Geology

National Atlas of Sweden

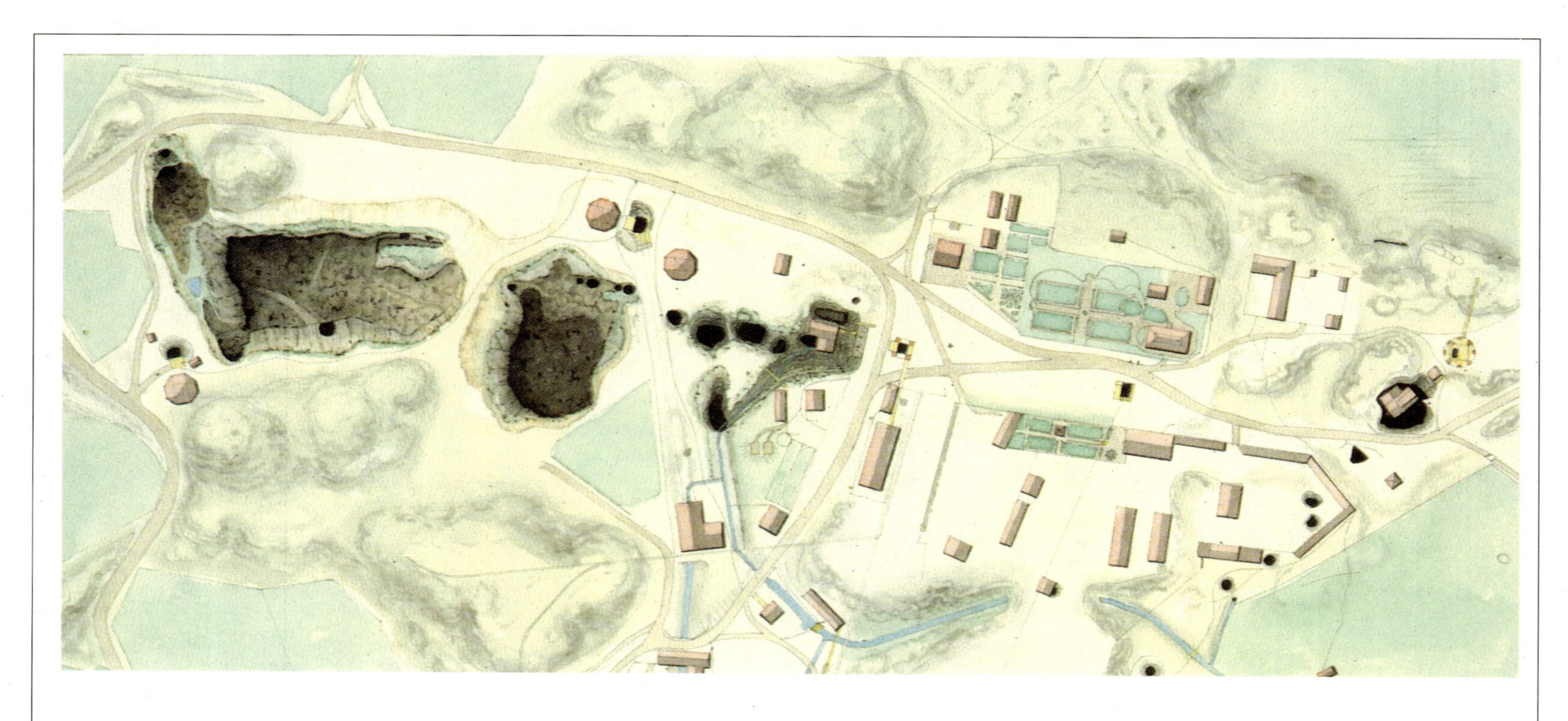

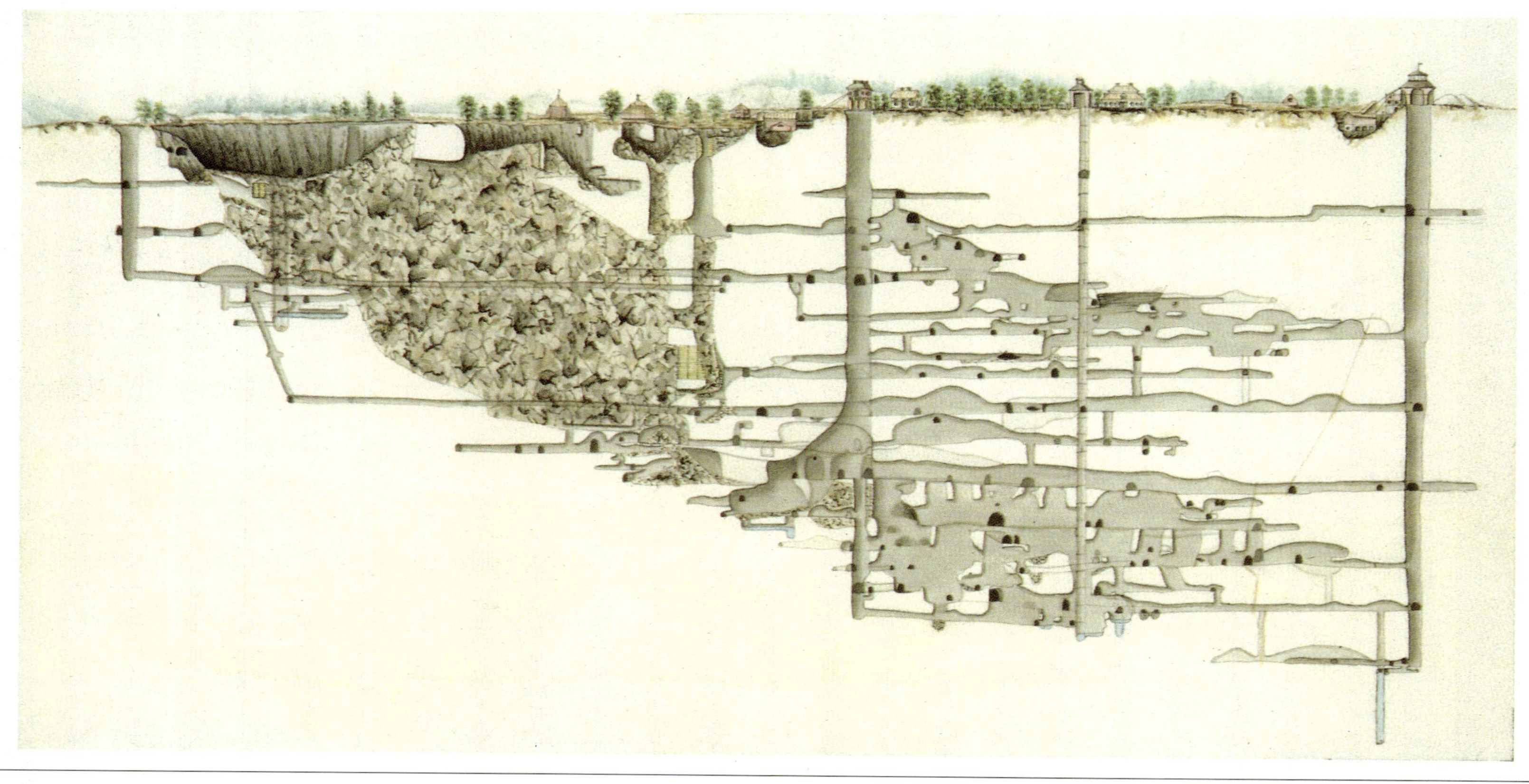

Sala Silver Mine, 1864–65. Wilhelm Georg Bergman

Geology

SPECIAL EDITOR

Curt Fredén

THEME MANAGER

Geological Survey of Sweden

National Atlas of Sweden

SNA Publishing will publish between 1990 and 1996 a government-financed National Atlas of Sweden. The first national atlas, *Atlas över Sverige*, was published in 1953–71 by *Svenska Sällskapet för Antropologi och Geografi, SSAG* (the Swedish Society for Anthropology and Geography). The new national atlas describes Sweden in seventeen volumes, each of which deals with a separate theme. The organisations responsible for this new national atlas are *Lantmäteriverket, LMV* (the National Land Survey of Sweden), *SSAG* and *Statistiska centralbyrån, SCB* (Statistics Sweden). The whole project is under the supervision of a board consisting of the chairman, Sture Norberg and Thomas Mann (LMV), Staffan Helmfrid and Åke Sundborg (SSAG), Frithiof Billström and Gösta Guteland (SCB) and Leif Wastenson (SNA). To assist the board and the editors there is a scientific advisory group of three permanent members: Professor Staffan Helmfrid (Chairman), Professor Erik Bylund and Professor Anders Rapp. For this theme Professor Jan Lundqvist has been coopted to the advisory group as a specialist. A theme manager is responsible for compiling the manuscript for each individual volume. The National Atlas of Sweden is to be published in book form both in Swedish and in English, and in a computer-based version for use in personal computers.

The English edition of the National Atlas of Sweden is published under the auspices of the *Royal Swedish Academy of Sciences* by the National Committee of Geography with financial support from *Knut och Alice Wallenbergs Stiftelse* and *Marcus och Amalia Wallenbergs Stiftelse*.

The whole work comprises the following volumes (in order of publication):

MAPS AND MAPPING
THE FORESTS
THE POPULATION
THE ENVIRONMENT
AGRICULTURE
THE INFRASTRUCTURE
SEA AND COAST
CULTURAL LIFE, RECREATION AND TOURISM
SWEDEN IN THE WORLD
WORK AND LEISURE
CULTURAL HERITAGE AND PRESERVATION
GEOLOGY
LANDSCAPE AND SETTLEMENTS
CLIMATE, LAKES AND RIVERS
MANUFACTURING, SERVICES AND TRADE
GEOGRAPHY OF PLANTS AND ANIMALS
THE GEOGRAPHY OF SWEDEN

CHIEF EDITOR	Leif Wastenson
EDITORS	Staffan Helmfrid, Scientific Editor Ulla Arnberg, Editor of *Geology* Margareta Elg, Editor Märta Syrén, Editor
PRODUCTION	LM Maps, Kiruna
SPECIAL EDITOR	Curt Fredén
TRANSLATOR	Nigel Rollison
GRAPHIC DESIGN	Håkan Lindström
LAYOUT	Typoform/Gunnel Eriksson, Stockholm
REPRODUCTION	LM Repro, Luleå
COMPOSITION	Bokstaven Text & Bild AB, Göteborg
DISTRIBUTION	Almqvist & Wiksell International, Stockholm
COVER ILLUSTRATION	Jan Peter Lahall/GreatShots

First edition

Printed in Italy 1994

ISBN 91-87760-04-5 (All volumes)
ISBN 91-87760-28-2 (Geology)

Contents

Geology—understanding Time

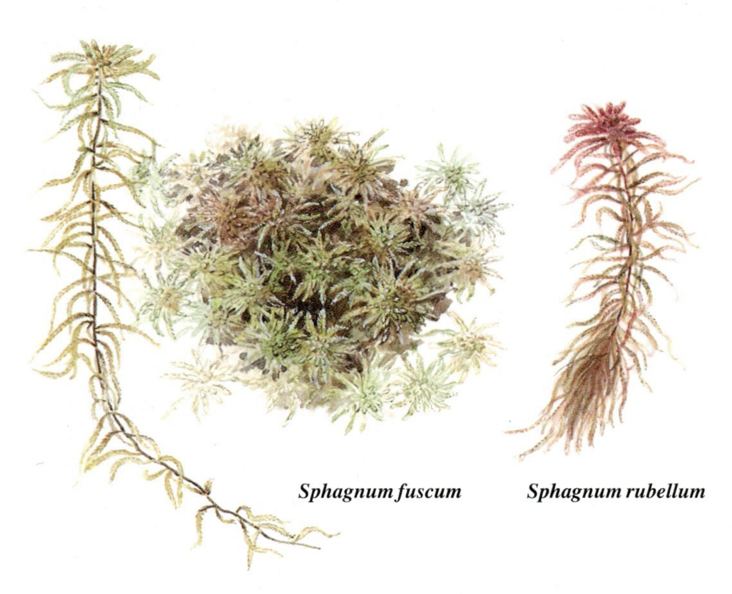

Sphagnum fuscum *Sphagnum rubellum*

—A brown sticky plug was removed from the bog. *Time* was slowly being drawn up.

An eight-year-old can appreciate the span of time encompassed by the summer holidays and a thirteen-year-old can certainly relate to the span of time that has elapsed since the reign of a king in the Middle Ages. But how can you understand time in a perspective of thousands, millions and billions of years?

My own first realisation of the geological time scale occurred on a rainy autumn day at Riddartorpsmossen Bog in Sörmland. I had just started my education as a geologist and we were out on our very first field course. We walked out to the middle of the bog in Indian file, the dwarf pines getting fewer and fewer, the bog myrtle being replaced by whortleberries and bog moss. The silence was total.

We were to practice using a corer to get a brown, sticky plug from the bog. Metre by metre we extracted the layers of peat. *Time* was slowly being drawn up out of the bog. The cores were laid out on the ground in a long row and a time axle gradually took form before our very eyes.

At the top was the fresh pink of today's Sphagnum moss, further down there were increasingly older decomposed residues of *Sphagnum* from ancient times. At a depth of four metres the core had the consistency of gyttja mud and just below that was the clay—the pure clay that had been deposited in a glacial sea. A time span of 10,000 years, extending from glacial times to the present-day, lay before us.

And there! At a depth of four and a half metres in the core there was a seed. A well-preserved pale yellow seed—about as large as a grain of pearl-sago—that had been deposited there more than two thousand years before the Egyptian pyramids were built. It was a seed of bogbean, that had sunk to the bottom of what was then a fen. It had never germinated, and had been found by a group of surprised students some 7,000 years later.

This is how geologists work and this is the way our knowledge of geology has developed. Small traces and remnants in the ground or in the rock can be used to create pictures of ancient environments. When all details in the core from Riddartorpsmossen had been analysed we had uncovered a complete process: The clay had been deposited in a glacial sea, the land had risen from the sea, an ice lake had formed, gyttja mud had been deposited in the lake, the lake had become clogged with vegetation and turned first into a fen and then gradually into the bog on which we stood, enveloped by the fragrance of bog myrtle and lulled by the splattering of a gentle shower of rain.

Knowledge of geology—a base to stand on

The slow but inexorable processes of nature are humbling. A peat bog perhaps takes a millennium to create but only a few short summer months to break up and transform into peat wafers. When you stand there on a bog, with time laid out in front of you—10,000 years of slow development—you realise how long it takes for a natural resource such as peat to become renewed.

The gravel and the sand deposited at the end of the glacial period as eskers and deltas is today being used in road-building and concrete-making. A new ice age is necessary to replenish and create new reserves of gravel and sand. The ore mined in Kiirunavaara is two billion years old and hundreds of millions of years are probably needed before new ore is formed in what we today call Sweden. Geological processes continuously create new natural resources but at an almost unbelievably slow rate, and thus peat, gravel and ore can be considered as non-renewable resources with which we must economise.

Understanding the term *non-renewable* resources also assumes awareness of the importance of keeping the groundwater clean, to develop methods of dealing with old mining wastes, and to decide on how, or if, nuclear waste can be safely stored. Knowledge of geology is essential in all work for a better environment.

Knowledge enables us to choose our path into the future. But the choice is by no means simple. The more I learn the better I can understand how complex the world is. New environmental problems are continuously being discovered, new scientific knowledge is emerging, and the "truth" of yesterday must be critically re-examined today. Our increased knowledge of the world will not make life easier, but it will give us better opportunities to influence development in a positive direction.

Geological knowledge is an important prerequisite for social development. This not only concerns solving problems of importance for the necessities of life and for our survival. Geological research involves the seeking for truths in unknown realms.

One must also bear in mind that we have no previous knowledge of benefits that will result from advances in basic research. Scientific information that may seem worthless today might be of infinite value in twenty years time. Thus, geological knowledge of rare earth metals was used in the hunt for high-temperature superconductors in the late 1980s. Another example is when the metal palladium suddenly became interesting as a replacement for platinum in catalysts for purification of car exhaust gases. Where could palladium be found? The geologists knew!

Peat—in practice a non-renewable natural resource. A peat bog takes many thousands of years to create but only a few short summer months to break up and transform into peat wafers. Tore Mosse, to the east of Hunneberg.

The magnitude of simplicity

The science of the Earth has developed during thousands of years. Periodically, knowledge has been lost or forgotten and has had to be rediscovered after many years. As early as in 500 BC there were clear ideas among the Ionian philosophers in Miletos that creation was a continuous process, that the planet Earth had received its shape as a result of slow processes, the same processes that are ongoing today. It took several hundred years for the same idea to surface again, but more than two thousand years were to pass before it became established.

Knowledge of the geological time perspective and geological development was effectively hindered for hundreds of years by the custodians of the church who wished to interpret the Bible by the letter. Nonetheless, Arabian scientists continued to study these fields and research was generously assisted by Muslim rulers of Baghdad and Samarkand. In about 1000 A.D., Avicenna presented his hypothesis on the genesis of mountain ranges, the hypothesis that the Christian world still regarded as both heretical and radical even as late as in the 19th century. He described how mountains could be formed in two ways; by rock movements in connection with earthquakes, and by the sculpturing effect of water on the landscape, whereby peaks and valleys are formed. Avicenna also realised that this took a long period to achieve. His theories were much closer to those of modern geology than the established views of that time in the west.

On occasions, knowledge has advanced in small jumps and text-books have had to be rewritten, an example occurring in the late 1970s when the *theory of plate tectonics* was presented. The German meteorologist and physicist Alfred Wegener had already in 1912 submitted his theory on continental drift, whereby all continents were considered to have once been part of a huge continent called *Pangea*, but that this had become split into smaller continents that then dispersed over the globe. The clear fit of the pieces in the jig-saw puzzle represented by Africa and South America strengthened his theory. Nonetheless, there was a lot that remained unexplained, particularly relating to the driving force behind the migration of the continents around the globe.

The theory of continental displacement was gradually replaced by a new theory on how the Earth's crust is divided into large plates that move in relation to each other. When the plates collide they form mountain chains, when they separate in the depths of the oceans the magma that emerges from the mantle leads to the formation of new crust. The theory of plate tectonics provided a simple and logical explanation of numerous processes that had formerly been described by far-fetched reasoning.

Good scientific theory is characterised by this kind of simple and logical approach. The theory of plate tectonics became of immense importance for all bedrock and ore geologists.

Small details and major theories

Theories are an essential part of the systematisation and structuring of knowledge. However, every theory must be based on observations of nature. Tiny bubbles in the rock at the small ore deposit at Östra Högkulla in the county of Västerbotten contain minute droplets of water that became trapped when two plates collided with each other two billion years ago. Drops of this kind reveal that metallic saline solutions with a temperature of almost 400°C were forced up through the sea floor in an ancient ocean. When the solutions were cooled by the seawater the ore was precipitated and small droplets of fluid became enclosed in crystals of quartz and sphalerite. Studies of these small bubbles in the rock allow geologists to create a picture of the geological environment prevailing at that time. Life on earth had not yet developed very far some two billion years ago—the first cells with a nucleus probably evolved at about the same time as the ore at Östra Högkulla was formed.

Deep down in the Greenland icecap a wafer-thin layer of ash has been found from the eruption of Vesuvius on August 24 in the year 79 A.D., which caused Pompeii to became buried in ash. This layer can be recognised and differentiated from other emissions of ash and in this way we can get a reliable age-determination of that layer in an ice core that is three thousand metres long. Ages of other horizons have been determined in a similar way. Such small details provide the evidence needed to establish the veracity of our knowledge.

Knowledge leading to insight

With better knowledge of our planet Earth I believe we will become wiser to the point of doing away with superstition. Instead of turning to charlatans for help in dispelling mystical forces and demons from our homes, we can devote ourselves to the true causes of the problem, for example, ventilating off the radon gas that may cause lung cancer, and the moisture that leads to moulds and allergies.

Scientific knowledge stimulates a critical attitude. With knowledge of the Earth's development following the "Big Bang", continuing through the birth of life, the dinosaurs and glacial eras, we are able to reject antiscientific and dogmatic doctrines that urge us to believe that the creation of the Earth took place some 6,000 years ago, and that fossils are remnants of the Great Flood.

The scientific view of the evolution of the Earth is based on the interpretation of traces left by massive geological processes. That picture is much more exciting than that offered by any pseudo-scientific doctrine!

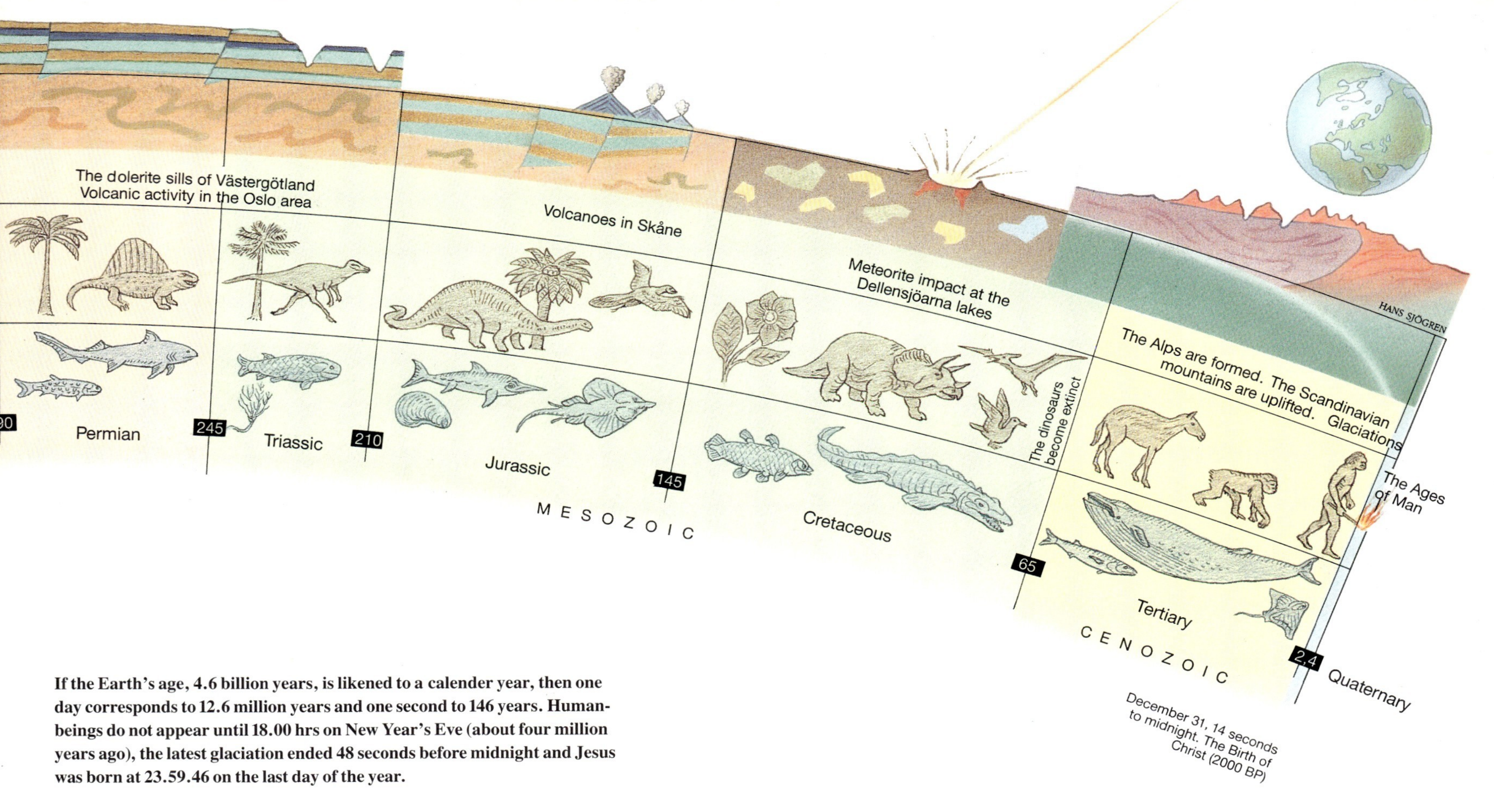

If the Earth's age, 4.6 billion years, is likened to a calender year, then one day corresponds to 12.6 million years and one second to 146 years. Human-beings do not appear until 18.00 hrs on New Year's Eve (about four million years ago), the latest glaciation ended 48 seconds before midnight and Jesus was born at 23.59.46 on the last day of the year.

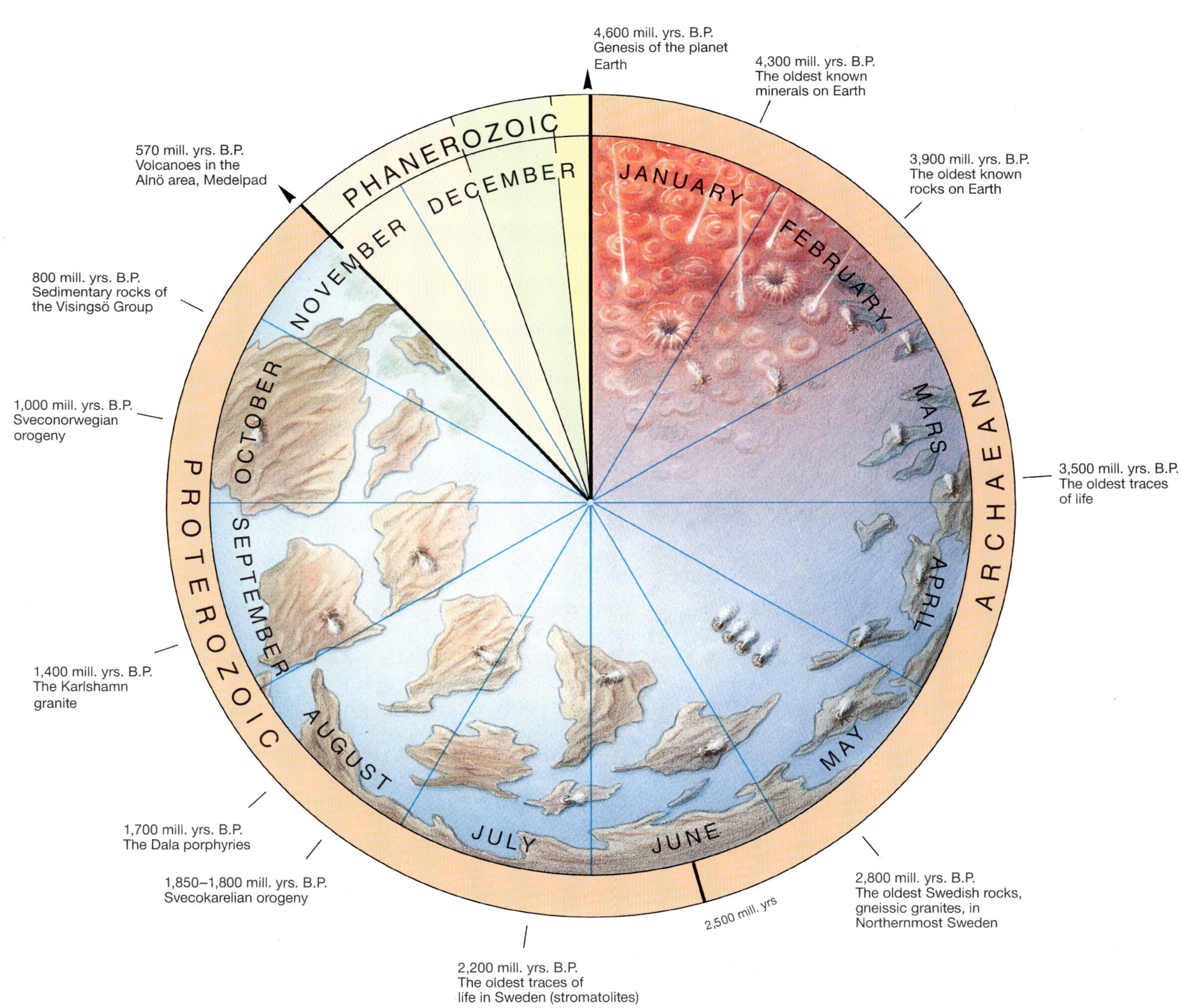

Structure of the Earth

Our knowledge of the planet *Earth* has increased dramatically during recent years. Technical developments that led to space probes and deep-sea exploration have enabled us to understand the relationships between a stone which we can hold in our hands and the planets in space.

The development we can see today started between 15 and 20 billion years ago with the Big Bang. However, our solar system is much younger, around 4.5 billion years. Since our planet is geologically active, the traces of the very oldest development have been destroyed by younger processes.

The Earth's radius is 6,370 km. The innermost parts, up to a depth of 2,900 km, consist of a core made up mainly of iron and nickel. The inner core is solid, the outer core is liquid.

The mantle is the next important unit and is bordered upwards by the so-called *Moho discontinuity* which is clearly seen in analysis of seismic waves. The mantle is solid but in a thin zone in the outermost parts it is partly fluid.

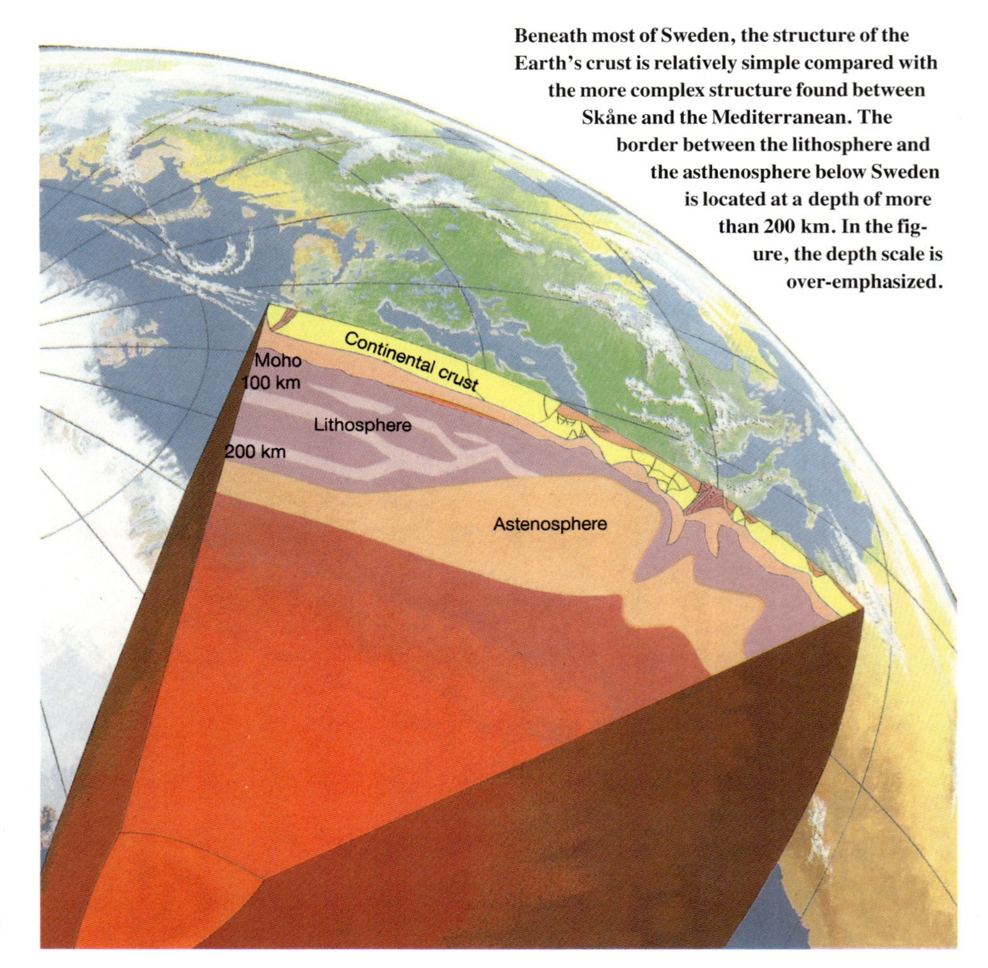

Beneath most of Sweden, the structure of the Earth's crust is relatively simple compared with the more complex structure found between Skåne and the Mediterranean. The border between the lithosphere and the asthenosphere below Sweden is located at a depth of more than 200 km. In the figure, the depth scale is over-emphasized.

The *crust* is the uppermost layer of the Earth and is the part that we can see. The crust is generally divided into a continental and an oceanic part, which have radically different properties. The thickness of the continental crust varies between 25 and 90 km, whereas the oceanic crust varies between 6 and 11 km.

Two other terms are also encountered in the vertical division of the Earth's different layers. One is the *lithosphere*, consisting of the crust and the upper parts of the mantle. The lithosphere is relatively solid and comprises the material in the various plates. The second term is the *asthenosphere* which is located below the lithosphere. The asthenosphere can become plastically deformed and most plate movements occur within this layer.

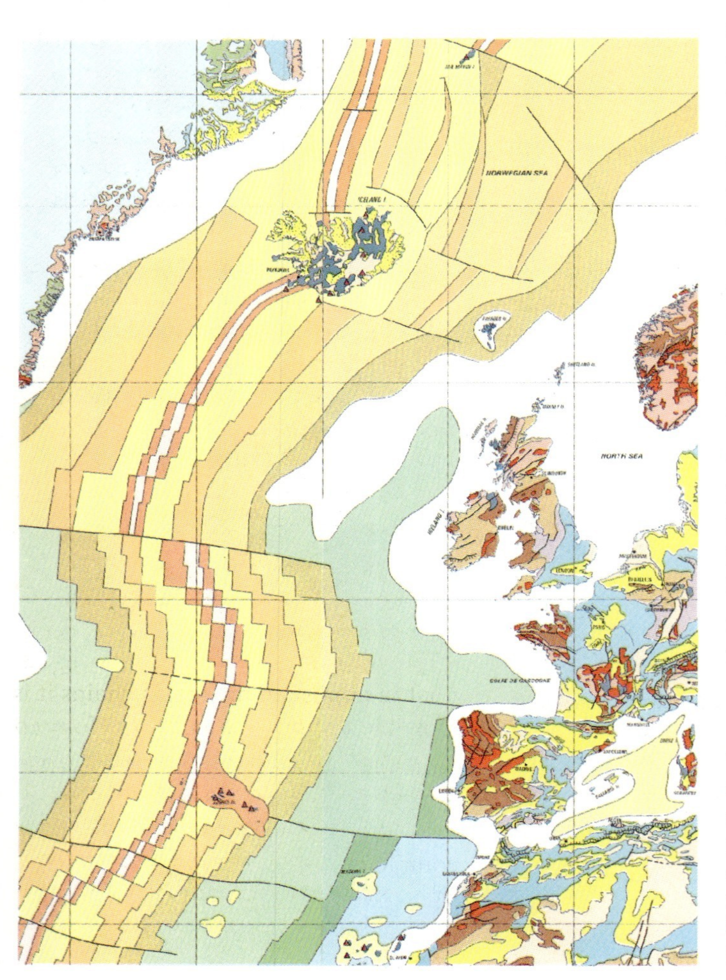

There are large differences between oceanic and continental crust, as illustrated by this bedrock map. The oceanic crust is relatively young, youngest in the spreading ridges and then gradually older further away. This contrasts with the mosaic pattern on the continents. The white areas illustrate the continental shelf which is underlain by continental crust.

Plate tectonics

In the early 20th century, the German meteorologist Alfred Wegener, among others, discussed the possibility that the continents were mobile. That this observation was principally correct was confirmed mainly during the 1960s by geophysical research on the structure of the ocean floors.

It was discovered that the ocean floors are continuously expanding along central spreading ridges of the same type as the Mid-Atlantic Ridge. Since the total area of the Earth does not increase, then this growth must be compensated in other places. This takes place mainly in the deep trenches of the oceans, where the oceanic crust and upper mantle, together with overlying water-saturated sediments, are drawn down and resorbed into the inner parts of the Earth. This circulation means that the bedrock of the ocean floors can never be particularly old in a geological context. The maximum age for bedrock of this kind is around 200 million years.

Information from the ocean floors

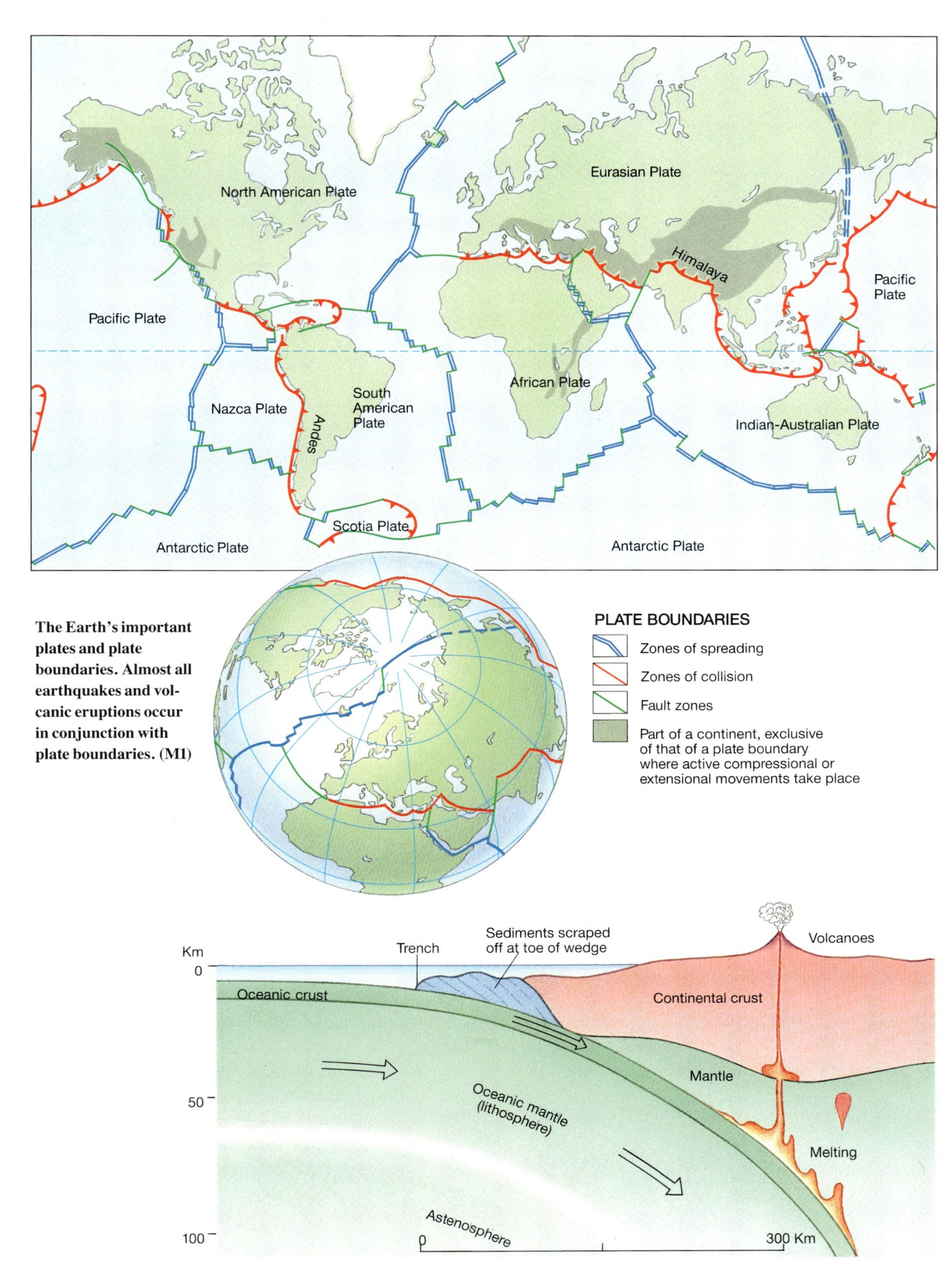

The Earth's important plates and plate boundaries. Almost all earthquakes and volcanic eruptions occur in conjunction with plate boundaries. (M1)

Along the island arcs in the western Pacific Ocean, oceanic crust is drawn down below continental crust. The type of volcanism that characterizes these island arcs has been proposed as a model for the volcanism that took place in Swedish ore districts around 1.9 billion years ago. The figure illustrates conditions in Sumatra today. In about 100 million years, the rocks formed here will probably be located in a mountain belt formed in connection with the movement of the Indian-Australian plate towards the north.

When two continents collide, the rocks that are present in the collision zone are folded and thrust over the neighbouring areas in the form of nappes. The example shows a profile through the Alps.

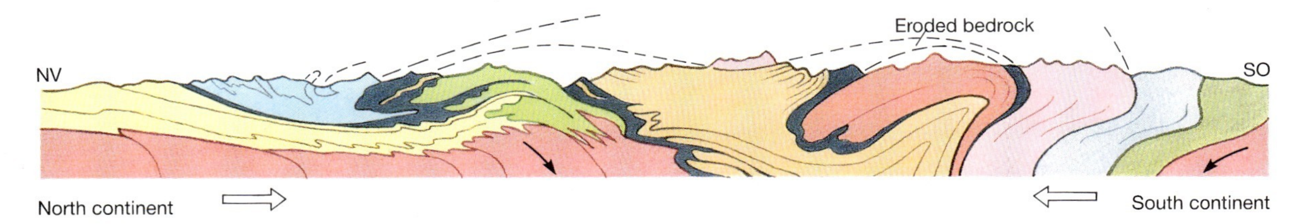

is systematised in *plate tectonic theory*. The outer part of the Earth, the lithosphere, is divided into a number of plates. Each plate moves around its own rotation axis. The boundaries between the plates are of three different types:

- *Constructive*, where the plates are created in spreading ridges.
- *Destructive*, where the plates collide and are destroyed.
- *Steep fault zones* where the plates slide past each other.

EARTHQUAKES AND VOLCANISM

Almost all geological activity in the form of volcanism and earthquakes is associated with processes along these boundaries. The character of volcanism is influenced, however, by the type of boundary in question. Volcanoes located along spreading ridges produce mainly lavas with basaltic composition (low in silica). Products from volcanoes above the destructive plate boundaries, for example around the Pacific Ocean, are much richer in silica, aluminium and alkali elements, *andesites*.

CONTINENTAL CRUST

The continents are composed of light silica-rich rocks and "float" on the underlying asthenosphere. They are linked together with oceanic crust but are not drawn down into the inner parts of the Earth at destructive margins on account of their low density. Rocks of all ages are found on the continents, some of which are extremely old. The oldest rocks found so far are in Canada and have been dated to 3,960 million years old.

New continental crust is formed in collisional zones of different kinds where mountain chains are developed. Where, for instance, the Nazca Plate in the Pacific Ocean is drawn down below the South American Plate, the Andes are formed; where a continental part of the Indo-Australian Plate moves under the Eurasian Plate, the Himalayas are formed.

When collisions take place, vast amounts of energy are released. An expression of these fluxes of energy is that in almost all mountain chains it is possible to find *granitic rocks* (formed as a result of melting processes). This applies to both young and old eroded mountain chains. Granitic rocks also completely dominate the structure of the continents even though they are covered by *sedimentary rocks* in many places.

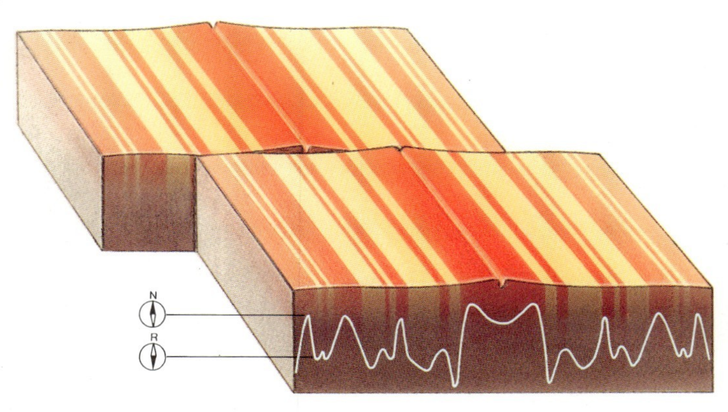

Results of magnetic measurements on the ocean floors show bands with different magnetic intensities. The pattern is symmetrical around the central ridges. Red indicates the normal direction of the magnetic field and yellow the reverse direction. The figure below on the left illustrates what the pattern actually looks like along the Reykjanes Ridge to the south of Iceland.

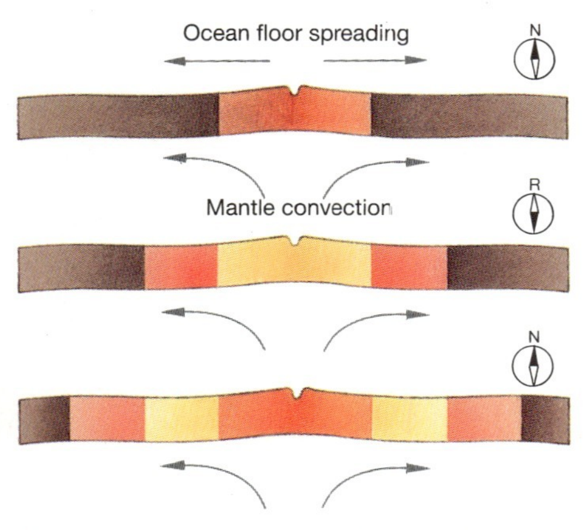

The oceanic crust grows around a spreading ridge. The rocks become older the further away from the ridge they are located. By using the reversals in the Earth's magnetic field, it has been possible to construct a time scale.

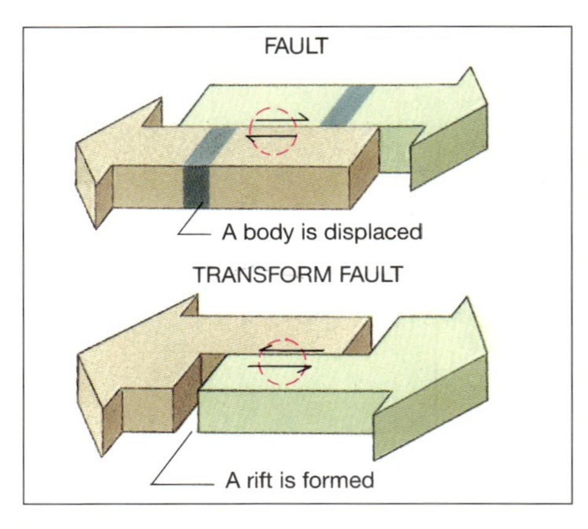

Two completely different fault mechanisms may result in similar structural patterns. Note the different, relative movements along the fault surface. The fissures formed in connection with the transformed faulting result in volcanism.

The continents make up a mozaic of different provinces. Each province is characterised by granites and folded rocks of about the same age.

How the plate tectonic theory developed

MAGNETIC PATTERN ON THE SEA FLOOR

One of the most important discoveries made during the development of the plate tectonic theory concerned the *magnetic pattern* on the ocean floors. When molten rock crystallises, the magnetic minerals orient themselves in the prevailing magnetic field in the same manner as small compass needles. These minerals subsequently retain the direction they received when the rock was formed. When maps were prepared of the magnetic field in the Earth's crust on the ocean floors, a stippled pattern appeared that provided evidence of periods with different directions of the magnetic field. It was discovered that the Earth's magnetic field totally reversed in direction at time intervals which varied between 50,000 and 3 million years. Within these intervals, the changes were small. Using the magnetic stripes, it was possible to establish a time scale that demonstrated that the ages became progressively older the further away from the mid-ocean ridges the measurements were made.

FAULT PATTERN IN THE SEA-FLOOR

Another discovery concerned the pattern of faults. When the mid-ocean ridges were studied in detail, for example the Mid-Atlantic Ridge, it was observed that they were displaced in a remarkable pattern across their longitudinal direction. Initially, it was believed that this was a case of normal lateral faulting. However, analyses of earthquakes that occurred along these faults demonstrated that the relative movement was the opposite. It was concluded that the faults are an active part of the sea-floor spreading process and it was now possible to schematically demonstrate how oceanic crust was created.

When a continental block starts to fracture and break apart, the upper parts of the Earth's mantle start to melt. The molten material rises up into the fracture system and builds basaltic lava at the surface. As this proceeds and oceanic material is developed, a symmetrical pattern is formed around the spreading ridge with the youngest rocks closest to the ridge.

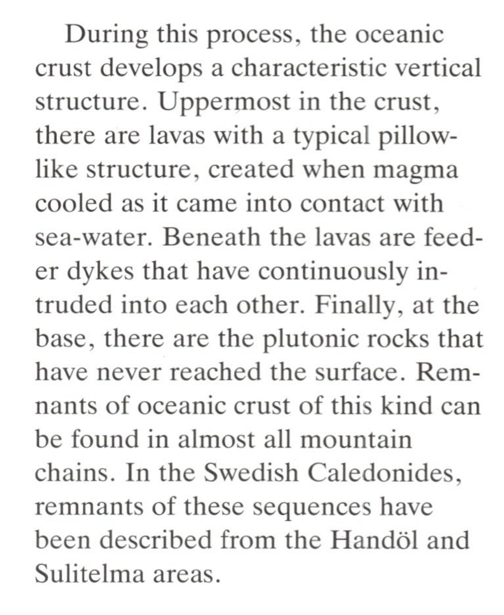

During this process, the oceanic crust develops a characteristic vertical structure. Uppermost in the crust, there are lavas with a typical pillow-like structure, created when magma cooled as it came into contact with sea-water. Beneath the lavas are feeder dykes that have continuously intruded into each other. Finally, at the base, there are the plutonic rocks that have never reached the surface. Remnants of oceanic crust of this kind can be found in almost all mountain chains. In the Swedish Caledonides, remnants of these sequences have been described from the Handöl and Sulitelma areas.

FOSSIL FAUNAS

During development of the plate tectonic theory, it was possible to explain a number of known yet puzzling facts from different disciplines within the geological sciences.

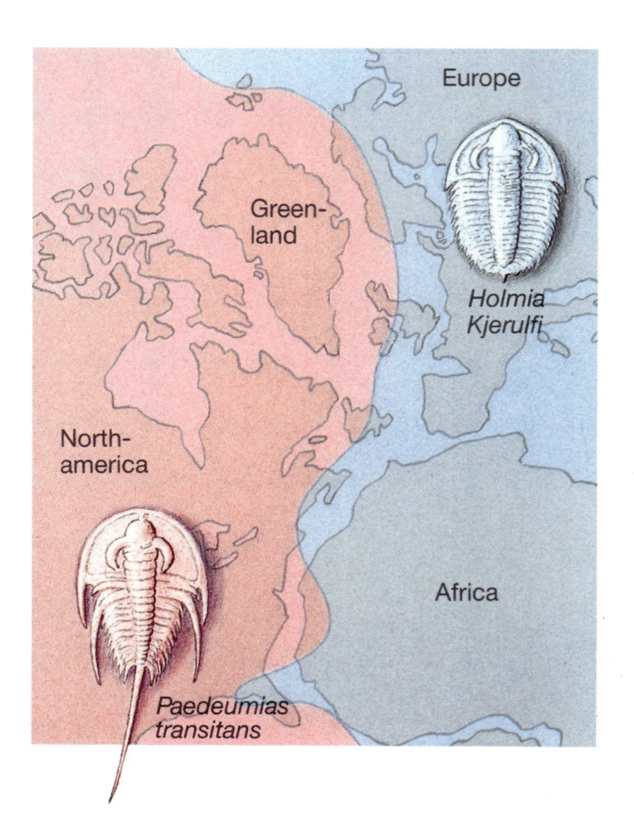

During the Cambrian, the fauna in Sweden was characterized by, for example, the trilobite *Holmia kjerulfi.* In the Trondheim area, Norway, there was a different fauna characterized by the trilobite *Paedeumias transitans,* which was largely identical with the one found in North America. The dispersal of the various types of fauna illustrates which land areas were joined at that time. When the Swedish mountain chain was formed by continental collision during the Silurian, the fossil faunas became more similar.

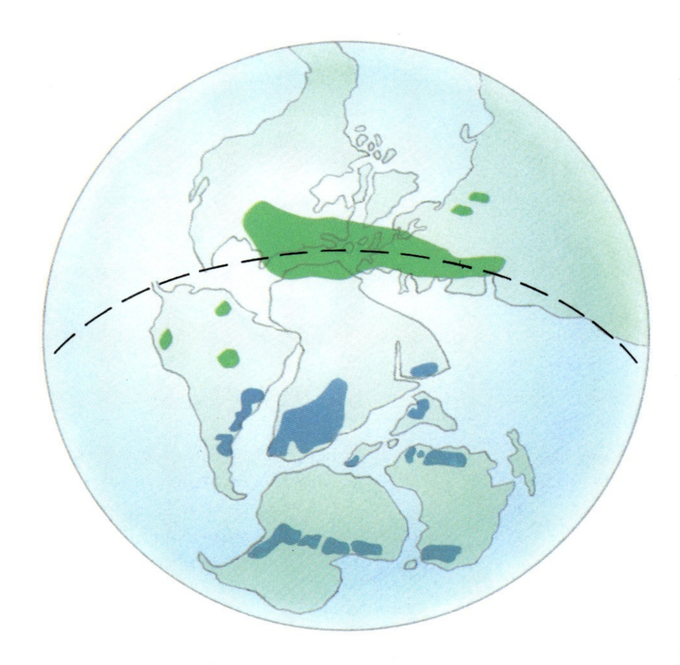

Most of the coal that is mined today has been formed from remnants of a tropical forest which grew around 300 million years ago. The tillites found in, for example, southern Africa and Brazil, are approximately of the same age. If we assemble the known occurrences of tropical forest (green) and areas covered by ice (blue) from this period and move the continents in the same way as pieces in a puzzle, the glaciated areas group together at the South Pole and the tropical forests around the equator.

A good indicator of climatic conditions are limestone reefs. Today, they are only formed in a zone around the equator. The limestone reefs present on, e.g., Gotland, about 420 million years ago, also formed when the area was located close to the equator. Stenkyrka, Gotland.

One such example is the distribution of different *fossil faunas* in time. When studying fossils from different time epochs, we can find different faunas in immediately neighbouring terrains and, conversely, similar faunas in different terrains. When continents collide with each other, the various living species are able to disperse, and the fauna will become similar. Conversely, a continent that is isolated will have its own development of different animal species. This has happened, for example, with the fauna in Australia.

PALAEOCLIMATE

When Alfred Wegener introduced his theory of continental drift in the early years of the present century, he based his theories on the different climatic conditions that could be identified from various rocks and how they were located in relation to different latitudinal positions. Wegener also believed that the continents floated around on the ocean floors in the same way as icebergs float on water. When this was found to be incorrect, his theories were rejected for many years. Today we know that the plates frequently consist of both continental and oceanic crust—the plates in the Pacific Ocean even consist solely of oceanic crust—and plate movement occurs instead in the upper parts of the Earth's mantle.

PALAEOMAGNETISM

In the same way as measurements were made of the magnetic poles on the ocean floors, analyses have also been made of rapidly-cooled magmatic rocks on the continents in order to find out the positions of the magnetic poles at different times. The frozen fossil magnetism, *palaeomagnetism*, reveals the latitude at which a rock was formed. By studying the palaeomagnetism of different rocks, it is possible to see how, for example, the position of Scandinavia on the Earth's surface has changed.

The latitudinal position and orientation of Fennoscandia during the past 2.7 billion years can be identified from paleomagnetic studies. The area has not had the same appearance throughout the entire development. The oldest parts are found in eastern Finland and on the Kola Peninsula. Subsequently, the continents have periodically expanded towards the south-west and west. The most important geological events are indicated on the diagramme.

The Bedrock

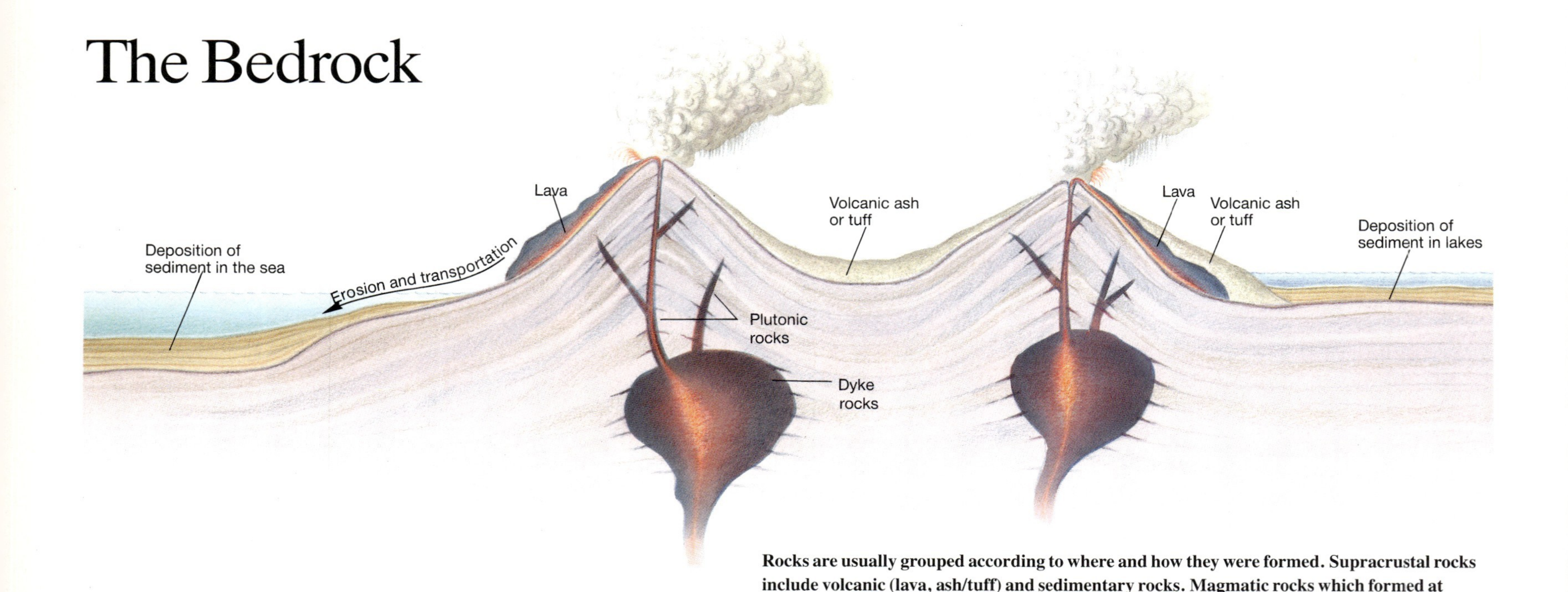

Rocks are usually grouped according to where and how they were formed. Supracrustal rocks include volcanic (lava, ash/tuff) and sedimentary rocks. Magmatic rocks which formed at greater or lesser depths in the crust are referred to as plutonic and hypabyssal rocks, respectively.

HOW ROCKS ARE FORMED

The elements in the Earth's crust have combined in different ways to form a number of *minerals*. Relatively few minerals comprise the main components of the most common *rocks*. Depending on how they have been formed, the rocks are divided into *igneous, sedimentary* and *metamorphic* types. These can be transformed into each other in different ways in the geological cycle. Igneous rocks are divided, in turn, into *volcanic, hypabyssal* and *plutonic rocks*. The volcanic and sedimentary rocks have been deposited on the surface of the Earth and are therefore usually called *supracrustal rocks*.

The Swedish bedrock has developed during a timespan of almost three thousand million years as a result of igneous, sedimentary and metamorphic processes, in many cases in connection with folding and other kinds of deformation. When the bedrock in Sweden started to be formed, the Earth was already more than one and a half thousand million years old and had a crust that was largely similar to the one we know today. Life was also present on Earth at that time, although in primitive form.

The bedrock we see in the countryside, or as presented on a geological map, is a two-dimensional surface mosaic of rocks. In order to understand how this pattern of rocks has arisen, it is necessary to have a good imagination and the capacity to think three-dimensionally. Most rocks have formed under completely different conditions than those prevailing today in Sweden. They may, for example, have been formed at great depths in the Earth's crust or in an ancient sea. Mountain-building processes may have placed originally horizontal layers in a vertical direction and magmas might have intruded older rock types. Weathering and erosion have subsequently exposed structures that have been hidden for millions of years. Another complication when attempting to interpret bedrock geology is that for long periods, Sweden was located at latitudes with completely different climatic conditions than the ones we experience today.

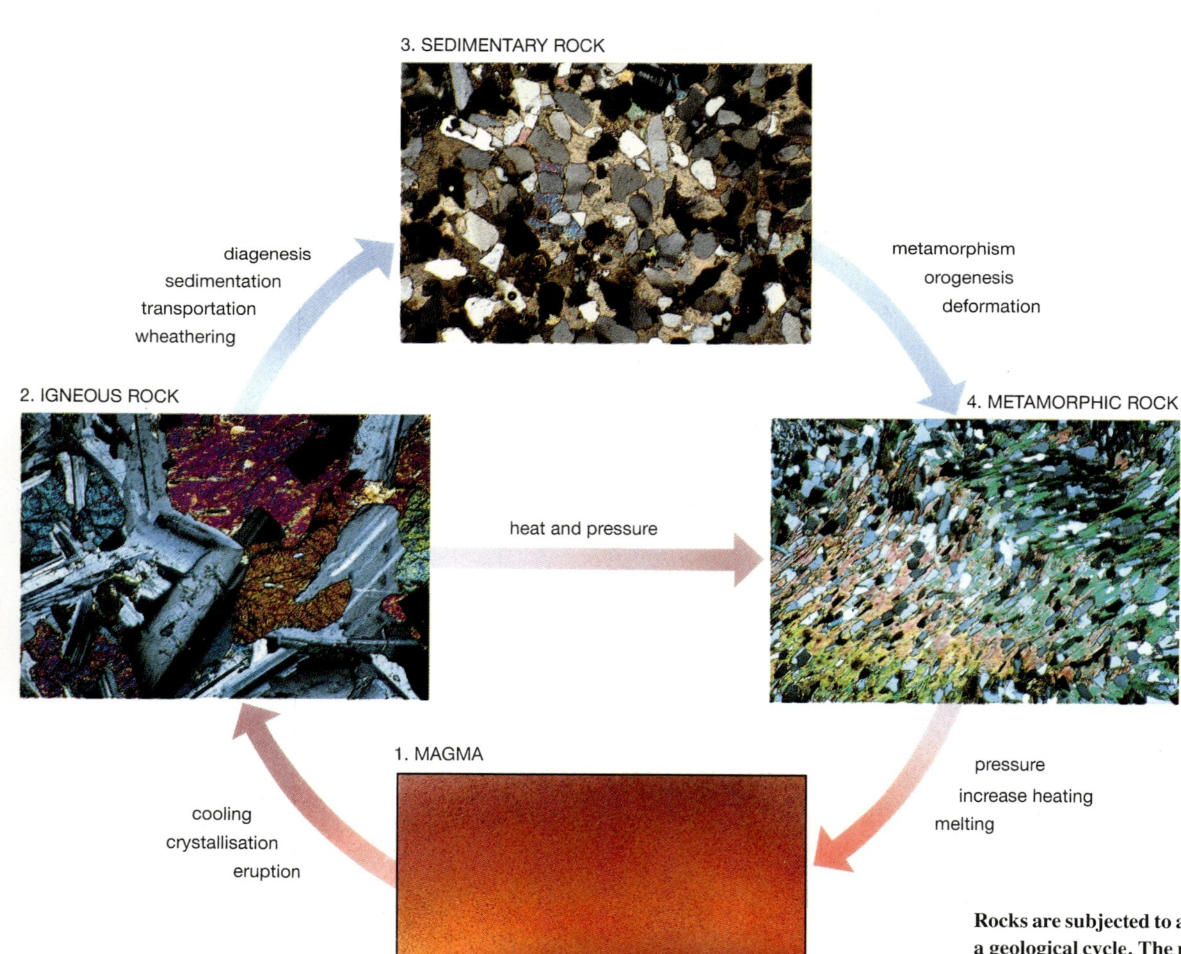

Rocks are subjected to a continuous but slow change—we speak of a geological cycle. The magmatic rocks are slowly broken down through weathering, etc., and are transformed into sedimentary rocks. Both magmatic and sedimentary rocks can be converted by heat, pressure and folding into metamorphic rocks, which in turn can melt when exposed to great heat and again be converted into magmatic rocks.

THE MOST ABUNDANT ELEMENTS IN THE EARTH'S CRUST . . .

Element	In upper continental crust, %	In oceanic crust, %
Oxygen (O)	46.8	44.0
Silicon (Si)	30.6	23.2
Aluminium (Al)	7.9	8.9
Iron (Fe)	3.3	6.8
Calcium (Ca)	3.0	8.4
Sodium (Na)	2.7	2.0
Potassium (K)	2.7	0.2
Magnesium (Mg)	1.3	4.3
Others*	1.7	2.2

*E.g. carbon (C) and hydrogen (H)

. . . FORM THE MOST COMMON MINERALS . . .

Quartz (Si, O)
Potassium feldspar (K, Al, Si, O)
Plagioclase (Na, Ca, Al, Si, O)
Mica (K, Mg, Fe, Al, Si, O, H)
Hornblende (Ca, Mg, Fe, Al, Si, O, H)
Pyroxene (Mg, Fe, Si, O, +/– Al, Ca)
Olivine (Mg, Fe, Si, O)
Carbonate (Ca, Mg, C, O)

. . . WHICH FORM THE MOST COMMON ROCKS

	METAMORPHIC	
SEDIMENTARY	Low grade ———>	High grade
Shale	Slate, mica schist	Gneiss, veined gneiss,
Sandstone	Fine-grained quartzite	Coarse-grained quartzite
Limestone	Fine-grained marble	Coarse-grained marble
Conglomerate		
MAGMATIC		
VOLCANIC		
Rhyolite (porphyry)	Hälleflinta, leptite	Leptite gneiss
Basalt	Greenschist	Amphibolite
DYKE ROCKS		
Granite porphyry		
Dolerite		
PLUTONIC ROCKS		
Granite	Foliated granite	Gneissic granite
Syenite		
Gabbro		

The magmatic rocks have names according to their content of pale-coloured minerals (quartz, microcline/orthoclase and plagioclase). The figure on the left shows the contents of minerals in common plutonic rocks, the figure on the right shows some volcanic rocks.

Granite is an example of a rock with approximately similar contents of quartz, microcline and plagioclase. Basalt, on the other hand, contains mainly plagioclase and very little or no quartz or orthoclase. Monzonite contains similar amounts of plagioclase and microcline but hardly any quartz.

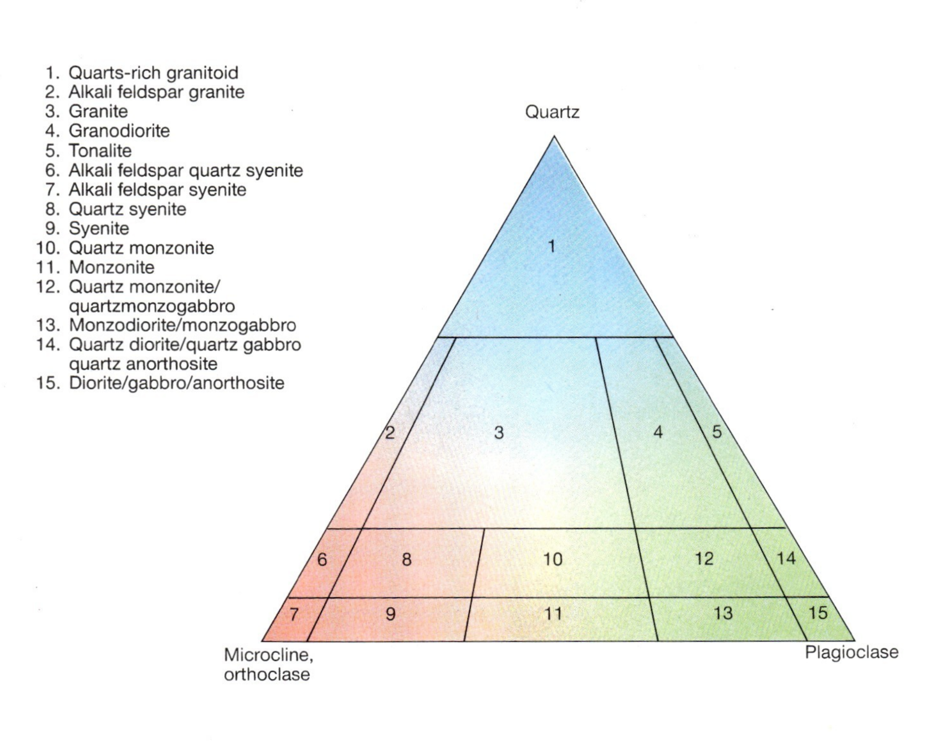

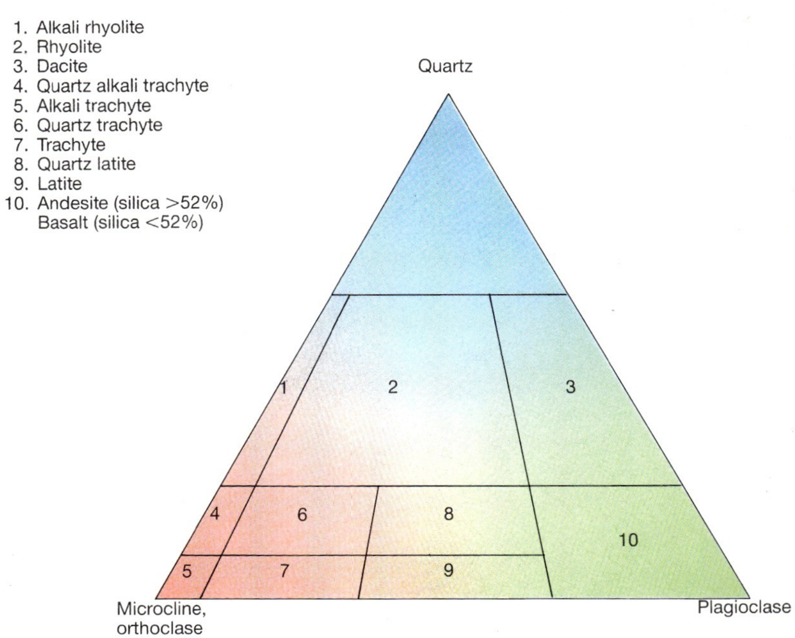

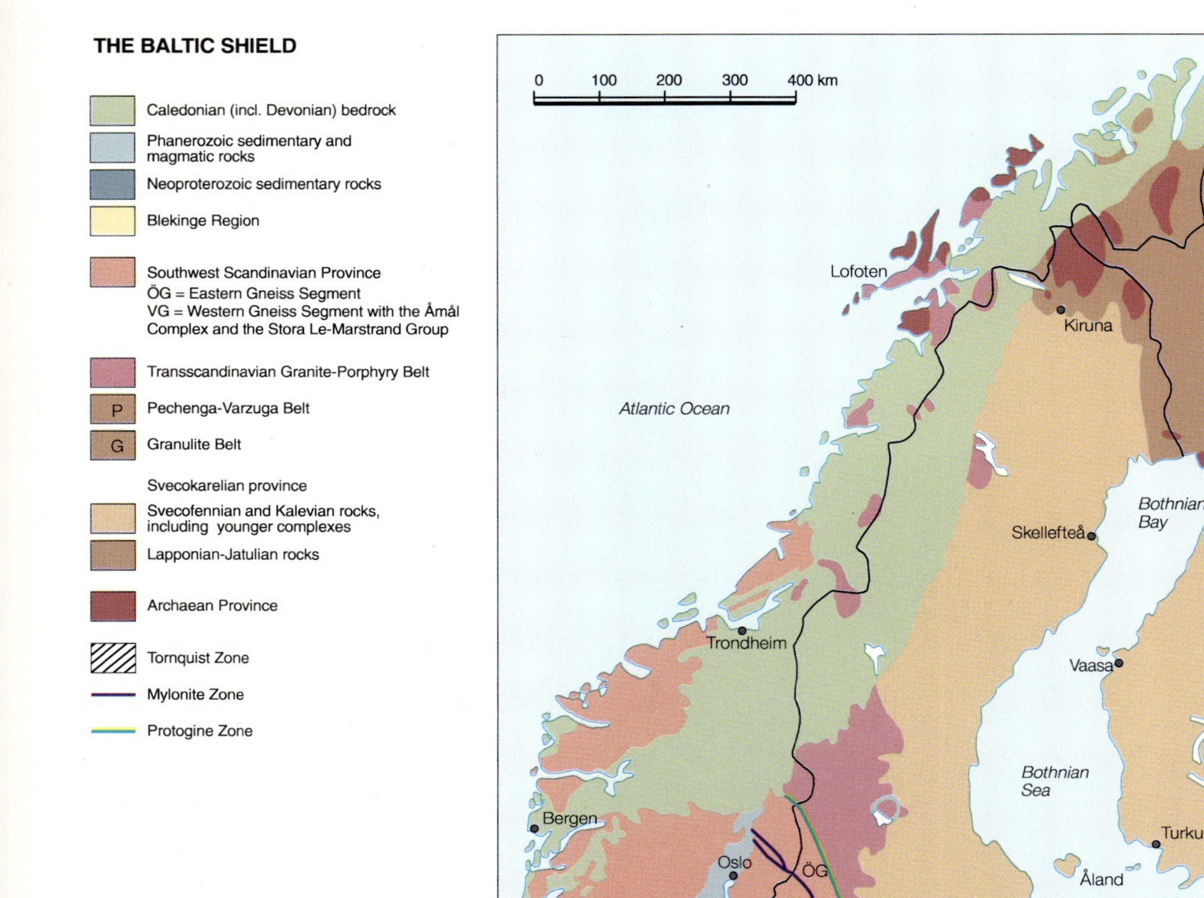

The Baltic Shield is divided according to the age of the bedrock and important geological events. Towards the southwest, the shield is bordered by the Tornquist Zone. Other important movement zones are the Protogine Zone that can be traced from Skåne to Värmland, and the Mylonite Zone that runs through Värmland and Västergötland to northern Halland. (M2)

BEDROCK IN SWEDEN

The Swedish bedrock consists of three major units: the Precambrian crystalline basement, the Caledonides, and the sedimentary bedrock outside the Caledonides.

The Swedish *Precambrian bedrock* is older than 570 million years. It belongs to the Baltic Shield that extends from the Kola Peninsula in the northeast to southwestern Norway.

The *Caledonides* are an ancient mountain chain that formed between 510 and 400 million years ago. They extend along the Swedish-Norwegian mountain region and continue into Great Britain and Ireland.

Most of *the sedimentary bedrock outside the Caledonides* was formed during the Cambrian–Tertiary, contains fossils and overlies the Precambrian basement in much the same way as a flat blanket.

The Swedish Precambrian is part of the Baltic Shield

The *Baltic Shield* is one of the Earth's Precambrian shields. Its bedrock has developed through several *orogenies* of mountain-building processes when sedimentary and volcanic rocks were folded and affected by metamorphic processes at different depths in the crust. Large amounts of granitic magma were added to the bedrock in connection with orogeny. The bedrock as we can see it today has subsequently been exposed through erosion.

The Baltic Shield is divided into different provinces with regard to age and important geological events.

THE ARCHAEAN PROVINCE

The Archaean province in northernmost Sweden contains the oldest bedrock in the country, between two and

THE TORNQUIST ZONE

The Tornquist Zone, named after a German geologist (1868–1944), is a zone of repeated movement in the continental crust. It extends from the Rumanian coast of the Black Sea in the southeast to the North Sea in the northwest. The zone forms the border between the East European Platform with the Baltic Shield to the northeast and a vast subsidence area in the western part of continental Europe.

In Skåne, the Tornquist Zone can be seen as faults which strike in a NW–SE direction. Basement horsts—the hilly ridges in Skåne—alternate with subsidence areas where younger fossiliferous strata frequently attain considerable thicknesses. Numerous dolerite dykes of Permo-Carboniferous age are also found in the zone, together with basalts from the Jurassic-Cretaceous periods.

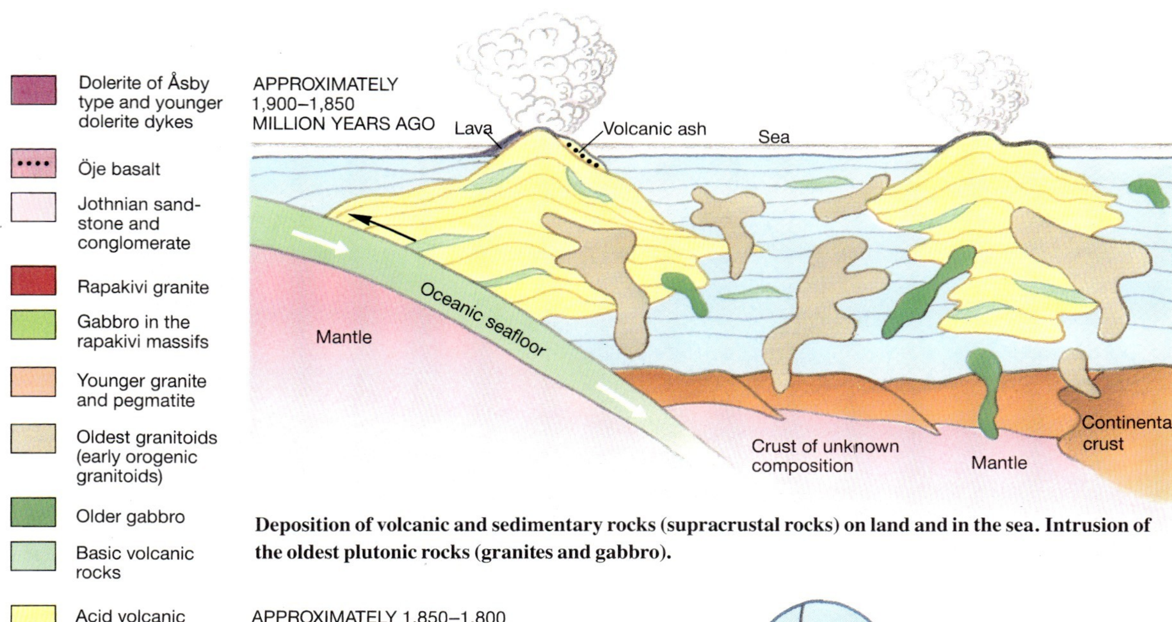

Deposition of volcanic and sedimentary rocks (supracrustal rocks) on land and in the sea. Intrusion of the oldest plutonic rocks (granites and gabbro).

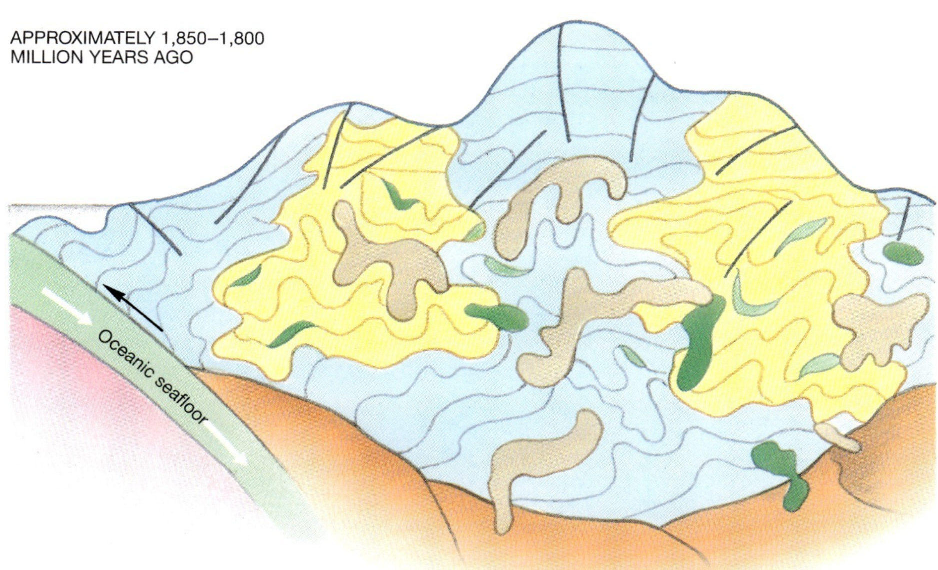

Folding and formation of a mountain chain during the Svecokarelian orogeny. Metamorphism increases in strength with depth and ultimately leads to melting (migmatisation). Thickening of the Earth's crust. Erosion of the mountain chain.

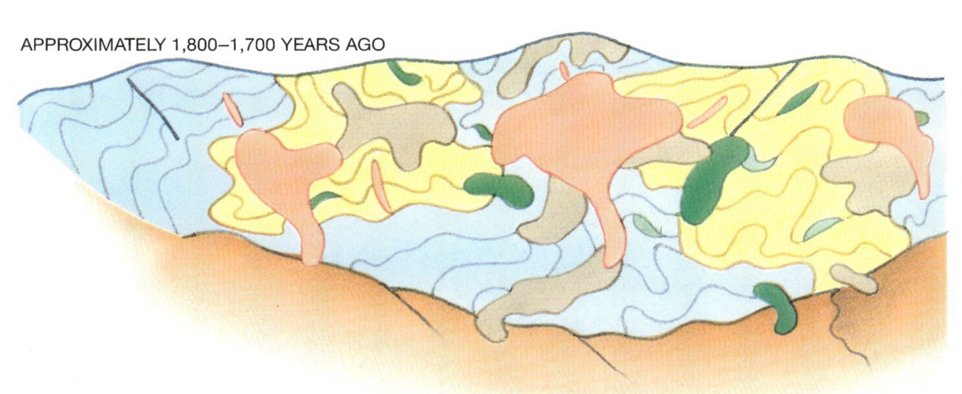

Following folding, the mountain chain becomes strongly eroded and younger granites penetrate into the Earth's crust. In contrast to the older granites, these are mainly undeformed and unmetamorphosed.

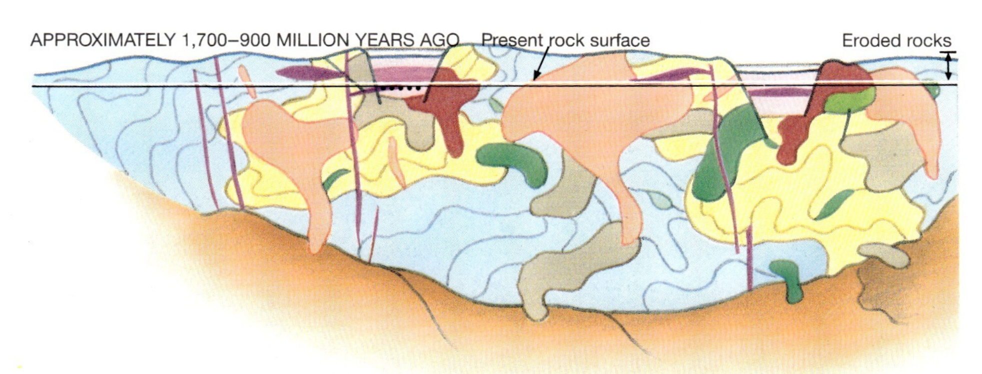

Rapakivi granites and associated gabbros are formed. Red Jotnian sandstones and conglomerates are deposited in deltas in fault depressions. Basalt lava covers the sandstones and subsequently dolerite magma of different generations penetrates along both flat-lying and steep fractures. During earlier folding, the section we can see today at the Earth's surface was at a depth of 10–15 km.

Illustration of proposed geological development in the Svecokarelian basement of central Sweden. The horizontal scale is about ten times smaller than the vertical scale.

a half and three thousand million years old. It forms an isolated inlier surrounded by younger rocks relative to the larger, coherent complexes in Finland, Norway and Russia. The Råstojaure and Kukkola gneiss complexes, which largely consist of different gneissic and migmatitic plutonic rocks, such as diorites, granodiorites and granites, mainly belong to these complexes.

THE SVECOKARELIAN PROVINCE

The bedrock of the Svecokarelian province largely developed before and in connection with the Svecokarelian orogenesis that occurred about 1,850–1,800 million years ago.

This bedrock is dominated by plutonic rocks, usually granitoids, that are between 1,900 and 1,600 million years old. These rocks are found throughout the entire province and belong to different generations. The earliest (early orogenic) granitoids, for example, have been exposed to metamorphism and major deformation, whereas the younger are largely unaffected. In addition, there are dolerites that are 1,500–900 million years old.

The supracrustal rocks of the province are divided into Lapponian–Jatulian rocks and Svecofennian rocks.

The Lapponian–Jatulian bedrock

The Lapponian-Jatulian bedrock in Sweden consists of sedimentary and volcanic rocks. The Lapponian rocks, which are roughly between 2,500 and 2,200 million years old, consist of mainly basaltic volcanic rocks such as those in the Kiruna and Vittangi areas. In many cases, volcanic eruptions have taken place on the sea floor and formed pillow lavas.

The slightly younger (2,200–2,000 million years) Jatulian units consist mainly of quartzite with phyllite, mica schist, dolomite and basaltic volcanic rocks. These can be found at several places in the county of Norrbotten.

The Svecofennian bedrock

The Svecofennian bedrock contains metamorphosed sedimentary and volcanic rocks that are usually between 1,880 and 1,870 million years old. The Kalevian sedimentary units in the province of Norrbotten can also be included here.

Metamorphosed greywacke and argillite predominate among the Svecofennian sedimentary rocks. Volcanic rocks have mainly silicic (acid) compositions, e.g., rhyolitic. In many cases, the metamorphism was of such a high grade that the supracrustal rocks have partly melted, leading to migmatitic gneisses. In the section we can see today, metamorphism has taken place at depths of ca. 10–15 km at temperatures of about 500–650°C. In connection with this metamorphic event, the original horizontal stratification in the supracrustal rocks has been modified to a more or less vertical dip and folded, frequently on several occasions.

Svecofennian sedimentary rocks are particularly abundant in central Norrland and in Södermanland. Large areas with acid volcanic rocks can be found in an area extending from Kiruna down towards Arvidsjaur and further to the Skellefte Field. The bedrock of the Skellefte Field has been interpreted as the remnants of a volcanic island, similar to the one forming Japan today. Another region where acid volcanic rocks dominate has its centre in the Bergslagen area. In places where acid volcanic rocks are found, there are often major ore deposits and mineralisations; most of the Swedish ore deposits are found in the Kiruna area, the Skellefte Field and in the Bergslagen district. Basic volcanic rocks occur in the southeastern part of the province.

Thin layers of carbonate rocks (limestone, dolomite) have been deposited in connection with volcanic activity. Such deposits are particularly common in Bergslagen and neighbouring areas. During metamorphism, the carbonate rocks have been converted into marble. They some-

Top. When lava flows out onto the sea floor, it forms large drops, the surfaces of which solidify rapidly and form pillow lavas. Viscaria, Kiruna.

Centre. Stromatolites in Svecofennian marble. Dannemora, Uppland.

Bottom. Light Härnö granite with inclusions of grey, gneissic granite containing white "eyes" of microcline. The gneissic granite is older and the Härnö granite younger. Husum, Ångermanland.

Top. Brownish carbonate rocks with folded layers of acidic tuff. The latter resist weathering better than the limestone and thus form ridges in the surface of the rock. Lake Usken, Lindesberg.

Centre. Cross-bedding in Västervik quartzite. This type of bedding formed in flowing water. Tallskär, Västervik.

Bottom. Conglomerate with pebbles of granitic (light) and volcanic (dark) rocks. Vallen, Norrbotten.

Top. Amphibolitic rocks, strongly mixed with granite and folded, are typical intercalations in acidic volcanic rocks in the archipelagos of northeastern Småland and Östergötland. Städsholmen, near Loftahammar.

Centre. Typical Revsund granite, which is grey or reddish, contains up to decimetre-large crystals, "eyes", of potassic feldspar. Västnoret, Åsele.

Bottom. Granite porphyry with scattered grains of pink orthoclase and grey-blue quartz. Känningen, Sundsvall.

AGE million years	EON	ERA	MAJOR GEOLOGICAL AND ROCK-FORMING EVENTS
65	PHANEROZOIC	Cenozoic	Uplift of the Scandinavian mountain chain
		Mesozoic	
250		Palaeo-zoic	The Caledonides are formed
570	PROTEROZOIC	Neoprotero-zoic	The Alnö massif Widespread peneplain formation Sedimentary rocks of the Visingsö Group
1,000		Meso-protero-zoic	Blekinge-Dala dolerites; Bohus granite Sveconorwegian folding and metamorphism Sedimentary and volcanic rocks of the Dalsland Group Formation of dolerite sills in Central Sweden Red sandstones; Öje basalt in Dalarna Karlshamns-Spinkamåla-Vånga granites
1,600		Palaeo-protero-zoic	Rapakivi granites with gabbro, anorthosite Folding and metamorphism in SW Sweden and Blekinge Dala porphyries Stora Le-Marstrand Group greywackes Småland-Värmland granites Svecokarelian folding and metamorphism in eastern and northern Sweden Svecofennian volcanic and sedimentary rocks; iron and sulfide ores; granites, gabbro Deposition of sedimentary and volcanic rocks in northernmost Sweden (Lapponian, Jatulian and Kalevian rocks)
2,600	ARCHAEAN		Oldest Swedish bedrock in northernmost Sweden
3,500			Oldest remnants of life on Earth
			Oldest known rocks on Earth
			Oldest known mineral on Earth
4,600			Birth of the Planet Earth

This Dala porphyry is an ignimbrite, i.e., formed by a hot ash flow. It contains pink and grey-white scattered grains of feldspar. The pink stripes are fragments of the original pumice. Rännåsarna, Dalarna.

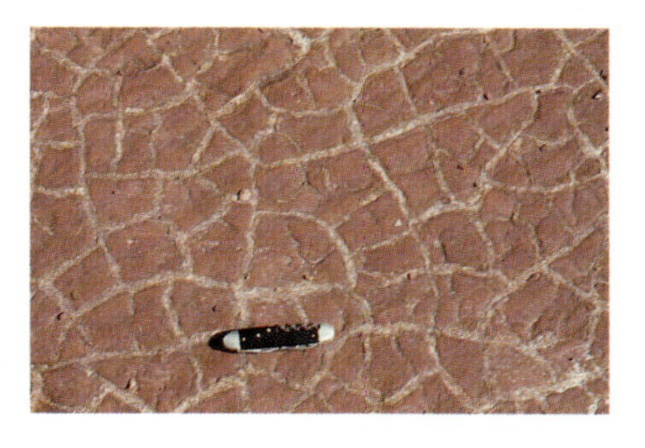

Mud cracks in a shale layer in Dala sandstone. The cracks are filled with sandstone. Mångsbodarna, Dalarna.

Several rare earth elements were discovered for the first time in gadolinite crystals from Ytterby near Vaxholm. Some of them have names related to the place where they were found, e.g., ytterbium, yttrium, terbium and erbium. The black gadolinite crystals are embedded in pale-coloured feldspar which is surrounded by black mica in a pegmatite that is 1,800 million years old.

times contain stromatolites, for example at Sala, Dannemora and Glanshammar.

Quartzite and arkose are not common among the Svecofennian supracrustal rocks, but occur mainly in the southern, southwestern and central parts of the Svecokarelian province.

Some conglomerates found, for example, in the Skellefte Field and near Luleå, contain pebbles of older, early orogenic, plutonic rocks. They are thus younger than the majority of Svecofennian supracrustal rocks.

THE TRANS-SCANDINAVIAN GRANITE-PORPHYRY BELT

This bedrock is dominated by granite and acid volcanic rocks (porphyries). Basic plutonic rocks such as gabbro and diorite also occur. The granites are usually about 1,800–1,700 million years old, i.e., mainly younger than the Svecokarelian orogeny. A number of small granite massifs in eastern Småland are even younger, about 1,400 million years old.

The granitic rocks in the southern and central parts of the area belong to the Småland and Värmland granites. These contain large (many km^2) inclusions of porphyritic volcanic rocks, referred to as Småland porphyries. The latter are slightly older than the oldest generation of granites, about 1,850–1,800 million years. Other inclusions belong to the Svecofennian supracrustal units. In the central and northern parts of the Belt, there are the Dala granites, as well as the volcanic Dala porphyries and the more basic Dala porphyrites. The 1,700–1,680 million years old Rätan granite, with megacrysts ("eyes") of feldspar, also belongs to this suite. The Dala porphyries, deposited approximately 1,690 million years ago, consist, as well as the Småland porphyries, largely of ignimbrite, a type

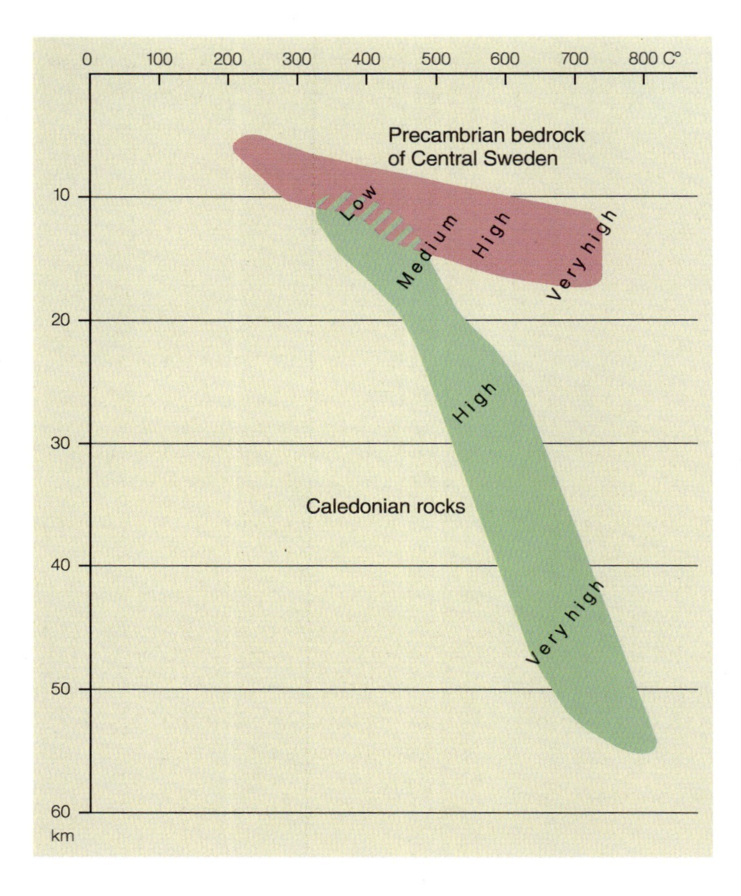

of volcanic rock formed by hot ash flows.

The ca. 800 m thick red Jotnian Dala sandstone with intercalated beds of basalt, Öje basalt, rests on top of the Dala porphyries and granites. Similar sandstones are found in Småland and in some places in the Svecokarelian province. Even younger sandstones, belonging to the Visingsö Group, occur in and around Lake Vättern.

Two generations of dolerite have cut the porphyries and the sandstone. These include the often gently dipping sheets of Åsby and Särna dolerites, about 1,250–1,200 million years old, and the steeply dipping Blekinge-Dala dolerites, about 900 million years old.

The diagram shows the temperatures and depths at which the Proterozoic basement rocks in central Sweden and the rocks in the Caledonides received their present mineral composition and structure. Large parts of the Caledonian bedrock were metamorphosed at greater depths than the Proterozoic basement.

THE SOUTHWEST SCANDINAVIAN PROVINCE

This bedrock was formed about 1,800–900 million years ago and consists largely of gneisses. Most of these are orthogneisses formed from quartz-bearing plutonic rocks (granite, granodiorite, tonalite). They are extensively migmatised. A minor part consists of metamorphosed sedimentary and volcanic rocks.

To the east, the province is restricted by a zone running from Skåne to Värmland, the *Protogine Zone*. Along this zone it is possible to find gradual transitional relationships between the gneisses of the province and the granites in the Trans-Scandinavian Granite-Porphyry Belt.

Along the Protogine Zone to the south of Lake Vättern, there are several generations of about 1,200–900 million year old hypabyssal and plutonic rocks. Monzonites, syenites and hyperite dolerites are present.

To the south, in Skåne, the gneisses of the province are covered by younger fossiliferous sedimentary rocks. Here, the border zone is cut by faults of the Tornquist Zone which strikes in a NW–SE direction. Basement horsts in this province alternate with grabens where the younger fossiliferous strata frequently have considerable thicknesses.

The *Mylonite Zone*, extending from the Norwegian border in Värmland towards the south along the eastern shore of the Värmlandsnäs peninsula down to the West Coast south of Göteborg, is an important zone of movement in the southwest Scandinavian province. Here, a thorough deformation in the bedrock took place approximately 1,000 million years ago, in connection with the Sveconorwegian orogeny. To the east of the Mylonite Zone, in the eastern gneiss segment, the bedrock is completely dominated by orthogneisses. There are also abundant basic hypabyssal and plutonic rocks, e.g., the 1,550–1,500 million year old Värmland hyperites. A special type of plutonic rock is the Varberg charnockite on the West Coast. Supracrustal rocks are seldom found, but one example is a remarkable quartzite with the minerals kyanite, lazulite and rutile in northern Värmland.

Orthogneisses also occur to the

The Stora Le-Marstrand greywackes contain intercalated acidic (light) and basic (dark) volcanic rocks. These have been folded in spectacular patterns that are clearly seen in rocks along the shoreline. Stigfjorden, Bohuslän.

When basic and acidic magma simultaneously penetrate the Earth's crust, the former may sometimes form "droplets" in the latter. The magmas have penetrated the greywackes of the Stora Le-Marstrand Group. To the right is a dyke of young, pale-coloured granite. Norra Testholmen, Marstrand.

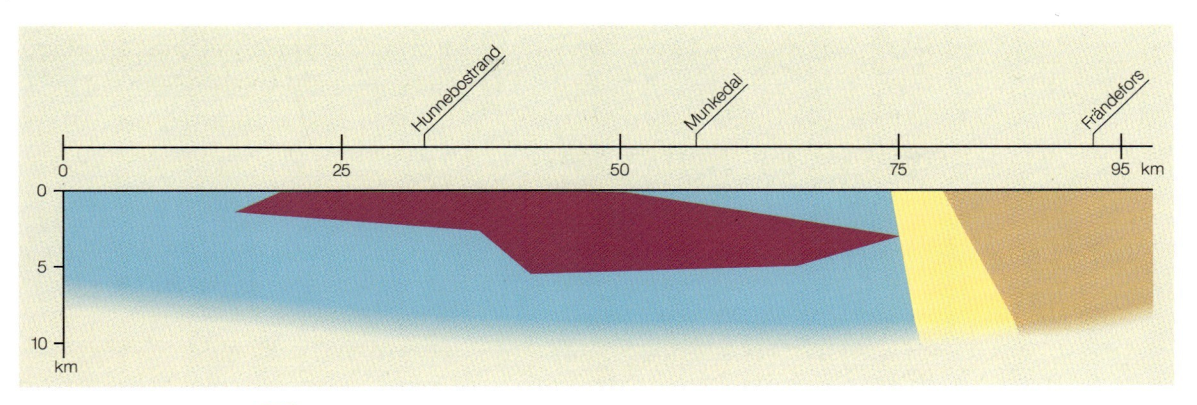

The form of the Bohus granite, seen in a profile between the sea off Hunnebostrand and Frändefors in the province of Dalsland, has been interpreted on the basis of gravity measurements.

west of the Mylonite Zone but here the supracrustal rocks are more extensive. In the northern-central parts of the Åmål Complex, the foliation is weak and both supracrustal and plutonic rocks are relatively well preserved. Examples are the ca. 1,650 million year old Åmål granite as well as porphyries and quartzites. There are also younger granites west of the Zone, e.g., the Askim granite near Göteborg (ca. 1,400 million years) and the granites in Dalsland (slightly more than 1,200 million years). The northern part of the Åmål Complex forms *nappes* resting on the gneissic bedrock. On top of the Complex lies a thick, folded and faulted sequence of slate, quartzite and spilite belonging to the more than 1,050 million year old Dalsland Group.

Further to the west are the gneissic and migmatised greywackes of the Stora Le-Marstrand Group with basic volcanic layers intruded by plutonic and hypabyssal rocks of four different generations. The 920 million year old massif of Bohus granite is the youngest granite of the Swedish Precambrian. Pegmatites of similar age are found all over the Province.

The coastal gneiss of Blekinge is a grey granitic gneiss, here with pink veins. Among other applications, it has been used in street paving. Listershuvud, Blekinge.

THE BLEKINGE REGION

The bedrock in the Blekinge region consists of quartzite, acid and basic volcanic rocks, and also the so-called Blekinge coastal gneiss, the Tving granite and the Småland granites. Ages vary between about 1,800 and 1,700 million years. The microcline-porphyritic Karlshamn granite and the related Spinkamåla granite are younger (ca. 1,400–1,350 million years). The foliated Vånga granite in northeastern Skåne, used in the stone industry, also belongs to this group.

The youngest rocks in Blekinge are the north-south striking, steeply dipping Blekinge-Dala dolerite dykes, with an age of approximately 900 million years. They can be followed all the way up to northern Dalarna.

Among the remarkable rocks found in the alkaline Alnö massif near Sundsvall, there is a white magmatic limestone that, in this case, has penetrated a black, pyroxene-rich rock.

ALKALINE COMPLEXES AND DYKES

Special igneous rocks of alkaline composition are locally found in Sweden. They are characterised by minerals with high contents of alkali (sodium and potassium) in relation to silicon and aluminium. Examples of such minerals are nepheline, cancrinite, alkali pyroxene and alkali amphibole. At Norra Kärr near Gränna, Almunge in Uppland (massifs), Särna in Dalarna, on Alnön Island off Sundsvall (massifs and dykes), and in the Kalix archipelago, northern Sweden (dykes only). On Alnön Island and in the Kalix archipelago, the alkaline rocks are associated with igneous carbonate-rich rocks, called carbonatites. Rocks belonging to the alkaline-carbonatitic group are often rich in rare elements and minerals.

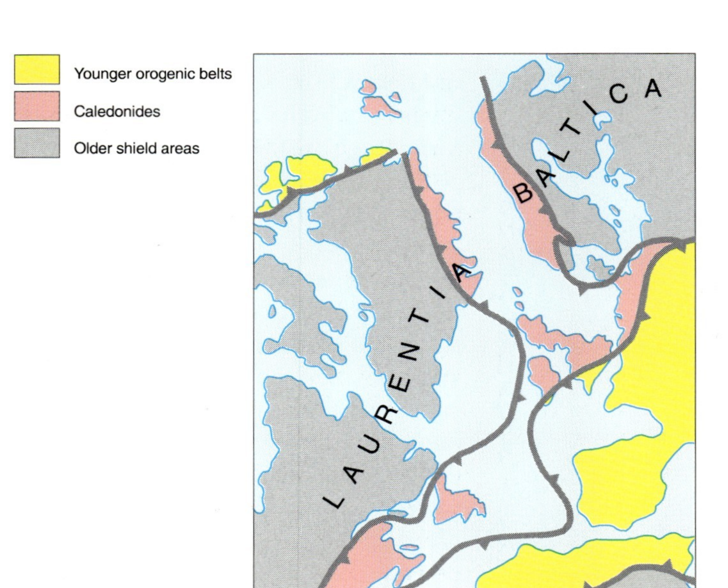

The Caledonides—after Caledonia, the Latin name for Scotland—in a reconstruction of what the North Atlantic area looked like around 250 million years ago. At that time, the Pangaea supercontinent had formed, but the present-day Atlantic had not yet opened. (M3)

The Caledonides

In Scandinavia, the *Caledonide orogen* extends in a 2,000 km long and 100–200 km wide belt from North Cape in the north to Stavanger in the south. Earlier, this area was connected with Caledonian bedrock in, e.g., eastern Greenland, Ireland, Great Britain and eastern North America before the entire chain was broken when the present North Atlantic opened about 65 million years ago.

The Caledonides were formed between 510 and 400 million years ago when the *Iapetus Ocean* that lay between the old continents *Baltica*, (present-day northern Europe) and *Laurentia* (North America and Greenland) started to contract and finally disappeared. Continent-continent collision resulted and the marginal zone of Baltica was pressed down beneath Laurentia and drastically shortened. Thrusting caused sedimentary and volcanic rocks deposited originally on the floor of the Iapetus Ocean and along the margin of the continent Baltica to be pushed up and onto Baltica, in the same way as the Himalayas were formed in modern times as a result of the collision between India and the remainder of Asia.

NAPPES AND THRUSTS

The dominant structure in the Caledonides is large transported sheets of rock, i.e., *nappes*. These are separated by flat-lying *thrusts*. Below the nappes, there is a thin zone of late Precambrian-Ordovician *autochthonous* rocks that have not moved from their original site of deposition.

The nappes have been transported several hundred kilometres to the east-southeast up onto each other and onto Baltica. They have been divided into the *Lower, Middle, Upper* and *Uppermost Allochthons*. Nappes high up in the sequence have been transported further relative to the lower ones. The Upper Allochthon is divided into lower and upper units, the *Seve* and *Köli* Nappes, respectively.

The degree of metamorphism generally increases upwards in the pile of nappes and, in individual units, from east to west across the orogen. In the lowermost units, sedimentary rocks are hardly affected, whereas, in the Seve Nappes, the degree of metamorphism is medium to high grade. In the Köli Nappes and in the Uppermost Allochthon, however, there is a wide variation in the degree of metamorphism. The minerals in some rocks in the Seve Nappes demonstrate metamor phism at temperatures ranging up to about 780°C and pressures up to 18 kbar. This suggests that these so-called *eclogites* were once present at a depth of approximately 60 km in the Earth's crust.

ROCK TYPES

The *Lower Allochthon* consists largely of weakly metamorphosed feldspathic sandstone, quartzite, greywacke, limestone and shale that were deposited during the late Precambrian–Silurian along the margin of the continent Baltica.

The *Middle Allochthon* contains a major component of deformed and metamorphosed, Precambrian, acid, intrusive and volcanic rocks. The ol-

The profile shows the Scandinavian Caledonides in cross-section along a line from Lofoten to Kvikkjokk (A-B on the map to the left). The Baltic Shield and its sedimentary coverrocks continue a long way beneath the overthrust Caledonian bedrock. In its lower part, this contains drastically shortened remnants of the marginal zone of the continent known as Baltica. (M4, M5)

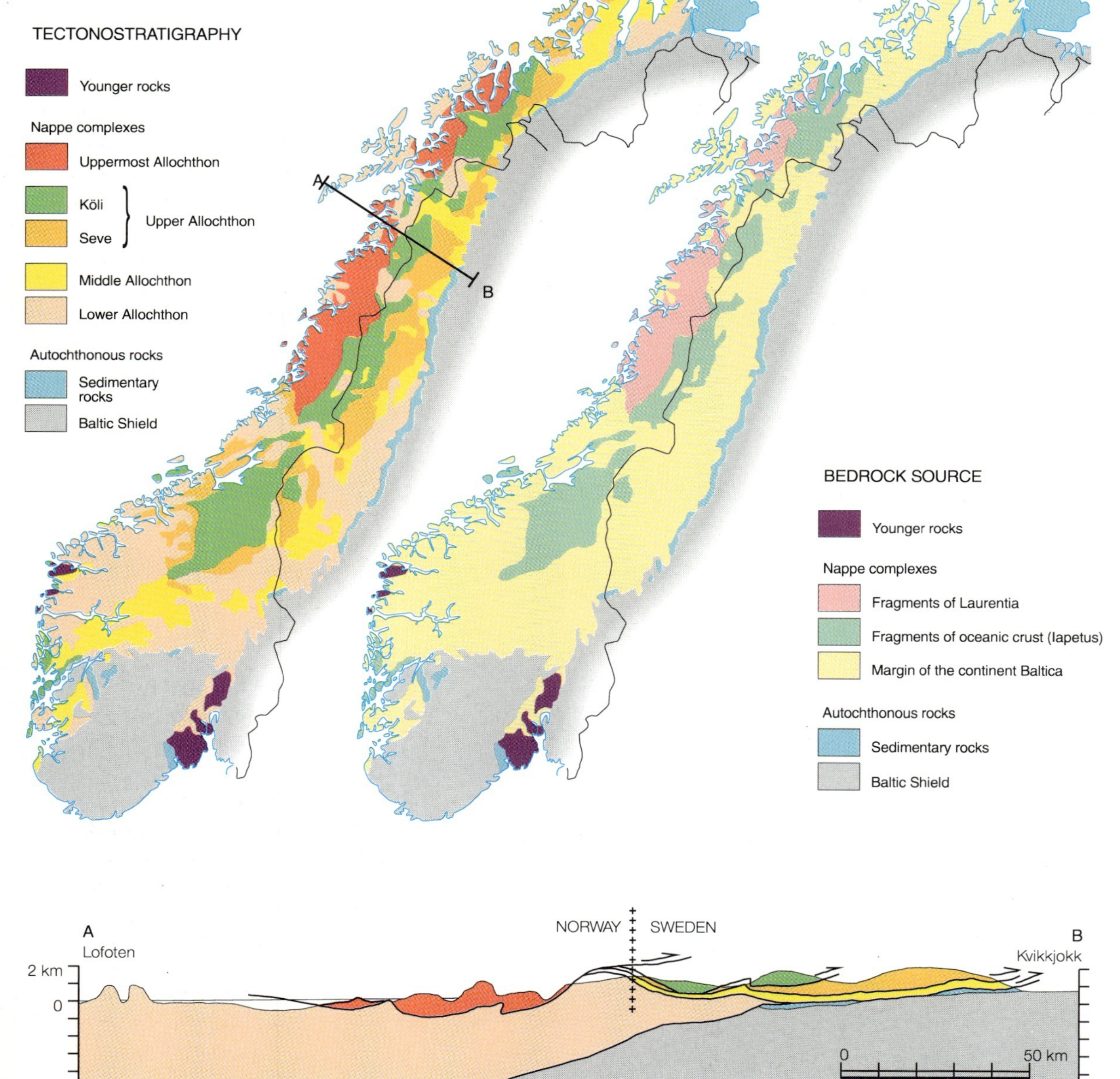

TECTONOSTRATIGRAPHY		SOURCE	METAMORPHIC GRADE
Uppermost Allochthon		Terraines derived from out-board of Baltica	Medium and high grade
Upper Allochthon	Köli Nappes		Low to high grade
	Seve Nappes	Tectonically shortened and folded margin of Baltica	Medium and high grade, locally high pressure
Middle Allochton			Low grade
Lower Allochthon			Low and very low grade
Baltic Shield and overlying sedimentary rocks = Continent Baltica			

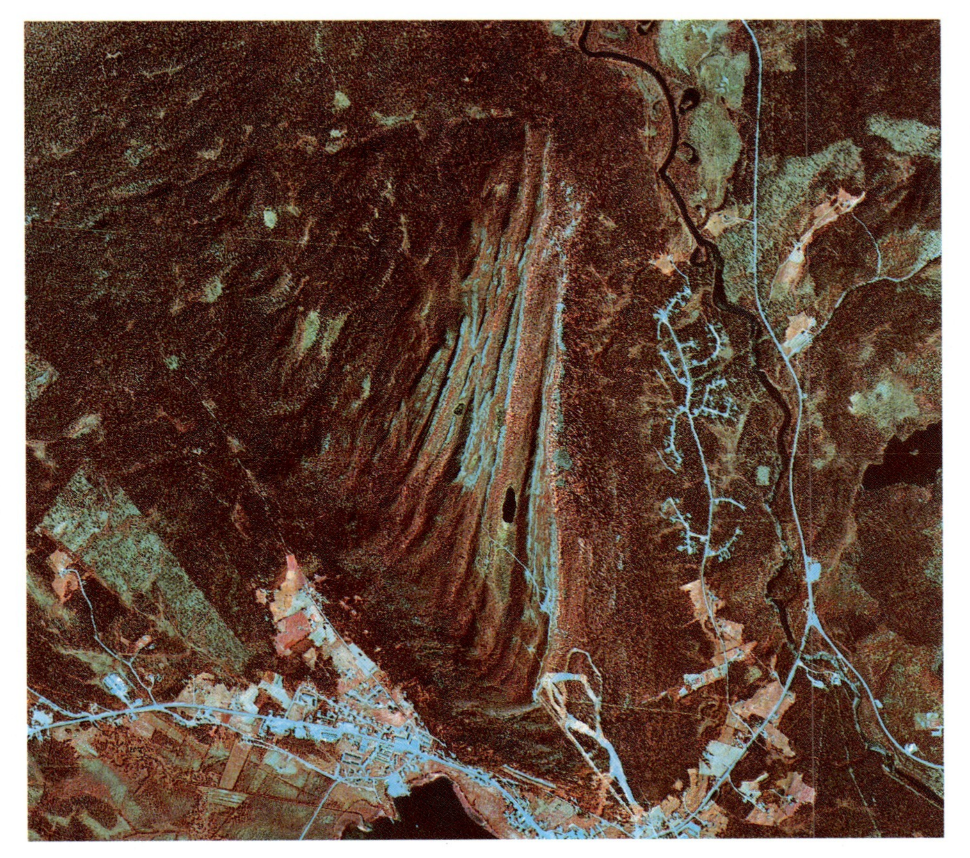

Aerial photograph of Funäsdalen. The well-defined topographic ridges north of Funäsdalen correspond to 600 million years old dolerite dykes in the Särv Nappe.

der parts of the sedimentary cover sequence are also included here. These were also deposited along the margin of the continent Baltica and include feldspathic sandstone, conglomerate and dolomite. In the uppermost part ot the Middle Allochthon, the so-called *Särv Nappe*, the sedimentary rocks are intruded by dolerites which are 600 million years old.

In the *Upper Allochthon*, the Seve Nappes represent the outermost part of the continent Baltica, i.e. the transition between continental crust and newly-formed oceanic crust. Amphibolite and eclogite, which represent strongly metamorphosed basalt and basic intrusive rocks, as well as ultrabasic rocks alternate with mica schist, gneiss, quartzite and marble. In the Köli Nappes, in the upper part of the Upper Allochthon, there are remnants of sedimentary and volcanic material deposited on the floor of the Iapetus Ocean. The sedimentary rocks were subsequently metamorphosed into different kinds of phyllite, mica schist and marble. Quartz keratophyre, greenschist and greenstone represent strongly metamorphosed acidic and basic volcanic rocks. These rocks were deposited during the Ordovician and Silurian in a series of basins which lay in the vicinity of island arcs similar to those we find today in the western Pacific Ocean.

In the *Uppermost Allochthon*, there are remnants of an exotic continent, probably Laurentia. This unit contains mica schist, gneiss, marble and granite and is considerably more extensive in Norway.

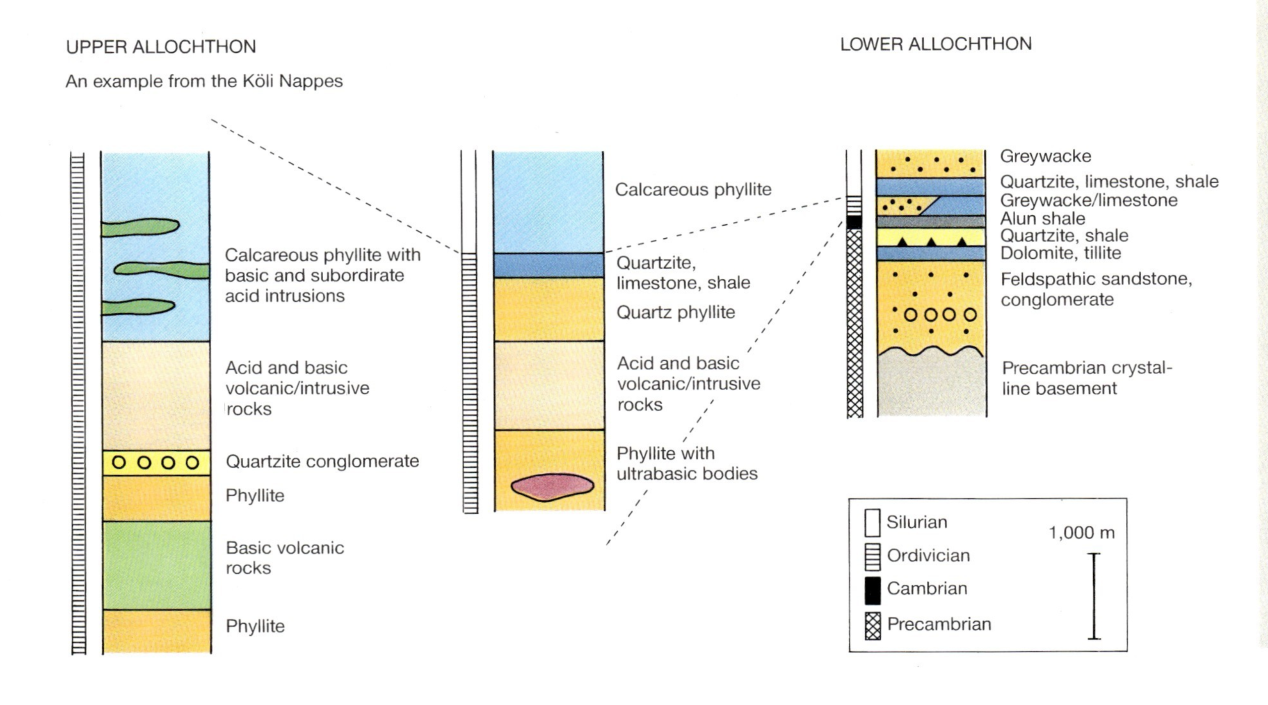

STRATIGRAPHY IN THE THRUST NAPPES

The distribution of different rocks can be seen on a bedrock map. In some areas it is possible to reconstruct the original stratigraphy. The three columns illustrate the stratigraphy of rock sequences deposited on the marginal zone of the continent Baltica during the Late Precambrian–Silurian, and on the sea-floor in connection with island arc development in the Iapetus Ocean during the Ordovician–Silurian. The Baltica sequence is based on relationships in the Lower Allochthon and the Iapetus sequences are based on two different Köli Nappes in the Upper Allochthon.

The mountain Akka in the Stora Sjöfallet National Park is composed of rocks belonging to the Seve Nappes (Upper Allochthon), thrust at least 80 km on top of rocks belonging to the Middle Allochthon.

600 mill. years

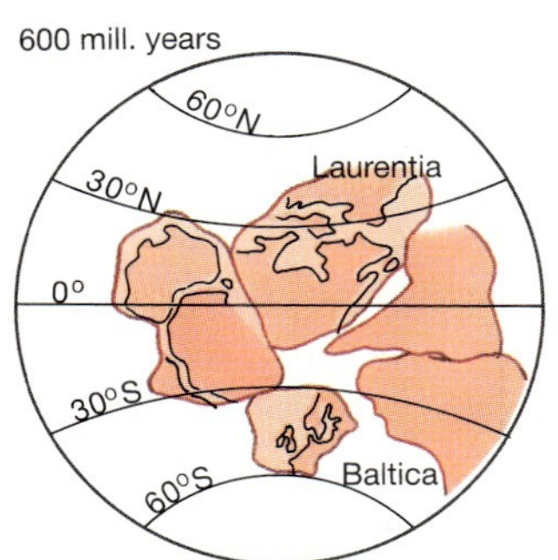

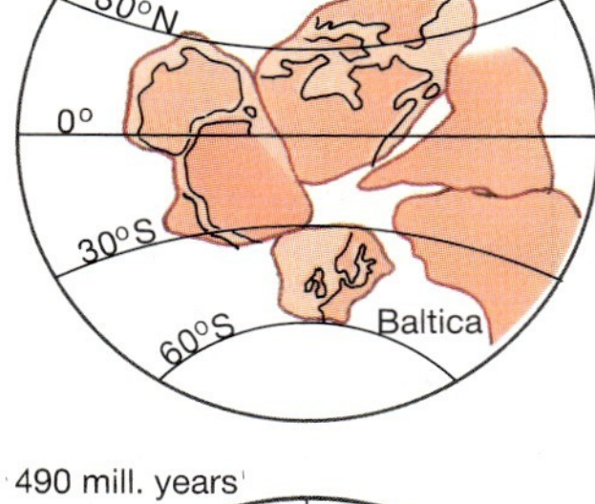

Cross-section through the Earth's crust and mantle at the time when the Caledonides were formed. The different positions of the continents at the different time intervals are illustrated to the left.

CALEDONIAN TECTONIC CYCLE

1. Rifting

The Caledonian tectonic cycle started about 800 million years ago when a major supercontinent started to rift. This continent was located at this time in the southern hemisphere. Continental rifting continued with the intrusion of dolerites ca 600 million years ago. Subsequently, a continental margin was formed with a transition zone into oceanic crust (Iapetus Ocean).

490 mill. years

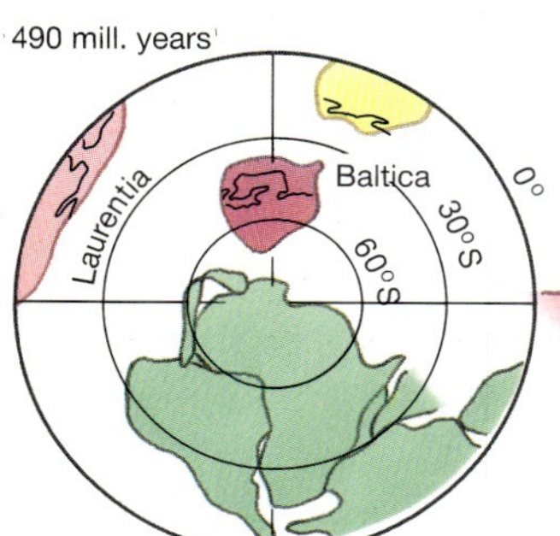

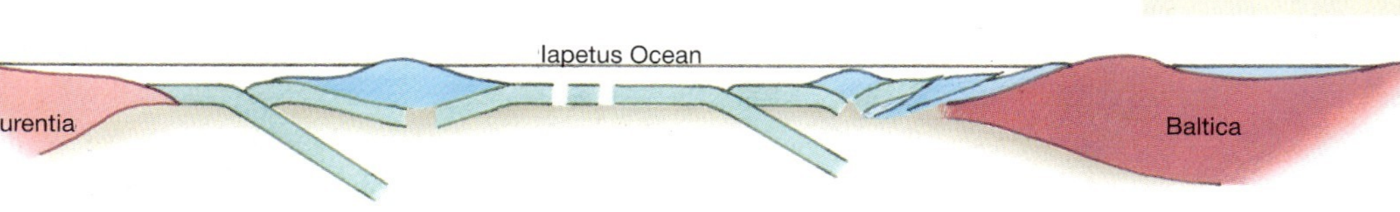

2. The first collisions

At the end of the Cambrian subduction zones formed and the Iapetus Ocean started to contract. The "western" continental margin of Baltica collided with an island arc and the outermost margin was pressed down, in some areas to a depth of about 60 km. This early collision gave rise, for example, to the eclogites in the Seve Nappes.

The Köli Nappes show that extensive sedimentary basins were present in the vicinity of volcanic island arcs at the start of the Ordovician. Most of the rocks in the Köli Nappes were deposited, however, on the other side of the Iapetus Ocean, where they collided with a continental land mass, probably Laurentia, about 470 million years ago. A subduction zone beneath Laurentia's continental margin was formed later during the Ordovician, a situation similar to that in the Andes in modern time.

470 mill. years

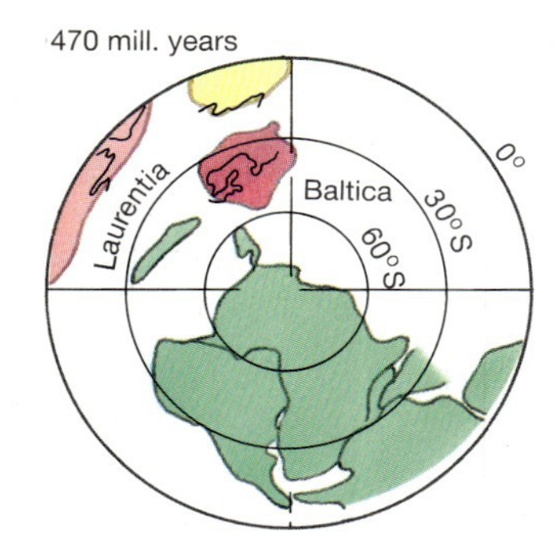

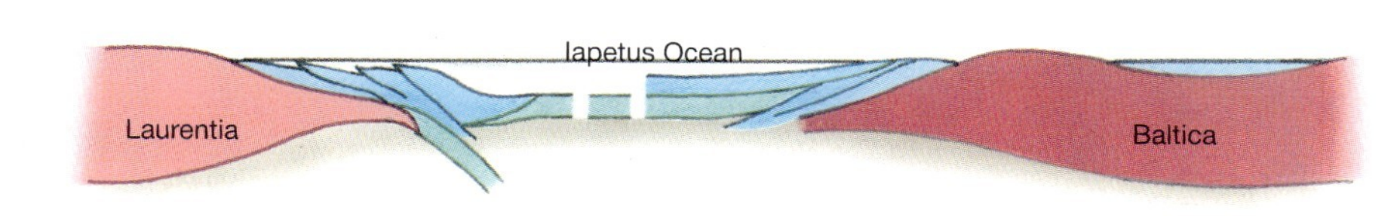

3. Continent—continent collision and subsequent collapse

During the Silurian, the continent Baltica was located at the equator. Subduction zones had consumed all the oceanic crust between Laurentia and Baltica and a continent-continent collision had resulted. At this time, Baltica was pressed down beneath Laurentia and the various nappes were pushed together and pressed on top of each other. About 400 million years ago, at the same time as thrusting continued in the east, the western part of the orogenic belt collapsed under its own weight and started to glide apart.

Formation of the Caledonian orogenic belt is the first stage in the creation of the supercontinent Pangaea. This process was completed approximately 280 million years ago.

425 mill. years

60°N
30°N
Laurentia
Baltica
0°
30°S
60°S

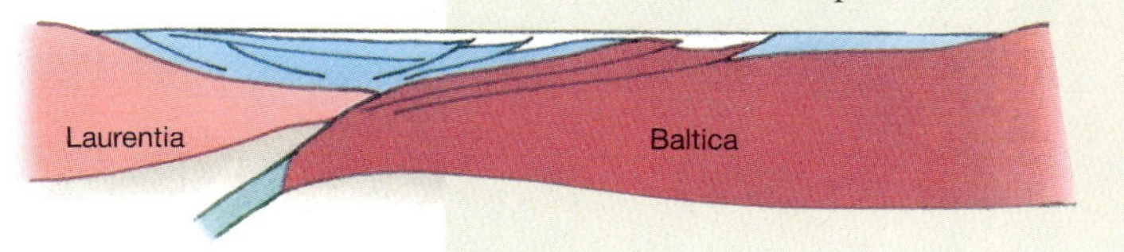

SEDIMENTARY ROCKS OUTSIDE THE MOUNTAIN CHAIN

1:10 000 000

Phanerozoic
Precambrian

Luleå
Umeå
Östersund
Falun
Stockholm
Göteborg
Malmö

(M6)

The Silurian bedrock of Gotland contains abundant fossils. Mainly brachiopods can be seen in the illustration.

The sedimentary rocks of the Visingsö Group are found in and around Lake Vättern. They are about 800 million years old, are relatively horizontal and have been preserved in a fault graben.

Reef limestone
Limestone
Alun shale
Shale
Sandstone
Breccia
Conglomerate

Stratification in areas with Palaeozoic bedrock in Sweden.

SERIES	JÄMTLAND (autochthon)	SILJANS-RINGEN	NÄRKE	ÖSTER-GÖTLAND	VÄSTER-GÖTLAND	ÖLAND	GOTLAND	SKÅNE
UPPER SILURIAN								
MIDDLE SILURIAN		*						
LOWER SILURIAN								
UPPER ORDOVICIUM								
MIDDLE ORDOVICIUM								
LOWER ORDOVICIUM								
UPPER CAMBRIUM								
MIDDLE KAMBRIUM								
LOWER CAMBRIUM								

* Orsa sandstone

Fossiliferous bedrock outside the Swedish mountain chain

Fossiliferous bedrock is normally associated with the Phanerozoic that started with the Cambrian about 570 million years ago. *Fossils*, remnants of living organisms, however, have a much longer history.

In Sweden, fossil algal colonies, *stromatolites*, occur in rocks that are more than two billion years old. In the Vättern Graben, there are fossiliferous sedimentary rocks, the Visingsö Group, which have been dated by means of microfossils to 850–700 million years. Remnants of highly-developed organisms, however, do not occur until the Phanerozoic in the form of skeletons, shells, imprints and tracks. The Phanerozoic also includes rocks without fossils.

Sweden has no Devonian, Carboniferous or Permian sedimentary rocks. However, in Jurassic rocks in Skåne, there are well-preserved, reworked microfossils from these periods. This suggests that sedimentary bedrock from the Devonian, Carboniferous and Permian were once present in southern Sweden.

Scandinavia is a peninsula of crystalline bedrock surrounded by sedimentary rocks. Of the supracrustal bedrock making up the Swedish land area, the fossiliferous rocks make up less than 20%. On the Swedish continental shelf, the situation is almost the reverse; 70–80% consists of sedimentary rocks.

PALAEOZOIC ROCKS

In Sweden, Palaeozoic rocks from the Cambrian, Ordovician and Silurian periods are present. Rocks from younger systems also occur on the continental shelf. The Cambro-Silurian bedrock outside the western mountain chain is dispersed in a number of isolated areas. The islands of Öland and Gotland form part of a large, continuous area of fossiliferous bedrock in the Baltic Depression. All these areas share a marine depositional environment for the Palaeozoic deposits. Most of these sedimentary masses were subsequently eroded away. However, in Jämtland, Närke, Östergötland and Skåne, for example, relics have been preserved in submerged blocks framed by faults. In Västergötland, a cover of resistant dolerites has provided protection against

Anticlinal structure in Ordovician limestone. Jämtland.

erosion. In the Siljan region, the deposits were preserved in a circular structure formed in connection with a gigantic meteorite impact during the Devonian.

Cambrian

The Cambrian is represented in all the areas mentioned above except the Siljan Ring. Using fossils in calcareous bedrock, mainly different species of *trilobites*, it has been possible to establish relative age, zonation and a stratigraphical correlation between the different areas.

The relatively rare remnants of organic life in Lower Cambrian rocks consist mainly of trilobites and traces of their movements. It is often possible to distinguish traces of creeping, digging and hunting. Remnants of organic life are much more common in the Middle Cambrian. At this time, life on Earth went through an almost explosive development; the arthropods, for example, occur in numerous different forms.

Solitary forms of rugose corals—a group that also formed colonies. Silurian. Gotland.

Shell of a fossilized cephalopod, *Lituites*. Ordovician, Öland.

Graptolites, a fossil group that is very important in stratigraphy. Silurian. Ovanmyra, Dalarna.

Traces of trilobites hunting for smaller animals. Lower Cambrian sandstone. Hjälmsäter, Kinnekulle, Västergötland.

Trilobite of the genus *Calymene*. Ordovician. Skogatorp, Västergötland.

Ordovician and Silurian

Deposits from the Ordovician are found in all the areas mentioned. Limestone is the most common rock except in Skåne, where shales dominate. From the youngest Cambrian until the oldest Devonian, a group of animals called graptolites lived in the sea. Numerous small skeletons of *graptolites* have been observed in marine clays and shales. These fossils are of great importance for fine zonation, relative age determinations and correlation of Ordovician and Silurian shales. On the island of Gotland, subsurface Upper Ordovician reef limestones have yielded commercially extractable amounts of oil.

On the island of Öland, Ordovician limestone constitutes the surface bedrock, whereas the entire bedrock area of the island of Gotland is of Silurian age. The Silurian rocks on Gotland contain large coral reefs that were formed in a tropical shallow sea when Scandinavia was located close to the equator. The Silurian shale deposits in Skåne, on the other hand, were formed at a considerably greater water depth from erosion products in connection with Caledonian mountain-building activity; within the Colonus Shale Trough they have a thickness of up to 2,000 m.

MESOZOIC

Deposits from the Mesozoic are limited to Skåne, Halland and Blekinge. It is in this area that the Tornquist Zone is located. This tectonic zone marks the border between the East European Platform with the Baltic Shield in the northeast and a large area of subsidence in the southwest, including southwestern Skåne, Denmark, the North Sea, and the western part of continental Europe.

Triassic

During the Triassic, sedimentary rocks up to 600 m thick were deposited in southwestern Skåne. The stratigraphy includes all three Triassic series. However, only the Upper

AGE million years	ERA	PERIOD	CHARACTERISTIC IMPORTANT EVENTS	FOSSIL IN THE GEOLOGICAL EVOLUTION
65	CENOZOIC	TERTIARY		TERTIARY The Caledonian mountain chain is uplifted. Early Tertiary marine conditions in southernmost Sweden are later replaced by continental. Finds of fossil birds and crocodiles. Limestones, formed by bryozoans and coccoliths (calc. algae), occur together with marlstones and glauconite sandstone.
145	MESOZOIC	CRETACEOUS		CRETACEOUS South Sweden transgrades by warm sea forming mainly limestones. Rich diversified marine fauna. Volcanic activity and inversion tectonic movements
210	MESOZOIC	JURASSIC		JURASSIC Block tectonic movements, volcanic activity in central Skåne. Moist subtropical climate, repeated sea level changes. Traces of dinosaurs. Rich vegetation resulting in coalseams.
245	MESOZOIC	TRIASSIC		TRIASSIC Mainly desert climate and continental sedimentation, in Late Triassic time replaced by humid conditions favourable for a rich vegetation in Skåne resulting in coal deposits.
290	PALAEOZOIC	PERMIAN		CARBONIFEROUS-PERMIAN Major volcanic activity in the Oslo Rift, rhombic porphyry formation in Bohuslän. Dolerite dykes in Skåne, plateau dolorite in Västergötland. No sedimentary rocks left in Sweden after long erosion periods.
360	PALAEOZOIC	CARBO-NIFEROUS		
400	PALAEOZOIC	DEVONIAN		DEVONIAN Gigantic meteor impact at Lake Siljan. No sedimentary rocks preserved in the Swedish mainland.
440	PALAEOZOIC	SILURIAN		CAMBRIAN-SILURIAN Caledonides formed (Scandinavian mountain chain), meteorite impacts at Granby, Tvären, Lockne and Hummeln. In Silurian time, Scandinavia was situated at the equator and was largely submerged. Accumulation of oil in the reef limestones, erosional products from the Caledonides resulted in thick shale deposits.
510	PALAEOZOIC	ORDI-VICIAN		
570	PALAEOZOIC	CAMBRIAN		

1. Paradoxides (Middle Cambrian)
2. Agnostus (Upper Cambrien-Ordovician)
3. Lituites (Ordovician)
4. Tretaspis (Upper Ordovician)
5. Monograptus (Silurian)
6. Cyrtograptus (Silurian)
7. Cycadeacae (tropical palm-like trees, U. Triassic-L. Jurassic)
8. Ginkgooites (Upper Triassic-Middle Jurassic)
9. Liostrea (Lower Jurassic)
10. Arnioceras (Lower Jurassic)
11. Hamites (Lower Cretaceous)
12. Gonioteuthis (Upper Cretaceous)
13. Globoconusa (Lower Tertiary)
14 and 15. Globigerina (Subbotina), (Lower Tertiary)

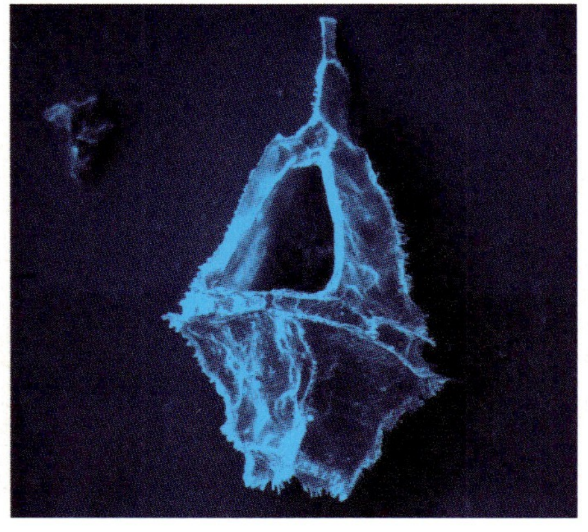

Left: Fossilized fern, *Lepidopteris ottonis*. Rhaetian Triassic. Skåne.

Right: Dinoflagellate, *Gonyolacysta*, Upper Jurassic. Skåne. Magnification x400.

Triassic is represented at the surface in northwestern Skåne with the *Kågeröd Formation*, continental deposits formed in a dry and warm climate.

A marked change from dry to warm, moist climatic conditions occurred during the latest part of the Triassic. At the same time, a tectonically quiet period was replaced by an active period, the *Kimmerian tectogenesis*, that lasted throughout the entire Jurassic and ceased during the earliest Cretaceous. These new conditions are reflected very clearly in the sedimentary record. From the latest Triassic and into the Jurassic, there was formation of delta deposits with coal-seams, clays, siltstones and sandstones. This formation, the *Höganäs Formation*, has been of economic importance for over 200 years not only for Skåne but also for the whole country. The total coal production, estimated to 30 million tonnes, has been used to provide energy in lighthouses and steam-engines, for heating purposes and in industry. In connection with the coal-mining, clays have also been exploited, some with considerable amounts of kaolin with good ceramic properties.

Jurassic

In Skåne, the Jurassic deposits are up to 800 m thick. They consist mainly of clastic sediments formed in limnic and marine environments.

In large parts of northwestern Skåne and along the border between the Danish Basin and the Fennoscandian Border Zone, the surface bedrock is of Early Jurassic age. *Plant fossils, ammonites* (cephalopods with spiral shells), *ostracods* (crustaceans) and *foraminifers* (tests of Protozoans) have been used for dating and stratigraphical correlation.

The Middle and Upper Jurassic in Skåne contains, for example, thin coal-seams, thick deposits of very pure quartz sand, claystones and sometimes thin limestone layers. The quartz sand has been used in manufacturing glass, fibre glass and abrasives. The quarries show traces of a periodically intensive tectonic activity. The strata have been tilted and faulted and in some cases inverted so that older strata have been placed on top of younger. In central Skåne, there are also traces of active *volcanism* in the form of eroded volcanic necks, basalts and tuffites.

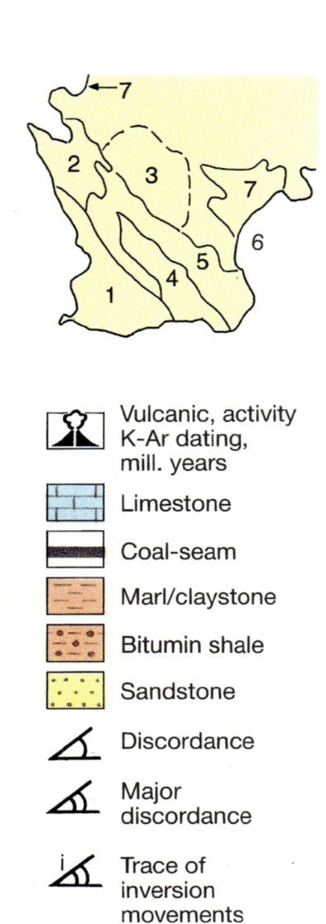

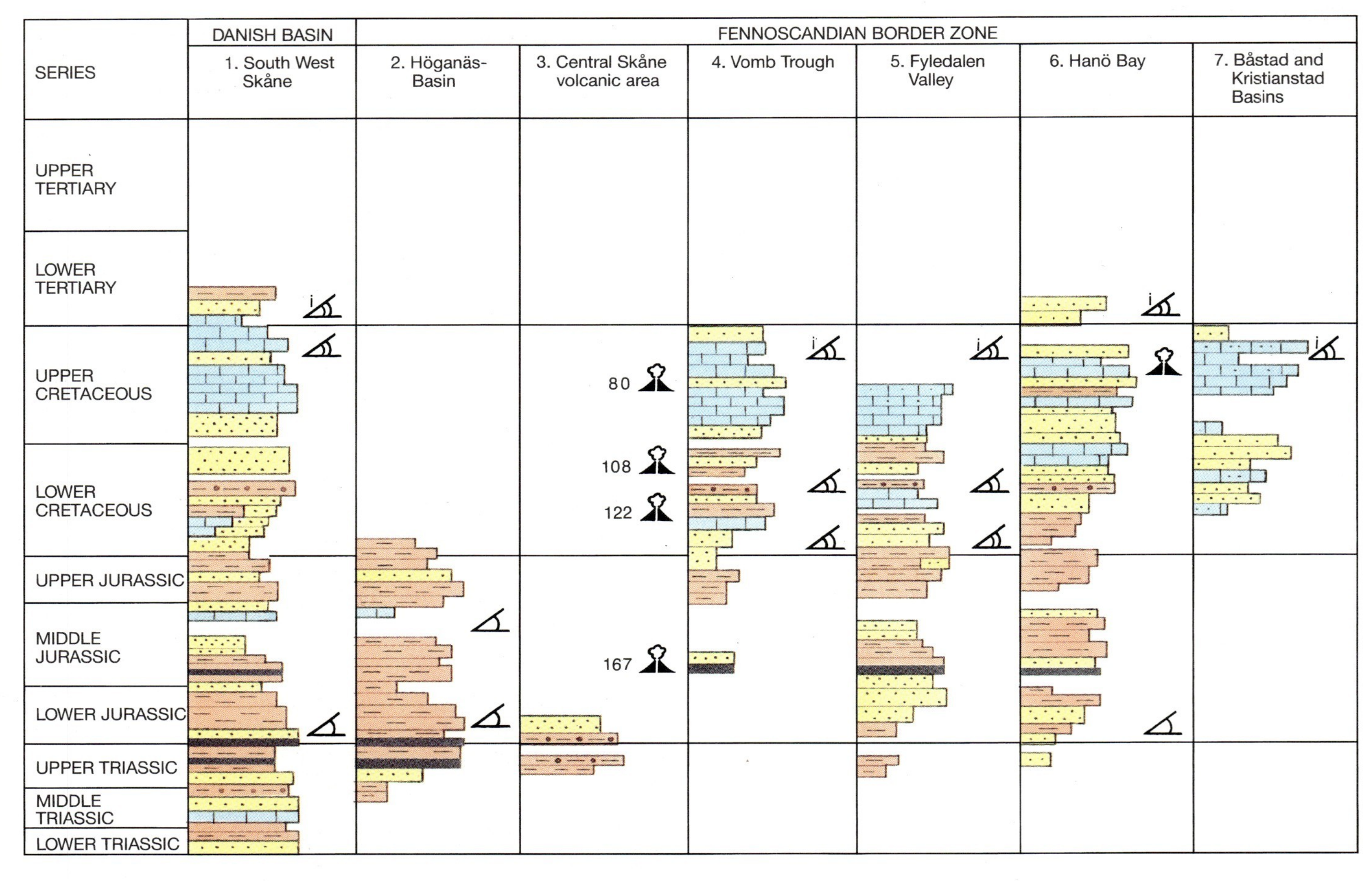

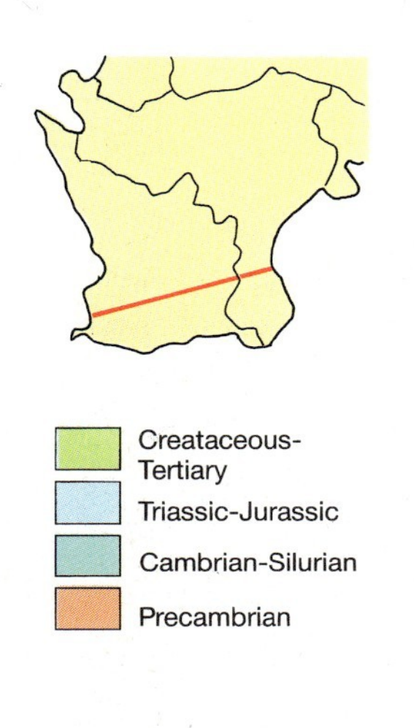

Cross-section through the Phanerozoic succession of Skåne.

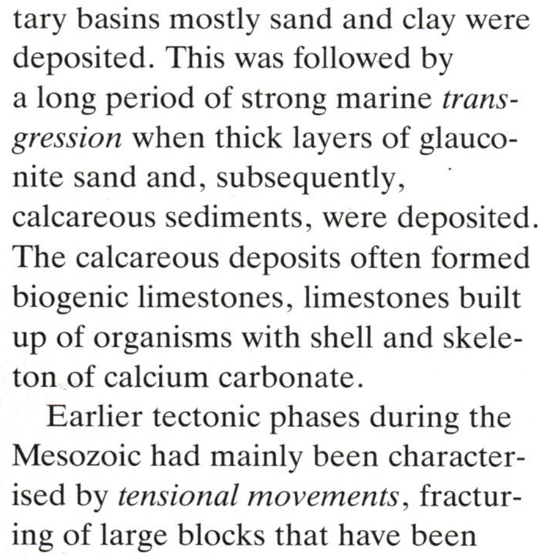

Recently discovered floral structure, *Silvianthemum suecicum*, from the Late Cretaceous in Skåne. The open flower in the illustration is, in fact, about 4 mm in diameter, whereas the photographed flower has a magnification of about x30.

Cretaceous

Deposits from the Early Cretaceous are similar to those from the Jurassic. In the northwest European sedimentary basins mostly sand and clay were deposited. This was followed by a long period of strong marine *transgression* when thick layers of glauconite sand and, subsequently, calcareous sediments, were deposited. The calcareous deposits often formed biogenic limestones, limestones built up of organisms with shell and skeleton of calcium carbonate.

Earlier tectonic phases during the Mesozoic had mainly been characterised by *tensional movements*, fracturing of large blocks that have been depressed under the weight of the sedimentary units. About 85 million years ago, a new tectonic regime, compressional in character, was initiated. Blocks were pushed together so that areas that had been sinking for millions of years to form sedimentary basins were now elevated and exposed to erosion. This *inversion tectonics* has highly characterised Skåne's geology and morphology with horsts and grabens. The Silurian block extending across Skåne in a NW–SE direction was once covered by the same type of Jurassic–Cretaceous sediments as found today in neighbouring blocks in the Vomb Basin and the Hanö Bay.

TERTIARY

The Tertiary sedimentary rocks in Skåne are considerably thinner than the several thousand metres thick deposits in the North Sea area. The thickness of the Danian limestone in Skåne rarely exceeds 65 metres. It forms the surface bedrock to the west of the line Landskrona–Ystad. At Limhamn, this limestone is quarried

METEORITE IMPACTS

1:10 000 000

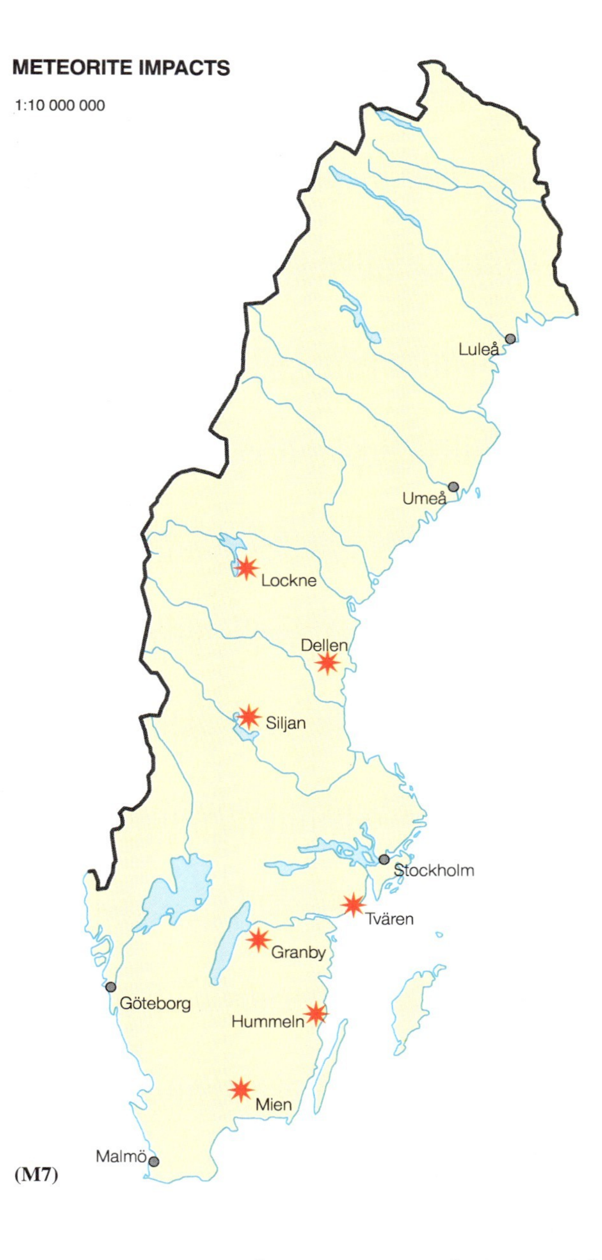

(M7)

for cement manufacture and for use in agriculture. Within small areas, the Danian limestone is covered by younger sedimentary rocks, residues that have been protected from erosion in, e.g., fault scarps and flexures.

This is where the Swedish Pre-Quaternary fossiliferous stratigraphy ends. Younger strata were eroded away during the Late Tertiary and in connection with the Quaternary glaciations.

Meteorite impact craters

Rounded bedrock structures, which vary in size from a few kilometres to about fifty kilometres in diameter, occur at several places in Sweden. These structures can more or less reliably be interpreted as craters formed by large meteorite impacts during earlier phases of the Earth's history. Among the more well-known craters are the Siljan Ring, Lake Mien in Småland, and the Dellensjöarna lakes in Hälsingland. These craters were formed during the Palaeozoic and Mesozoic periods.

IMPACT	DIAMETRE, KM	AGE
Lockne	8	Middle Ordovician
Dellen	19	Upper Cretaceous
Siljan	55	Upper Devonian
Tvären	6	Middle Ordovician
Granby	3	Lower Ordovician
Hummeln	2	Cambrian
Mien	7	Lower Cretaceous

Older, Precambrian craters are probably also present. However, these have become more or less difficult to identify as a result of folding, weathering and erosion.

Phanerozoic igneous rocks

Most of the Phanerozoic bedrock outside the Swedish mountain belt consists of sedimentary rocks. However, igneous rocks from the Phanerozoic eon can be found both here and in the surrounding basement rocks. These consist of both volcanic and intrusive units.

Examples of these igneous rocks include the alkaline massif at Alnön island (Precambrian—Cambrian) and at Särna (Carboniferous—Permian), the flat plateau dolerites overlying the Cambro-Silurian rocks of Västergötland (Carboniferous—Permian), and dolerite dykes with NW—SE strike in Skåne (Late Carboniferous). The dolerite and rhomb-porphyry dykes which intrude the Bohus granite in a N–S direction, and which formed in connection with volcanic activity in the Oslo area, also date from the transition between the Carboniferous and the Permian. To the north of the lake Ringsjön in Skåne, there are also basaltic fillings of volcanic pipes from the Jurassic and Cretaceous.

Top. Columnar jointed Mesozoic basalt, Juskushall, Ljungbyhed. When the basalt has cooled, cracks have formed that divide the rock into hexagonal pillars.
Bottom. Dolerite dyke in gneiss running in a northwestern direction. Arild, Kullen, Skåne.

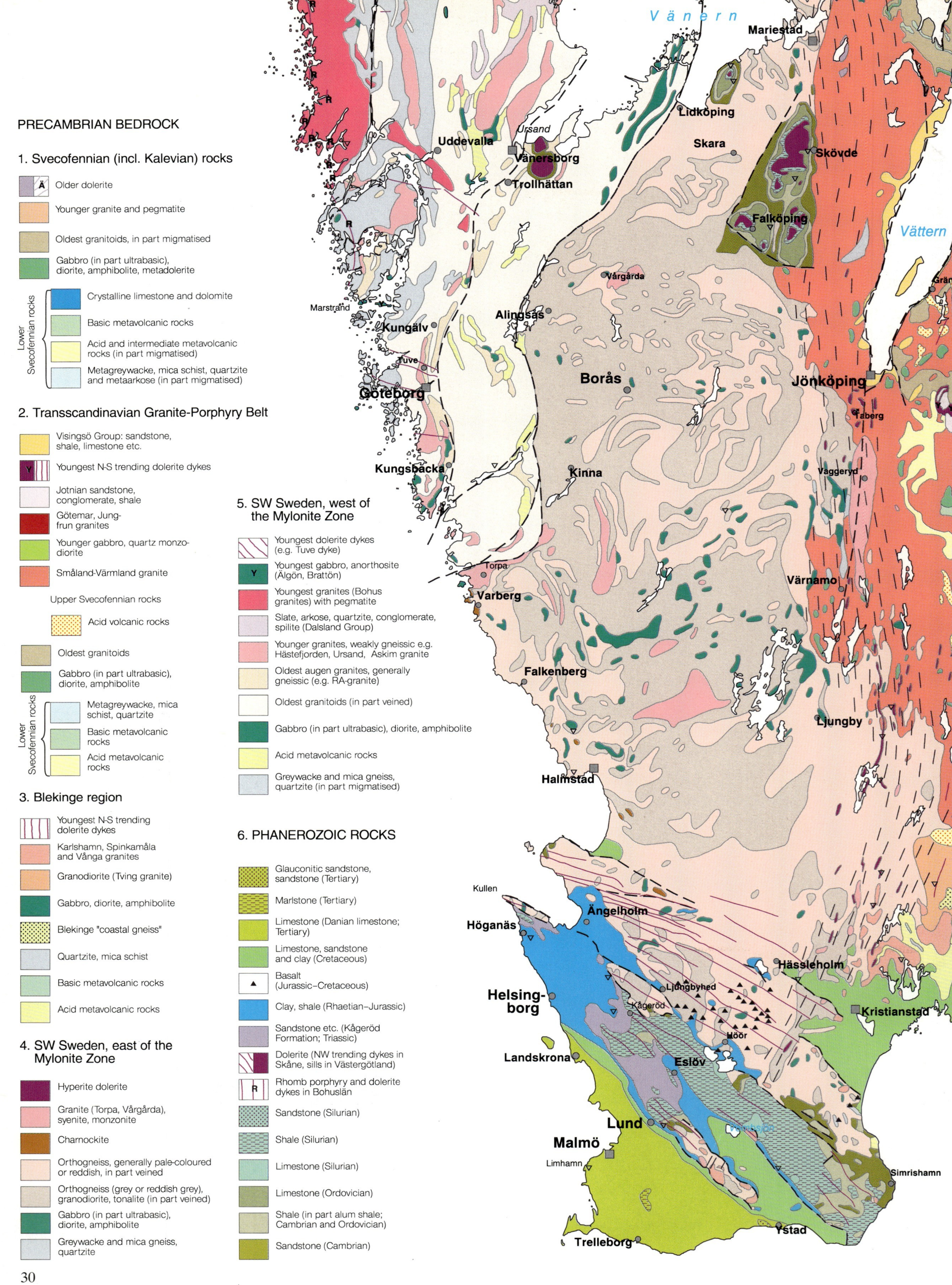
PRECAMBRIAN BEDROCK
1. Svecofennian (incl. Kalevian) rocks
Ä
Older dolerite
Younger granite and pegmatite
Oldest granitoids, in part migmatised
Gabbro (in part ultrabasic), diorite, amphibolite, metadolerite
Lower Svecofennian rocks
Crystalline limestone and dolomite
Basic metavolcanic rocks
Acid and intermediate metavolcanic rocks (in part migmatised)
Metagreywacke, mica schist, quartzite and metaarkose (in part migmatised)
2. Transscandinavian Granite-Porphyry Belt
Visingsö Group: sandstone, shale, limestone etc.
Y
Youngest N-S trending dolerite dykes
Jotnian sandstone, conglomerate, shale
Götemar, Jungfrun granites
Younger gabbro, quartz monzodiorite
Småland-Värmland granite
Upper Svecofennian rocks
Acid volcanic rocks
Oldest granitoids
Gabbro (in part ultrabasic), diorite, amphibolite
Lower Svecofennian rocks
Metagreywacke, mica schist, quartzite
Basic metavolcanic rocks
Acid metavolcanic rocks
3. Blekinge region
Youngest N-S trending dolerite dykes
Karlshamn, Spinkamåla and Vånga granites
Granodiorite (Tving granite)
Gabbro, diorite, amphibolite
Blekinge "coastal gneiss"
Quartzite, mica schist
Basic metavolcanic rocks
Acid metavolcanic rocks
4. SW Sweden, east of the Mylonite Zone
Hyperite dolerite
Granite (Torpa, Vårgårda), syenite, monzonite
Charnockite
Orthogneiss, generally pale-coloured or reddish, in part veined
Orthogneiss (grey or reddish grey), granodiorite, tonalite (in part veined)
Gabbro (in part ultrabasic), diorite, amphibolite
Greywacke and mica gneiss, quartzite
5. SW Sweden, west of the Mylonite Zone
Youngest dolerite dykes (e.g. Tuve dyke)
Y
Youngest gabbro, anorthosite (Älgön, Brattön)
Youngest granites (Bohus granites) with pegmatite
Slate, arkose, quartzite, conglomerate, spilite (Dalsland Group)
Younger granites, weakly gneissic e.g. Hästefjorden, Ursand, Askim granite
Oldest augen granites, generally gneissic (e.g. RA-granite)
Oldest granitoids (in part veined)
Gabbro (in part ultrabasic), diorite, amphibolite
Acid metavolcanic rocks
Greywacke and mica gneiss, quartzite (in part migmatised)
6. PHANEROZOIC ROCKS
Glauconitic sandstone, sandstone (Tertiary)
Marlstone (Tertiary)
Limestone (Danian limestone; Tertiary)
Limestone, sandstone and clay (Cretaceous)
Basalt (Jurassic–Cretaceous)
Clay, shale (Rhaetian–Jurassic)
Sandstone etc. (Kågeröd Formation; Triassic)
Dolerite (NW trending dykes in Skåne, sills in Västergötland)
R
Rhomb porphyry and dolerite dykes in Bohuslän
Sandstone (Silurian)
Shale (Silurian)
Limestone (Silurian)
Limestone (Ordovician)
Shale (in part alum shale; Cambrian and Ordovician)
Sandstone (Cambrian)
Vänern
Vättern
Mariestad
Lidköping
Skara
Skövde
Falköping
Uddevalla
Ursand
Vänersborg
Trollhättan
Vårgårda
Alingsås
Marstrand
Kungälv
Tuve
Göteborg
Borås
Jönköping
Taberg
Gränna
Kungsbacka
Kinna
Vaggeryd
Torpa
Varberg
Värnamo
Falkenberg
Ljungby
Halmstad
Kullen
Höganäs
Ängelholm
Hässleholm
Ljungbyhed
Helsingborg
Kågeröd
Kristianstad
Höör
Landskrona
Eslöv
Lund
Malmö
Limhamn
Simrishamn
Ystad
Trelleborg

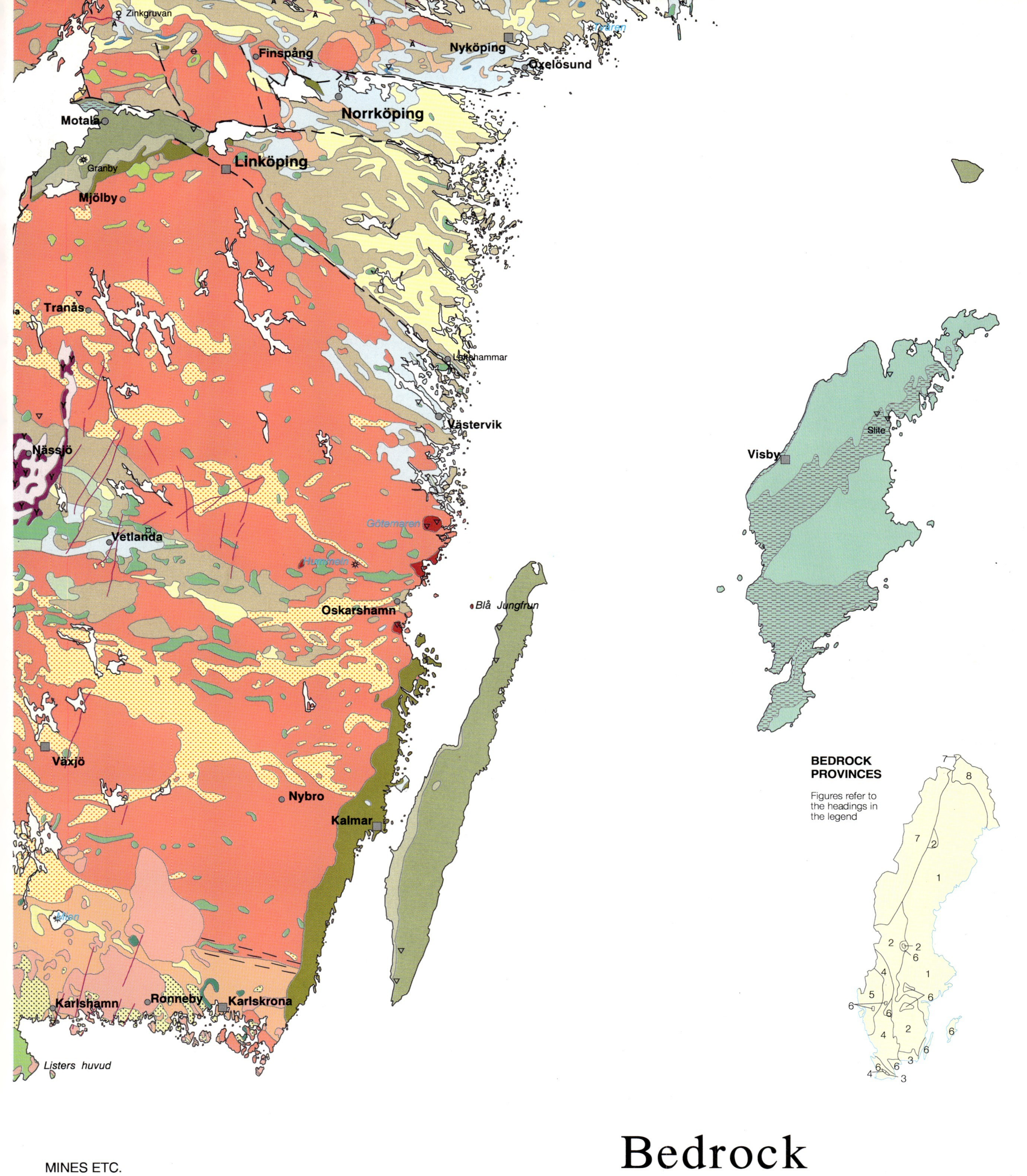

Bedrock

MINES ETC.

¤	Nickel
♀	Copper, zinc, lead
♂	Iron, titanium-vanadium
▽	Quarry
✷	Impact crater, centre
– –	Fault or ductile deformation zone
\ \	Protogine Zone, Blekinge-Småland Zone
A	Alkaline plutonic rocks

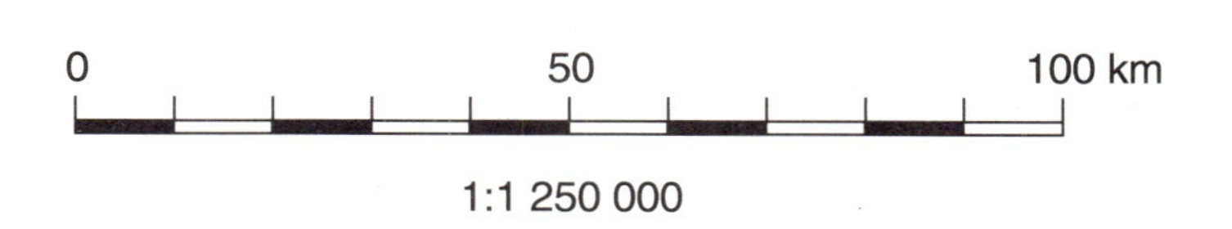

1:1 250 000

(M8)

BEDROCK

1:1 250 000

PRECAMBRIAN BEDROCK

1. Svecofennian (incl. Kalevian) rocks

- Youngest, NNW trending dolerite dykes
- Tuna dolerite
- Dolerite of Åsby-Ulvö type
- Jotnian sandstone, conglomerate, shale
- Granite, monzonite, syenite of rapakivi type
- Older dolerite
- Younger granite and pegmatite
- Oldest granitoids, in part migmatised
- Gabbro (in part ultrabasic), diorite, amphibolite, metadolerite

Lower Svecofennian rocks
- Crystalline limestone and dolomite
- Basic metavolcanic rocks
- Acid and intermediate metavolcanic rocks (in part migmatised)
- Metagreywacke, mica schist, quartzite and metaarkose (in part migmatised)

2. Transscandinavian Granite-Porphyry Belt

- Visingsö Group: sandstone, shale, limestone etc.
- Youngest N-S trending dolerite dykes
- Dolerite of Åsby type
- Öje basalt
- Jotnian sandstone, conglomerate, shale
- Dala granite
- Sandstone, conglomerate (intercalated in Dala porphyrites and porphyries)
- Dala porphyry
- Dala porphyrite
- Younger gabbro, quartz monzodiorite
- Småland-Värmland granite, Rätan granites
- Gabbro (in part ultrabasic), diorite, amphibolite, metadolerite

Upper Svecofennian rocks
- Quartzite, conglomerate, phyllite
- Basic volcanic rocks
- Intrusive Venjan porphyrite
- Acid volcanic rocks

4. SW Sweden, east of the Mylonite Zone

- Hyperite
- Orthogneiss, generally pale-coloured or reddish, in part veined
- Orthogneiss (grey or reddish grey), granodiorite, tonalite (in part veined)
- Gabbro (in part ultrabasic), diorite, amphibolite
- Acid metavolcanic rocks
- Basic metavolcanic rocks
- Greywacke and mica gneiss, quartzite

5. SW Sweden, west of the Mylonite Zone

- Youngest granites (Bohus, Blomskog granites) with pegmatite
- Slate, arkose, quartzite, conglomerate, spilite (Dalsland Group)
- Younger granites, weakly gneissic e.g. Hästefjorden, Ursand, Askim granite
- Oldest granitoids (in part veined)
- Gabbro (in part ultrabasic), diorite, amphibolite
- Gneiss of mixed origin (in part veined)
- Acid metavolcanic rocks
- Greywacke and mica gneiss, quartzite (in part migmatised)

7. CALEDONIDES

Middle Allochthon

Late Precambrian
- Quartzite, feldspathic sandstone, arkose, "hardschist", mylonite

Precambrian
- Gneiss, augen gneiss, mylonitic granite

Lower Allochthon

Late Precambrian–Cambrian
- Limestone, dolomite
- Quartzite with shale interbeds
- Feldspathic sandstone

Precambrian
- Acid intrusive rock

Autochthon

Late Precambrian–Ordovician
- Limestone
- Alum shale
- Quartzite with shale interbeds
- Tillite
- Conglomerate

Särna
Rännåsarna
Älvdalen
Sälen
Mångsbodarna
Malung
Hälsjöberget
Torsby
Långban
Filipstad
Arvika
Karlstad
Gillberga
Skoghall
Kristinehamn
Stora Le
Säffle
Åmål
Strömstad
Värmlandsnäs
Vänern
Mariestad

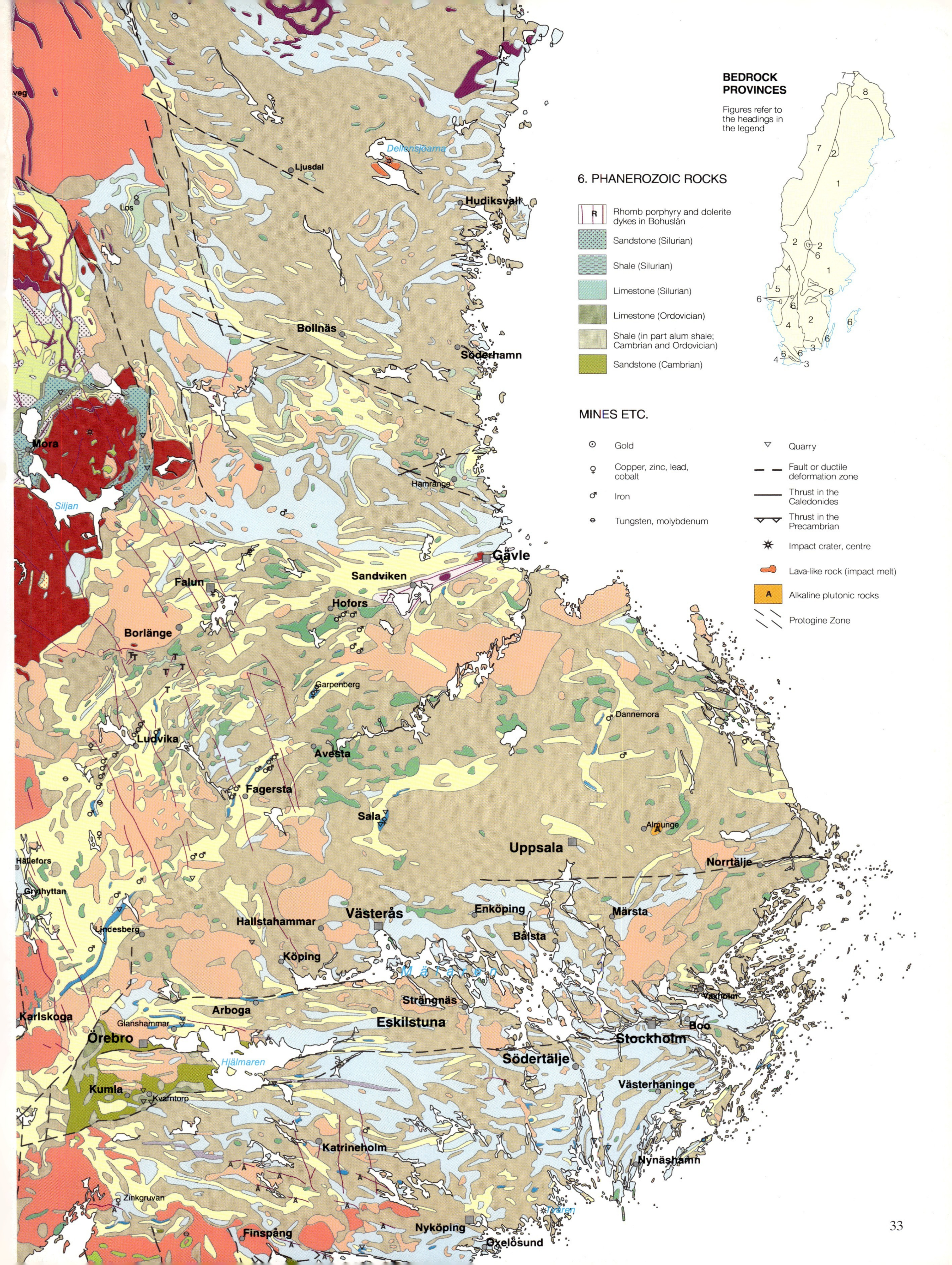
BEDROCK PROVINCES
Figures refer to the headings in the legend
6. PHANEROZOIC ROCKS
Rhomb porphyry and dolerite dykes in Bohuslän
Sandstone (Silurian)
Shale (Silurian)
Limestone (Silurian)
Limestone (Ordovician)
Shale (in part alum shale; Cambrian and Ordovician)
Sandstone (Cambrian)
MINES ETC.
Gold
Copper, zinc, lead, cobalt
Iron
Tungsten, molybdenum
Quarry
Fault or ductile deformation zone
Thrust in the Caledonides
Thrust in the Precambrian
Impact crater, centre
Lava-like rock (impact melt)
Alkaline plutonic rocks
Protogine Zone
Dellensjöarna
Ljusdal
Los
Hudiksvall
Bollnäs
Söderhamn
Mora
Siljan
Hamrånge
Gävle
Sandviken
Falun
Hofors
Borlänge
Garpenberg
Dannemora
Ludvika
Avesta
Fagersta
Sala
Almunge
Uppsala
Norrtälje
Hällefors
Grythyttan
Lindesberg
Hallstahammar
Västerås
Enköping
Märsta
Bålsta
Köping
Mälaren
Strängnäs
Vaxholm
Arboga
Eskilstuna
Karlskoga
Glanshammar
Örebro
Boo
Stockholm
Hjälmaren
Södertälje
Västerhaninge
Kumla
Kvarntorp
Katrineholm
Nynäshamn
Zinkgruvan
Tvären
Finspång
Nyköping
Oxelösund

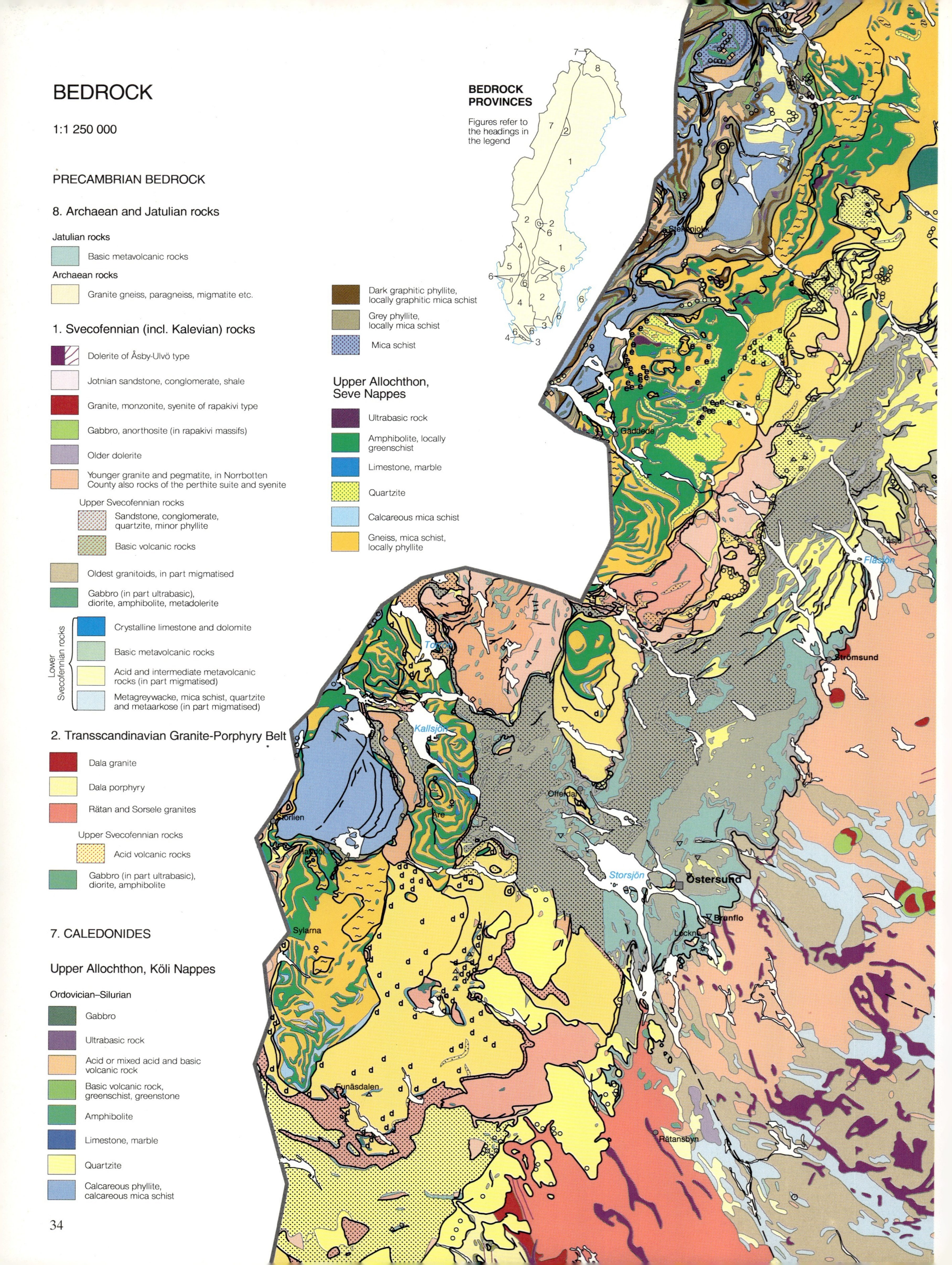
BEDROCK
1:1 250 000
PRECAMBRIAN BEDROCK
8. Archaean and Jatulian rocks
Jatulian rocks
Basic metavolcanic rocks
Archaean rocks
Granite gneiss, paragneiss, migmatite etc.
1. Svecofennian (incl. Kalevian) rocks
Dolerite of Åsby-Ulvö type
Jotnian sandstone, conglomerate, shale
Granite, monzonite, syenite of rapakivi type
Gabbro, anorthosite (in rapakivi massifs)
Older dolerite
Younger granite and pegmatite, in Norrbotten County also rocks of the perthite suite and syenite
Upper Svecofennian rocks
Sandstone, conglomerate, quartzite, minor phyllite
Basic volcanic rocks
Oldest granitoids, in part migmatised
Gabbro (in part ultrabasic), diorite, amphibolite, metadolerite
Lower Svecofennian rocks
Crystalline limestone and dolomite
Basic metavolcanic rocks
Acid and intermediate metavolcanic rocks (in part migmatised)
Metagreywacke, mica schist, quartzite and metaarkose (in part migmatised)
2. Transscandinavian Granite-Porphyry Belt
Dala granite
Dala porphyry
Rätan and Sorsele granites
Upper Svecofennian rocks
Acid volcanic rocks
Gabbro (in part ultrabasic), diorite, amphibolite
7. CALEDONIDES
Upper Allochthon, Köli Nappes
Ordovician–Silurian
Gabbro
Ultrabasic rock
Acid or mixed acid and basic volcanic rock
Basic volcanic rock, greenschist, greenstone
Amphibolite
Limestone, marble
Quartzite
Calcareous phyllite, calcareous mica schist
Dark graphitic phyllite, locally graphitic mica schist
Grey phyllite, locally mica schist
Mica schist
Upper Allochthon, Seve Nappes
Ultrabasic rock
Amphibolite, locally greenschist
Limestone, marble
Quartzite
Calcareous mica schist
Gneiss, mica schist, locally phyllite
BEDROCK PROVINCES
Figures refer to the headings in the legend
Tärnaby
Stekenjokk
Gäddede
Täsjö
Flåsjön
Torrön
Kallsjön
Strömsund
Storlien
Åre
Offerdal
Handöl
Storsjön
Östersund
Brunflo
Sylarna
Lockne
Funäsdalen
Rätansbyn

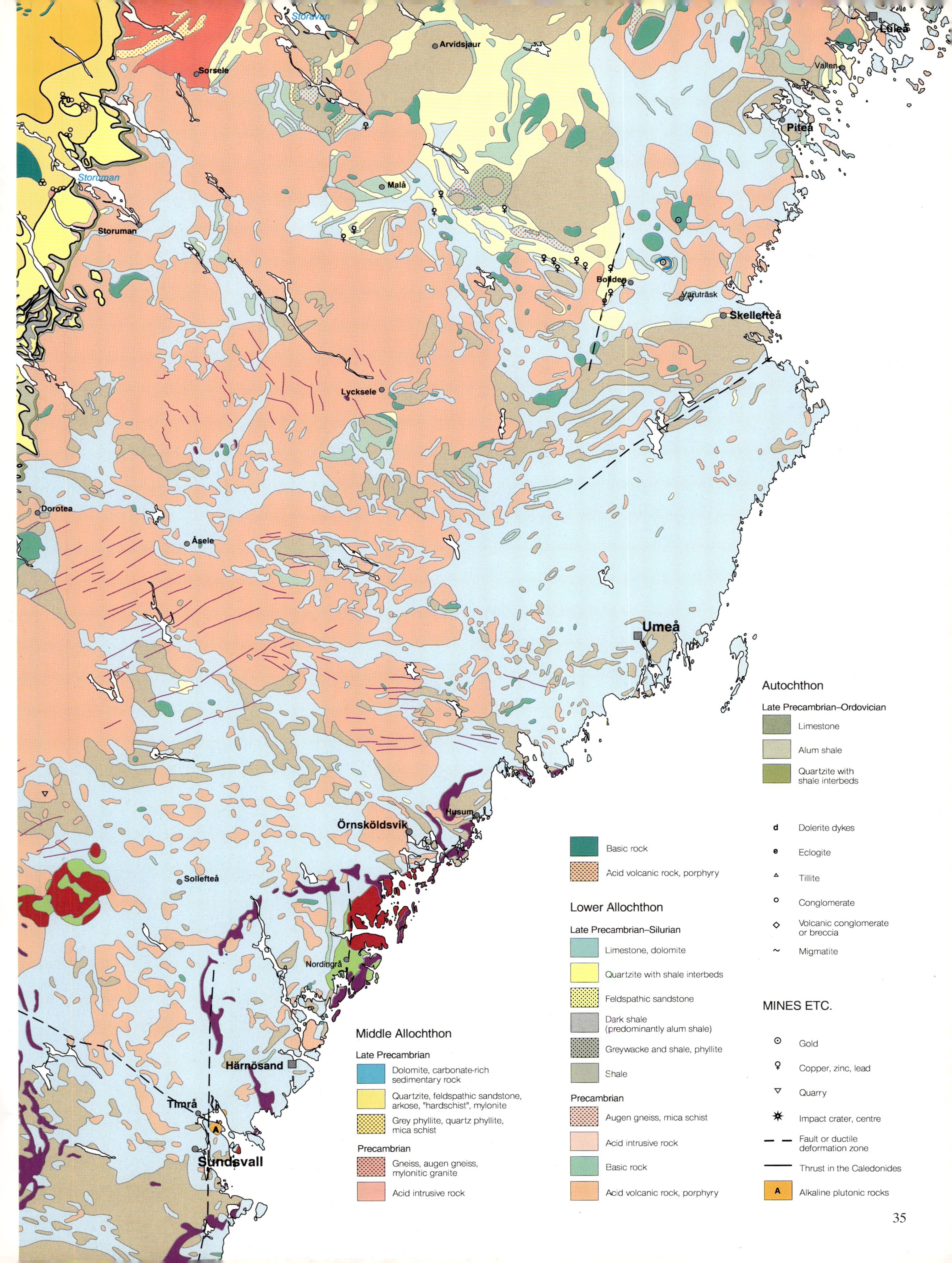
Storavan
Arvidsjaur
Luleå
Sorsele
Vallen
Piteå
Storuman
Malå
Storuman
Boliden
Varuträsk
Skellefteå
Lycksele
Dorotea
Åsele
Umeå
Husum
Örnsköldsvik
Sollefteå
Nordingrå
Härnösand
Timrå
Sundsvall
Autochthon
Late Precambrian–Ordovician
Limestone
Alum shale
Quartzite with shale interbeds
Dolerite dykes
Eclogite
Tillite
Conglomerate
Volcanic conglomerate or breccia
Migmatite
Basic rock
Acid volcanic rock, porphyry
Lower Allochthon
Late Precambrian–Silurian
Limestone, dolomite
Quartzite with shale interbeds
Feldspathic sandstone
Dark shale (predominantly alum shale)
Greywacke and shale, phyllite
Shale
Precambrian
Augen gneiss, mica schist
Acid intrusive rock
Basic rock
Acid volcanic rock, porphyry
MINES ETC.
Gold
Copper, zinc, lead
Quarry
Impact crater, centre
Fault or ductile deformation zone
Thrust in the Caledonides
Alkaline plutonic rocks
Middle Allochthon
Late Precambrian
Dolomite, carbonate-rich sedimentary rock
Quartzite, feldspathic sandstone, arkose, "hardschist", mylonite
Grey phyllite, quartz phyllite, mica schist
Precambrian
Gneiss, augen gneiss, mylonitic granite
Acid intrusive rock

BEDROCK

1:1 250 000

PRECAMBRIAN BEDROCK

8. Archaean and Lapponian-Jatulian rocks

Lapponian and Jatulian rocks

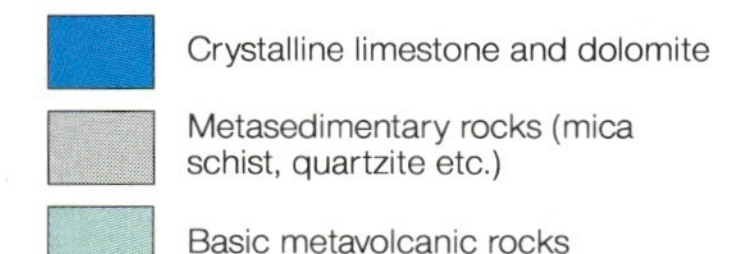

- Crystalline limestone and dolomite
- Metasedimentary rocks (mica schist, quartzite etc.)
- Basic metavolcanic rocks
- Gabbro (in part layered)

Archaean rocks

- Granite gneiss, paragneiss, migmatite etc.

1. Svecofennian (incl. Kalevian) rocks

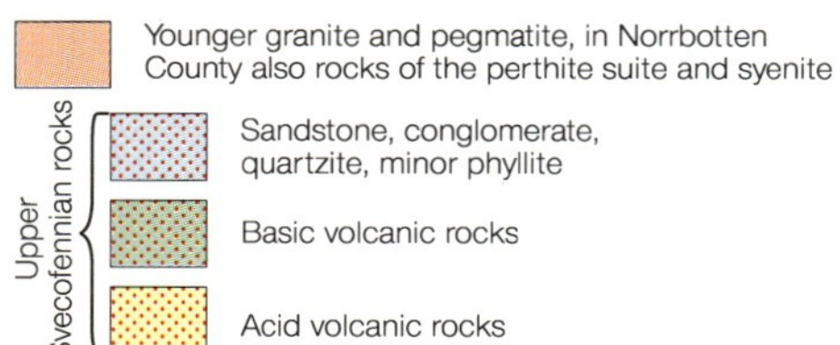

- Younger granite and pegmatite, in Norrbotten County also rocks of the perthite suite and syenite

Upper Svecofennian rocks

- Sandstone, conglomerate, quartzite, minor phyllite
- Basic volcanic rocks
- Acid volcanic rocks

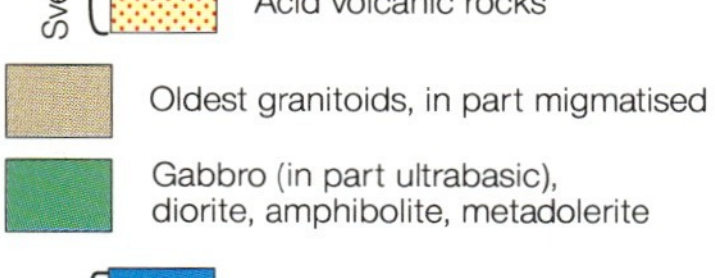

- Oldest granitoids, in part migmatised
- Gabbro (in part ultrabasic), diorite, amphibolite, metadolerite

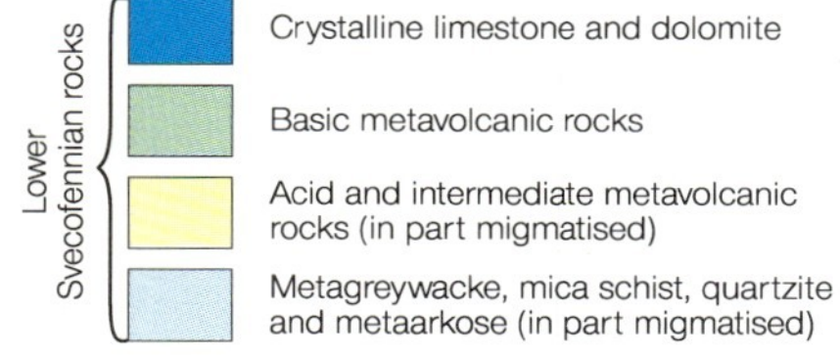

Lower Svecofennian rocks

- Crystalline limestone and dolomite
- Basic metavolcanic rocks
- Acid and intermediate metavolcanic rocks (in part migmatised)
- Metagreywacke, mica schist, quartzite and metaarkose (in part migmatised)

2. Transscandinavian Granite-Porphyry Belt

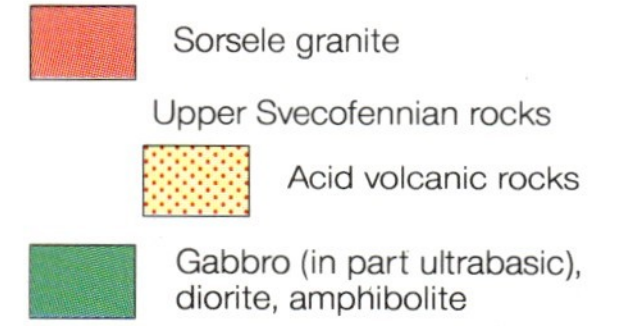

- Sorsele granite

Upper Svecofennian rocks

- Acid volcanic rocks
- Gabbro (in part ultrabasic), diorite, amphibolite

7. CALEDONIDES

Uppermost Allochthon

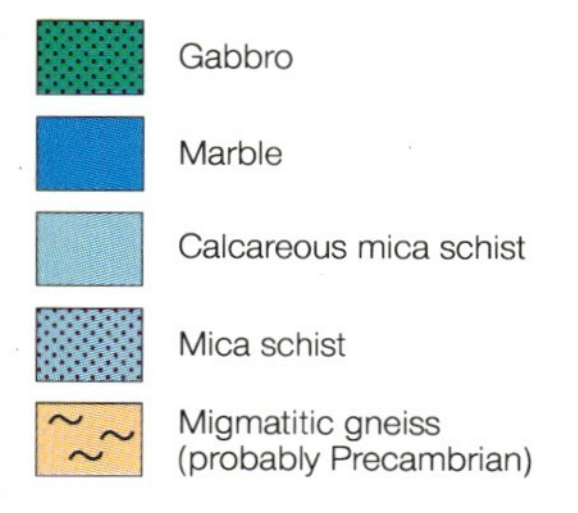

- Gabbro
- Marble
- Calcareous mica schist
- Mica schist
- Migmatitic gneiss (probably Precambrian)

Upper Allochthon, Köli Nappes

Ordovician–Silurian

- Granite
- Gabbro
- Ultrabasic rock
- Acid or mixed acid and basic volcanic rock
- Basic volcanic rock, greenschist, greenstone

- Amphibolite
- Limestone, marble
- Quartzite
- Calcareous phyllite, calcareous mica schist
- Dark graphitic phyllite, locally graphitic mica schist
- Grey phyllite, locally mica schist
- Mica schist

Upper Allochthon, Seve Nappes

- Ultrabasic rock
- Amphibolite, locally greenschist
- Limestone, marble
- Quartzite
- Gneiss, mica schist, locally phyllite
- Anorthosite-gabbro-ultrabasic rock complex (probably Precambrian)

Middle Allochthon

Late Precambrian

- Dolomite, carbonate-rich sedimentary rock
- Quartzite, feldspathic sandstone, arkose, "hardschist", mylonite
- Grey phyllite, quartz phyllite, mica schist

Precambrian

- Gneiss, augen gneiss, mylonitic granite
- Acid intrusive rock
- Basic rock

Lower Allochthon

Late Precambrian–Ordovician

- Limestone, dolomite
- Quartzite with shale interbeds
- Feldspathic sandstone
- Greywacke and shale, phyllite
- Shale

Precambrian

- Acid intrusive rock
- Basic rock
- Acid volcanic rock, porphyry
- Quartzite

Autochthon

Late Precambrian–Ordovician

- Alum shale
- Quartzite with shale interbeds

- t Trondhjemite dykes
- d Dolerite dykes
- e Eclogite
- o Conglomerate
- ◇ Volcanic conglomerate or breccia
- ~ Migmatite

MINES ETC.

- ♀ Copper, zinc, lead
- ♂ Iron, titanium-vanadium
- ▽ Quarry
- – – Fault or ductile deformation zone
- —— Thrust in the Caledonides

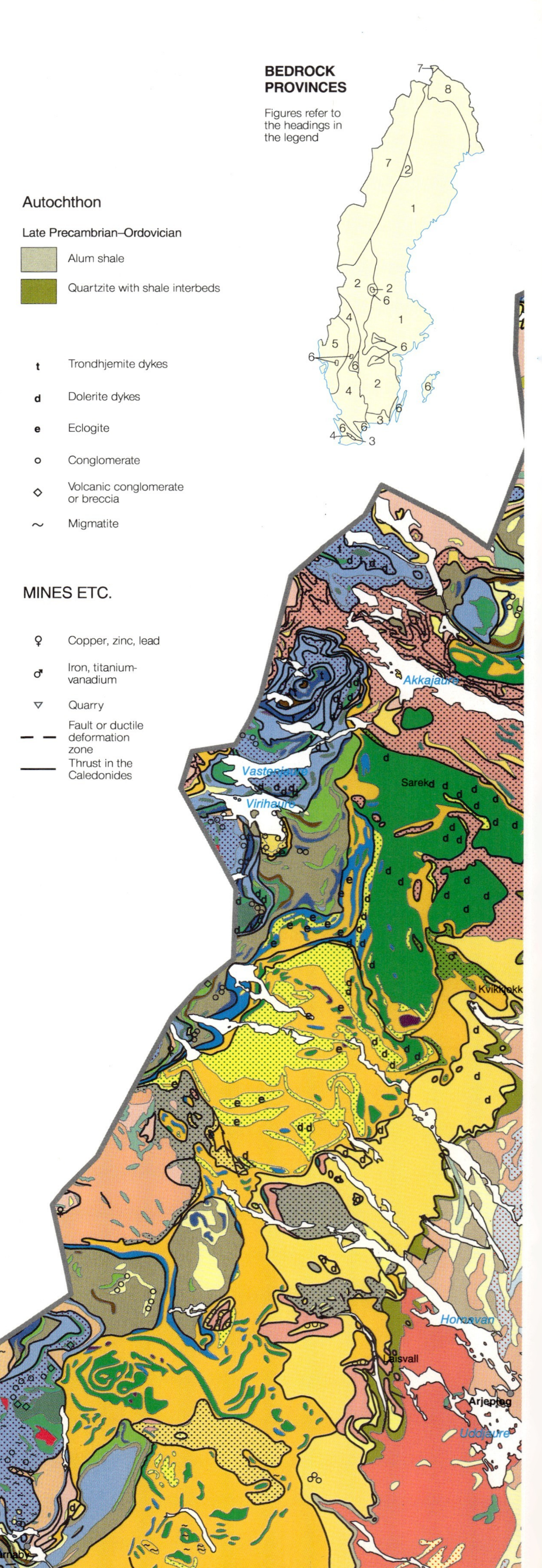

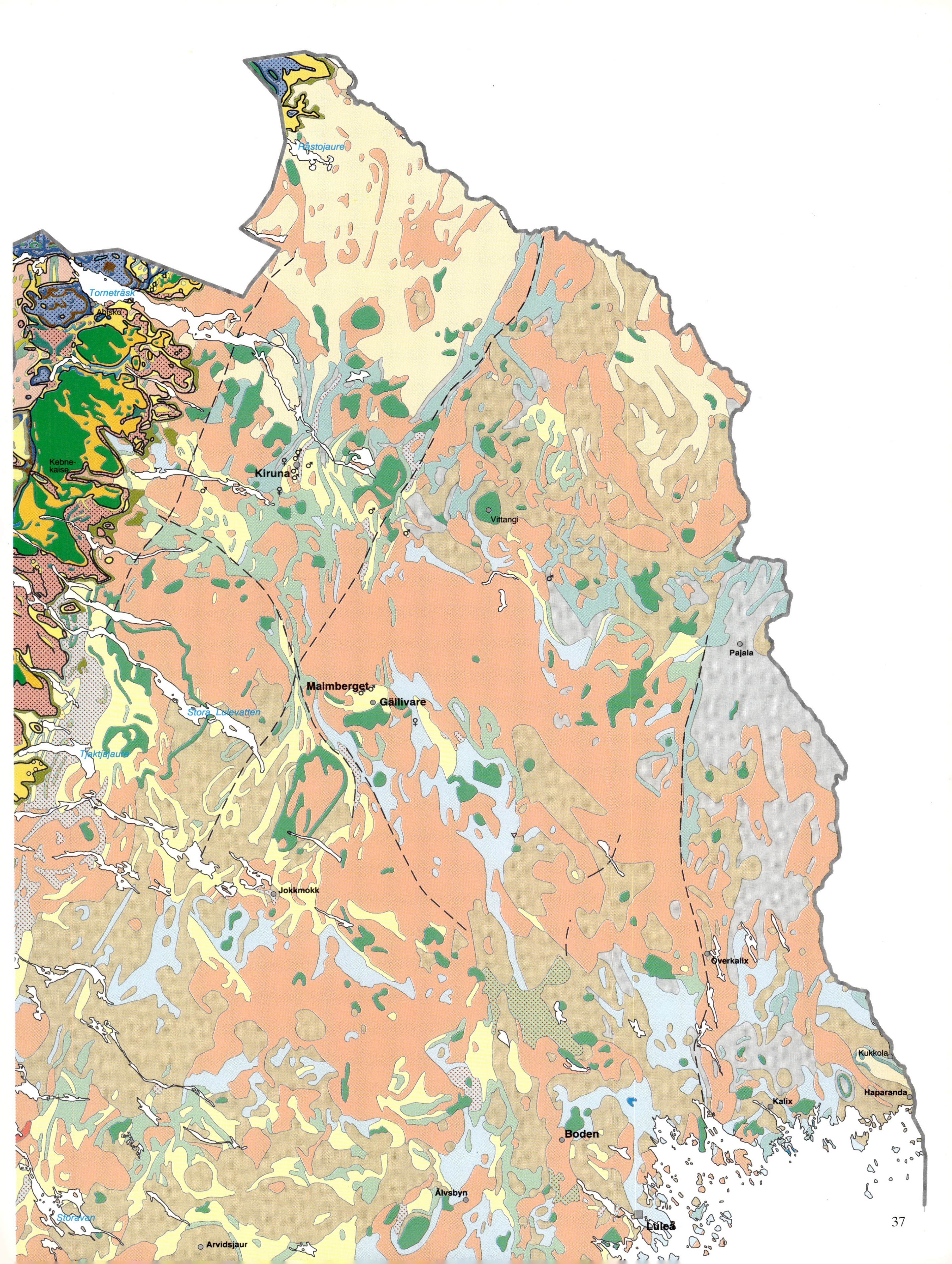
Råstojaure
Torneträsk
Abisko
Kebne-
kaise
Kiruna
Vittangi
Pajala
Malmberget
Gällivare
Stora Lulevatten
Tjaktjajaure
Jokkmokk
Överkalix
Kukkola
Haparanda
Kalix
Boden
Älvsbyn
Luleå
Storavan
Arvidsjaur

Bedrock of the Swedish continental shelf

The contintents of the Earth are surrounded by a border of shallow sea that gradually becomes deeper away from the land. On average the slope is two metres per kilometre. At a given depth, rarely exceeding 200 m, the slope increases radically. This is the outermost border of the shelf. Naturally, the depth for the change in slope can vary, but for practical reasons the limit of the *continental shelf* is placed at the 200 m depth curve.

The width of the shelf varies from a few kilometres to 40–50 kilometres. The steep slope of the sea floor outside the continental shelf is called the *continental slope*. Geologically, the shelf is part of the continents, a continuation under the sea of the coastal plains and bedrock of the land areas. The shelf is generally covered by Ouaternary sediments and young sedimentary rocks which are often quite thick. The oldest, unfolded sedimentary bedrock on the Swedish continental shelf is more than one thousand million years old and the youngest is about fifty million years old.

The sedimentary rocks preserved in the Baltic Depression provide evidence that for very long periods this area was part of a larger subsidence area in the Baltic Shield. Periods of subsidence and sedimentation were followed by land formation and erosion. Whereas the Swedish mainland is completely dominated by Precambrian crystalline rocks, the Swedish marine areas show a completely different picture. Here, sedimentary bedrock dominates, from the Bothnian Bay farthest in the north in the Baltic Depression down to the Kattegat in the southwest.

The sedimentary strata of shallow-sea areas are often an important environment for accumulation of hydrocarbons (gas and oil). They may also be favourable for other natural resources such as coal, metallic ores and industrial minerals.

BOTHNIAN BAY, BOTHNIAN SEA AND ÅLAND SEA

The crystalline bedrock in the Bothnian Bay and Bothnian Sea probably connects directly to the Svecokarelian bedrock in the surrounding land areas. However, certain other types of rock have been distinguished on the sea floor, e.g., younger dolerites, that also occur in the land areas.

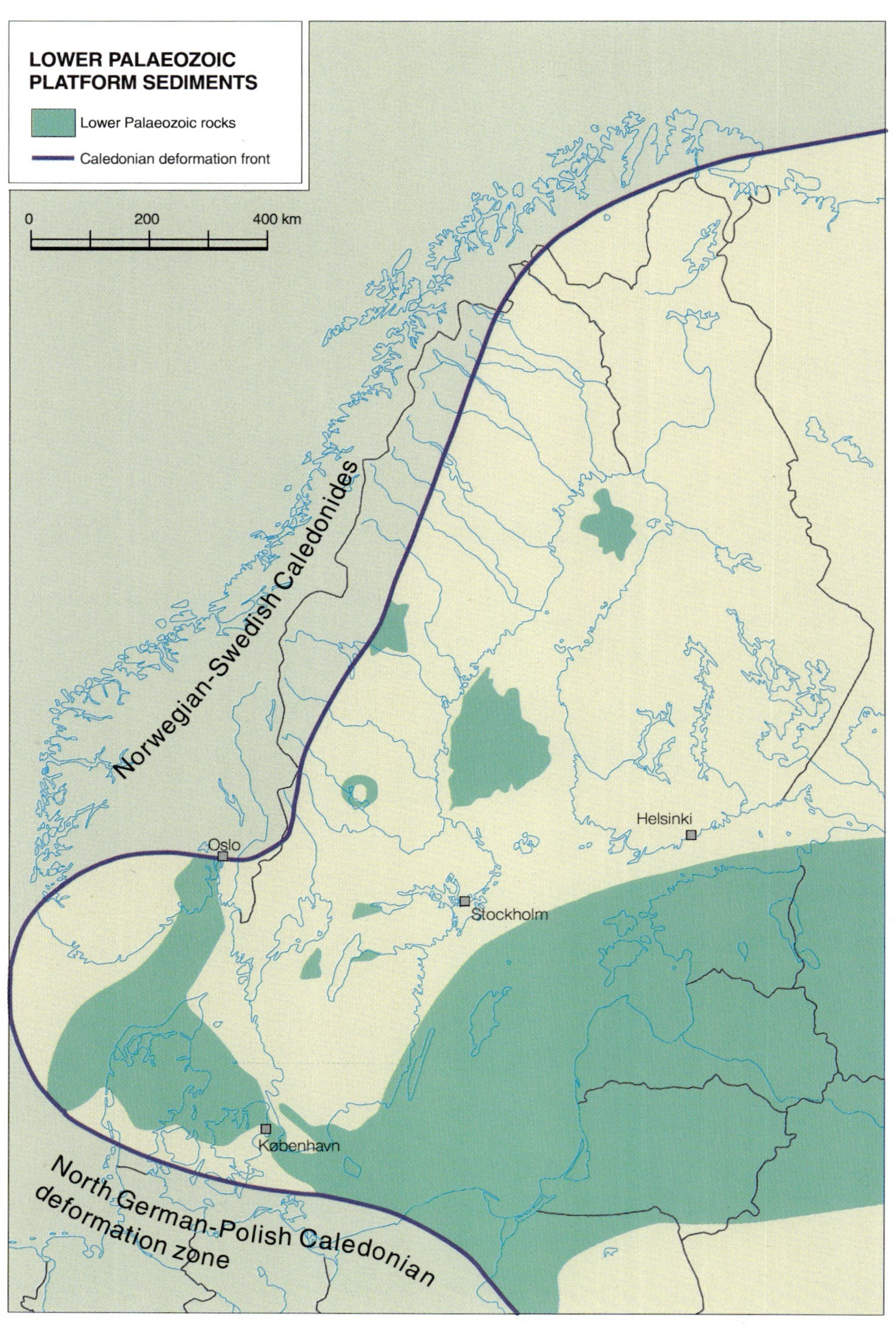

Lower Palaeozoic rocks are widely spread throughout the Baltic Basin, as well as to the west, southeast and south of the basin. The Cambro-Silurian strata are marine whereas Devonian rocks, located to the south and southeast of Gotland, are continental formations. The dispersed occurrences of older Palaeozoic strata within Swedish territory suggest that almost the entire area of Scandinavia was submerged at this time. (M9)

Precambrian sedimentary rocks

In the Bothnian Bay and Bothnian Sea, the sedimentary bedrock consists mainly of 1,600–1,200 million year old Jotnian sandstones. Sedimentary rocks from this time are found both to the east and to the west of the Baltic Depression. Examples in Sweden occur at Nordingrå, in the Kramfors (Ångermanland) district, Dalarna and Småland. This suggests that the Precambrian sedimentary basins of this age extended over large parts of Scandinavia.

The Precambrian sedimentary bedrock in the Bothnian Bay has been correlated with the *Muhos Formation* in Finland, which has long been considered to be of Jotnian age. Occurrences of microfossils indicate, however, that at least parts of this bedrock are much younger, 610–570 million years.

Cambro–Ordovician rocks

The earliest studies of bedrock composition in the Bothnian Bay and Bothnian Sea were carried out on loose blocks in coastal areas in both Sweden and Finland. Later, comprehensive geophysical and marine geological studies have increased our

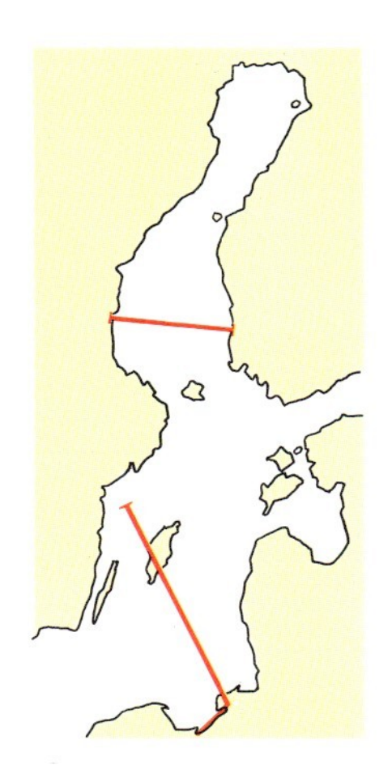

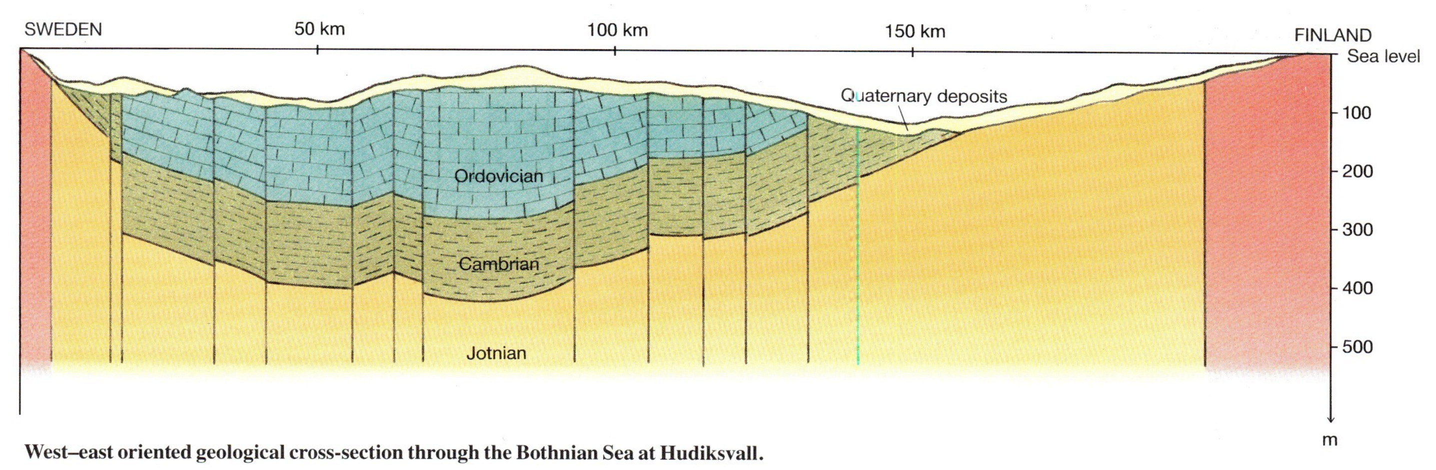

West–east oriented geological cross-section through the Bothnian Sea at Hudiksvall.

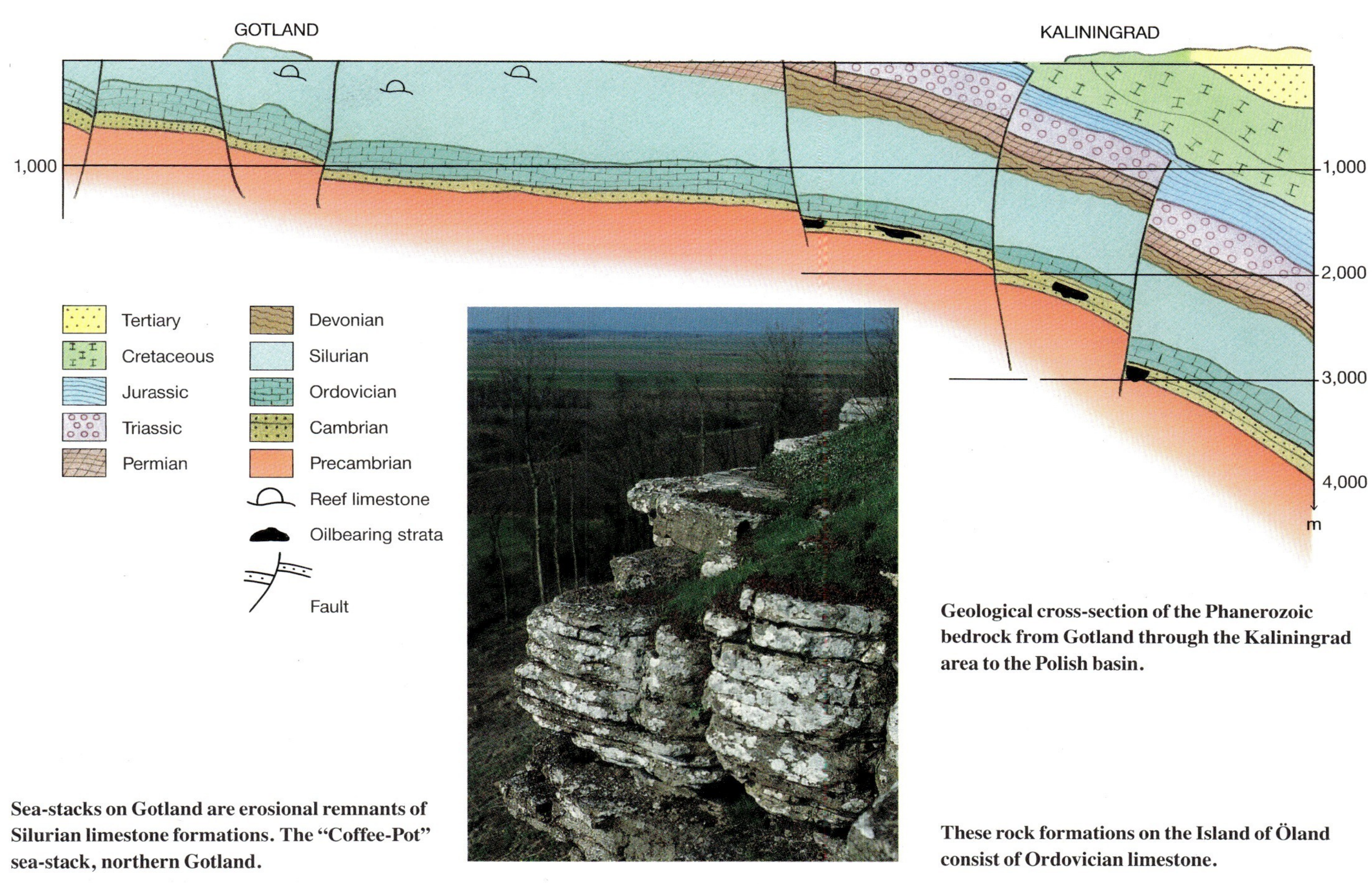

Geological cross-section of the Phanerozoic bedrock from Gotland through the Kaliningrad area to the Polish basin.

These rock formations on the Island of Öland consist of Ordovician limestone.

Sea-stacks on Gotland are erosional remnants of Silurian limestone formations. The "Coffee-Pot" sea-stack, northern Gotland.

knowledge of submarine bedrock geology. Drilling to the northeast of Gävle, on Västra Banken and on Finngrundet, has provided detailed information on stratigraphy from the Cambrian and Ordovician. Strata from these periods are represented in two separate areas. In the Bothnian Bay, only Cambrian rocks are present. In the Bothnian Sea, Ordovician limestone overlies Cambrian and Jotnian sedimentary rocks.

BALTIC SEA, ÖRESUND STRAIT, SKAGERRAK AND KATTEGAT

Knowledge of bedrock in the Baltic Sea Proper is based on comprehensive studies, commissions and oil prospecting. Information from the Baltic States has also been used.

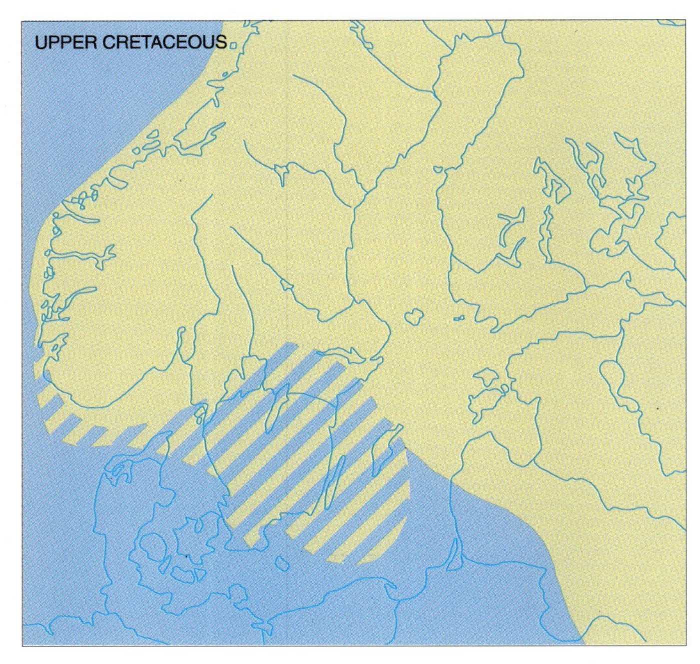

Denudation area
Limnic and deltaic deposits
Brackish-marine deposits
Marine deposits

Distribution of sedimentary deposits in the Late Cretaceous. The bands mark the probable extent of the marine deposits that have now been eroded away. (M10)

Early Jurassic deposits in Sweden and neighbouring areas. (M11)

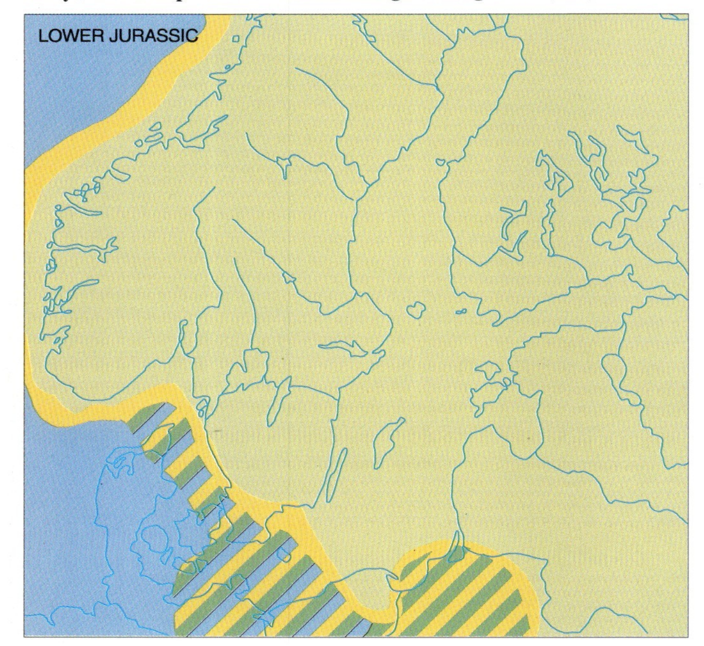

Younger Triassic deposits in Sweden and neighbouring areas. (M12)

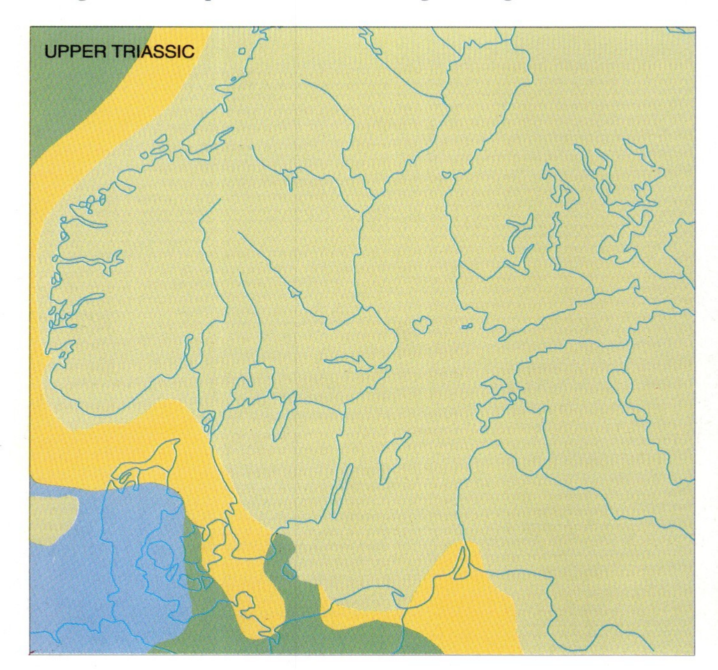

Precambrian sedimentary rocks

The southernmost known occurrences of Precambrian sedimentary rocks belonging to the Jotnian sandstone comprise the surface bedrock in the Landsort Deep and under the Island of Gotska Sandön.

Cambro-Silurian bedrock

In the Baltic Sea proper, the dominant surface bedrock on the Swedish continental shelf consists of layers formed during the Cambrian, Ordovician and Silurian. The strata dip towards the southeast and, consequently, gradually younger units occur in this direction. In most of the Baltic area, this bedrock consists of sedimentary material that was deposited on the East European Platform. The southwest margin of this platform corresponds to the Tornquist Zone. To the west of this major tectonic zone, there is a large area of subsidence including southwestern Skåne, Denmark, the North Sea and the western part of continental Europe.

Mesozoic and Tertiary bedrock

The southern and southwestern parts of the Baltic Sea which lie west of the Tornquist Zone, belong to a separate geological regime than the rest of the Baltic Depression. In this area, the Palaeozoic bedrock is largely overlain by formations from the Mesozoic and Tertiary. These younger strata form the bedrock surface in the Hanö Bay (mainly Cretaceous limestone and sandstone), to the south of Skåne (Cretaceous—Lower Tertiary), in the Bornholm-Gat (Lower Palaeozoic and Triassic), in the Öresund Strait (mainly Lower Tertiary), in the western Kattegat, and in the Swedish part of the Skagerrak (mainly Cretaceous).

Oil-drilling platform on the Swedish continental shelf. The picture was taken in November 1987 when Treasure Seeker was drilling the Yoldia–1 Well in the outer Hanö Bay for Oljeprospektering AB.

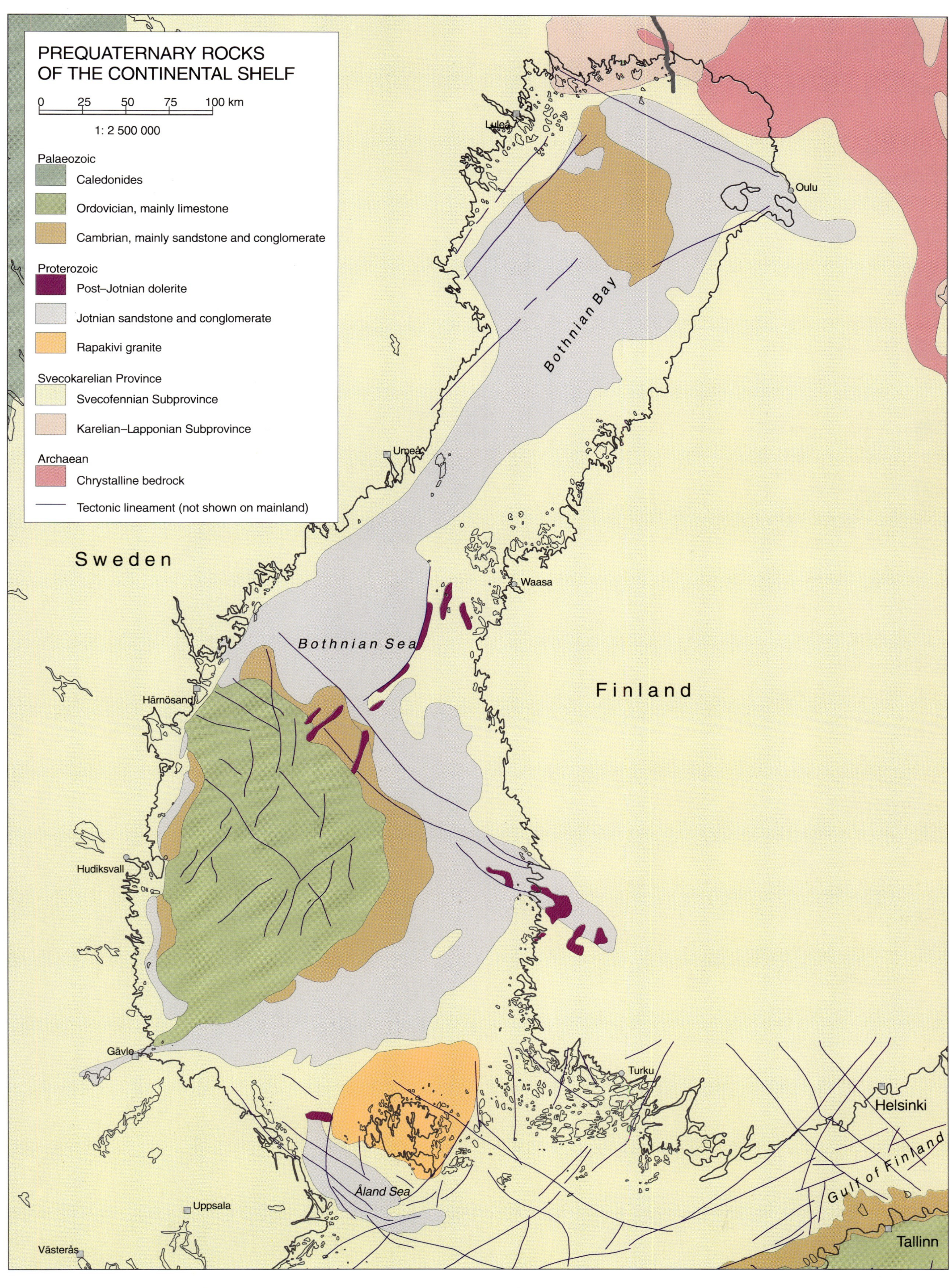
PREQUATERNARY ROCKS
OF THE CONTINENTAL SHELF
0
25
50
75
100 km
1: 2 500 000
Palaeozoic
Caledonides
Ordovician, mainly limestone
Cambrian, mainly sandstone and conglomerate
Proterozoic
Post–Jotnian dolerite
Jotnian sandstone and conglomerate
Rapakivi granite
Svecokarelian Province
Svecofennian Subprovince
Karelian–Lapponian Subprovince
Archaean
Chrystalline bedrock
Tectonic lineament (not shown on mainland)
Luleå
Oulu
Bothnian Bay
Umeå
Sweden
Waasa
Bothnian Sea
Finland
Härnösand
Hudiksvall
Gävle
Turku
Helsinki
Gulf of Finland
Åland Sea
Uppsala
Västerås
Tallinn

PREQUATERNARY ROCKS OF THE CONTINENTAL SHELF

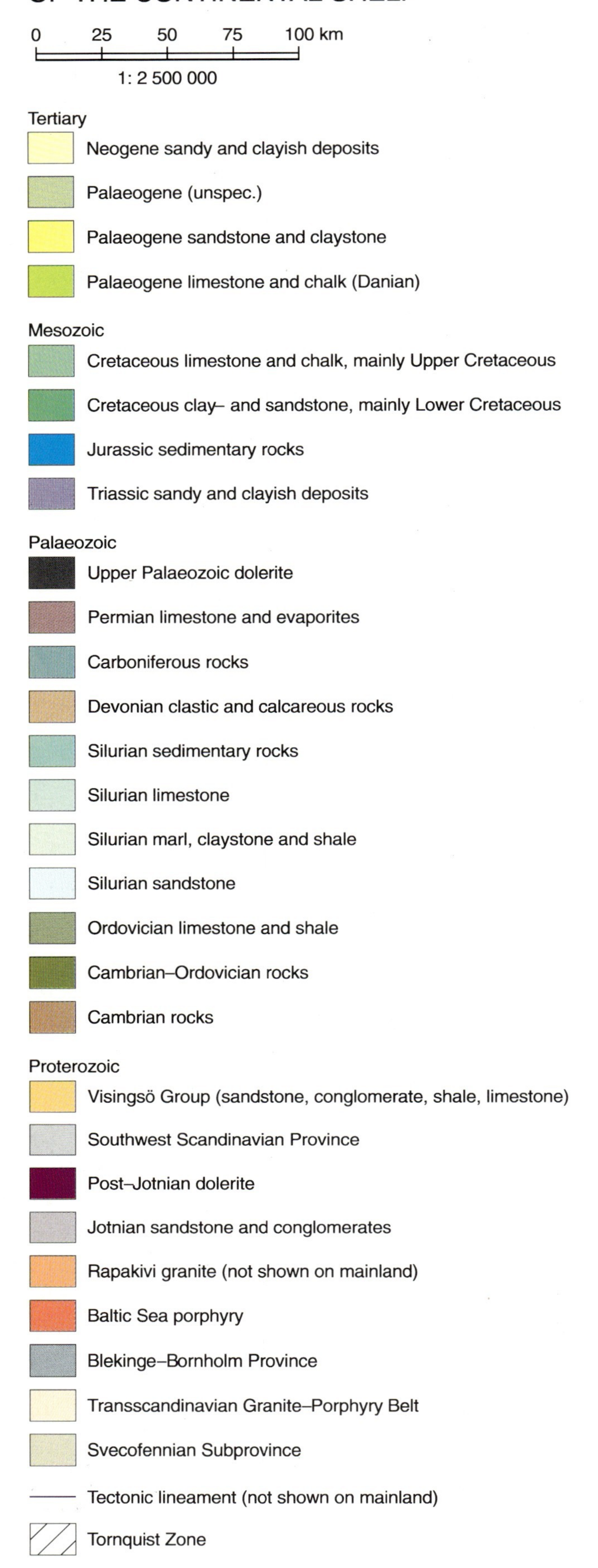

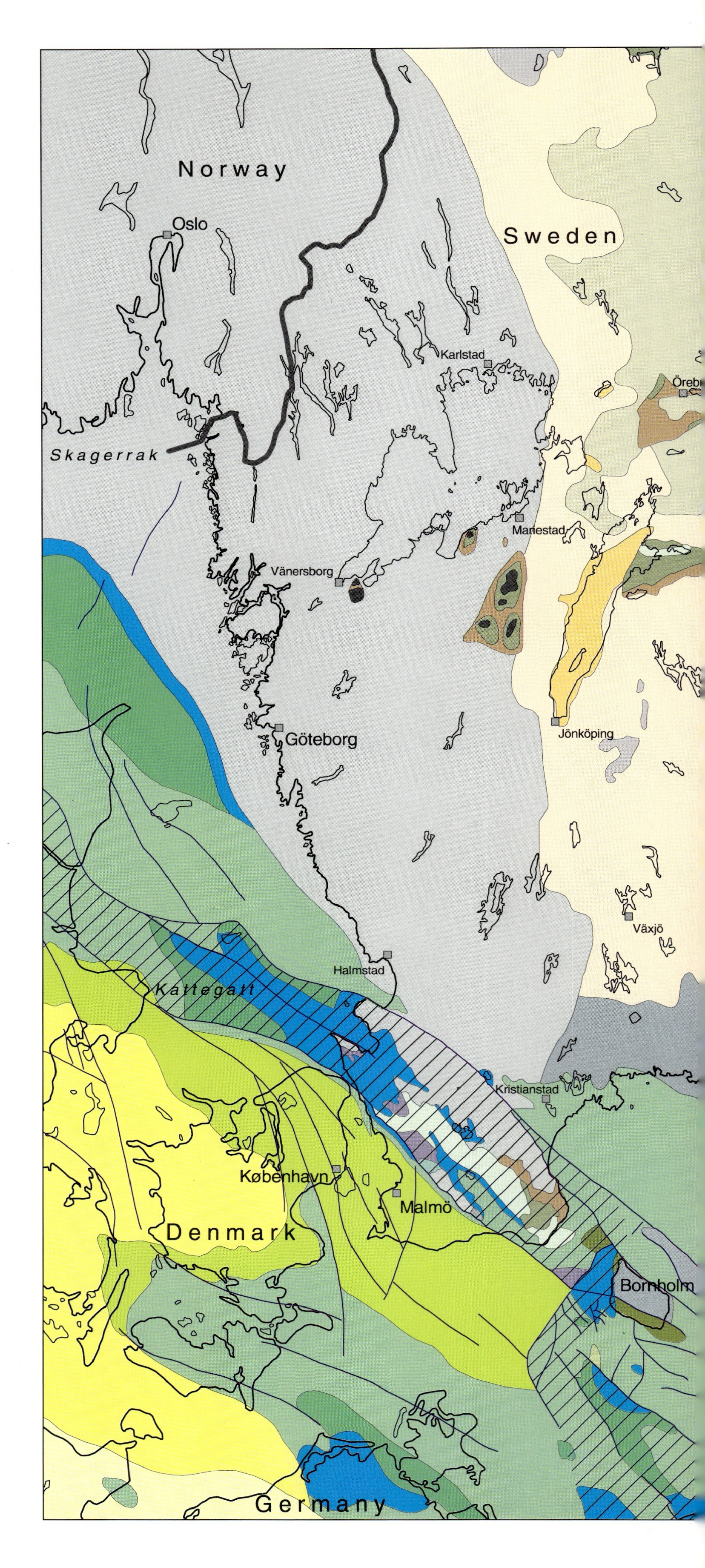

The map of marine bedrock is based on both published and unpublished material from comprehensive regional geological surveys, commissioned activities and prospecting, particularly from the hunt for oil and gas within the borders of the respective countries. There is a variable degree of accuracy in the material. (M13)

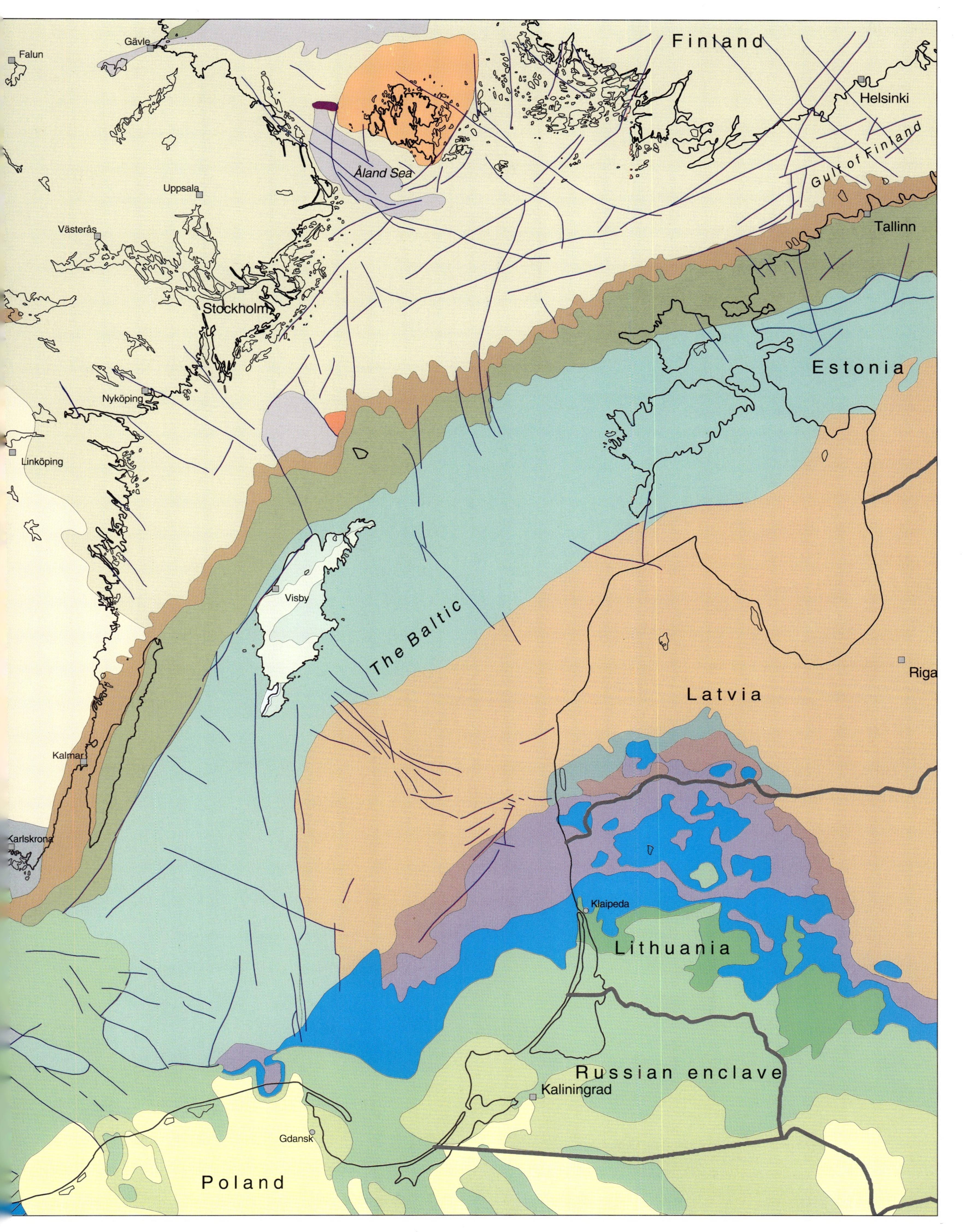
Falun
Gävle
Finland
Helsinki
Gulf of Finland
Åland Sea
Uppsala
Västerås
Stockholm
Tallinn
Estonia
Nyköping
Linköping
Visby
The Baltic
Riga
Latvia
Kalmar
Karlskrona
Klaipeda
Lithuania
Russian enclave
Kaliningrad
Gdansk
Poland

Morphology of the bedrock surface

FORM GROUPS

The various forms of the bedrock surface are presented in two different maps, the *height layer map* showing heights above sea level and the *relief map* showing whether the landscape is hilly or flat. Both maps show the forms of the land surface, not of the bedrock surface. However, these forms largely coincide since the thickness of the Quaternary cover rarely exceeds 25 metres.

Five main form groups can be distinguished: *Mountains, elongate upland areas, plains with or without residual hills, valleys*, and *slopes*. The height layer map clearly shows mountains, elongate upland areas and valleys. The relief map distinguishes plains, plains with residual hills and individual mountains and hills. It also depicts how relief can be related to the structure of the bedrock. The *morphotectonic* map shows how valleys and elongate upland areas depend on fractures and faults.

Mountains

In the present context, the term *mountains* refers to heights above 700–800 metres above sea level (m a.s.l.). These are the fjelds or "fjällen" in Swedish. The mountain chain is the result of the interaction between uplift and denudation of the land surface since the Mesozoic. Valleys were formed by river erosion and the areas in between gradually evolved into isolated mountains as a result of weathering and slope processes.

The northern part of the mountain chain has developed both within areas of Precambrian shield and Caledonian bedrock. In the south, the mountains are almost entirely found within the Caledonian nappes. The two southernmost mountains, Transtrandsfjällen and Fulufjället, are formed in Jotnian sandstone.

The effects of weathering and erosion depend largely on the rock types of the Caledonian Nappes. An east–west zonation of topography with ridges and depressions is the result. Some rock types were rapidly worn down whereas resistant parts re-

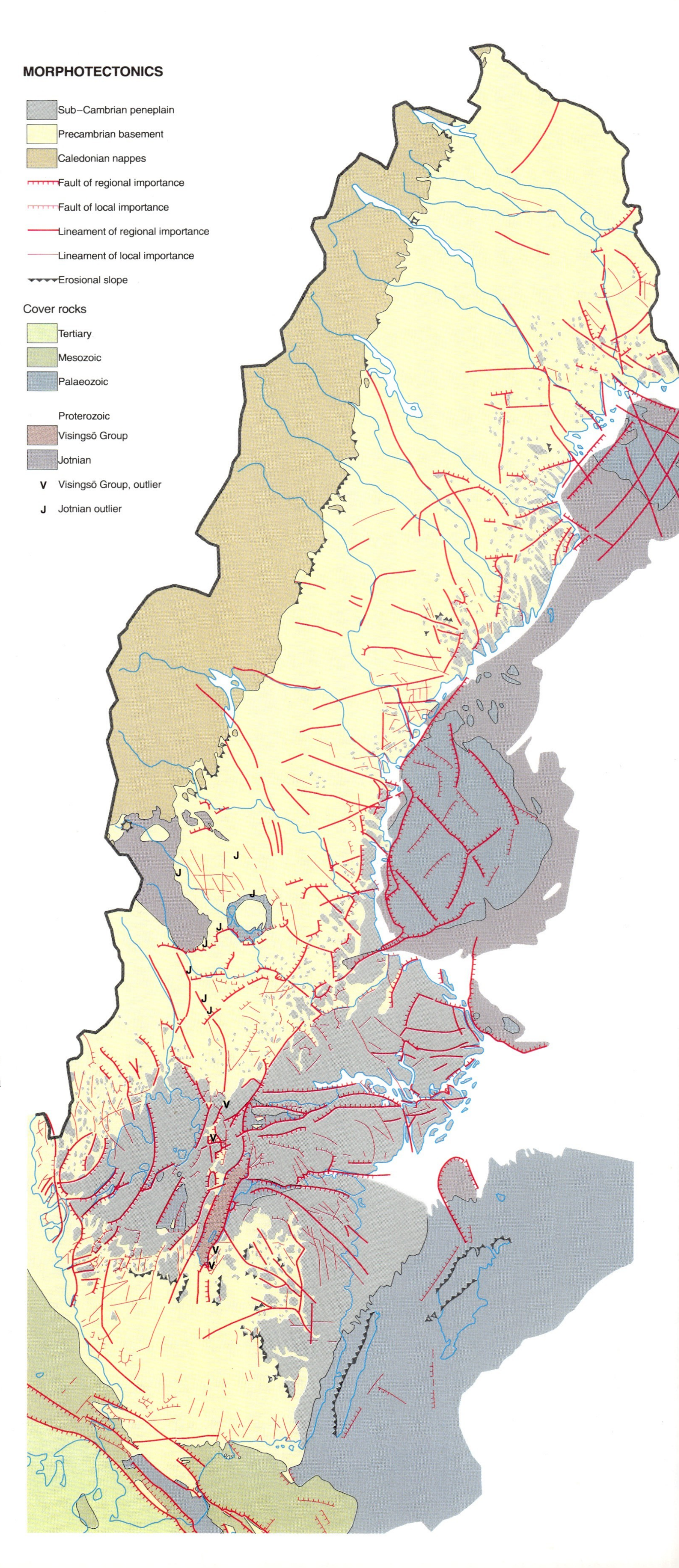

Tectonic features as reflected in the morphology in the shape of joint aligned valleys and faults. The latter have been identified in relation to denudation surfaces and cover rocks of different ages. (M14)

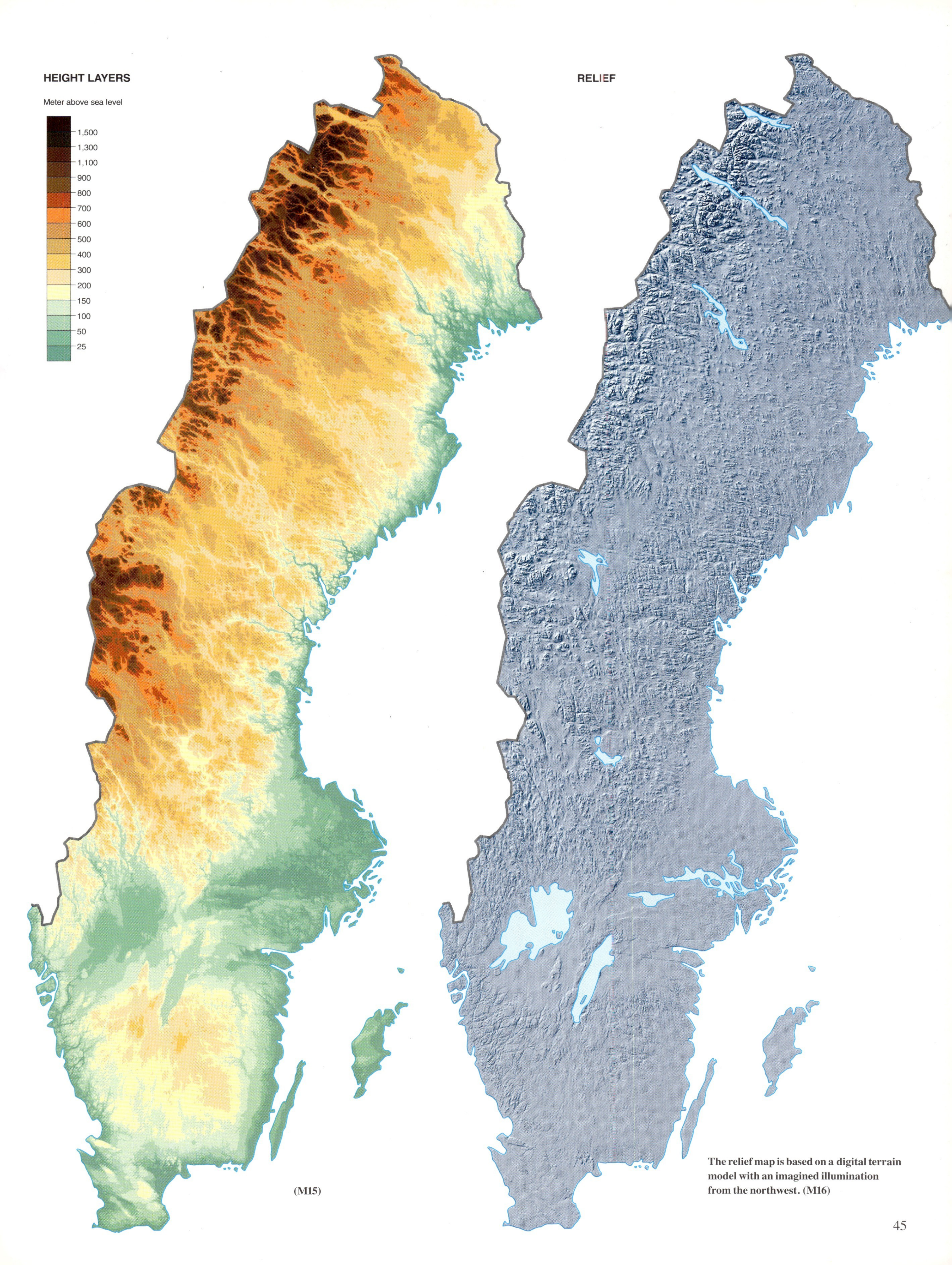

The relief map is based on a digital terrain model with an imagined illumination from the northwest. (M16)

mained as isolated mountains or hills, e.g., the quartzite ridges in Jämtland. The highest mountains consist of amphibolite, e.g., Sarek and Kebnekaise, or of mica schists.

Elongate upland areas

Elongate upland areas extend throughout the areas with Precambrian rocks in Norrland. They may be some 100 kilometres long and rise a couple of hundred metres above the surroundings. Several of these areas extend up to mountain altitudes, e.g., Ultevis to the west of Jokkmokk and the low mountains to the east of the Caledonian front at Lake Torneträsk. Between the rivers Skellefteälven and Umeälven, there is a triangular block area with the northwestern corner at Malå with Hornberget (561 m a.s.l.) as the highest point. In southern Lappland, an elongate upland area extends from Stöttingfjället (717 m a.s.l.) as far to the southeast as Ångermanbalen (488 m a.s.l.), only 50 km from the coast. In Ångermanland, Medelpad and in eastern Jämtland, the basement surface forms a broad elongate upland area which follows the coast in a 150 km broad zone.

Areas with altitudes above 200 m a.s.l., surrounding the southern part of Lake Vättern, are referred to as the *South Swedish Uplands*. They extend towards the southeast. The highest summit is Tomtabacken (377 m a.s.l.) to the southwest of Nässjö.

The summit surface mountains are remnants of old denudation surfaces at different levels. View from Säkok in the southern Sarek Mountains towards the south over the Njåtsojåkka valley.

The river Indalsälven runs from Liden towards the sea in a large joint aligned valley with steep sides. The river reached the zone of jointing at a late stage and has cut through the wide north–south ridge along the coast. The valley is remarkably narrow in comparison with those of the rivers Ljungan, Ljusnan and Ångermanälven.

Plains and bedrock hills

There are extensive *plains* within the area of Precambrian rocks. They comprise the scenery in southern Småland, along the east coast, within the Central Swedish Lowlands, in the inner parts of Norrland, and along parts of the Norrland coast. Plains located on sedimentary cover rocks occur, e.g., in Skåne and Östergötland, on the islands of Öland and Gotland, and in the provinces of Dalarna and Jämtland.

The plains often contain *residual hills*. Within the shield area, these hills may rise up to 400 m above the surroundings but usually not more than 50–200 m. Northernmost Norrland is characterized by such plains with residual hills.

Table hills (mesas and buttes) are residual hills that have developed through erosion of horizontally stratified sedimentary rocks. They are found mainly in Västergötland where they have been preserved due to protective caps of dolerite. Dolerite also gives rise to residual hills in Dalarna.

A landscape with closely spaced hills is encountered in northeastern Skåne and in the central and inner parts of southern Halland. *Hilly terrain* also characterizes Värmland, northwestern Västmanland, southern Dalarna, the inner parts of Gästrikland and southern Hälsingland.

Valleys

Well-developed river valleys are much more common in northern than in southern Sweden. In Norrland and in northern Svealand, three types of valleys can be distinguished: Well-developed valleys within the mountain chain, narrow and straight coastal valleys, and wide and irregular valleys within the hilly terrain.

Areas of flat land extend between the mountains and coastal valleys. Across these plains, rivers usually run in shallow valley channels. Rivers do not always follow their pre-glacial

Västra väggar—fault line scarp on Omberg, Östergötland.

channels, which might have been filled with till or glaciofluvial sediments. Consequently, it is sometimes uncertain where the original valley in the bedrock surface is located. The River Luleälven differs from other rivers by running in a distinct valley throughout all its course. Several large valleys, e.g., Ströms Vattudal and Indalsälven, break through the higher parts of the mountain chain with resistant rocks, and form *water gaps*.

Certain valley types follow structures in the bedrock. Examples of these are the *joint-aligned valleys*—short and shallow valleys that are found along parts of the east coast, in Blekinge, northern Halland and Bohuslän. A landscape consisting of more large-scale joint-aligned valleys occurs inland from the Höga Kusten (the High Coast) in central Norrland.

Slopes

Terrain slopes are of two completely different kinds and origin, namely *erosional slopes* and *slopes caused by faulting*.

The islands of Öland and Gotland are built up of southeast-dipping sedimentary strata of Cambrian/Ordovician and Silurian age, respectively. In areas where the outcropping layers consist of resistant limestone, erosional slopes have been formed, e.g. along the western coast of Öland.

A special type of erosional slope is the *glint*, the eastern erosional front of the Caledonian thrust nappes. Along the southern part of the Caledonian front, the bedrock consists of resistant quartzite forming the glint line with mountains such as Nipfjället, Städjan, Sömlingshågna, Sånfjället and Klövsjöfjällen. In Jämtland, the glint is absent and the landscape around Lake Storsjön is lower than the Baltic Shield area to the east. The glint with quartzite ridges reoccurs at Strömsund. Between Storuman and Kvikkjokk, the surface of the Precambrian basement attains mountain altitudes and merges with the mountains in the Caledonian bedrock. The glint is not found here but reoccurs in northern Lappland.

Slopes as a result of faulting are common in central Sweden. One example is the eastern slope of Kilsbergen. Hills located in rows may disclose old fault-line scarps. Steep coasts caused by faulting are found along Höga kusten in Ångermanland and at Hovs hallar and Kullaberg in Skåne.

AGE AND EVOLUTION OF THE LANDFORMS

The present-day scenery carries traces of a much longer history than most people are aware of. The bedrock landforms were established during warm climatic conditions, long before the glaciations. Apart from the most recent million years, warm climates have predominated in Sweden during the Phanerozoic.

The primary peneplain

The Swedish landscape is largely formed in metamorphic and igneous rocks, e.g., gneisses and granites, of Precambrian age. They constitute the *basement rock* upon which younger *sedimentary cover rocks* were deposited. Where remnants of cover rocks still remain, we can obtain knowledge concerning the nature of the old landforms and associated saprolites (weathering mantles) by studying the contact surface between the cover and basement rocks. The old land surfaces can be traced far out from the cover rocks. Many present landforms belong to extremely old landscapes that have been preserved more or less intact owing to the widespread persistance of a protective cover. Based on their positions with respect to the eroded cover rocks, these exhumed ancient landscapes are called sub-Cambrian, sub-Jurassic, etc.

During the Proterozoic, the oldest rocks in the Baltic Shield area were already severely eroded and a *denudation surface* was formed, on which the Jotnian sediments were deposited. In southwestern Sweden, this old land surface with its sedimentary cover was folded and the rocks were subject to metamorphism and transformed into a new basement. In other parts of Sweden, the old land surface with its Jotnian cover was broken up by faults. A new, almost level land surface, a *peneplain*, was formed across new and old basement as well as the sedimentary Jotnian bedrock. This is the *primary peneplain* from which all younger relief in the basement rock has formed.

The primary peneplain was slowly covered by younger sedimentary strata and, during the Ordovician, prob-

At Nordkroken in Västergötland the sub-Cambrian peneplain extends perfectly flat from beneath the Cambro-Silurian cover rocks that form Halleberg. The sub-Cambrian peneplain forms the substratum for the large plains in the Central Swedish Lowlands and in eastern Sweden.

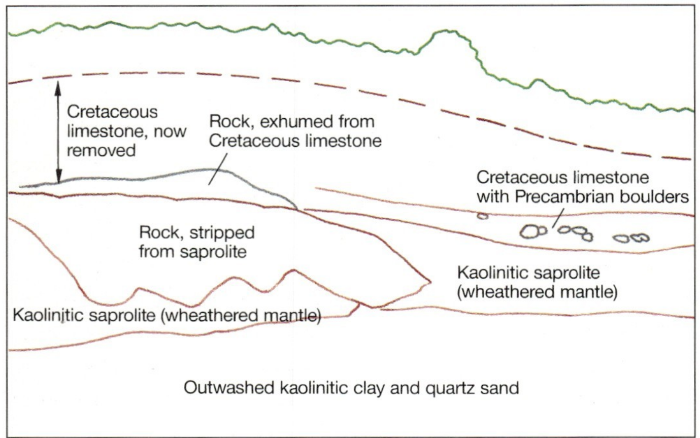

The kaolin and limestone quarry at Ivön in northeastern Skåne

The rock on the left shows an unweathered part of the basement that, in its lower part, is enveloped by kaolinitic saprolite, i.e., completely weathered basement rock. The border between unweathered (fresh) and weathered rock is sharp and has caused a steep rock-face. The upper part of the rock and the kaolinitic saprolite has been exhumed from beneath a cover of Cretaceous limestone by quarrying.

Correspondingly, depressions in the surrounding landscape have once been filled by weathered basement rock and Cretaceous sediments. The rolling forms of the fresh basement surface have appeared as a result of natural erosion of the Cretaceous limestone and the kaolin. This quarry is an important key to parts of the Swedish landscape.

View from Kjugekull over Lake Ivösjön, with Ivön and Ivöklack in the middle of the picture. Deep weathering and subsequent erosion of the weathering mantle created the rolling relief—undulating hilly land—during the Jurassic and Cretaceous.

ably all of Sweden was covered by such strata. Where the primary peneplain has been preserved up to the present, it is mainly a sub-Cambrian surface. It is not known how thick the sedimentary cover was, but calculations suggest that it might have been up to a couple of kilometres in Central Sweden. Apart from Cambrian, Ordovician and Silurian strata, the cover may also have included Devonian and Carboniferous layers.

At the end of the Silurian and during the Early Devonian, Caledonian thrust nappes were emplaced from the west on top of the basement rocks to the east. After uplift erosion gradually cut deeply into the Caledonian bedrock, and the Palaeozoic rocks covering the rest of the country were almost completely eroded. Different types of relief were formed from the primary peneplain depending on when and for how long a certain part of the peneplain surface was exposed.

Mesozoic deep weathering

Large parts of the primary peneplain were already uncovered and exposed in the Mesozoic. From the Late Triassic until the end of the Cretaceous, the climate was not only warm but also humid. This led to extensive *deep weathering* of the exposed basement and a total re-shaping of the relief.

Weathering mainly followed fracture zones in the bedrock. The depth of the weathering gradually increased with time. The longer the weathering processes were in operation, the smaller became the unweathered parts between the weathering zones. The saprolite was removed (stripped) during certain periods of accelerated erosion. If the bedrock was exposed to deep weathering and stripping only during a short time, then a *joint-valley landscape* was formed when the saprolite was removed. If the deep weathering was able to work for longer periods, the result was a *hilly terrain* once the saprolite was stripped.

During the Jurassic the sea had transgressed over Skåne, but it was not until the Late Cretaceous that it was to cover large parts of the country. A protective veneer of Cretaceous sediments was deposited over both hilly terrain and joint-valley landscapes. The Cretaceous cover remained over large areas of Sweden well into the Tertiary and remnants are still found in southern Sweden.

LATE PRECAMBRIAN

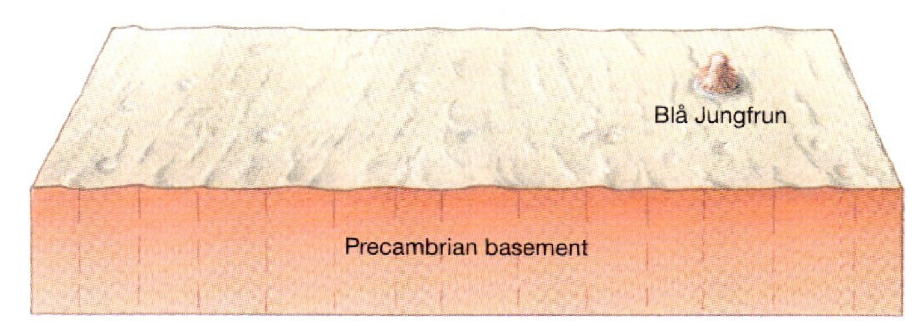

The basement surface has been eroded to a peneplain—an almost flat surface—with occasional residual hills.

SILURIAN

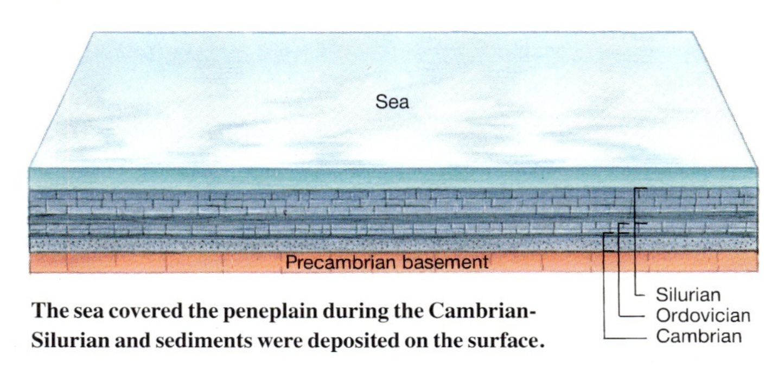

The sea covered the peneplain during the Cambrian-Silurian and sediments were deposited on the surface.

JURASSIC/CRETACEOUS

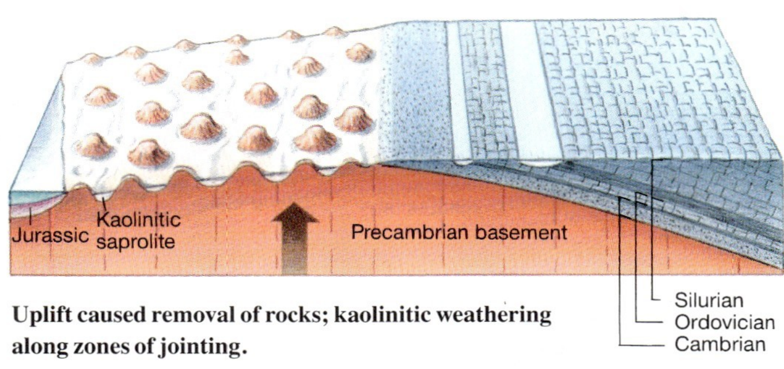

Uplift caused removal of rocks; kaolinitic weathering along zones of jointing.

LATE CRETACEOUS

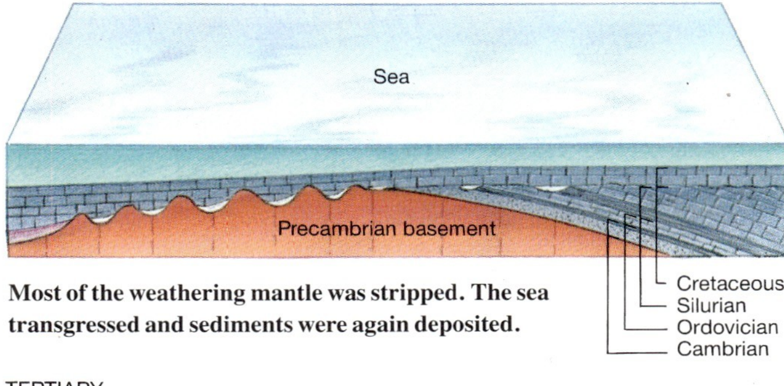

Most of the weathering mantle was stripped. The sea transgressed and sediments were again deposited.

TERTIARY

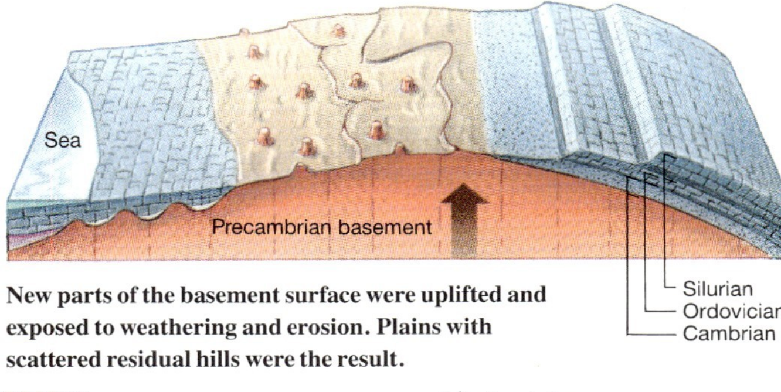

New parts of the basement surface were uplifted and exposed to weathering and erosion. Plains with scattered residual hills were the result.

RECENT

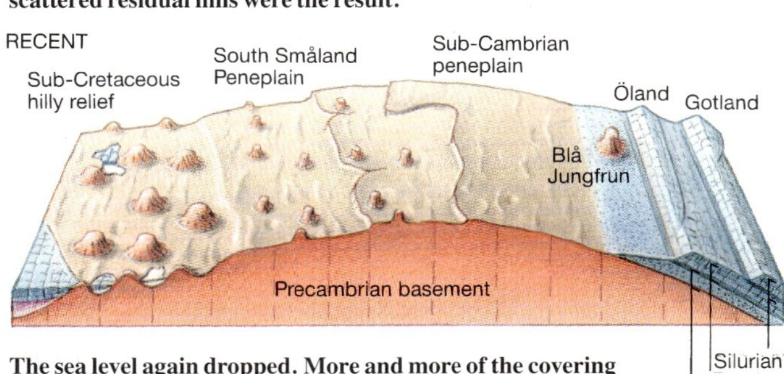

The sea level again dropped. More and more of the covering rocks were eroded and the old landscapes appeared again.

Tertiary plains with residual hills

The Cretaceous cover was gradually removed and the basement surface re-exposed when the sea level dropped. Weathering mantles of the same thickness and character as during the Mesozoic were not formed during the more variable climates of the Tertiary. Arid and humid periods alternated and erosion was periodically intense. Stripping of the weathered material exposed hills that were further denuded by surface weathering and slope processes. Plains with residual hills were developed.

New polycyclic landforms and ancient relief

Weathering, erosion and mass movements start to attack sections of the ground surface when it is exposed above sea level. Valleys and denudation surfaces are frequently formed step-wise below each other. The mechanism behind this is still not clear but the most commonly accepted theory is that the sea was the original base level for each step and that the steps are indications of new, successive uplift of the land. We speak of *polycyclic relief* when surfaces and valleys follow each other in a *step-wise manner*. The steps may be of very different dimensions and, besides tectonic events, may also reflect changes in climate and sea level.

Within Sweden, the Caledonian bedrock has hardly had any cover rocks. All relief within this tectonic complex may therefore be considered as new. By contrast, the old erosion surfaces in the Precambrian basement were conserved for long periods of time beneath a cover of sedimentary rocks. Relief evolution could start again when uplift of the Earth's crust caused erosion of these cover rocks. Sometimes, the ancient relief has been completely obliterated. However, the present-day relief at lower levels frequently consists of a mixture of the ancient surfaces and newly-formed polycyclic relief.

GEOMORPHOLOGICAL MAP

The map of the morphology of the bedrock on the next page shows the exhumed surfaces in their relationship to cover rocks of different ages, polycyclic relief, scarps and valleys. Tectonically controlled landforms as joint-aligned valleys and fault scarps are shown in a special map on morphotectonics (p. 44). Faults, cutting through the surrounding cover rocks, are marked for comparison.

Series of block diagrams illustrating the development of bedrock relief in southern Sweden between the Baltic and the Kattegat. The lowermost block diagram, showing the situation at present, illustrates how the landscape of today is composed of surface facets from widely different periods.

Exhumed ancient surfaces

The exhumed surfaces have been identified by means of the occurrence of remnant cover rocks.

Large parts of the plains of southern Sweden belong to an exhumed *sub-Cambrian peneplain*, a flat surface with a relief of only a few metres and which originally was formed more than 550 million years ago. With some exceptions, e.g., Blå Jungfrun in Kalmarsund, this surface lacks residual hills. The sub-Cambrian peneplain makes up the summit level in southeastern Uppland, Södermanland and parts of eastern Östergötland and northeastern Småland. The peneplain is here dissected by joint-aligned valleys, which were probably formed after the uncovering of the peneplain along ancient zones of weakness.

Parts of the Norrland coast are also thought to be a sub-Cambrian peneplain. The border towards the interior of the country is irregular but distinct and consists of an old fault-line scarp. Sometimes the peneplain surface has been tilted and younger erosion has cut into the dipping sub-Cambrian surface. The latter can be loosely identified on the basis of the highest summits. Consequently, the coastal plain here constitutes a new formation and the border to the interior an erosional slope. A particularly good example is the coastal plain between Piteå and Kalix, with residual hills rising up to 100 metres above the plain.

Sub-Jurassic hilly terrain occurs in central Skåne, sub-Cretaceous in northeastern Skåne and in the central and inner parts of southern Halland. The joint-valley relief in Blekinge, in northern Halland and the neighbouring parts of Västergötland, and in Bohuslän is also sub-Cretaceous. Along the west coast to the south of Falkenberg, the primary sub-Cambrian peneplain often occurs on the summits of the landscape. The hilly terrain in northern Svealand and southern Norrland is also considered to be sub-Cretaceous in its main features, even though no remnants of the cover rock remain to confirm this.

The map shows the extent of old exhumed bedrock landforms in their relationship to cover rocks of different ages, as well as polycyclic relief, scarps and valleys. There are two kinds of scarps, erosional scarps and fault line scarps. (M17)

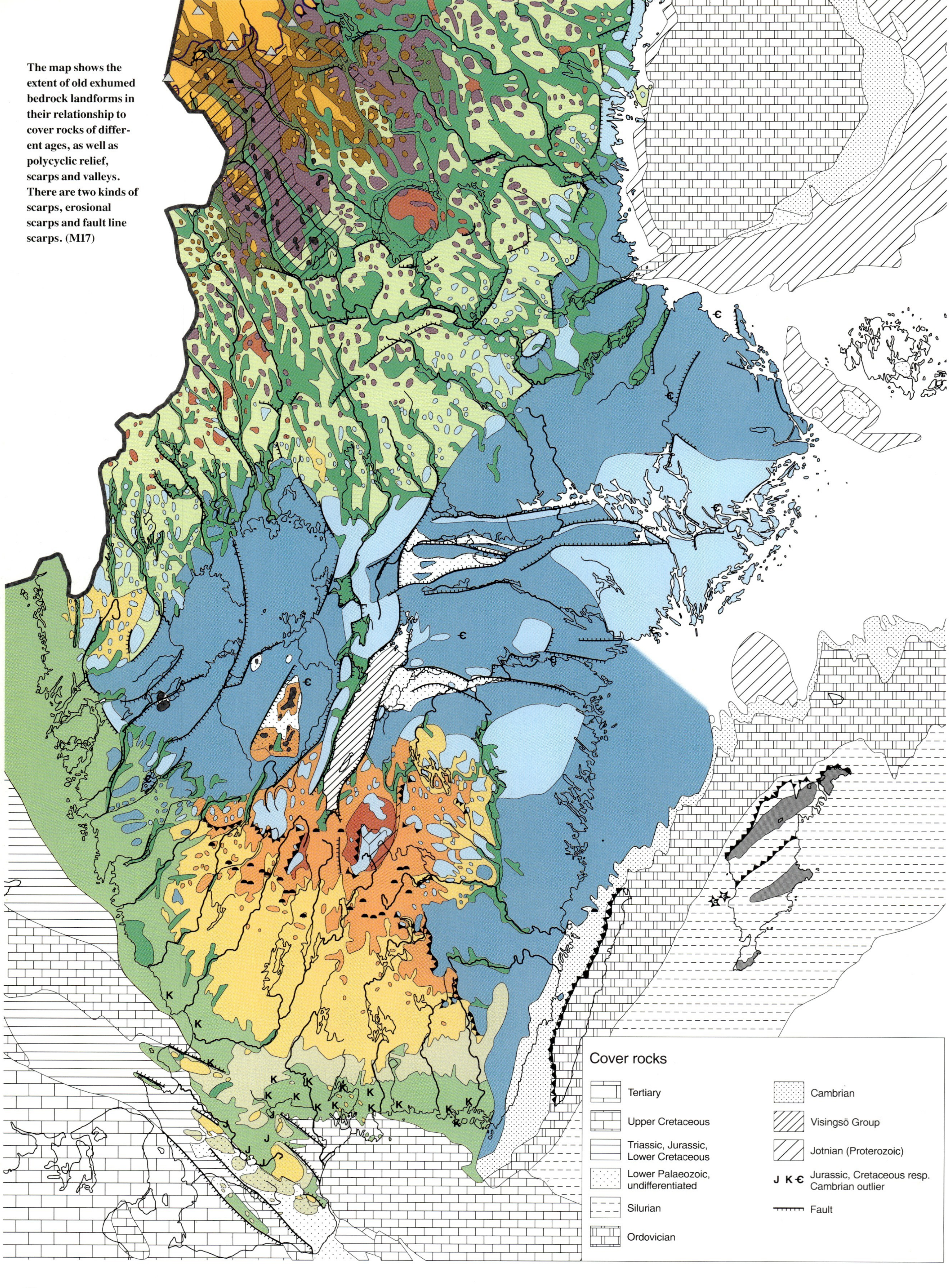

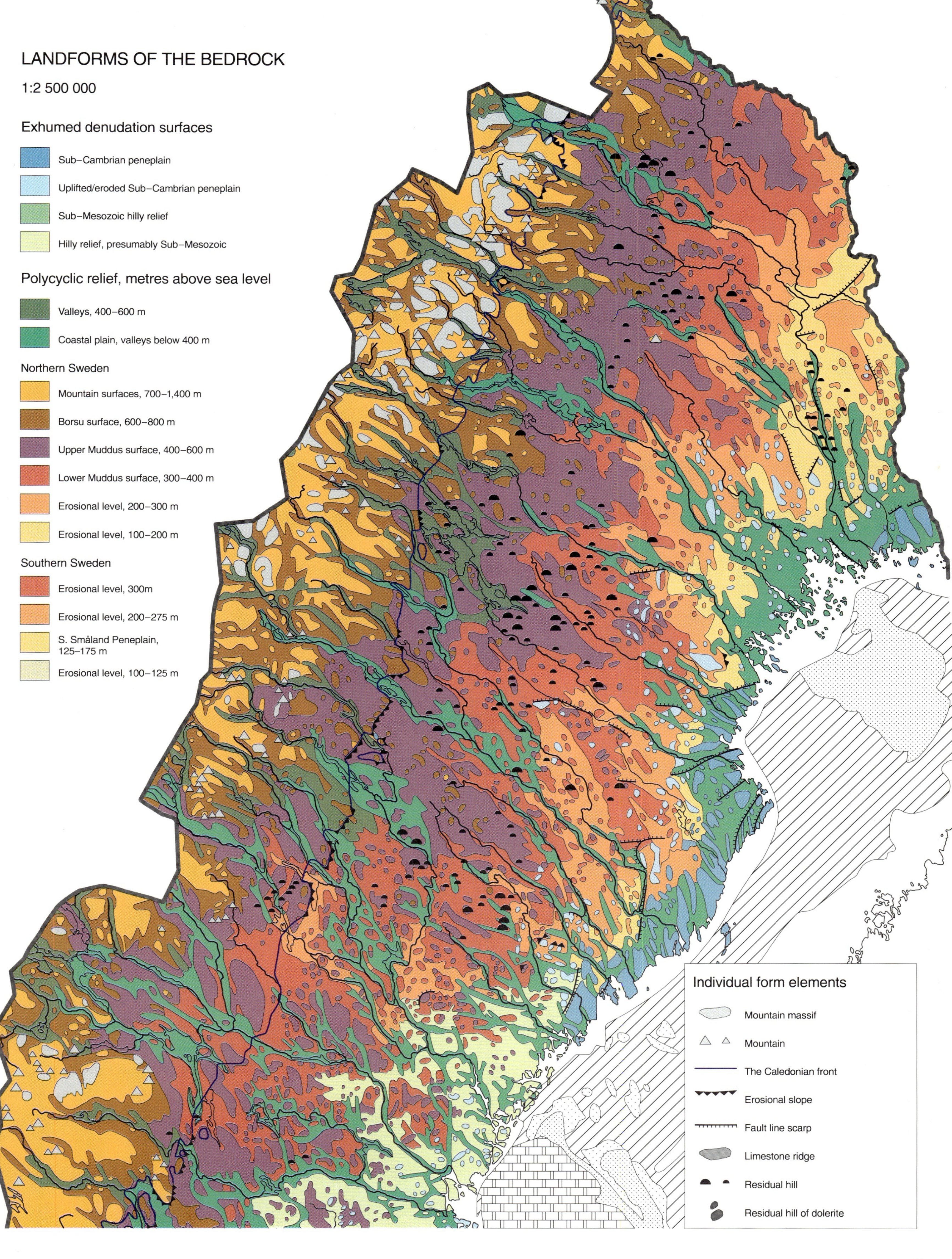

LANDFORMS OF THE BEDROCK
1:2 500 000
Exhumed denudation surfaces
Sub–Cambrian peneplain
Uplifted/eroded Sub–Cambrian peneplain
Sub–Mesozoic hilly relief
Hilly relief, presumably Sub–Mesozoic
Polycyclic relief, metres above sea level
Valleys, 400–600 m
Coastal plain, valleys below 400 m
Northern Sweden
Mountain surfaces, 700–1,400 m
Borsu surface, 600–800 m
Upper Muddus surface, 400–600 m
Lower Muddus surface, 300–400 m
Erosional level, 200–300 m
Erosional level, 100–200 m
Southern Sweden
Erosional level, 300m
Erosional level, 200–275 m
S. Småland Peneplain, 125–175 m
Erosional level, 100–125 m
Individual form elements
Mountain massif
Mountain
The Caledonian front
Erosional slope
Fault line scarp
Limestone ridge
Residual hill
Residual hill of dolerite

After the sub-Cambrian peneplain had been covered by Cambro-Silurian rocks, the Earth's crust was faulted. This can be seen as clear scarps in the surface of the peneplain after erosion of most of the Cambro-Silurian cover. The photo shows the escarpment of the Vånga fault to the north of Lake Roxen in Östergötland.

Faults in the ancient surfaces

The ancient erosion surfaces with their remnants of cover rocks are important surfaces of reference for identification and rough dating of faults. On this basis, it has been possible to construct a completely new tectonic map illustrating late Proterozoic and Phanerozoic brittle tectonics (p. 44).

The largest area with preserved Jotnian sandstone occurs in Dalarna. This area is bordered to the southeast by a ring of hills that probably shows the position of an old fault scarp below which the sandstone has been preserved. Small remnants of Jotnian sandstone and rows of hills suggest that the plains at Sveg, Dala-Järna and Stora Tuna can be interpreted as old fault basins where Jotnian strata have been present. At Gävle, the Jotnian sandstone also lies in a basin surrounded by faults. When considered as a whole, this gives a pattern of old faults in a ENE–WSW direction which have broken the sub-Jotnian surface. The faults can be followed out into the Bothnian Sea.

Younger faults in a north-south direction have preserved the sedimentary rocks of the Visingö Group in the Vättern Graben and neighbouring areas. They belong to a fracturing of the Earth's crust approximately 800 million years ago. The fault scarps we see today probably also reflect later movements that have taken place partly along the old faults and partly along completely new lines.

Numerous *horsts* occur in the sub-Cambrian peneplain in Central Sweden, e.g., Kroppefjäll and Kilsbergen. Faults in a north–south direction delimit low ridges in Uppland. Faults along intersecting systems have led to the formation of the Roslagen and Södertörn horsts. From here, horsts can be followed to the west, e.g., Mälarmården, Käglan and Tiveden. The Kolmården horst occurs further to the south at Norrköping. East of the Vättern Graben there is a small horst, Omberg, and a larger one to the west, Hökensås. Faults along the coast of Norrland and further inland have also been identified in relation to remnants of the sub-Cambrian peneplain.

Skåne is situated on the border between the Baltic Shield and the younger bedrock of central Europe. This border consists of a tectonic zone, the Tornquist Zone. At the end of the Cretaceous, faulting occurred in the sub-Mesozoic surfaces along this zone whereupon the present horsts and basins were formed, e.g., Hallandsåsen and Romeleåsen, the Kristianstad Plain and the Vomb Basin.

Polycyclic relief

In large parts of Sweden, polycyclic relief in the form of surfaces and valleys in steps below each other have been identified. The steps vary considerably in size. The most important steps in northern Sweden were described in 1908 and named by Walter Wråk.

The stepped relief in Norrland can be divided into three zones in a west-east direction: *Steps within the mountain chain, the Muddus plains* and the *mixed relief of the coastal zone.*

The relief in the mountain chain consists of steep high mountains, fjeld plains and valleys. In general, the high mountains have been re-shaped glacially but remnants of stepped relief can be found. The high mountains reach altitudes between 1,200 and 2,000 m a.s.l. Plains and valleys are common at altitudes between 800 and 1,000 m a.s.l., e.g., the *Tuipal Plains*. A clear level in front of the mountains, the *Borsu level*, can be followed

The sub-Cambrian peneplain extends from under the Cambrian cover rocks in the east, whereas in the west a hilly relief is being exhumed from Cretaceous cover rocks. From altitudes of about 300 m a.s.l. in the east, the basement cover becomes lower in two steps, the first at about 200 m a.s.l. and the second at 125 – 175 m a.s.l., towards the dipping exhumed sub-Cretaceous hilly relief in the west.

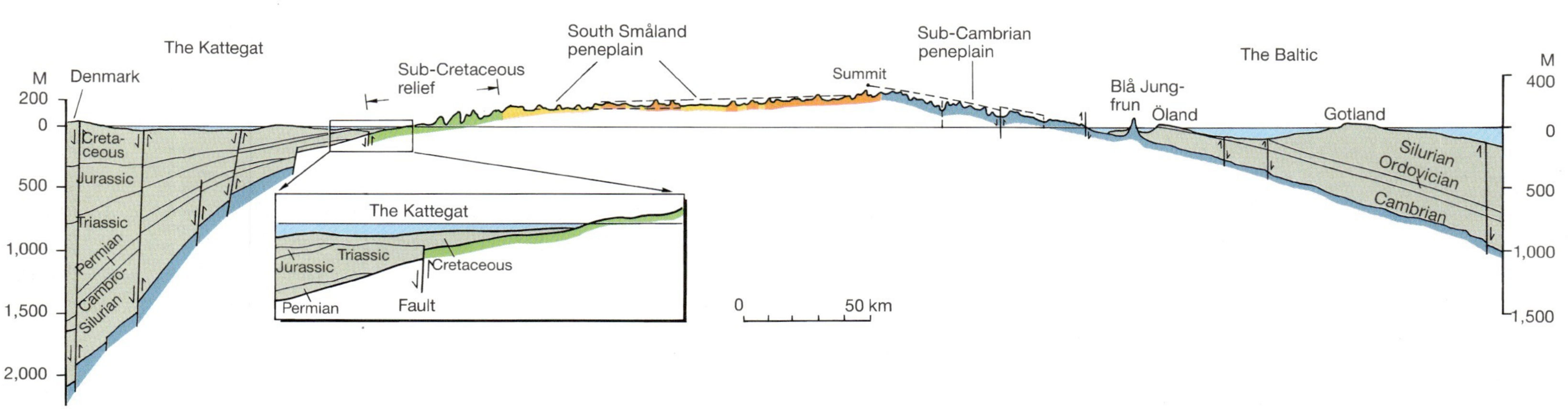

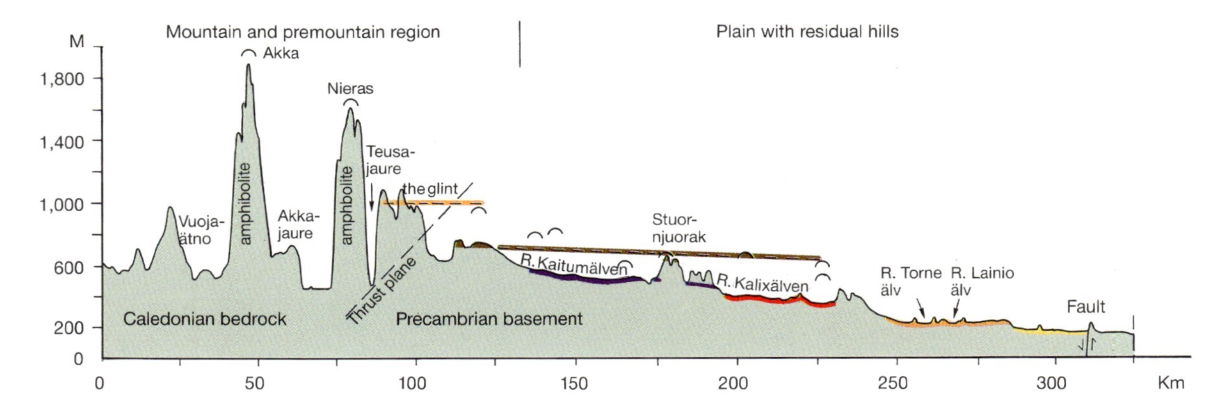

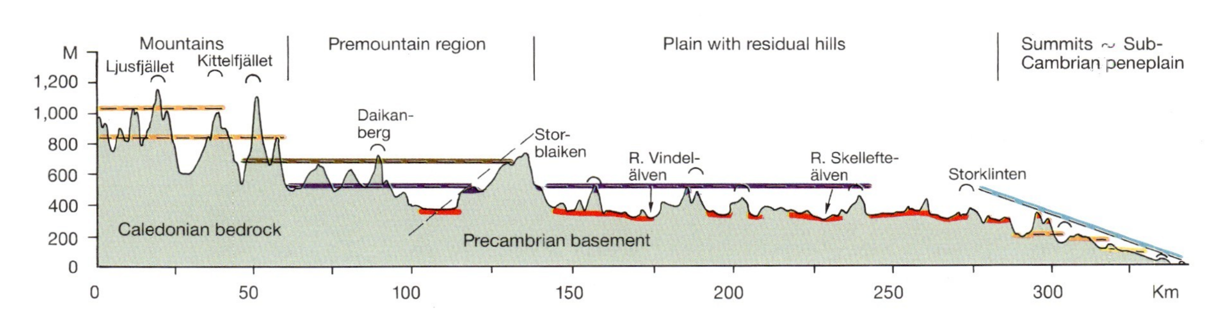

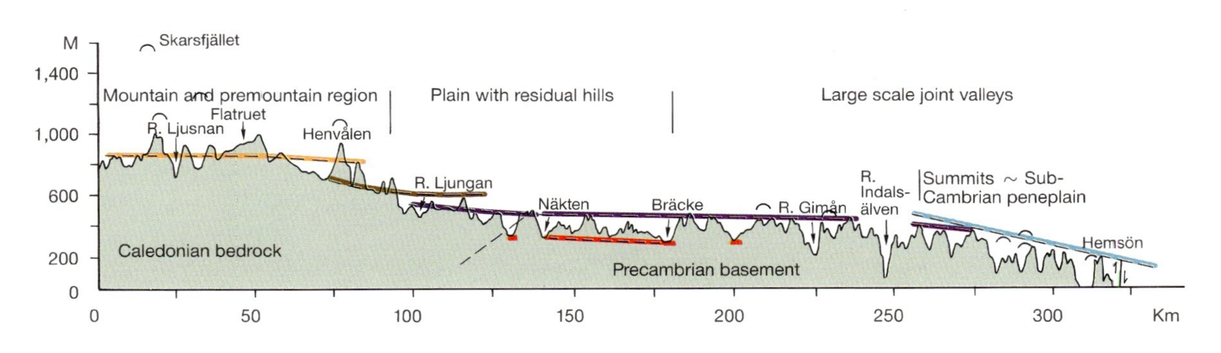

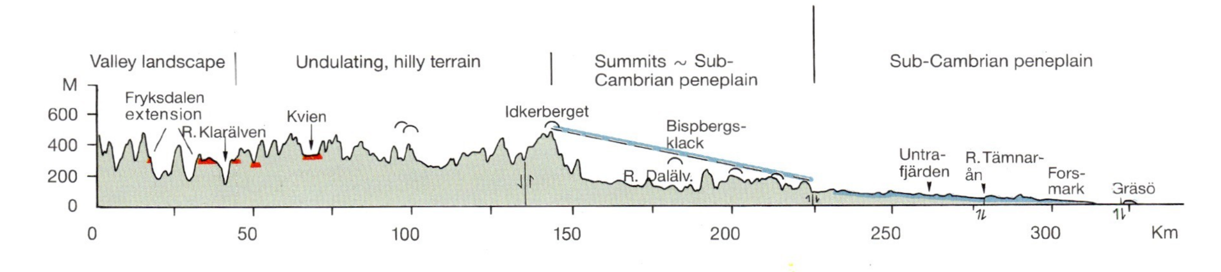

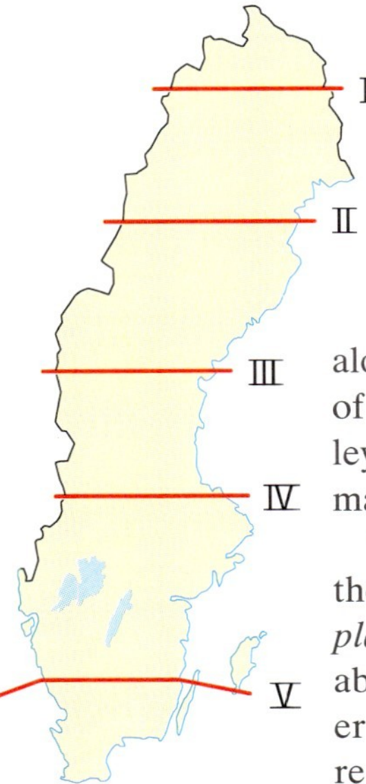

A series of east-west profiles showing different types of terrain and denudation surfaces. Along the east coast, the summit surface is interpreted as the sub-Cambrian peneplain. Denudation surfaces in a polycyclic pattern have been identified in the inland areas of northern Sweden.

along the mountain chain at a height of about 600 metres, from which valleys extend into the high mountain massifs.

To the east of the mountain chain, there are large flat areas, the *Muddus plains*. The Upper Muddus level is about 400–500 m a.s.l. A slightly lower level, the Lower Muddus level, reaches down to 300 metres. The Muddus Plains are charachterised by numerous residual hills. The plains extend over the Caledonian bedrock to the south of Storuman and reaches as far as the Norwegian border in western Jämtland. In Dalarna, the upper plain cuts across both the Precambrian basement and the Jotnian sandstone.

The *relief at lower levels* contrasts with the monotony of the Muddus plains. Here we find a mixture of ancient exhumed surfaces and young stepped relief. In northern Norrland, the primary peneplain slopes to the southeast and here a simple stepped relief has been formed. In central Norrland, on the other hand, the peneplain has been elevated along faults following the coastline. This has caused erosion to cut deep into the bedrock and, for example, the rivers Ångermanälven and Indalsälven run in huge water gaps. The valleys largely follow the fracture systems. A landscape of large-scale joint-aligned valleys is encountered. Locally, it is possible to distinguish indications of stepped relief. In southern Norrland and northern Svealand, the landscape consists of a typical undulating hilly terrain, probably an exhumed sub-Cretaceous surface. The summits are cut by the Muddus surface but it is not possible to distinguish any stepped relief below it.

The large extent of the Muddus surfaces in comparison with steps at higher altitudes and their almost horizontal position suggests that they were formed during a very long period of time. The Muddus surfaces are characterized by plains with residual hills and were probably formed during the Tertiary.

The South Swedish Uplands consist of elevated, broken, and partly eroded parts of the sub-Cambrian peneplain. The peneplain ends in the south and west in an erosional scarp. This leads down to younger surfaces at levels of about 300 and 200 metres that have cut into the surface of the peneplain. The 200 m surface has high residual hills that are interpreted as remnants of an older erosional slope. The South Småland Peneplain, 125–175 m a.s.l., distinctly cuts off the dipping sub-Cretaceous surfaces and is therefore younger. Remnants of the old weathering mantle (kaolin) and old cover rocks (Cretaceous), which occur in northern Skåne and southern Halland on the sub-Cretaceous surface, disappear at the lower margin of the Småland plain.

River valleys, e.g., along Emån and Stångån, have formed along fracture zones in the broken sub-Cambrian peneplain. At the uplifted western margin of the sub-Cambrian peneplain, the valley of the River Göta Älv forms a water gap.

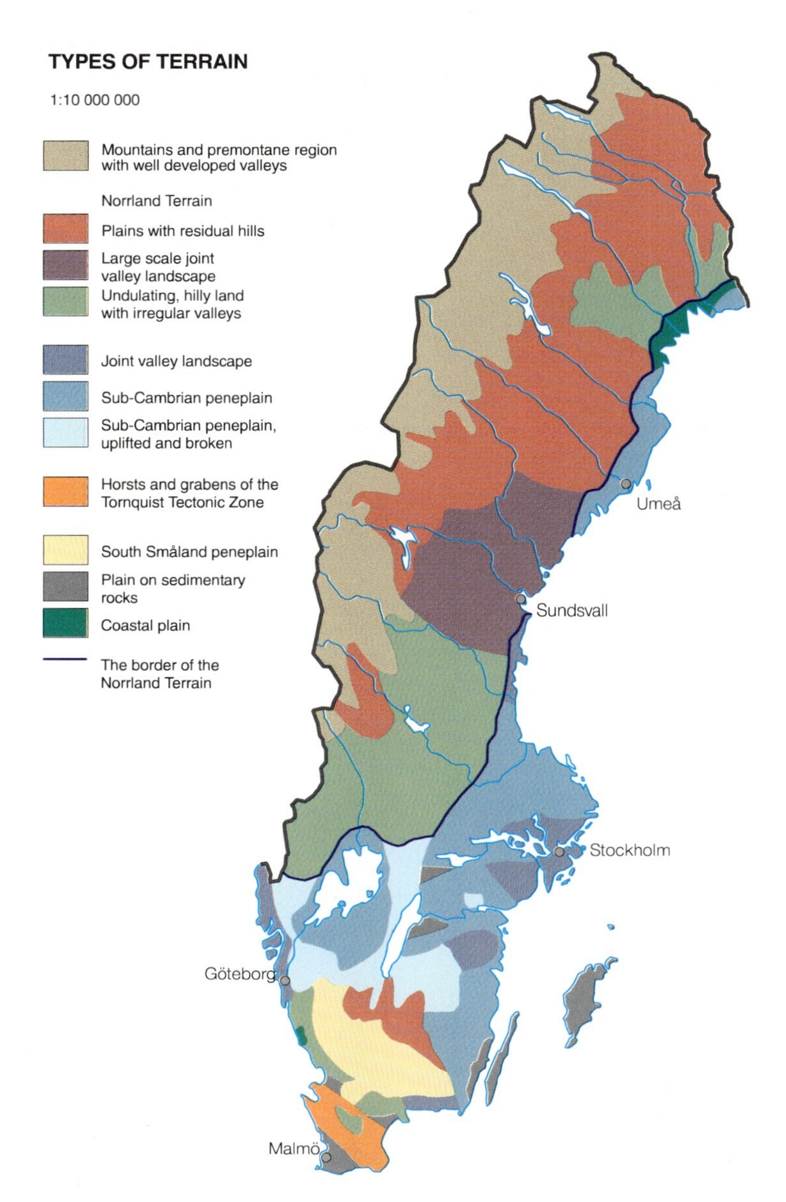

The present landforms are a mosaic of forms from different times. The sub-Cambrian peneplain is more than 600 million years old. The joint aligned valleys and the undulating hilly land probably date from Mesozoic times, whereas the plains with residual hills are probably from the Tertiary. Also the mountain plateaus are probably mainly Tertiary but may have residual relief from earlier times. The formation of valleys has probably been affected by the uplift of the western border of Fennoscandia related to with the formation and development of the Atlantic Ocean. (M18)

As a result of glacial lateral erosion, V-shaped river valleys have been transformed into U-valleys. Conversely, valleys have preserved their old shape below a bottom-frozen ice sheet. This is apparent, for example, along the valleys at Söderåsen in Skåne.

Glacial erosion on protruding bedrock hills occurs largely through plucking, mainly on the leeside in relation to the movement of the ice, but also along the sides.

Despite several glaciations, there are still considerable remnants of the pre-glacial weathering mantles preserved under the glacial deposits, particularly in northern Norrland, Östergötland, Småland and Skåne.

Valleys formed by flowing water have a V-shaped cross section. When a glacier has flowed through a valley and eroded the bottom and sides, the valley will become distinctly U-shaped, as here at Syterskalet in southern Lappland.

GLACIAL EROSION

The erosional ability of an ice sheet in motion depends, among other things, on the temperature of its basal parts. If the ice is frozen to the ground, the motion takes place at higher levels. An ice sheet of this kind does not cause erosion but instead has a preserving effect on old landforms. On the other hand, if the ice sheet is melting at its base, erosion will result. The erosion is of a minor amount over smooth relief but may be very large in valleys. Several of the lakes along the Swedish mountain chain have great depths caused in this manner, e.g., Lake Hornavan, which is 221 m deep.

Ores and mineral deposits

In Sweden, metals started to be extracted from ore probably several centuries B.C., when iron was obtained from bog iron ore and marsh ore. The deposits were small and frequently had low iron contents. Thus, they decreased in importance as richer "rock ore" became available.

Mining of ore from rock started in Bergslagen. New research has demonstrated that copper-mining at Falun might have started as early as 500 A.D. Mining for iron ore probably started in northern Västmanland—the blast-furnace at Lapphyttan, to the north of Norberg, has been dated to the 12th century. The copper mine at Falun and the silver mine at Sala were of major importance for Sweden's development during the Great Power Period.

Until the end of the 19th century, mining activities outside the Bergslagen area were of minor importance. At this time, mining started near the towns of Kiruna and Malmberget in Norrbotten. At the present time, these are the only two areas in Sweden where mining of iron ore is in progress.

The Skellefte Field is another important ore province, where complex sulphide ores have been mined since the middle of the 1920's. In all three provinces—Bergslagen, Norrbotten and the Skellefte field—the ores are hosted by Precambrian rocks, particularly volcanic and sedimentary rocks, approximately 1880 million years old.

The deposits of sulphide ore in the Caledonides are much younger. At, for example, Laisvall and Vassbo, lead ore is hosted by sandstone approximately 600 million years old. The rocks in the nappes of the Caledonides are slightly younger and contain complex ores, e.g., the deposit at Stekenjokk.

The youngest ore deposits in Sweden are the Mesozoic iron ores in Skåne that were mined during 1938–39 in Fyledalen.

Among the thousands of mineralisations known in Sweden, about 420 have been utilised for commercial mining on a large scale; 400 of these have been in operation since 1890. During the summer of 1993, only 18 mines were in production.

Iron ores at Kiirunavaara and Malmberget, together with the copper ore at Aitik, are the largest ore deposits in Sweden. During 1991, they produced more than 40 million tonnes of ore. This corresponds to 83.5% of the total ore production in Sweden for that year.

Flogberget Mine, to the west of Smedjebacken is one of the many mines in Bergslagen that closed at the end of the First World War.

Some of the old mine holes are filled with water, whereas others are open shafts.

Limonite ore was used in Sweden as late as during the 1920's. From Lake Torrvarpen to the south of Hällefors.

The number of mines in production has decreased since the 1950's but ore production is increasing. Note that the scale on the x-axle is not linear.

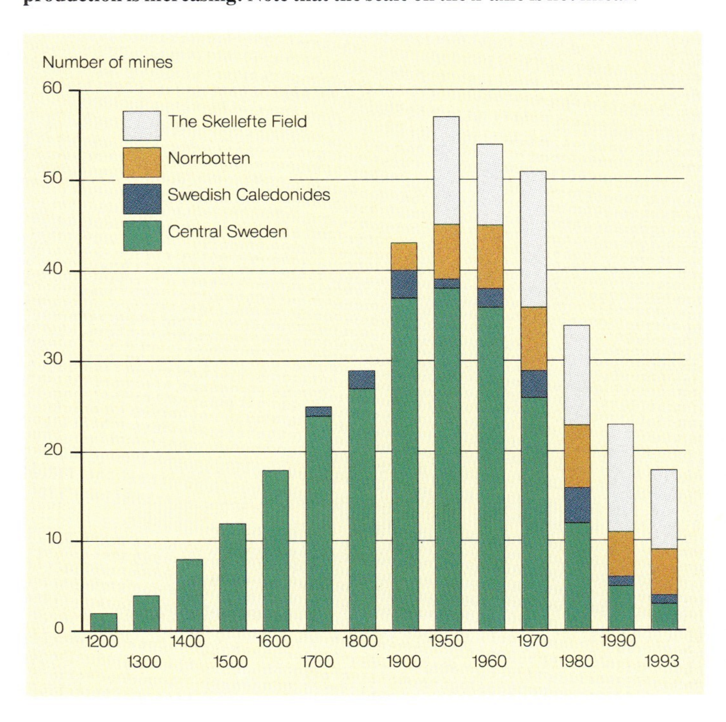

Shaft-head and dressing-house, Långban, Värmland. Seventy type-minerals have been discovered at Långban, i.e., minerals described for the first time, which is unique.

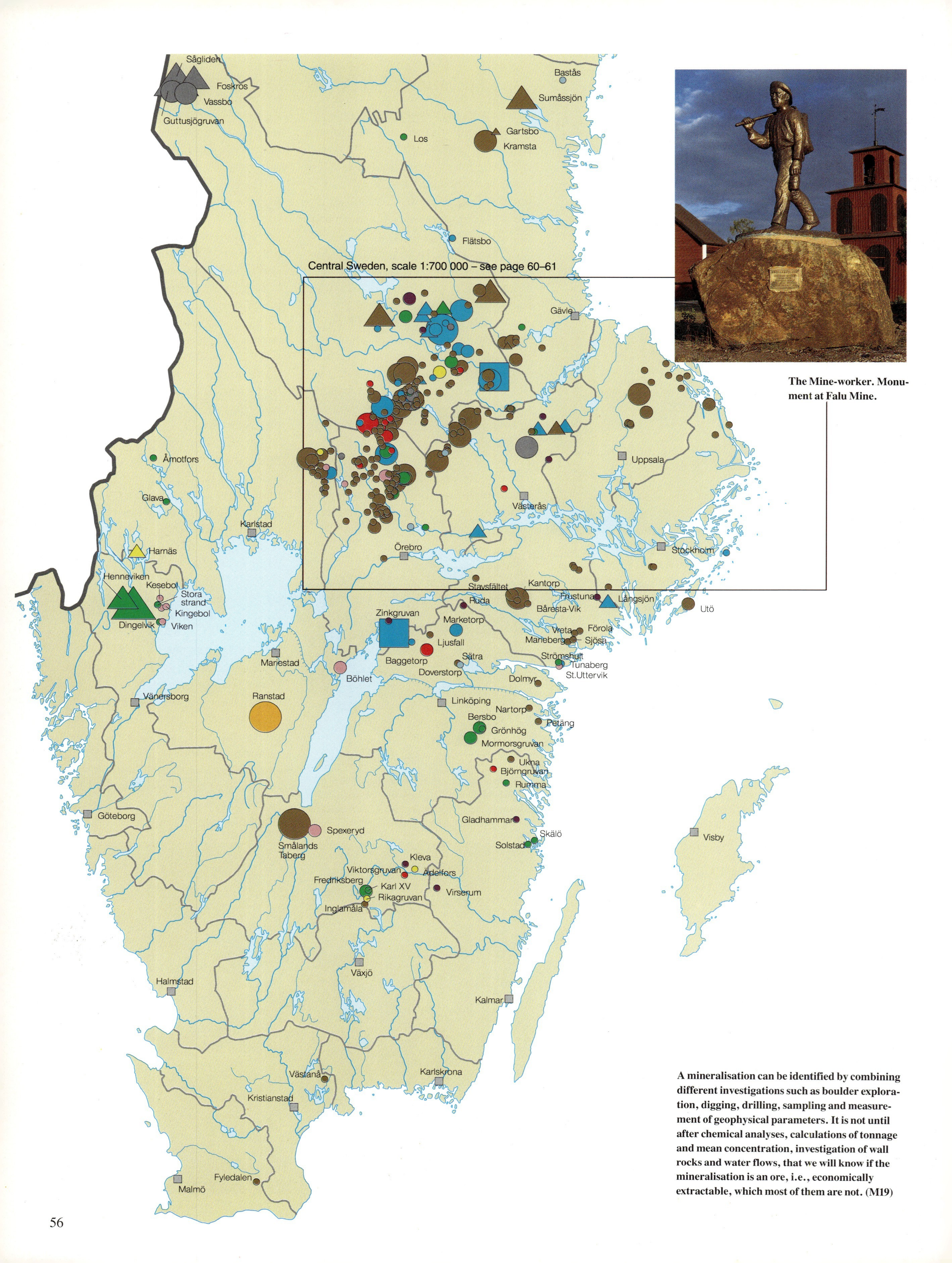

The Mine-worker. Monument at Falu Mine.

A mineralisation can be identified by combining different investigations such as boulder exploration, digging, drilling, sampling and measurement of geophysical parameters. It is not until after chemical analyses, calculations of tonnage and mean concentration, investigation of wall rocks and water flows, that we will know if the mineralisation is an ore, i.e., economically extractable, which most of them are not. (M19)

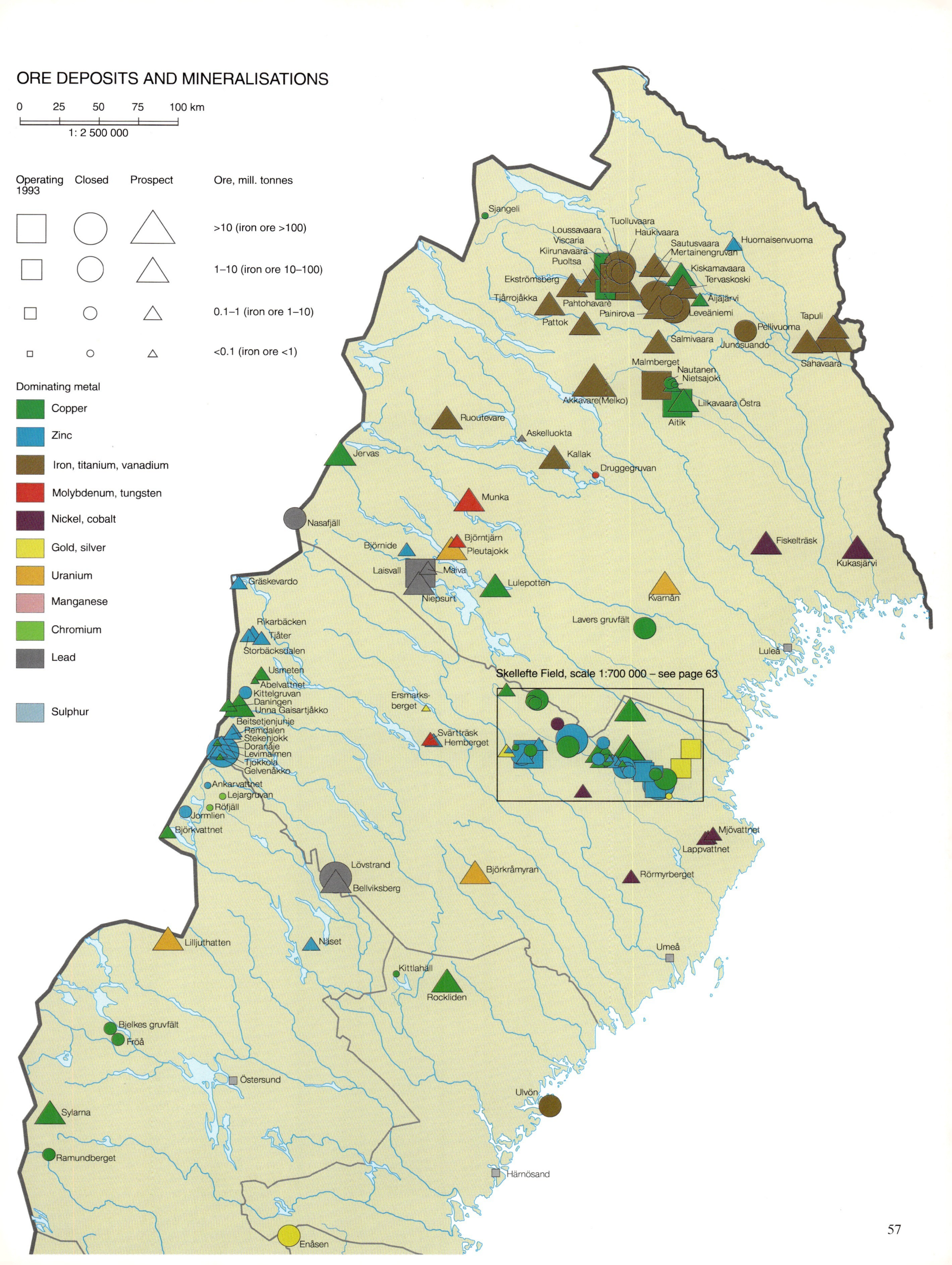
ORE DEPOSITS AND MINERALISATIONS
0 25 50 75 100 km
1: 2 500 000
Operating 1993
Closed
Prospect
Ore, mill. tonnes
>10 (iron ore >100)
1–10 (iron ore 10–100)
0.1–1 (iron ore 1–10)
<0.1 (iron ore <1)
Dominating metal
Copper
Zinc
Iron, titanium, vanadium
Molybdenum, tungsten
Nickel, cobalt
Gold, silver
Uranium
Manganese
Chromium
Lead
Sulphur
Sjangeli
Tuolluvaara
Loussavaara
Haukivaara
Viscaria
Sautusvaara
Kiirunavaara
Mertainengruvan
Huornaisenvuoma
Puoltsa
Kiskamavaara
Ekströmsberg
Tervaskoski
Tjårrojåkka
Pahtohavare
Äijäjärvi
Painirova
Leveäniemi
Pattok
Tapuli
Pellivuoma
Salmivaara
Junosuando
Sahavaara
Malmberget
Nautanen
Nietsajoki
Akkavare(Melko)
Liikavaara Östra
Aitik
Ruoutevare
Askelluokta
Jervas
Kallak
Druggegruvan
Munka
Nasafjäll
Björntjärn
Björnide
Pleutajokk
Fiskelträsk
Kukasjärvi
Laisvall
Maiva
Gräskevardo
Lulepotten
Niepsurt
Kvarnån
Lavers gruvfält
Rikarbäcken
Tjåter
Storbäcksdalen
Luleå
Usmeten
Skellefte Field, scale 1:700 000 – see page 63
Abelvattnet
Kittelgruvan
Ersmarks-berget
Daningen
Unna Gaisartjåkko
Beitsetjenjunje
Remdalen
Svärtträsk
Stekenjokk
Hemberget
Doranåje
Levimalmen
Tjokkola
Gelvenåkko
Ankarvattnet
Lejargruvan
Röfjäll
Jormlien
Björkvattnet
Mjövattnet
Lappvattnet
Lövstrand
Björkråmyran
Rörmyrberget
Bellviksberg
Lilljuthatten
Näset
Umeå
Kittlahäll
Rockliden
Bjelkes gruvfält
Fröå
Östersund
Ulvön
Sylarna
Ramundberget
Härnösand
Enåsen

In Bergslagen, mining today only takes place at Garpenberg (foreground) and Garpenberg norra (background). Zinc-lead ore is mined at both places.

More iron and manganese ore remains in the rock at Dannemora than the total amount mined during the entire history of the mine—but today there are no buyers.

DEPOSITS IN CENTRAL SWEDEN

In early 1992, five mines were operating in central Sweden. On 1 April 1992, work ceased at the Dannemora Mine, thereby ending more than 500 years of iron-ore mining in northern Uppland. On 8 December 1992, the final charge was detonated at the Falu Mine. This signalled the closure of one of the world's oldest mines. The three remaining mines in central Sweden are Zinkgruvan, Garpenbergs Odalfält and Garpenberg Norra.

Iron ore

Iron ores are divided into three types: *Quartz-banded iron ores, skarn ores* and *calcareous iron ores,* and *apatite iron ores*. Quartz-banded iron ores, with iron contents of 30–50%, usually consist of hematite and/or magnetite layers which alternate with quartz-rich layers containing small amounts of skarn minerals. Sometimes the ore has a beautiful red colour (jaspilite) which is related to the occurrence of small flakes of hematite in the quartz-rich layers. The largest quantities of quartz-banded iron ore have been mined in the Håksberg, Norberg, Riddarhytte and Stripa areas and at Stråssa.

The Norberg mining area consists of a horizon of tuffs and lavas, up to 2.5 km wide and 8 km long, with layers of quartz-banded iron ores as well as skarn and calcareous iron ores. In the central parts of the area the ores are beautifully folded.

A belt of quartz-banded iron ores can be followed from Ludvika approximately 17 km to the north along the eastern side of a marked thrust zone. Within the Håksberg mining area, this zone forms the boundary between felsic volcanic rocks ("leptites") and iron ores in the east, and granites in the west. The Håksberg area has a total of more than 150 mines. At Iviken, excellent quartz-banding has been preserved. Otherwise, the structure has been destroyed in connection with later tectonic processes.

At Stråssa, the iron content of the ore was about 30%. In connection with intensive folding and recrystallisation to homogeneous or banded hematite and magnetite ores the originally quartz-banded hematite ores were metamorphosed.

The best-preserved quartz-banded iron ores in Bergslagen have been mined in the Stripa mining area, where the typical ore is a finely and

LARGE MINES IN CENTRAL SWEDEN (PRODUCTION UP TO 1993)

Deposit	Opened*	Closed	Ore production, mill. tonnes**	Main metals
Bastkärnsfältet	1854	1978	6.24	iron, manganese
Bispbergsfältet	1380	1967	2.61	iron
Blötbergsfältet	1768	1979	18.76	iron
Dalkarlsbergs Odalfält	1200	1960	4.25	iron
Dannemora gruvor	1481	1992	24.0	iron, manganese
Falu gruva	1150	1992	28.15	zinc, lead, copper
Garpenberg Norra	1967	producing	6.15	zinc, lead, silver
Garpenbergs Odalfält	1300	producing	7.82	zinc, lead, copper, silver
Grängesbergsfältet	1500	1989	150	iron
Herrängsfältet	1833	1961	2.53	iron
Hillängsgruvan	1917	1950	0.98	iron, manganese, lead, zinc
Håksbergsfältet	1550	1979	18.2	iron
Idkerbergsfältet	1901	1977	10.73	iron
Intrångets gruvor	1934	1969	5.23	iron
Kantorps gruvfält	1855	1967	6.13	iron
Kaveltorps gruvor	1857	1971	1.10	zinc, lead, copper
Lekombergsfältet	1860	1945	2.89	iron
Långbansfältet	1710	1958	1.28	iron, manganese
Norbergsfältet	1350	1980	34.67	iron
Nordmarksfältet	1600	1973	6.53	iron
Nybergsfältet, Smedjebacken	1600	1965	3.71	iron
Persbergs odalfält	1400	1977	7.73	iron
Pershyttefältet	1600	1967	4.75	iron
Ramhälls gruvor	1832	1975	4.22	iron
Riddarhyttefältet	1700	1979	16	iron, copper
Ryllshyttefältet	1512	1944	1.0	zinc, lead, silver, iron
Sala silvergruva	1500	1962	1.6	lead, zinc, silver
Saxbergsfältet	1886	1988	6.43	zinc, copper, lead
Semlafältet	1600	1967	2.89	iron
Sköttegruve-Mossgruvefälten	1873	1972	3.34	iron
Smältarmossgruvan	1873	1979	3.66	iron
Stollbergsfältet	1300	1982	4.0	lead, zinc, iron, manganese
Stribergs gruvfält	1500	1967	5.22	iron
Stripafältet	1400	1977	16.13	iron
Stråssa odalgruva	1857	1982	28	iron
Ställbergs-Haggruvefälten	1868	1977	6.5	iron, manganese
Tuna-Hästbergsfältet	1648	1968	6.38	iron, manganese
Utö gruvor	1624	1879	1.81	iron
Vingesbackefältet	1640	1980	8.5	iron
Vintjärnsfältet	1731	1978	7.36	iron
Yxsjöbergsfältet	1728	1989	5.15	tungsten, copper
Zinkgruvan	1700	producing	17.91	zinc, lead, silver

* Dates referring to when the oldest mines started production are approximate.
** Systematic registration of sulphide ore production started during the 1630s and of iron ores during the 1830s. The production volumes of the oldest mines are based on estimates and aproximations.

Quartz-banded iron ore of alternate magnetite and hematite. Stora Jakobsgruvan, Hästefältet.

Folded quartz-banded red iron ore from the 175 m level in Nygruvan Mine, Norberg.

regularly quartz-banded hematite ore with 40–50% iron.

Skarn and calcareous iron ores with low contents of manganese were mined in the Herräng, Vingesbacke, Nordmark and Vintjärn Mines, as well as in the Persberg and Dalkarlsberg mining areas. The iron contents were 35–50% and manganese contents less than 1%.

In the Persberg mining area, which is the most important ore zone in Filipstad's Bergslag, there are numerous irregularly formed ore bodies in a 800 m long and 400 m wide zone of limestone, dolomite, skarn and iron ore. The most important ore type consists of magnetite in garnetiferous pyroxene skarn. Epidote and hornblende occur closer to the felsic volcanic rocks in the wall rock.

Examples of manganese-rich iron ores are the Ställberg, Haggruvan, Bastkärn and Tuna-Hästberg mining areas together with the Hilläng Mine and the mines at Dannemora. The ores contain 1–10% manganese and 30–50% iron. Most of the manganese is bound to the minerals knebelite and dannemorite. Some 2% of manganese is also included in magnetite.

The ore zone at Dannemora is 3 km long and consists of about 25 magnetite concentrations in a stratified series of fine-grained felsic volcanic rocks (hälleflinta) and carbonate rocks. In one drill hole, ore with 30–50% iron and 2–3% manganese occurs to a depth of 1,000 m. The ore sometimes lies directly in carbonate rocks but is usually surrounded by skarn of knebelite, dannemorite and serpentine. Manganese-poor skarn iron ores are also present.

The Hilläng Mine occurs east of Ludvika. The mined ore contained 8–10% manganese and 30–36% iron together with sphalerite and galena, sometimes with up to 6% lead and zinc.

The deposits in the Långban mining area consist of alternating iron ores and manganese oxide ores. In the latter, the percentage of iron normally did not exceed 2%, whereas the manganese content amounted to 38% in first quality ore and 28% in second quality ore. Långban is well-known internationally for its richness in minerals.

The economically most important iron ores are the apatite iron ores, in which magnetite and/or hematite are accompanied by apatite and small amounts of quartz and skarn. The contents of iron and phosphorus are 45–63% and 0.3–1.3%, respectively. Five mining areas have been of importance, ranging from Grängesberg in the south to Idkerberget 37 km to the north.

The largest concentrations of ore in central Sweden occur at Grängesberg. The largest individual body of ore was mined in Grängesberg's Exportfält; 126 million tonnes of ore with, on average, 60% iron and 1% phosphorus were extracted. The ore body, 20–90 m wide, is dominated by magnetite ore in the main central part.

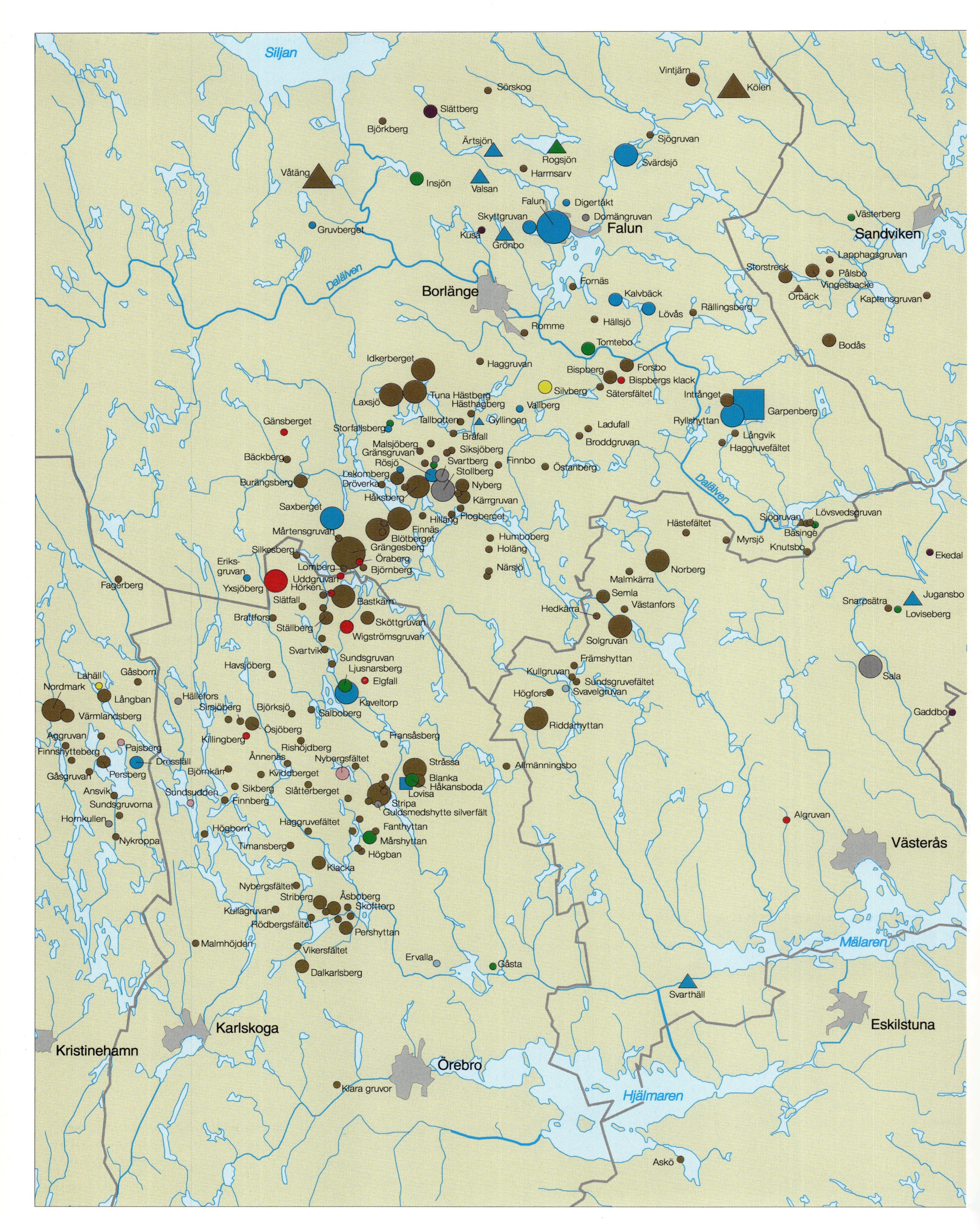

Siljan
Vintjärn
Kölen
Sörskog
Slättberg
Björkberg
Sjögruvan
Ärtsjön
Rogsjön
Svärdsjö
Våtäng
Harmsarv
Insjön
Valsan
Falun
Digertäkt
Gruvberget
Skyttgruvan
Domängruvan
Falun
Kusa
Grönbo
Västerberg
Sandviken
Dalälven
Lapphagsgruvan
Storstreck
Pålsbo
Fornäs
Borlänge
Vingesbacke
Örbäck
Kalvbäck
Kaptensgruvan
Lövås
Rällingsberg
Hällsjö
Romme
Tomtebo
Bodås
Idkerberget
Haggruvan
Bispberg
Forsbo
Bispbergs klack
Silvberg
Sätersfältet
Laxsjö
Tuna Hästberg
Intrånget
Hästhagberg
Vallberg
Garpenberg
Gänsberget
Storfallsberg
Tallbotten
Gyllingen
Ryllshyttan
Ladufall
Bråfall
Långvik
Broddgruvan
Haggruvefältet
Malsjöberg
Siksjöberg
Gränsgruvan
Bäckberg
Svartberg
Finnbo
Rösjö
Östanberg
Stollberg
Lekomberg
Dröverka
Buråsberg
Nyberg
Håksberg
Kärrgruvan
Saxberget
Dalälven
Lövsvedsgruvan
Sjögruvan
Hillang
Flogberget
Hästefältet
Båsinge
Mårtensgruvan
Finnäs
Myrsjö
Blötberget
Humboberg
Knutsbo
Silkesberg
Grängesberg
Holäng
Ekedal
Eriks-
gruvan
Öraberg
Norberg
Lomberg
Björnberg
Närsjö
Fagerberg
Uddgruvan
Malmkärra
Yxsjöberg
Hörken
Semla
Slätfall
Bastkärn
Juganbo
Västanfors
Snarpsätra
Loviseberg
Hedkärra
Brattfors
Skottgruvan
Ställberg
Wigströmsgruvan
Solgruvan
Svartvik
Främshyttan
Sundsgruvan
Havsjöberg
Ljusnarsberg
Sala
Kullgruvan
Gåsborn
Elgfall
Lahäll
Sundsgruvefältet
Nordmark
Svavelgruvan
Hälleforsc
Långban
Högfors
Kaveltorp
Gaddbo
Sirsjöberg
Björksjö
Salboberg
Riddarhyttan
Värmlandsberg
Ösjöberg
Aggruvan
Fransåsberg
Killingberg
Finnshytteberg
Pajsberg
Rishöjdberg
Dressfäll
Ånnenäs
Nybergsfältet
Stråssa
Allmänningsbo
Björnkärr
Persberg
Gåsgruvan
Kviddberget
Blanka
Håkansboda
Ansvik
Sikberg
Slätterberget
Lovisa
Sundsudden
Stripa
Sundsgruvorna
Finnberg
Guldsmedshytte silverfält
Hornkullen
Haggruvefältet
Algruvan
Fanthyttan
Högborn
Nykroppa
Mårshyttan
Timansberg
Västerås
Högban
Klacka
Nybergsfältet
Åsboberg
Striberg
Skottorp
Kullagruvan
Rödbergsfältet
Pershyttan
Mälaren
Malmhöjden
Vikersfältet
Ervalla
Gästa
Dalkarlsberg
Svarthäll
Eskilstuna
Karlskoga
Kristinehamn
Örebro
Klara gruvor
Hjälmaren
Askö

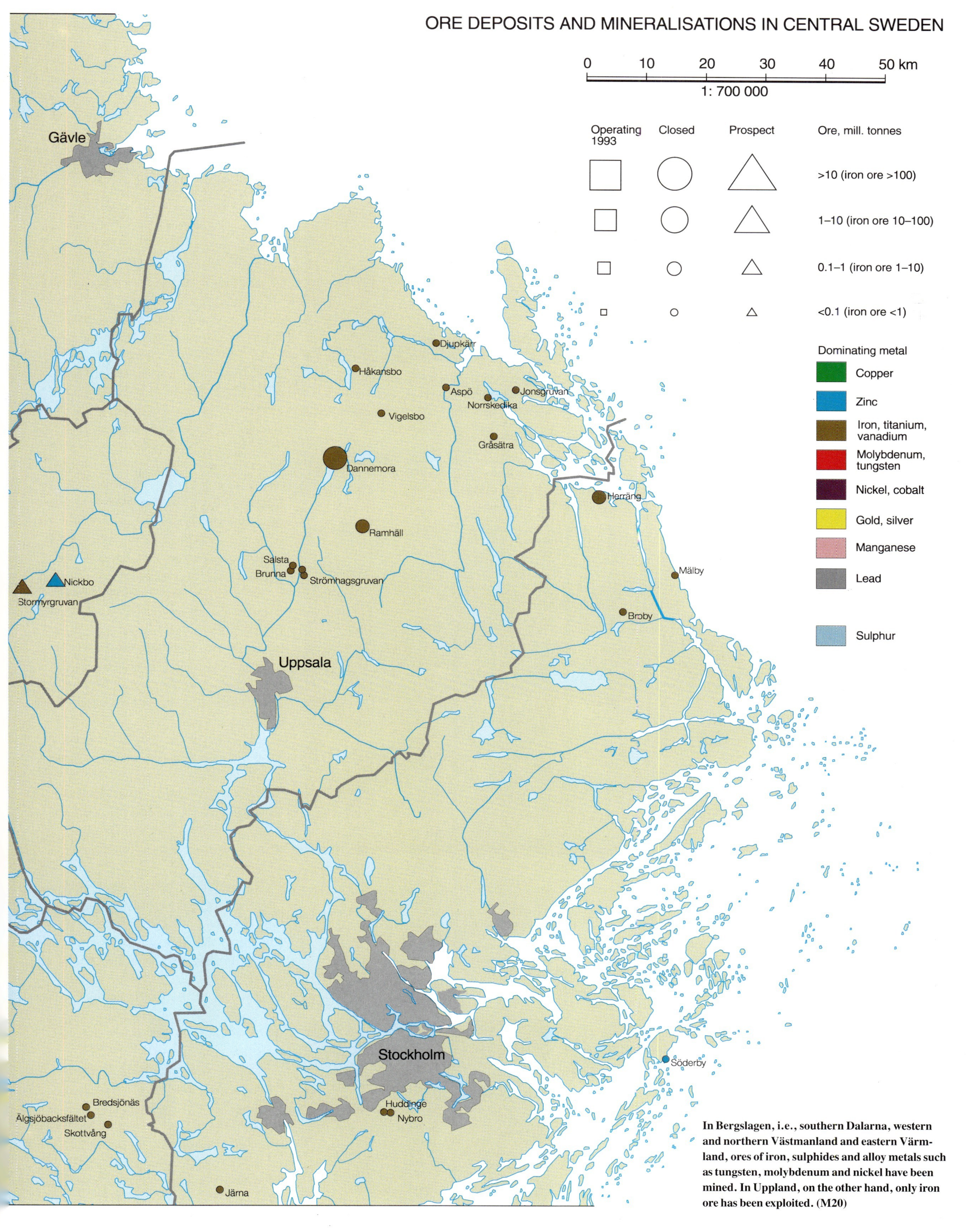

In Bergslagen, i.e., southern Dalarna, western and northern Västmanland and eastern Värmland, ores of iron, sulphides and alloy metals such as tungsten, molybdenum and nickel have been mined. In Uppland, on the other hand, only iron ore has been exploited. (M20)

The eastern part of the open-pit Stora Stöten at the Falu Mine that was formed by a massive fall in 1687 and which today measures about 350 m in diameter. Despite the mining activities at Stora Kopparberg being responsible for two-thirds of the world's copper production during the 17th century, zinc was the dominating metal in Falu Mine.

In order to be able to mine all the ore at Långdal Mine in the Skellefte Field, Boliden Mineral has temporarily changed a short stretch of the bed of the river Skellefteälven. About one million tonnes of complex sulphide ore with 5.7% zinc, 1.79% lead and 1.9 grammes of gold and 160 grammes of silver per tonne can now be obtained.

Chalcopyrite in quartzite. Falu Mine.

Visible grains of gold in quartz—the first find of gold in the Falu Mine, 1881.

Other ore types

Among the sulphide ore deposits, the mines at Falu Gruva, Zinkgruvan, the Garpenberg, Saxberg and Stollberg mining areas, together with the Sala Silver Mine, have been the most productive. Typical for the sulphide ores in central Sweden is that they occur together with iron ores and also have low copper contents in comparison with the contents of zinc and lead.

Ore of "Falu type" refers to the massive pyrite ore with zinc, lead and copper sulphides that was mined at Falun. The mining occurred originally in open-pit operation and was subsequently followed by underground work in three lenses that could be followed up to 500 m downwards. The massive ore which forms the central part of Falu Gruva is 500 m long and up to 130 m wide. It contains 5% zinc, 1.7% lead, 0.7% copper and 35% sulphur. This part of the mine also yields the weathering product of pyrite which is used as raw material in manufacturing Falu red paint. To the east and west of the Stora Stöten open-pit, "hard ores" consisting of disseminated chalcopyrite and small amounts of other sulphides are hosted by an "ore quartzite". The copper content was about 2.5%. In addition, fully visible grains of gold have been found in the eastern hard ore sector.

At Zinkgruvan (Åmmeberg) the ore is hosted by grey, felsic volcanic rocks with intercalations of carbonate rocks, skarn and quartzite. Massive zinc-lead-silver ore can be followed for 5 km as two layers ranging from 2 to 20 m thick in abedded tuffite. The deposit is known to a depth of 1,300 m and is considered to contain more than 30 million tonnes of ore. The extracted ore has contained 6–10% zinc.

The Garpenberg mining area comprises a 4 km long sulphide-mineralised horizon and several iron ores (including Smältarmossen). Zinc-lead ore containing silver is today mined at Garpenbergs Odalfält and Garpenberg Norra. Between these two mines, there is the Dammsjö Mine with 5 million tonnes of ore that has only been test-mined.

Among tungsten mineralisations in Sweden, only the deposit at Yxsjöberg has been of economic importance. At Yxsjöberg, copper ore was originally mined but, since 1918, the mining has extracted tungsten ore. Scheelite, chalcopyrite and fluorite comprise the ore minerals.

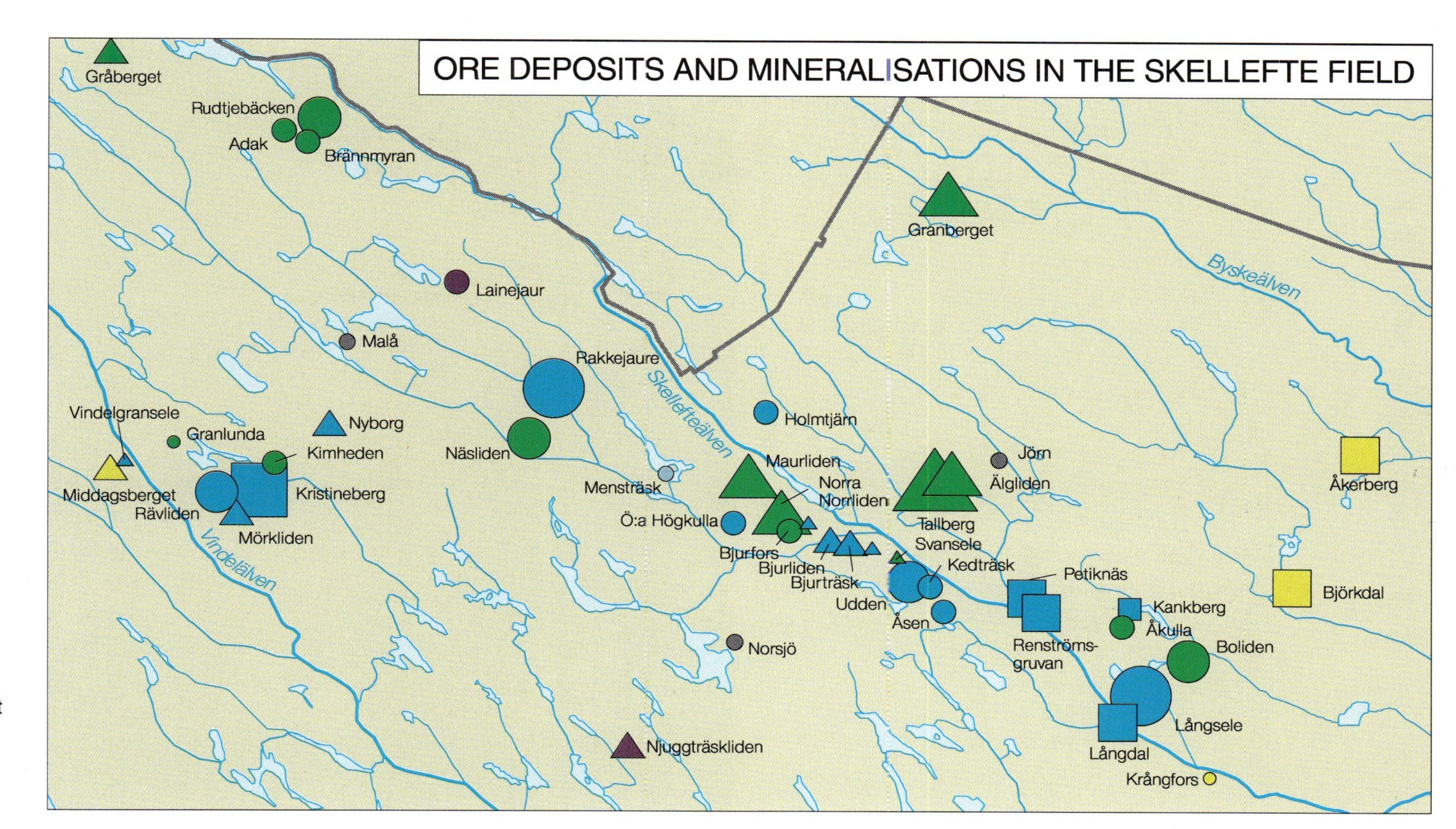

1: 700 000

Dominating metal

- Copper
- Zinc
- Nickel, cobalt
- Gold, silver
- Sulphur

Operating 1993	Closed	Prospect	Ore, mill. tonnes
			>10
			1–10
			0.1–1
			<0.1

In the Skellefte Field, the deposits consist of sulphide ore and a number of quartzite dykes containing gold. No iron ores are present. Zinc and copper dominate particularly in the central part of the field. Nickel deposits consist of gabbros and ultrabasic rocks containing pyrrhotite, pentlandite and chalcopyrite. (M21)

LARGE MINES IN THE SKELLEFTE FIELD (PRODUCTION UP TO 1993)

Deposit	Opened	Closed	Ore production, mill. tonnes	Main metals
Adakfältet	1941	1977	6.33	copper
Bjurfors gruvfält	1941	1945	0.20	copper, zinc, lead
Björkdalsgruvan	1989	producing	3.1	gold
Bolidens gruvfält	1925	1967	8.28	copper, gold, zinc, lead, silver
Brännmyrangruvan	1960	1976	0.98	copper
Holmtjärnsgruvan	1924	producing	0.46	zinc, copper
Kankbergsgruvan	1966	producing	0.52	zinc, gold, lead, copper, silver
Kedträskgruvan	1969	1991	0.16	zinc, copper, lead
Kimhedengruvan	1965	1975	0.13	copper
Kristinebergs gruvfält	1940	producing	16.29	zinc, copper, gold, silver
Lainijaurs gruvfält	1941	1945	0.12	nickel, copper
Långdalsgruvan	1967	producing	3.23	zinc, lead, gold, silver
Långselegruvan	1956	1991	11.18	zinc, copper, lead, gold
Näslidengruvan	1969	1989	3.71	copper, zinc, gold, silver
Petiknäs	1992	producing	0.24	zinc, copper, gold, silver
Rakkejaure	1965	1988	0.73	zinc, gold, silver
Renströmsgruvan	1952	producing	7.16	zinc, lead, copper, gold, silver
Rudtjebäcksfältet	1947	1975	4.74	copper, zinc
Rävlidenfältet	1936	1991	9.41	zinc, copper, lead
Uddengruvan	1971	1990	5.95	zinc, copper, lead, gold, silver
Åkerberg	1989	producing	0.38	gold
Åkullagruvan	1938	1957	0.98	copper
Åsengruvan	1988	1991	0.22	zinc, copper
Östra Högkullagruvorna	1951	1959	0.11	zinc, silver

THE SKELLEFTE FIELD

The Skellefte Field is approximately 150 km long and 50 km wide, and follows the river Skellefteälven. In June 1992, mining was in progress at 8 locations, whereas 21 deposits had been closed and 15 were subeconomic and dormant. The deposits consist of sulphide mineralisations which are mostly complex ores, with zinc, copper, lead etc., requiring selective enrichment. Complex ores with gold and silver form the Skellefte Field's important deposits. Examples include the mining areas at Kristineberg, Rävliden and Boliden as well as the Långsele and Renström mines. Pyrite is the dominant sulphide, whereas the main ore mineral is sphalerite, followed by chalcopyrite. Arsenopyrite is commonly found and is sometimes abundant. The average ore contains 2.3% zinc, 0.8% copper, 0.8% arsenic and 0.2% lead together with 40 grammes per tonne of silver and 1.5 grammes per tonne of gold.

The Boliden ore was previously Europe's largest gold deposit and the largest arsenic deposit in the world. The mean concentrations in the ore were 15.5 grammes per tonne of gold and 6.8% of arsenic. The gold occurs in quartz-tourmaline veins and in arsenopyrite. Altogether 128 tonnes of gold and 566,000 tonnes of arsenic were extracted from the Boliden Mine, together with other products.

The Kristineberg mining area has provided the largest ore tonnage. It is estimated that approximately 6 million tonnes of ore reserves are present, containing 5.2% zinc and 1.2% copper together with precious metals and lead.

Two deposits, Björkdalsgruvan and Åkerberg, consist of quartz veins containing gold in granodiorite and gabbro, respectively. The average gold content of the ore, in both cases, is 3 grammes per tonne.

THE ORE DISTRICTS IN NORRBOTTEN

Five deposits are in operation in the Gällivare and Kiruna regions at the present time. These include iron ores at Malmberget and Kiirunavaara, and copper ores at Aitik, Pahtohavare and Viscaria.

Iron ore

Skarn iron ores and quartz-banded iron ores are well represented in the county of Norrbotten, but only apatite iron ores are economically important. These occur as 10–200 m wide sheets in felsic to intermediate lavas.

Kiirunavaara has one of the world's largest concentrations of iron ore. The ore sheet is more than 4 km long and dips approximately 60° towards the east. The ore contains 55–67% iron, in magnetite and to some extent in hematite, together with apatite, pyroxene, amphibole and calcite. Depending on the distribution of apatite, some of the ore is phosphorus-rich (ca. 2% P) and some is phosphorus-poor (0.02% P).

After Kiirunavaara, the largest deposit of iron ore in Sweden is at Malmberget, 5 km to the north of Gällivare. Here, the volcanic rocks are metamorphosed to red and grey gneisses. Intrusion of granite and pegmatite accompanied folding of the ore beds and several separate sheets of ore are present. In one of these, Stora Malmlagret, a more or less continuous mineralisation of magnetite and hematite can be followed for 6.5 km in an east-west direction. Other ore bodies (Kaptenslagret, Koskullskulle, etc.) consist almost entirely of magnetite, containing 49–65% iron and 0.03–1.2% phosphorus.

Other ore types

The largest copper mine in Europe and Sweden's largest gold mine is at Aitik, 15 km southeast of Gällivare. Although the mean concentrations of ore are less than 0.5 per cent copper and 0.5 grammes of gold per tonne,

Aerial photograph of Kiirunavaara with Luossavaara in the background. The terraces consist of rock waste.

Illustration of how the mining and the extraction of iron ore is done in Kiruna and Malmberget.

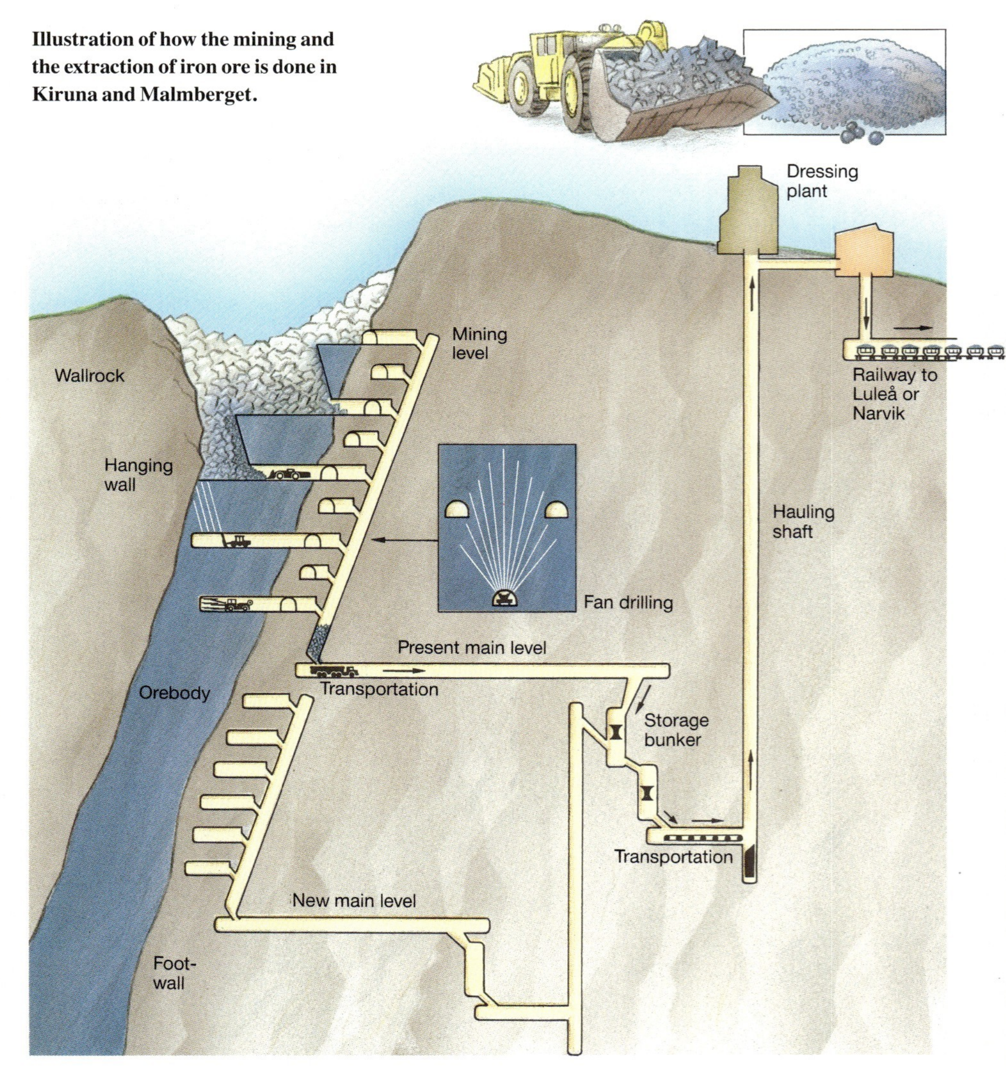

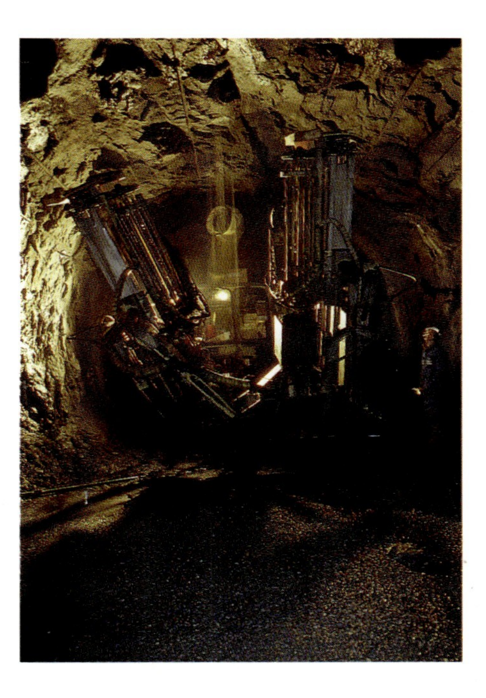

Drill for underground operation in one of the world's most modern mines, Kiirunavaara. This machine is capable of drilling 300 m per day.

large amounts of pure metal are extracted as a result of modern technology and open-cast mining. During 1991, atotal of 13.9 million tonnes of ore were mined, from which 47,098 tonnes of copper, 34.3 tonnes of silver and 1,645 kg of gold were extracted. Chalcopyrite and pyrite form disseminations, schlieren and veins in a 3 km long and 400 m wide mineralised zone. Within this zone, the ore is extracted in an open-cast mine which extends for more than 2 km. The wall rocks consist of paragneisses.

The copper deposits at Viscaria and Pahtohavare, to the west and southwest of the town of Kiruna, were discovered relatively recently. The ores are associated with intercalations of sedimentary rocks in mafic volcanic rocks (Kiruna greenstone). At Pahtohavare, gabbro and quartz diorite are also present.

At Viscaria, three ore layers have been identified consisting of rich impregnations and veins of chalcopyrite in limestone and graphite schist. Pyrrhotite, magnetite and some sphalerite also occur in the ore.

At Pahtohavare, open-pit mining is in operation. The copper content has been 2–3% and the gold content about 1 gramme per tonne. In one area the chalcopyrite is oxidised down to a depth of 150 m and modified into a colourful collection of minerals consisting of copper-rich silicates, oxides, carbonates and sulphides as well as native copper.

Iron ore from Malmberget is enriched and transformed into, e.g., pellets in the ore processing plant at Vitåfors.

Oxidized copper ore from Pahtohavare. The copper minerals malachite, chrysocolla and azurite with green to blue colours, dominate together with the iron oxide limonite (brown).

LARGE MINES AND MINERALISATIONS IN NORRBOTTEN COUNTY (PRODUCTION UP TO 1993)

Deposit	Opened	Closed	Ore production, mill. tonnes	Reserves. mill. tonnes	Main metal
Aitikgruvan	1968	producing	202	500	copper
Haukivaara-Nukutusfältet	1926	1986	10.4	140	iron
Kiirunavaara malmfält	1898	producing	968	800	iron
Lavers gruvfält	1934	1946	1.54	–	copper
Leveäniemi malmfält	1964	1982	56.65	250*	iron
Loussavaara malmfält	1925	1985	20.49	0	iron
Malmberget	1888	producing	287	400	iron
Mertainengruvan	1956	1959	0.43	166	iron
Pahtohavare	1990	producing	0.73	1*	copper, gold
Tuolluvaara malmfält	1902	1982	25.23	0	iron
Viscaria	1982	producing	10.04	2*	copper

* Estimated value

DEPOSITS IN THE CALEDONIDES

Two types of stratabound sulphide ores can be distinguished in the Swedish Caledonides. One type consists of disseminated lead-zinc ores in Late Precambrian and Cambrian sandstones along the Caledonian front, Laisvall is one example. The other type consists of partly massive complex ores which lie parallel with the layering in the surrounding Cambrian-Silurian volcano-sedimentary rocks. The most important deposits, e.g., Stekenjokk, Levimalmen, Remdalen and Ankarvattnet, occur in the Köli Nappes.

Lead ore at Laisvall is 5 km long, 3 km wide and up to 24 m thick. It consists of disseminated galena and sphalerite which, together with smaller amounts of pyrite, fluorite, baryte, calcite, etc., form a cement to the clastic particles in a quartz-rich sandstone. The ore is mined in a lower

The open pit at Stekenjokk has been filled and the area has been planted with grass. The photograph was taken in 1982. The ore mined was of a complex type with zinc and copper as the most important metals.

and an upper sandstone, with mean concentrations of 4% lead and 0.5% zinc.

The complex ore at Stekenjokk is 5 km long, 100–300 m wide and 2–10 m thick. The deposit contains, on average, 3.2% zinc and 1.4% copper. The extremely persistent ore body lies on the boundary between metamorphosed felsic volcanic rocks and graphitic schist. The volcanic rocks located close to the ore have been altered to sericite-rich quartzite and schist.

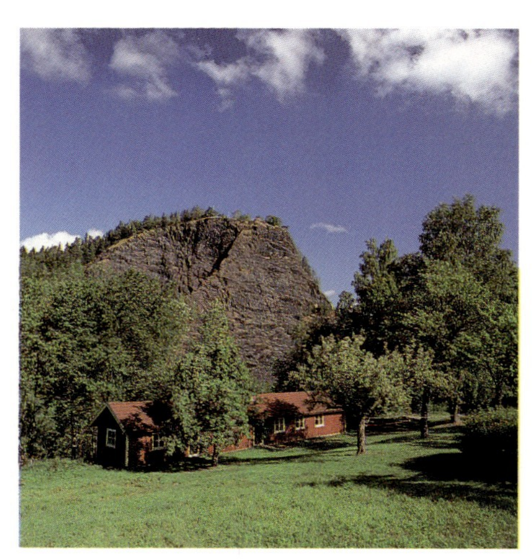

Smålands Taberg has been the classical deposit for extraction of titanium. Today, it is a nature conservation area.

LARGE MINES AND MINERALISATIONS IN THE CALEDONIDES (PRODUCTION UP TO 1993)

Deposit	Opened	Closed	Ore production, mill. tonnes	Reserves, mill. tonnes	Main metals
Ankarvattnet	1916	1917	0.01	0.75	zinc, copper, lead
Beitsetjenjunje	–	–	–	0.22	zinc, copper
Bellviksbergfältet	–	–	–	1*	lead, zinc
Bjelkes gruvfält	1739	1911	0.16	–	copper, zinc
Daningen	–	–	–	0.1–1	copper
Foskros	–	–	–	1–10	lead
Fröå/Gustavs-fältet	1742	1919	0.24*	–	copper, zinc
Guttusjögruvan	1978	1982	0.14	2*	lead
Jervas	–	–	–	10	copper
Jormlien	1919	1919	0.01	0.61	zinc, copper
Laisvall	1943	producing	48.38	60	lead
Levimalmen	–	–	–	5.26	zinc, copper
Lövstrandsgruvan	1956	1957	0.02	12	lead
Maiva	–	–	–	1	lead
Niepsurt	–	–	–	1.75	lead
Näset	–	–	–	0.1–1	zinc, lead
Ramundberget	1690	1756	0.01*	0.1–1	copper
Remdalen	–	–	–	0.74	zinc, copper
Routevare	–	–	–	20	iron, titanium, vanadium
Stekenjokk	1975	1988	8.08	7*	zinc, copper, lead
Sylarna	–	–	–	9.88	copper
Sågliden	–	–	–	13	lead
Unna Gaisartjåkko	–	–	–	1	copper, zinc
Vassbogruvan	1960	1982	4.84	0	lead, zinc

* Estimated value

DEPOSITS IN OTHER PARTS OF SWEDEN

The Enåsen gold mine lies to the south of the border between the provinces of Medelpad and Hälsingland. The ore body, 350 m long and 5–25 m wide, is found in a 10–20 m thick quartzite in veined gneisses. The gold mainly occurs in pure form associated with chalcopyrite, pyrrhotite and pyrite. Since the opening of the mine in 1984 until its closure in 1991, a total of 1.5 million tonnes of ore were extracted with 2.2 grammes of gold and 4 grammes of silver per tonne together with certain amounts of copper.

Iron ores with economically interesting concentrations of titanium and vanadium are found at eight places in Sweden. None of these deposits has been in operation during recent years. In the Smålands Taberg mining area more than one million tonnes of ore were extracted up to 1960, containing 28–32% iron, 5–10% titanium dioxide and 0.12–0.17% vanadium.

About 20 nickel deposits were investigated during the 1970's and 1980's in the county of Västerbotten. At Lappvattnet, for example, a mineralisation was found with one million tonnes containing 1% nickel and 0.2% copper.

At least 30 uranium mineralisations are known in Sweden. They are found in Cambrian-Ordovician sedimentary rocks as well as along faults and fractures in both plutonic and supracrustal rocks. The largest concentration of uranium has been found at Ranstad in Västergötland, in a Cambrian alum shale, where there is a 3.6 m thick uranium-rich zone consisting of black shale, anthraconite and kolm. The uranium content is, on average, 300 grammes per tonne. During 1965–1969, the production of uranium at Ranstad amounted to 200 tonnes.

At Pleutajokk, 30 km northwest of Arjeplog, pitchblende has been formed by hydrothermal solutions along fissures in felsic volcanic rocks. The mineralisation has been dated to 1,750 million years old. To the north of Östersund, a fracture zone with pitchblende and secondary uranium minerals has been investigated at Lilljuthatten. The mineralisation, dated to 420 million years old, is found in a hydrothermally metamorphosed granite. Uranium deposits at Pleutajokk, Lilljuthatten, Björkråmyran (northeast of Åsele) and Kvarnån (NW of Älvsbyn) contain atotal of 10 300 tonnes of uranium.

Numerous, and sometimes surprising, fields of use for industrial minerals and rocks are found in modern society.

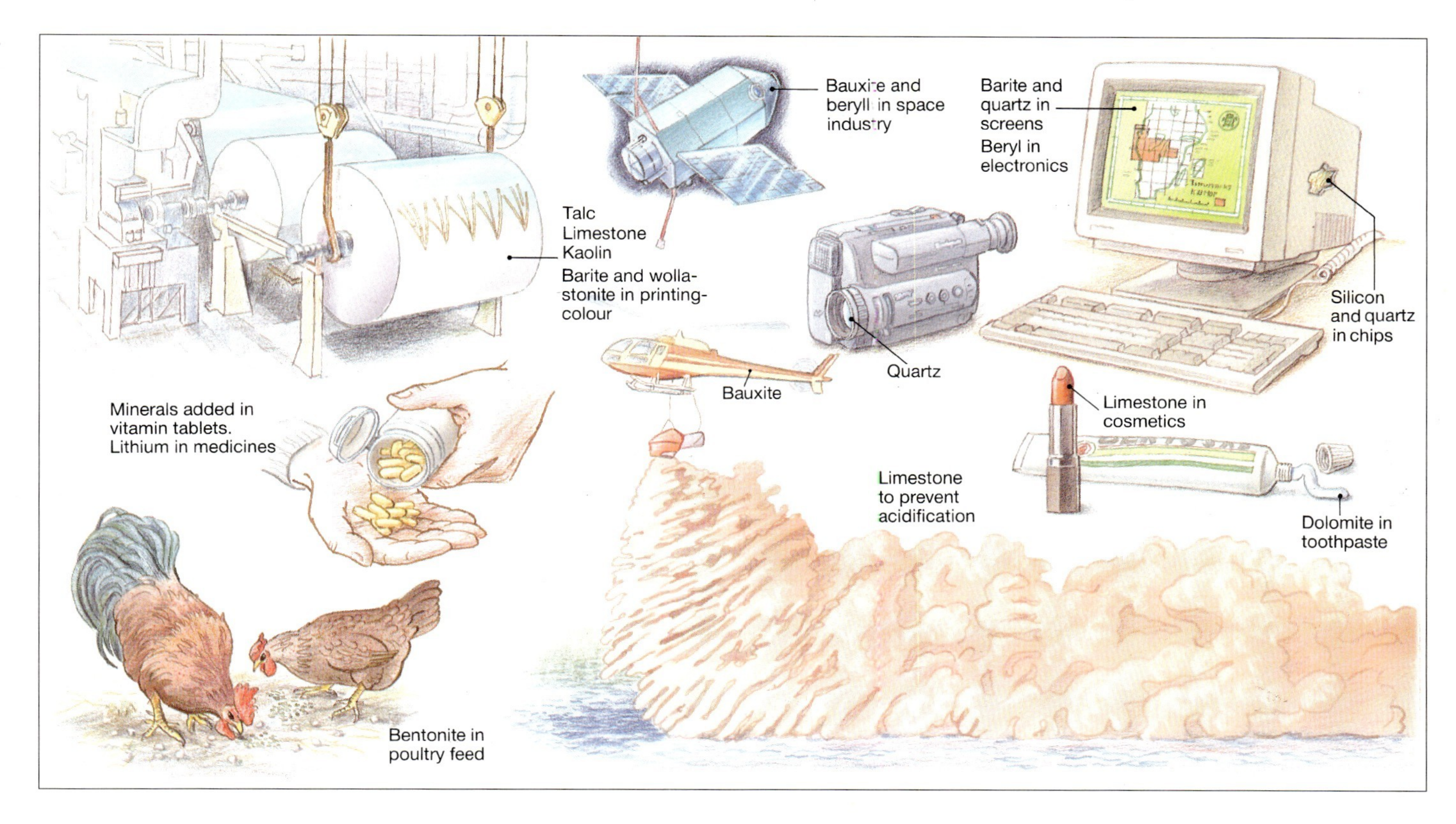

Industrial minerals and rocks

The term industrial minerals and rocks refers to minerals and rocks that are extracted for reasons other than for their metallic content or fuel value. The expression industrial minerals also sometimes includes elements utilized in metal production, e.g., bauxite and ilmenite.

SOME INDUSTRIAL MINERALS

Andalusite	Mica	Olivine
Anorthosite	Ilmenite	Pegmatite
Anthophyllite	Limestone	Rutile
Apatite, phosphorite	Kaolin	Sandstone
	Chromite	Sillimanite
Baryte	Quartz	Rare Earth minerals
Bauxite	Quartzite	Talc, soapstone
Diatomite	Kyanite	Heavy mineral sands
Dolomite	Lithium minerals	
Fluorite	Magnesite	Vermiculite
Feldspar	Monazite	Wollastonite
Garnet	Nepheline syenite	Zirconium minerals

STRENGTH AND RESISTANCE TO WEATHERING OF SOME ROCK TYPES

Hard and strong	Granite, mica-poor gneiss, quartzite, dolerite, porphyry and leptite
Moderate and low abrasion resistance	Limestone, mica schist, mica-rich gneiss, migmatised rocks of the minerals mentioned above
Easily weathered and erosive	Shale, chalk and argillised rocks

Industrial minerals are characterised by their diversified range of uses. They can rarely be re-cycled. They have a very wide variation as regards geological occurrences, mining and enrichment technology, purity requirements and pricing. This is thus a wide field, and includes numerous different minerals ranging from quartz sand, a bulk product costing approximately 50 SEK per tonne, to industrial diamonds sold per carat (0.2 g) at a cost of about 100 million SEK/tonne.

Industrial minerals are used within almost all industrial activities, for example, oil-drilling, the chemical-technical industry and the fertilizer industry. They are also important in the space and computer industries, in the drug industry, as fillers in the paper and plastics industries, in certain foods and as filters when purifying liquids.

Swedish bedrock offers good conditions for finding industrial minerals and rocks. However, by no means all deposits have been exploited. Those that have not been utilised may become important resources in the future and must be considered in the planning of environmental and natural resources.

APATITE

Apatite is a phosphate mineral that is found in very small quantities in most crystalline rocks. It is usually green to grey-green and occurs as hexagonal crystals. *Phosphorite* is a finely crystalline form of apatite. It is found in sedimentary rocks, locally in considerable amounts, and is often formed by fossil parts of skeletons and other organic material.

There are few occurrences of economically extractable phosphorite in Sweden. This mineral is found within the very widespread glauconite-phosphorite limestone of Ordovician age in Västergötland, Östergötland and Närke. Cambrian sandstone in Närke and Östergötland also contains glauconite-phosphorite in a deposit that is up to 3.5 m thick. Apatite for use as a raw product in the manufacturing of fertilizers and detergent phosphates was extracted until fairly recently as a secondary product of the iron ore mining at Kiruna and at Grängesberg. The latter mine is now closed.

BARYTE

Baryte is a heavy mineral with high density. It is often colourless or white and usually crystallizes in the form of small plates or prismas. It usually occurs as a dyke mineral together with sulphide minerals or iron and manganese ores. Baryte is resistant to weathering and, thus, high concentrations are often found in *residual deposits*, i.e., the minerals that previously occurred together with baryte have weathered away. In Sweden, commercially interesting

Barite (barium sulphate) is used in contrast liquid when X-raying the stomach and intestinal tract.

amounts of baryte have only been found at Pottäng and Hartung in the Alnö massif near Sundsvall. In the early 1940's, at least three baryte dykes were found here with a known length of about 100 m and a width of up to 3 m. In the late 1940's, the dyke at Pottäng was worked and a total of about 6,000 tonnes of baryte were mined.

Baryte is a very important ingredient in the lubricant used when drilling for oil and gas. It is mixed with water and clay, after which the liquid is pumped down into the bore hole where it cools the drill crown, flushes away cuttings and prevents the hole from collapsing. Baryte is also used as a filler in paint, paper, playing-cards, linoleum carpets, and as an additive when manufacturing glass. The mineral is also used in contrast liquid in X-rays where its ability to absorb X-rays is utilised.

When beryllium occurs in perfect hexagonal crystals they are considered as gems. Usually they have a greener shade than in the picture. The lengths of the large crystals are 57 and 61 mm. Hoting, Jämtland.

BERYL

The hexagonal shape of the crystals and its usually green colour are characteristics of *beryl*. The crystals are usually prismatic and more or less transparent. The colour varies and transparent perfect crystals are regarded as precious stones; green stones are called *emeralds*, blue-greens are called *aquamarine*, yellow stones are called *heliodor*, and rose-coloured stones are called *morganite*.

Beryl usually occurs as an accessory mineral in pegmatites but can also sometimes be found in cavities in granite. In Sweden, beryl has been found in several pegmatite dykes. In connection with mining of quartz and/or feldspar, it has also been possible to extract limited amounts of beryl. The largest known deposits where beryl has been utilised are in the Godegård district of Östergötland, in the Kolsva feldspar mine WNW of Köping, Angsta in Västernorrland, Varuträsk to the northwest of Skellefteå, and at Sörhällan to the north of Luleå.

The element *beryllium* is used within the space and nuclear energy industries, and within health care in X-ray equipment. Copper with an addition of a few per cent of beryllium gives a very important alloy, beryllium copper, which has properties of being non-magnetic, non-rusting, and highly durable. The alloy is used in special implements, in material for the aircraft industry, and in different components to be used at sea.

DOLOMITE

The mineral *dolomite* is usually found as rhombohedric white crystals. When these occur in massive, granular aggregates, they form the rock-forming mineral in dolomitic limestone. Apart from dolomite, there are several other carbonate-rich minerals including calcite, magnesite, siderite and aragonite. Dolomite is used within several sectors, e.g., in iron and steel manufacturing and within the building industry. Dolomite is spread on fields, in lakes and in forests in order to counteract acidification. In the chemical-technical industries, it has a wide range of uses, for example, as an in-

Mining of blocks of marble at Ekeberg, Glanshammar near Örebro.

The mineral fluorite (lilac in the illustration) is used to provide fluorides used in high-tension industry on account of their good insulating capacity. Carbon-fluorine compounds are considered to damage to ozone layer.

Finely-ground feldspar is used, for example, in the manufacture of sanitation ware.

Muscovite, or mica, has the characteristic of being separated into thin, transparent layers. It is used, for example, in the paint and electronic industries.

gredient in toothpaste and paint, and in the manufacturing of mineral-wool.

Dolomite is found at numerous places throughout Sweden, and there is quarrying at several places, e.g., Glanshammar and Fanthyttan in the province of Örebro, and near Sala.

FLUORITE

Fluorite frequently forms cubic crystals of varying colour, including yellow, green, blue and lilac. The relative softness of the mineral enables it to be processed mechanically and fluorite was used very early by the Greeks and Romans for different decorative purposes. The Indians also used large crystals of fluorite to make decorations. The properties of fluorite in connection with the melting of metals were discovered already in the mid–16th century.

In Sweden, fluorite has been found in different geological environments and has been mined at several places. During the Second World War, this mineral was mined at Gladsax in the neighbourhood of Simrishamn. When this production ceased, work continued until 1955 in the nearby Onslunda Mines. During recent years, fluorite has also been obtained as a secondary product in tungsten mining at Yxsjöberg in the county of Örebro. These activities ceased at the end of the 1980's. The lead ores in the eastern frontal part of the Swedish Caledonides locally contain significant amounts of fluorite.

Fluorite is mined and processed largely for use when smelting iron ore and also within the glass and welding electrode industries. A large part of the production is also used within the chemical industry and a smaller part for manufacturing ornamental objects and for production of special optical equipments. Fluorite is the raw material for processing of fluorides, which are used in the high-tension industry on account of their good insulating capacity. Fluorine has also become of major importance in the manufacture of organic fluorine compounds used, for example, in teflon frying pans.

FELDSPAR

Feldspar used at the present time is obtained from pegmatites and coarse-grained granites. Feldspar contains different concentrations of potassium, sodium and calcium and is classified according to its chemical content in *potassium feldspar* and *plagioclase*. Its colour may vary from white, grey and light red to red.

Feldspar has been used for many years in the manufacture of ceramic goods and production today is mainly used for the manufacture of sanitation ware. This mineral has been mined at several different places in Sweden but only two are still in operation today. Both occur in the neighbourhood of Riddarhyttan, in Bergslagen.

MICA

There are two main groups of mica: Dark mica, consisting of *biotite* and *phlogopite*, is rich in iron and magnesium, whereas white mica, *muscovite*, is rich in aluminium. A light red variant is called *lithium mica*. A characteristic of mica is its formation in thin sheets or layers that are often found in thick packs. Individual layers are more or less transparent and may vary in colour between white, pink and dark brown to black.

Mica has been well-known for many years and has played an important role in the development of modern society. The name *vitrum moscoviticum*, Russian glass, indicates the origin of mica that was imported during the 18th century from Russia. In connection with the introduction of electronics, the mineral became essential owing to its high electrical insulation capacity. Particularly during the Second World War, mica production was of strategic importance because the mineral was used in the electrical industry and in radiotechnology. The importance of the mineral decreased after the war due to the discovery of alternative minerals. However, mica is still an important industrial mineral in the electronic industries. In addition, large amounts of ground mica are used as fillers in roofing paper, asphalt, rubber, paint and plaster.

Economically interesting concentrations have been found at several places in Sweden. The most profitable production took place during the Second World War when mica deposits in the provinces of Göteborg & Bohus and Jämtland were exploited. At the present time, there is no mining of mica in Sweden. However, this mineral is utilised as a secondary product when mining quartz and feldspar near Riddarhyttan in Bergslagen.

GRAPHITE

Graphite is the soft, black, metallic, naturally-occurring form of coal with the same chemical symbol, C, as diamond. The word graphite comes from the Greek and means "to write".

The largest known deposits of graphite in Sweden are at Vittangi, Masugnsbyn and Övertorneå in Norrbotten. Graphite has earlier been mined to a smaller extent at various places in Sweden, for example, southeast of Masugnsbyn at Nybrännan. In Västmanland, a small graphite deposit a few kilometers NNW of Halvarsbenning, to the north of Fagersta, was worked during the 19th century and up to 1920. A total of 3,800 tonnes of elutriated graphite was obtained.

Large amounts of limestone have been quarried at Limhamn near Malmö. This photograph was taken in 1988.

Welding can be done on account of graphite in the electrodes.

Limestone that has been deposited on the ground vegetation, here mainly on mosses, forms tufa. This sample has been taken from the lower part of a ca. 5 m thick deposit of tufa at Ekedalen in the province of Västergötland.

Graphite has also been mined at Skälsta, to the west of Sigtuna, and on the island of Härnön at Gånsvik, southeast of Härnösand.

Graphite has a very wide range of use owing to its chemical and physical properties. Its melting-point is high, 3,650°C, and the mineral sublimes at 4,500°C. Graphite is mainly used in welding electrodes and in the iron and steel industries. Its low coefficient of friction makes it excellent as a lubricant. The good electrical conductivity of graphite means that it is of major importance in the manufacture of batteries, in the metallurgic and chemical industries, and when manufacturing starting cables for electric motors. Blacklead is probably the most well-known field for graphite. It is a mixture of clay and amorphic graphite that can be heated and then formed into lead for use in pencils. Good-quality blacklead contains 85% coal.

LIMESTONE

Limestone is a sedimentary rock containing more than 50% carbonate minerals, mainly *calcite*. Other carbonate minerals found in limestone are *dolomite, magnesite, ankerite* and *siderite*.

Limestone occurs in layers of different geological age. It has been formed, and is still being formed, on the sea floor as a result of chemical precipitation of calcium carbonate from sea-water or through accumulation of calcareous shells or parts of skeletons of dead organisms. *Chalk* is a special kind of limestone that is formed through accumulation of shells of ca. 0.004 mm large microorganisms (coccolites). *Travertine* is the name of a carbonate stone that was formed on land as a result of calcium carbonate being precipitated from calcareous water. During metamorphism of limestone, there is a redistribution and recrystallisation of the mineral so that limestone is transformed into *marble* (crystalline limestone).

The calcite concentration of carbonate rocks may vary strongly, from rocks with 100% calcite content to rocks where calcite makes up only a minor part, e.g. marls of various kinds.

During the Cretaceous, large parts of northern Europe were covered by a shallow, warm sea in which calcium carbonate was precipitated and limestone gradually formed. Such limestones are today found in areas around, for example, Malmö, Kristianstad and Båstad, where intensive quarrying previously took place. Older limestones, formed 570–410 million years ago, are found at numerous places, for example, on the islands of Öland and Gotland, in Västergötland and Östergötland, on the Närke plain, in the Siljan area, and in the

Quartzite is a rock that resists weathering. In Dalsland, quartzite is found in marked ridges. Ulerudstjärn, to the north of Håverud.

Quartz is used in many domestic utilities, in glass, for example, quartz is the main constituent.

Swedish Calidonides. At several places, quarrying is still in progress and the total production in 1990 was more than 7.2 million tonnes.

Limestone is used in the manufacture of cement, in the chemical industries, as a filler in paper, paint and rubber, in ceramics, in glass manufacturing, etc. Limestone is also an ingredient in toothpaste and lipstick and is used to purify water and beverages. Large amounts of limestone are used in agriculture and environmental protection to counteract acidification, in purification of combustion gases, etc. The building industry also uses large amounts of marble.

QUARTZ

The mineral *quartz* is extremely common in the Earth's crust and occurs in most rocks. It usually has a milky-white colour and semi-transparent irregular shapes. However, sometimes other colour variants are found with hexagonal crystals. These are regarded as semi-precious stones. Pure quartz is extracted from fissure fillings where it has crystallized from silica-rich residual solutions.

The main fields of use of quartz are within china and pottery manufacturing, glazing and enamel production, and in silicon production. Very pure quartz is used, for example, in camera lenses, in mirrors for technical applications and in manufacturing silicon metals. The technical properties of quartz are utilised within semi-conductor technology, in computer chips and in solar cells. Quartz sand is mainly used in the glass industry, in foundries, in the manufacture of fibre glass, etc.

Quartzite, suitable for the manufacture of heat-resistant stones and sheets, as well as ferro-silicon and silicon alloys, is mined in Dalsland.

CLAY

Clay is one of Nature's most difficultly-defined raw materials. The term "clay" in fact simply means a group of fine-grained sediments with particle diameters less than 0.002 mm.
A property that all commercially interesting clays have in common is, however, that most of them become plastic when they are mixed with small amounts of water.

Clays can be divided into three groups: Kaolinitic clays, smectitic clays and structural clays.

The term "kaolin" originates from the place "Kau-ling" in the Jiangxi Province of China. Mineralogically, kaolinite is an aluminium hydrosilicate and the term "kaolin" refers to a clay that largely consists of the mineral kaolinite. *Kaolin* is a weathering product of granite, gneiss and other feldspar-rich rocks. Its composition and colour vary considerably between different deposits. Kaolin is mainly used industrially in the manufacture of paper, ceramics, rubber and paint. Kaolin clay has earlier been quarried at a couple of places in Skåne for manufacturing heat-resistant bricks, floor-paving and china.

The *smectitic clays* largely consist of *bentonite*, which is mainly made up of the mineral montmorillonite. The physical properties of bentonite, mainly its ability to swell, make it an important industrial mineral. It is frequently used in drilling fluids in oil-prospecting, as a binding agent in foundry sand, in pelleting of iron ore, etc. Bentonite is also used in animal feeds, for example, in poultry feed in order to increase the size of the egg and the hardness of the shell. In the foodstuff industries, bentonite is used, for example, to clarify beer, wine and fruit-juice. The most well-known deposit of bentonite in Sweden are the ca. 1.8 m thick layer at Kinnekulle in Västergötland. However, bentonite is not commercially mined today in Sweden.

Structural clays have a highly variable content of minerals and may sometimes contain only a minor part of the clay minerals kaolinite or smectite. However, this has not prevented them being used and, since ancient time, they have found applications in the manufacture of building-material such as bricks or tiles for flooring, roofing and walls, or as clay vessels for household needs. However, in the modern society of today, we are not only restricted to the mineralogical

The mineralogical composition of clay is of decisive importance when making pottery.

content in the local clay but can create variable properties to satisfy the needs of consumers by adding different types of material. In northwestern Skåne and at Linköping, the clays obtained are used both for technically qualified purposes as well as for making bricks.

LITHIUM

Lithium occurs mainly in pegmatites or together with the minerals quartz, feldspar and mica. Lithium is found in a large number of minerals but commercially the most important are spodumene, lepidolite, petalite and amblygonite.

Lithium is used within the glass and ceramic industries, in metal alloys, in aluminium manufacturing, in batteries and in tranquillizers, as well as within the nuclear power industry. In Sweden, lithium has been found in a large number of pegmatite dykes. The most well-known is the deposit at Varuträsk, near Skellefteå, where mainly petalite, but also spodumene, has been mined. At the present time, however, there is no mining of lithium in Sweden.

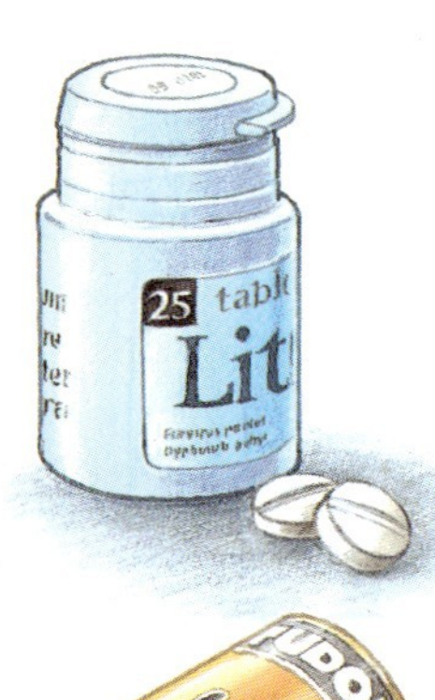

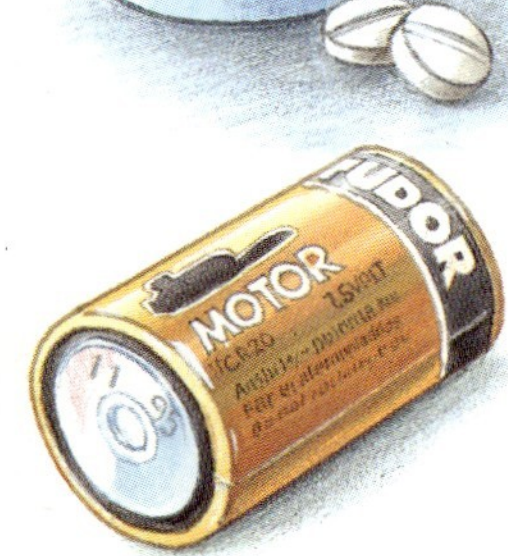

MAGNESITE

Magnesium is a very common element in the Earth's crust. It is found in a large number of minerals of which the most important commercially is *magnesite*. *Brucite* and *dolomite* are other minerals with high magnesium contents.

The most important field of use for magnesite is as a heat-resistant lining material in ovens. Magnesium oxide, produced by combustion of magnesite, is used in the paper and chemical industries and also as filler in rubber and plastics. To a minor extent, magnesium is also used in electrical components. Magnesite deposits in Sweden occur predominantly in the Swedish Caledonides. The largest known body of magnesite is in the Sarek National Park. Several magnesite bodies have also been identified in the mountains to the west of Kvikkjokk. In Sweden, commercial production of magnesite has never taken place, but at some places the mineral has been subjected to test-mining.

NEPHELINE SYENITE

Nepheline syenite is a relatively uncommon rock, similar to granite in texture and hardness.

The low melting-point and good flux properties of nepheline syenite make it an ideal crude product in glass manufacturing. When finely ground, this rock is used as a filler in paint, rubber and plastics. It has also been used as a cleaning and polishing agent, a market that today is dominated by volcanic ash or finely-ground silicon. The high aluminium content of nepheline syenite makes it an important crude product in aluminium production and is used for this purpose in Russia. Nepheline syenite is also used in manufacturing ceramic products such as glazed tiles and different types of china.

In Sweden, nepheline syenite and other nepheline-containing rocks have been found at four places: On the island of Alnön close to Sundsvall, near Särna, near Almunge in Uppland, and to the north of Gränna. No mining of nepheline-containing rocks takes place in Sweden.

OLIVINE

Olivine is a magnesium-rich mineral consisting of a mixture of forsterite and fayalite with fusing temperatures of 1,890°C and 1,205°C, respectively.

In the early 1930's, olivine started to be used in the manufacture of heat-resistant materials. Limited amounts were earlier used as a soil ameliorant and as fertilizer. Today, olivine is used mainly to form slag in the manufacture of crude iron where the magnesium content of olivine is important. Olivine is also used in foundry sand. Raw blocks of olivine have been used as heat reservoirs on account of the mineral's ability to release heat uniformly for a long period after being heated.

Most olivine deposits in Sweden are found in the Swedish Caledonides. In Jämtland, an olivine-rich rock with a characteristic red-yellow colour on the weathered surface was mined from the 1950's until the mid–1980's. These rocks were mainly used in manufacturing heat-resistant bricks and floor tiles. In the Pre-Cambrian shield area east of the Caledonides, deposits of olivine are uncommon. However, one occurrence at Purnu in Norrbotten is mined for use as an additive in the manufacture of iron ore pellets.

SILLIMANITE

The minerals *andalusite, sillimanite* and *kyanite* are grouped under the common heading sillimanite. All these minerals consist of aluminium silicate with a similar chemical composition. They differ in the way they have been formed, in their crystal structure, and in their physical properties.

Andalusite and sillimanite are found at a large number of places in Sweden but not in concentrations of economic interest. Kyanite, however, which is only found in a few deposits, is mined at the present time at Hålsjöberget in the county of Värmland. It contains about 30% kyanite and is the only deposit to be exploited in western Europe.

Sillimanite is used as a raw material in producing heat-resistant products that can be used in processes requiring high temperatures and pressures.

ROCKS USED IN THE STONE INDUSTRY

Rocks used in the stone industry refer to rocks that can be used for the manufacture of paving-stone, curbstone, building-stone, facing stone, ornamental stone and monument stone. *Granite, gneiss, diorite, hyperite dolerite, limestone, marble, sandstone* and *slate* are used for these purposes in Sweden. A decisive factor for the use of such rocks outdoors is mainly their ability to withstand the strain of climate and wear-and-tear.

Quarrying of natural stone for building purposes places great demands on the frequency of fissures in the rock, its splitting properties and the content of sulphide minerals. The rock must be homogeneous and must not contain cracks or dykes of other rock types. At the present time, quarrying of rock for the stone industry takes place mainly in the form of large raw blocks and is ongoing at

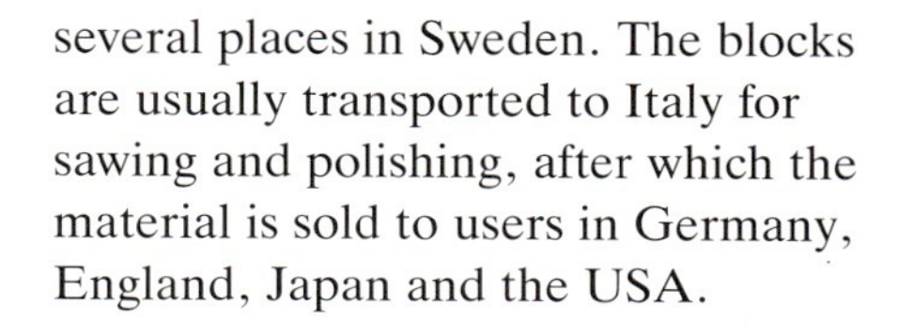
several places in Sweden. The blocks are usually transported to Italy for sawing and polishing, after which the material is sold to users in Germany, England, Japan and the USA.

TALC AND SOAPSTONE

Talc is a water-containing, magnesium-rich silicate mineral that constitutes the main component in soapstone which is found in metamorphosed ultrabasic rocks. In Sweden, it is mainly found in the Caledonides but also occurs in the counties of Värmland and Dalsland in the Precambrian shield area.

The most important fields of use for talc are as a filler and coating material in paper, as a filler in plastics, paint and cosmetics, and within the fertilizer and ceramic industries.

The *soapstone* quarried today at Handöl in the county of Jämtland is used for the manufacture of roofing paper, open fireplaces and within the pulp and paper industry.

The streets of Stockholm were paved with granite from Stenhamra, in the islands of Lake Mälaren.

The eastern quarry at Grythyttan in about 1900. The slates produced at Grythyttehed were of hard clay shale and mainly used for roofing.

Soapstone is so soft that it can be carved. This sculpture is about 19 cm high.

Roughly hewn pink granite from Vätö near Norrtälje is used as building material, etc. Here it is used on the facade of the Stockholm Opera House.

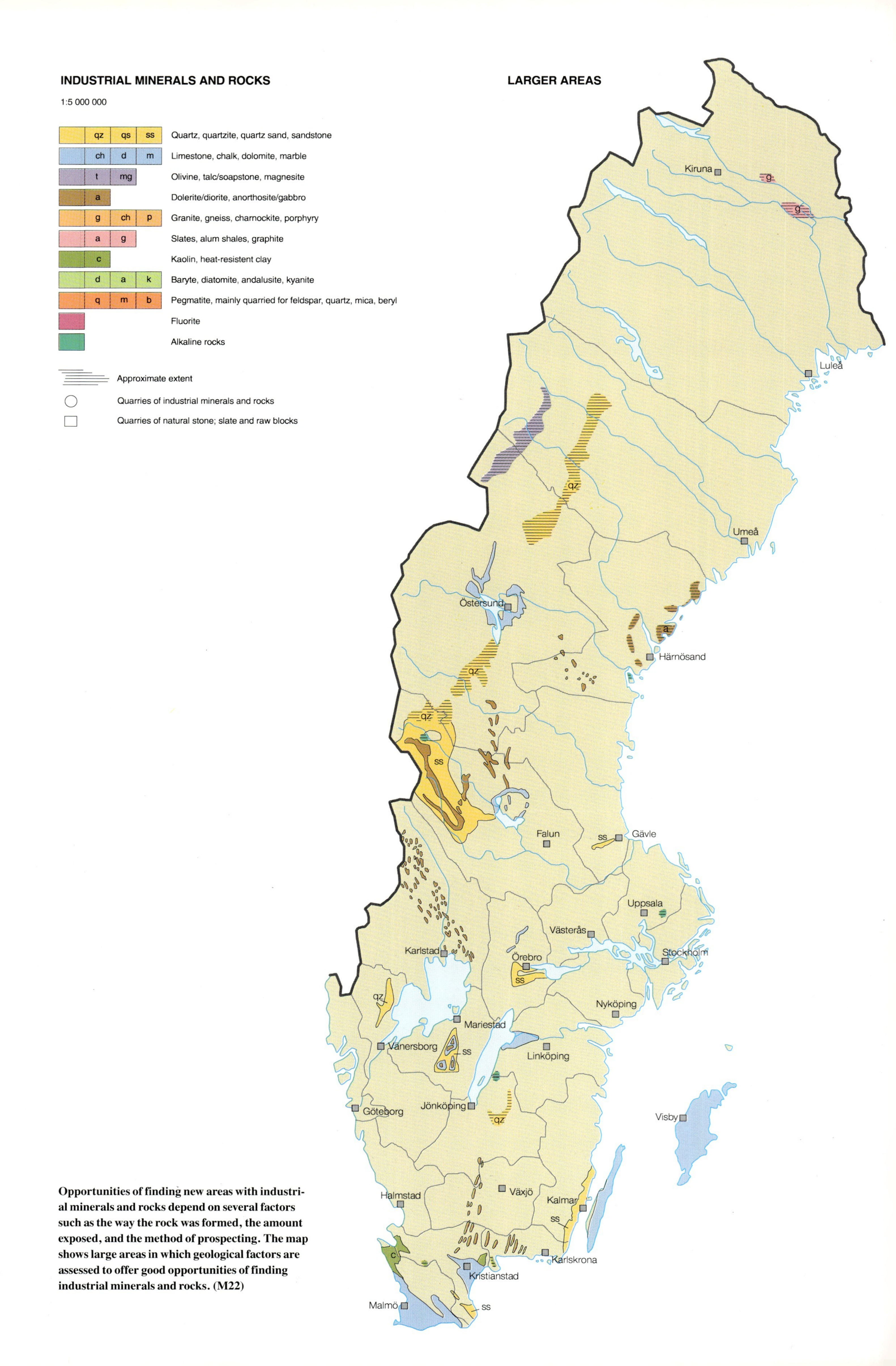

Opportunities of finding new areas with industrial minerals and rocks depend on several factors such as the way the rock was formed, the amount exposed, and the method of prospecting. The map shows large areas in which geological factors are assessed to offer good opportunities of finding industrial minerals and rocks. (M22)

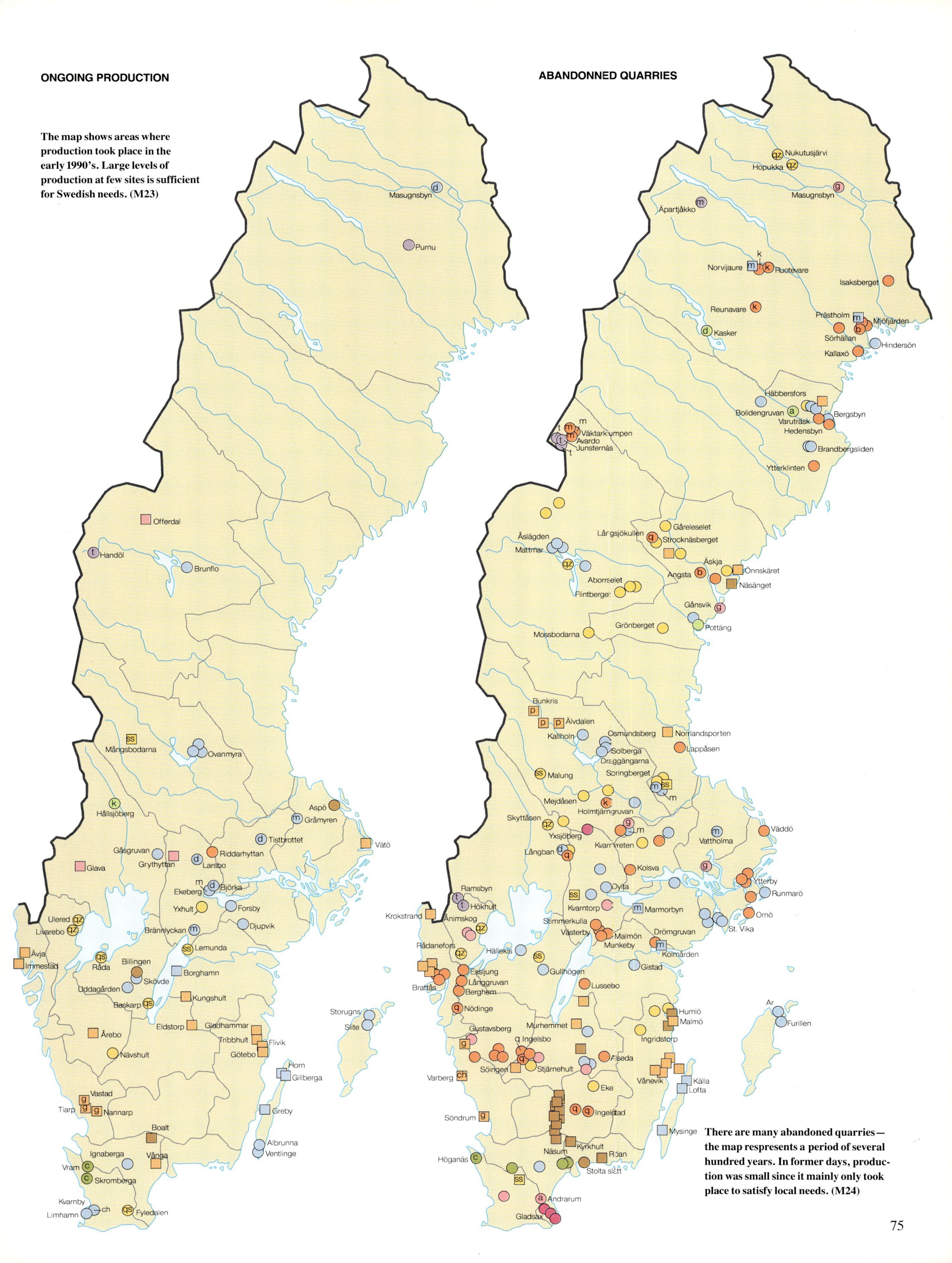

ONGOING PRODUCTION

The map shows areas where production took place in the early 1990's. Large levels of production at few sites is sufficient for Swedish needs. (M23)

ABANDONNED QUARRIES

There are many abandoned quarries—the map respresents a period of several hundred years. In former days, production was small since it mainly only took place to satisfy local needs. (M24)

Geophysics

Our knowledge of the bedrock, Quaternary deposits and water resources is mainly based on direct observations and analyses. With this knowledge, we can attempt to predict the conditions at greater depths. What does the overburden consist of? What kinds of bedrock are below it? How thick is the Earth's crust? What does it consist of? When and how was it formed? How has it been deformed and altered?

Answers to these questions require measurements and calculations based on the physical properties of the bedrock and the Quaternary deposits. Sensitive instruments are capable of measuring differences in the distribution of these properties even when they are hidden at great depths.

Differences in density of the bedrock cause small perturbations in the Earth's gravity field, magnetic rocks cause perturbations in the Earth's magnetic field, and differences in electrical properties of rocks and Quaternary deposits affect electric currents or radio-waves transmitted through the Earth.

Aircraft equipped for geophysical measurements. The nose and tail of the aircraft have been equipped with sensors for different measuring methods.

By making systematic measurements on the ground, in the air or in drillholes several kilometres deep, geophysicists can map how measured properties diverge from the normal situation. These anomalies depend on differences in the physical properties of bedrock and Quaternary deposits and can be used to create a third dimension on a geological map. The various measurement techniques give us an opportunity to "see" through water and the overburden and down into the bedrock. They also allow us to determine types and thicknesses of the Quaternary deposits.

EMANUAL SWEDENBORG (1688–1772): ON MAGNETISM:

"In the former part of our Principia, we arrived at that element of the world which may be denominated the magnetic; the first in which elementary nature presents herself as visible to the eye. Here it is that she begins to emerge out of her hiding place, and from darkness to issue forth into light: that she presents an image of herself as a whole: that at the very first glimpse of herself, she overwhelms as it were and confounds our senses by the forces with which she has invested a rough, dark, vile and heavy corpuscle. This corpuscle is the magnetic stone."

(From: *Opera Philosophica et Mineralia,* originally published in Latin, Dresden and Leipzig, 1734.)

Emanuel Swedenborg. Painting by Per Krafft the Elder.

Potential fields of the Earth

Our planet is enveloped by force fields that depend on different physical conditions and processes. Studies of these fields are one of the keys to understanding the basic composition of the Earth.

The gravity and magnetic fields are the most prominent potential fields. They both have size and direction, i.e., they are vectorial and their

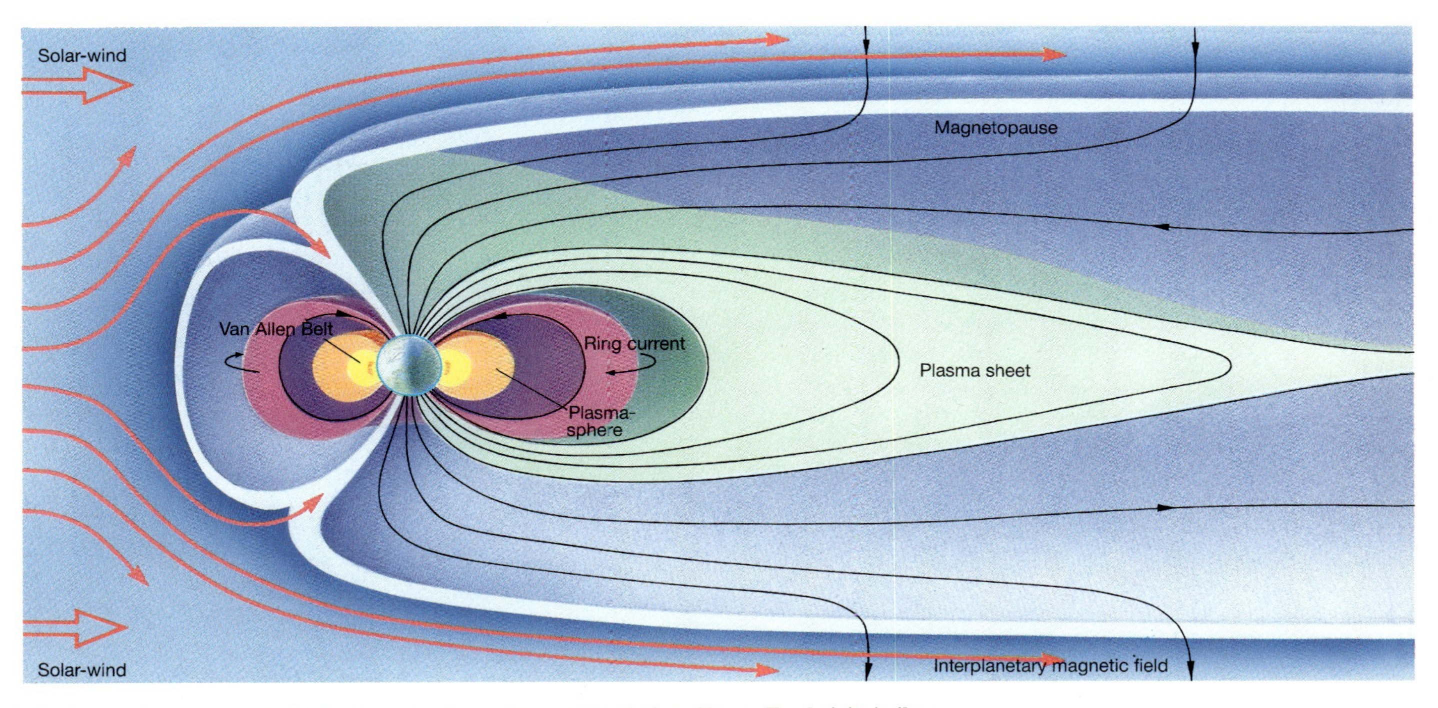

The shape of the geomagnetic field is strongly influenced by solar radiation. Close to Earth, it is similar to a dipole field. The magnetosphere protects the surface of the Earth from charged particles.

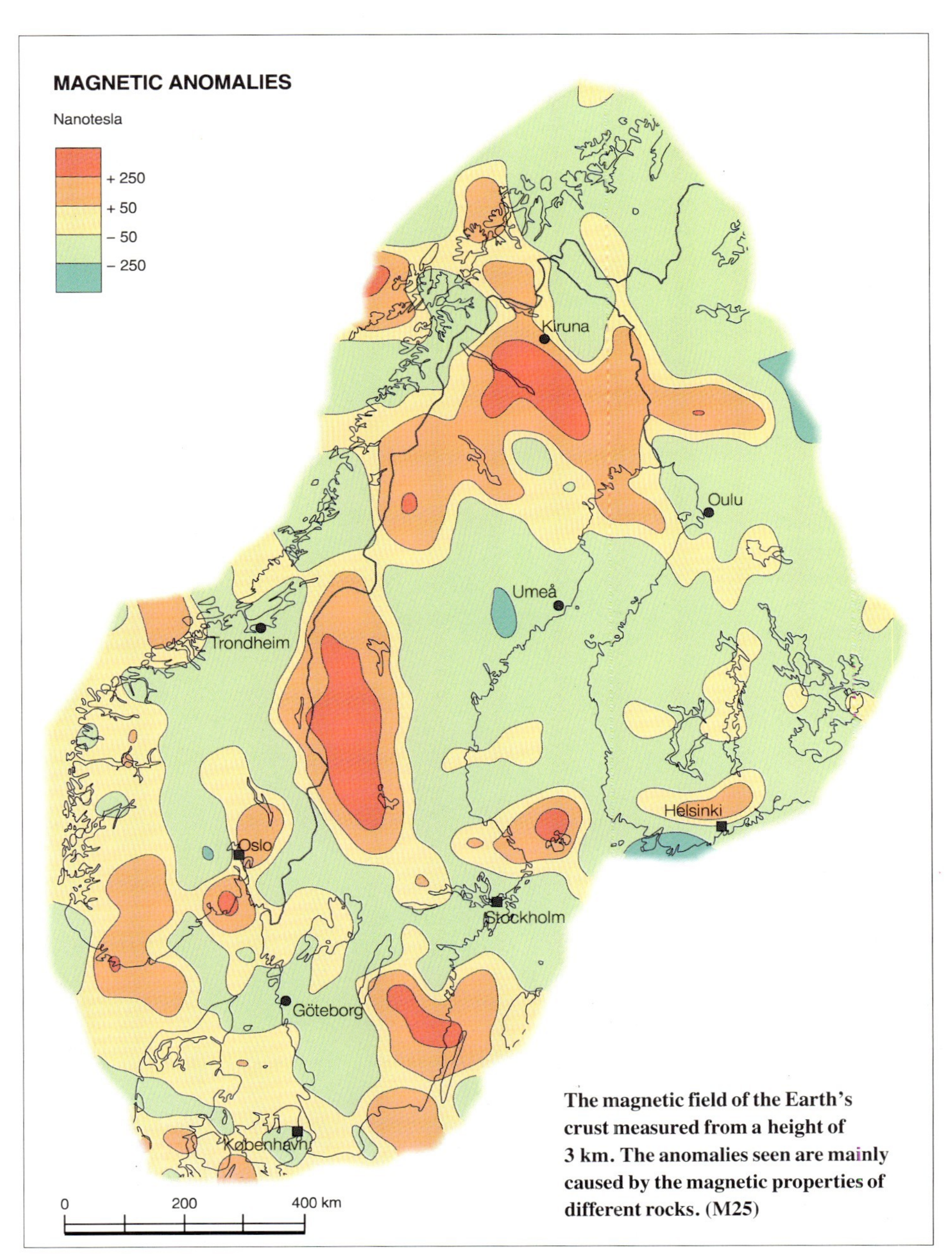

The magnetic field of the Earth's crust measured from a height of 3 km. The anomalies seen are mainly caused by the magnetic properties of different rocks. (M25)

strength decreases with distance. They are also continuous and exist both within and outside the Earth. The Earth's magnetic field extends far into space. The same applies to the Earth's gravity field, which keeps both the moon and various satellites in their trajectories. Despite the great distance, the moon's gravity field is fully noticeable on the Earth's surface, where it causes the ebb and flow of the tides, and also causes the Earth's crust to rise or fall several decimetres.

The geopotential fields at the surface of the Earth reflect the sum of both mass distribution and magnetic effects. The variations caused by rocks in the Earth's crust are fairly small. In the case of the magnetic field, they do not exceed 10%, and for the gravity field to about 0.01% of the normal field at the Earth's surface. Nonetheless, it is these variations that are used in the measurements of gravity and magnetism as a basis for mapping the bedrock.

MAGNETIC FIELD

The Earth's magnetic field is caused by electrical currents in the fluid core of the Earth. At the surface, it is similar to the field from a bar magnet located at the centre of the Earth. The electrical currents that cause the magnetic field are constant for long periods of time—several thousand years—but sometimes sudden changes occur, leading to changes in the direction and strength of the field. At present, we are in a period where the magnetic field is directed towards the centre of the Earth in the northern hemisphere and outwards from the

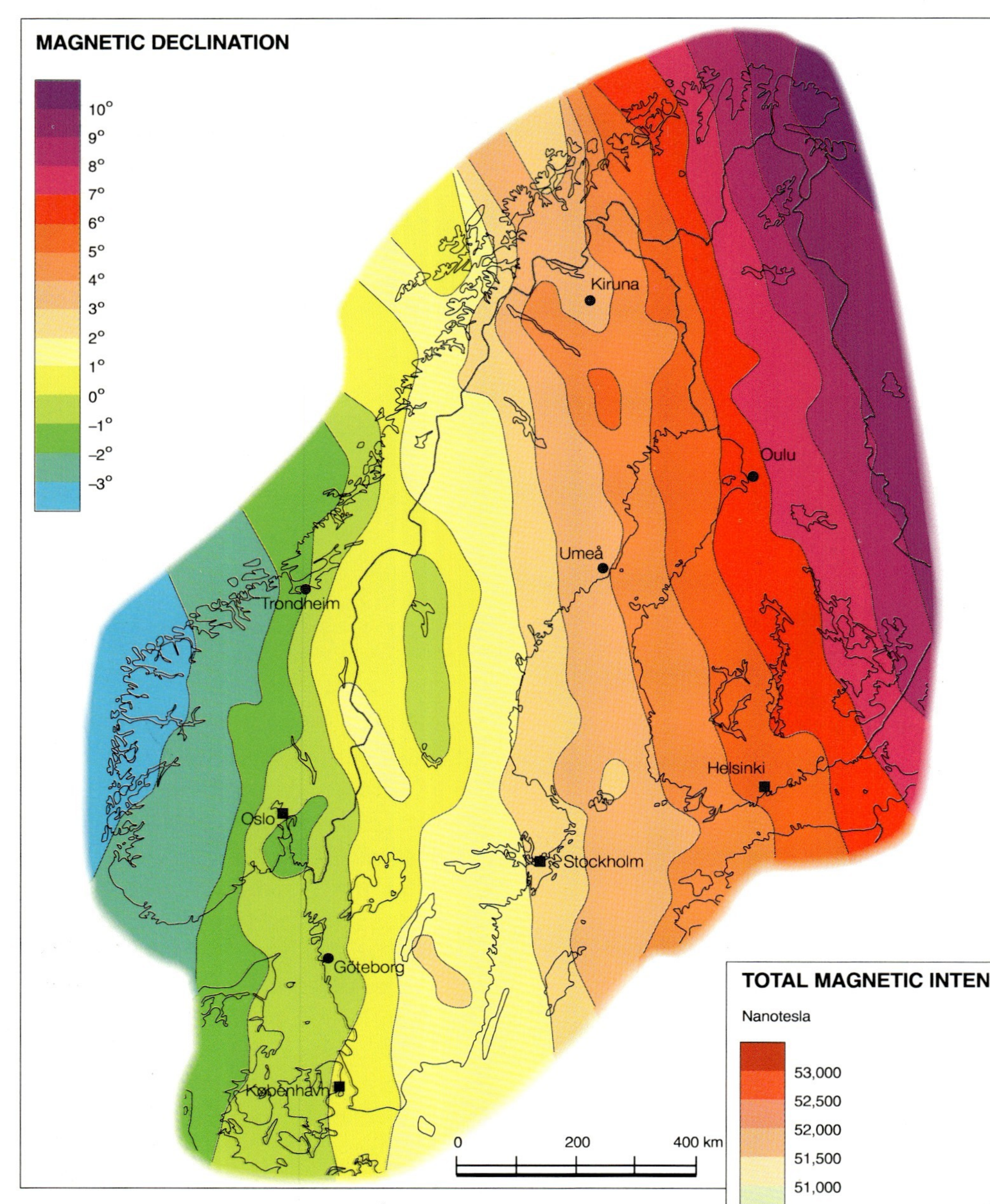

The magnetic field has both magnitude and direction. The direction is given by means of two angles, the magnetic declination, which is the angle in the horizontal plane measured from the north towards the east (corresponds to compass error), and inclination which is the angle downwards from the horizontal plane. The map above shows the declination, which in Sweden varies between –1 and +6 degrees (1990). The inclination of the magnetic field varies between 70 degrees in the south and 78 degrees in the north.

The unit for magnetic flux density is tesla (T). The magnitude of the magnetic field varies between 49,600 nT in the south and 52.600 nT in the north. (M26, M27)

Earth in the southern hemisphere. The magnetic field is directed vertically and is strongest at the magnetic poles. At the magnetic equator, the field is horizontal and much weaker. The magnetic poles almost coincide with the geographic poles, i.e., the Earth's axis of rotation.

Magnetic fields are measured with magnetometers which may have a sensitivity of one millionth of the normal geomagnetic field. The unit for magnetic flux density is *tesla* (T), in honour of the Croatian physicist Nikola Tesla (1856–1943).

The Earth's magnetic field influences all materials but certain elements are affected particularly strongly, e.g., iron and nickel. Minerals containing these elements may thus become magnetic. This effect is particularly noticeable in the minerals magnetite and magnetic pyrrhotite and results in entire rock complexes becoming magnetic. Generally, the amount of these minerals is much less than 10% by volume.

Locally, the Earth's magnetic field is irregular, and strongly magnetic rocks cause compass errors. If the magnetic variations caused by bedrock are to be studied, then a geomagnetic standard field is subtracted from the measured values. The difference is the *magnetic anomaly field*.

The magnetic field is also influenced by the *solar wind*—a continuous stream of charged particles from the sun—and, at some distance out in space, the magnetic field is shaped

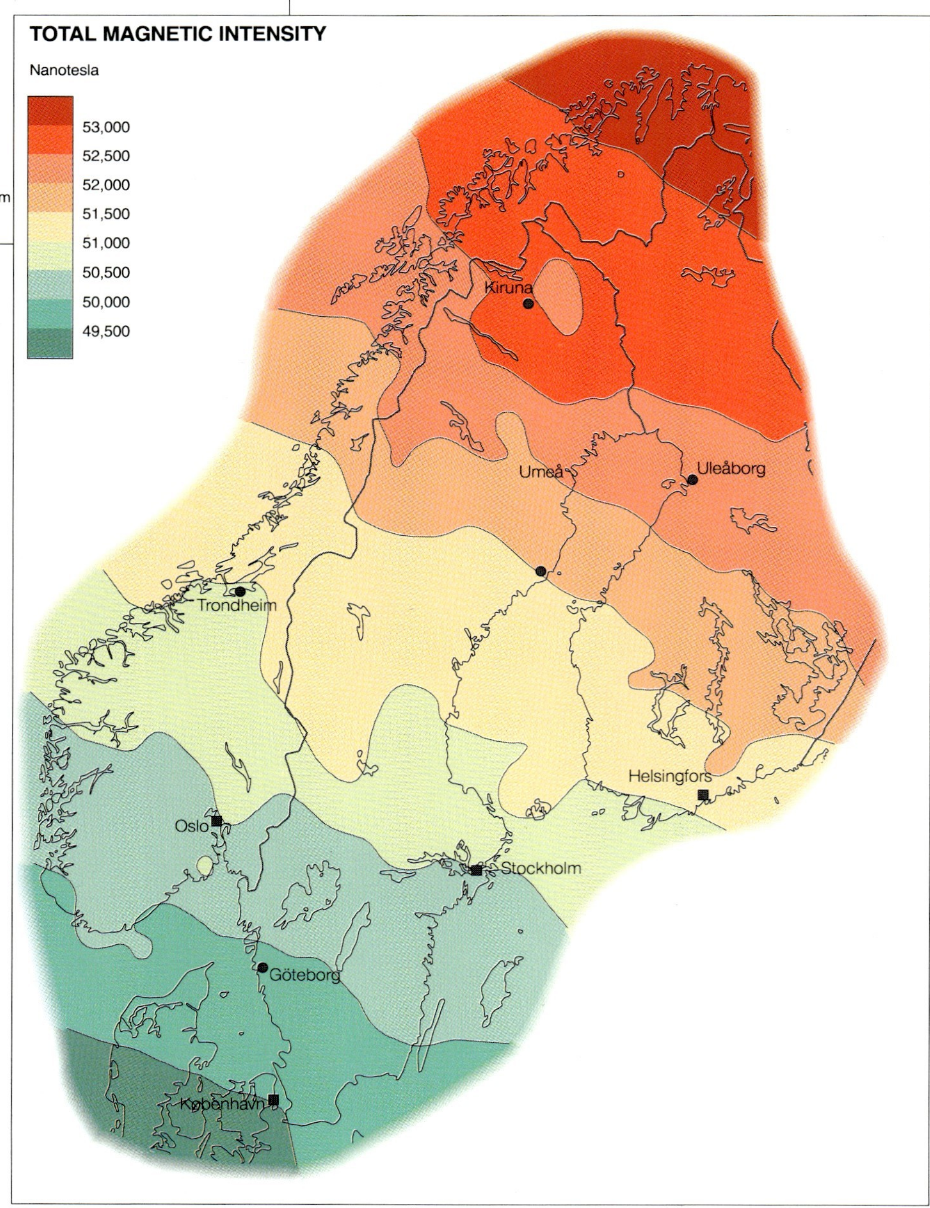

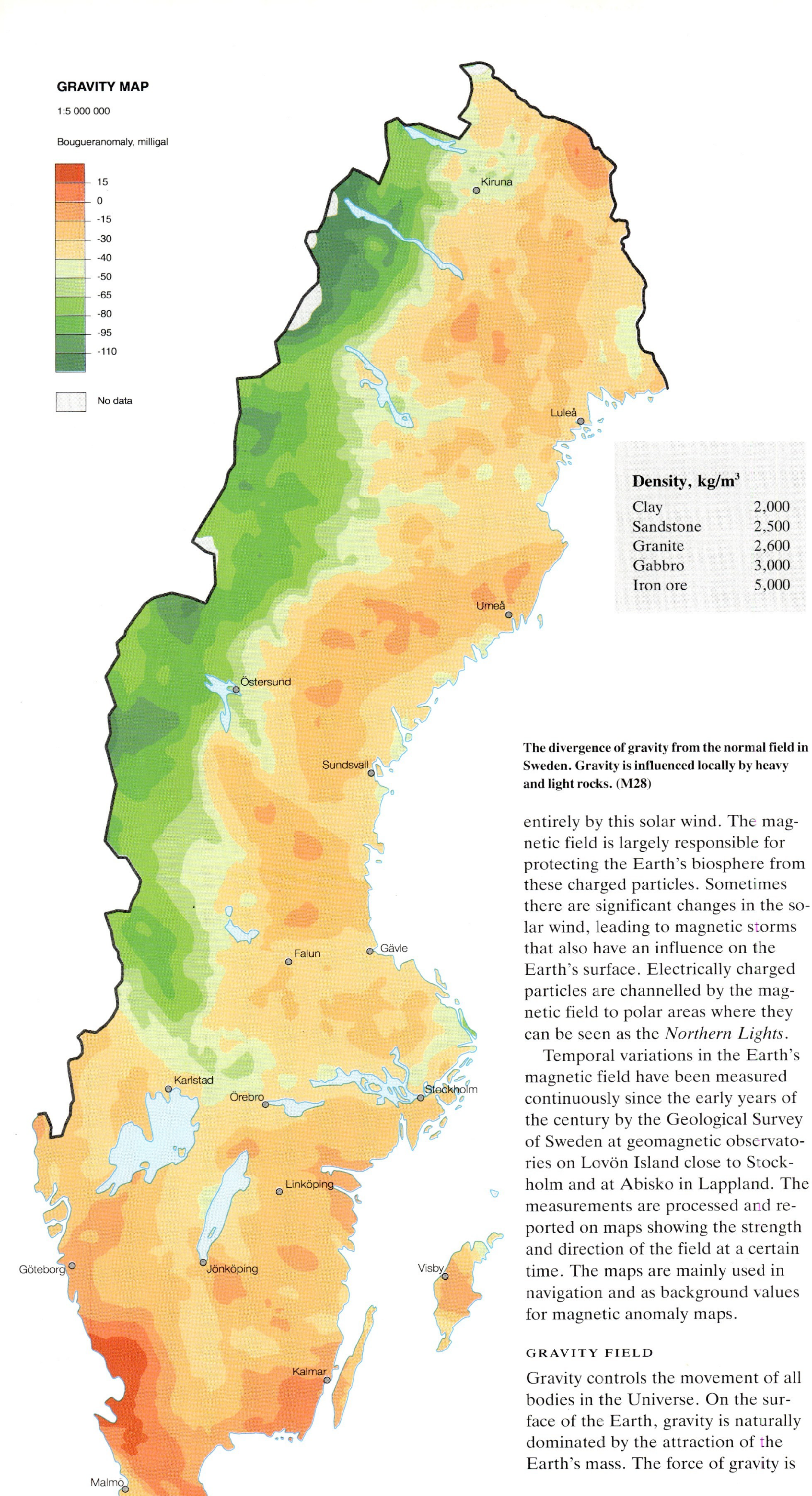

Density, kg/m³	
Clay	2,000
Sandstone	2,500
Granite	2,600
Gabbro	3,000
Iron ore	5,000

The divergence of gravity from the normal field in Sweden. Gravity is influenced locally by heavy and light rocks. (M28)

entirely by this solar wind. The magnetic field is largely responsible for protecting the Earth's biosphere from these charged particles. Sometimes there are significant changes in the solar wind, leading to magnetic storms that also have an influence on the Earth's surface. Electrically charged particles are channelled by the magnetic field to polar areas where they can be seen as the *Northern Lights*.

Temporal variations in the Earth's magnetic field have been measured continuously since the early years of the century by the Geological Survey of Sweden at geomagnetic observatories on Lovön Island close to Stockholm and at Abisko in Lappland. The measurements are processed and reported on maps showing the strength and direction of the field at a certain time. The maps are mainly used in navigation and as background values for magnetic anomaly maps.

GRAVITY FIELD

Gravity controls the movement of all bodies in the Universe. On the surface of the Earth, gravity is naturally dominated by the attraction of the Earth's mass. The force of gravity is the result of the Earth's gravity and the centrifugal force caused by the rotation of the Earth around its axle. Gravity measurements on the ground are carried out using a gravimeter, an extremely sensitive balance that can measure differences of one ten-millionths of the Earth's normal force of gravity. The unit for the force of gravity is *Gal*, after the Italian scientist Galilei.

Since the force of gravity decreases the further we move away from the centre of the Earth, all measured values must first be reduced to one and the same reference area, the so-called *geoid,* in order to be comparable with each other. If the distribution of material within the Earth was completely uniform, the mean level of the sea would assume a uniform oval shape, the earth ellipsoid. The undisturbed surface of the sea and its theoretical continuation below the continents is called the geoid. Since the distribution of material in the Earth is not uniform, in some places the geoid will lie above and in others below the earth ellipsoid, the greatest divergence being almost 100 m.

When *free-air correction* is made, consideration is taken only to the change in gravity caused by height differences between the point of observation and the geoid. When making a *Bouguer reduction* (named after the French physicist Bouguer), the value obtained is additionally reduced to compensate for the effect of material existing between the Earth's surface and the geoid, which has been given a standard density of 2,670 kg/m³.

Locally, gravity is influenced by the distribution of heavy and light rocks. Large volumes of rock types with high density give a surplus of mass that strengthens the force of gravity. A limited body with diverging density will cause an anomaly that decreases with the square root of the distance. If we know the difference in density between the rocks present, then measurements of gravity can be used to calculate their total masses and volumes. Rock density is determined by the amount of minerals present and the porosity. Quaternary deposits and sedimentary rocks have low density owing to their high porosity. Crystalline rocks, which dominate the Swedish bedrock, have relatively high density. In these rocks, the density mainly reflects the chemical composition. Ores have the highest densities.

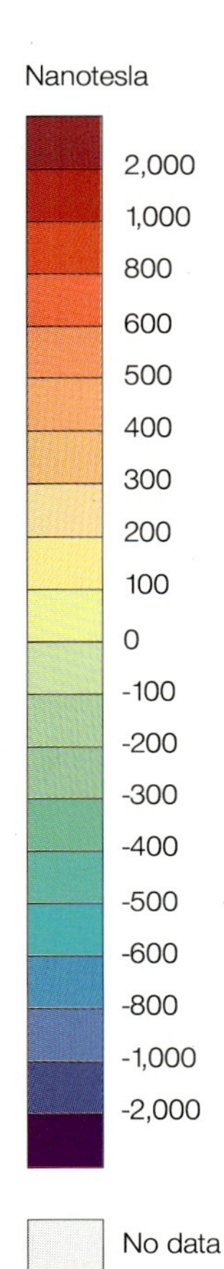

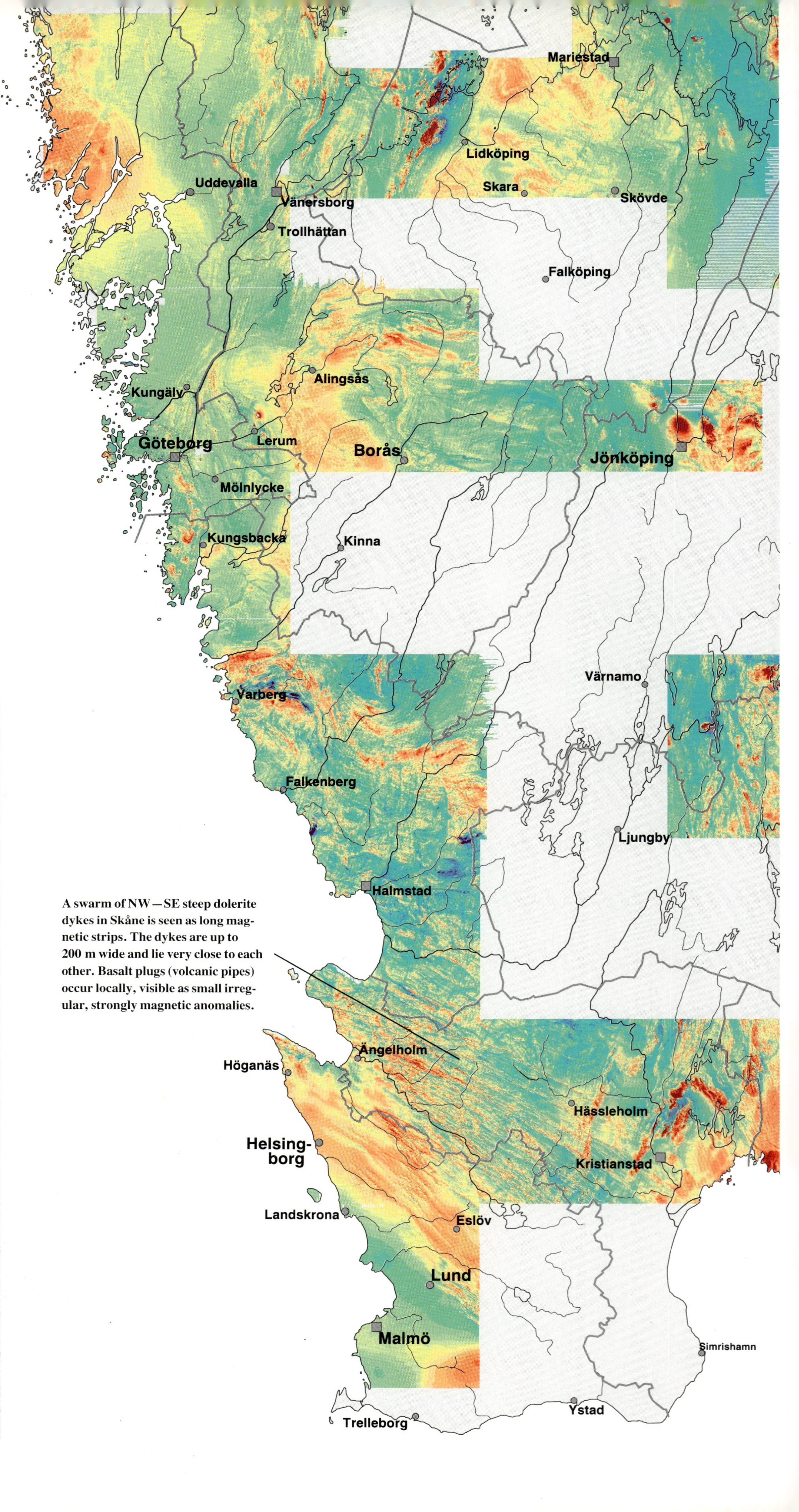

A swarm of NW—SE steep dolerite dykes in Skåne is seen as long magnetic strips. The dykes are up to 200 m wide and lie very close to each other. Basalt plugs (volcanic pipes) occur locally, visible as small irregular, strongly magnetic anomalies.

The map shows magnetic variations caused by the bedrock. The Earth's normal magnetic field has been subtracted from the measured values.

The crust's magnetic properties depend mainly on the content of the mineral magnetite in the rocks, which may vary from almost nothing to about 10% or more in gabbro and up to almost 100% in certain iron ores.

Measurements made from aircraft flying at an elevation of 30 m and with 200 m between the measuring lines, provide a picture of the magnetic anomaly field—a pattern that reflects the distribution of the rocks on the surface and at greater depths. In the mountains however, the distance between measuring lines is 1,000 m. The orientation of the magnetic rock types in relation to the Earth's magnetic field can be determined. In that way we can also obtain information on strikes and angles of dip, as well as the shape and character of the bodies of rocks.

Breaks in the magnetic pattern may reveal faults and their relative movements. The magnetic map provides important information of the bedrock and its third dimension. This is essential information for, and a complement to, the map of bedrock geology, and particularly in parts of the country where the bedrock is covered by water or a thick layer of Quaternary deposits. (M29)

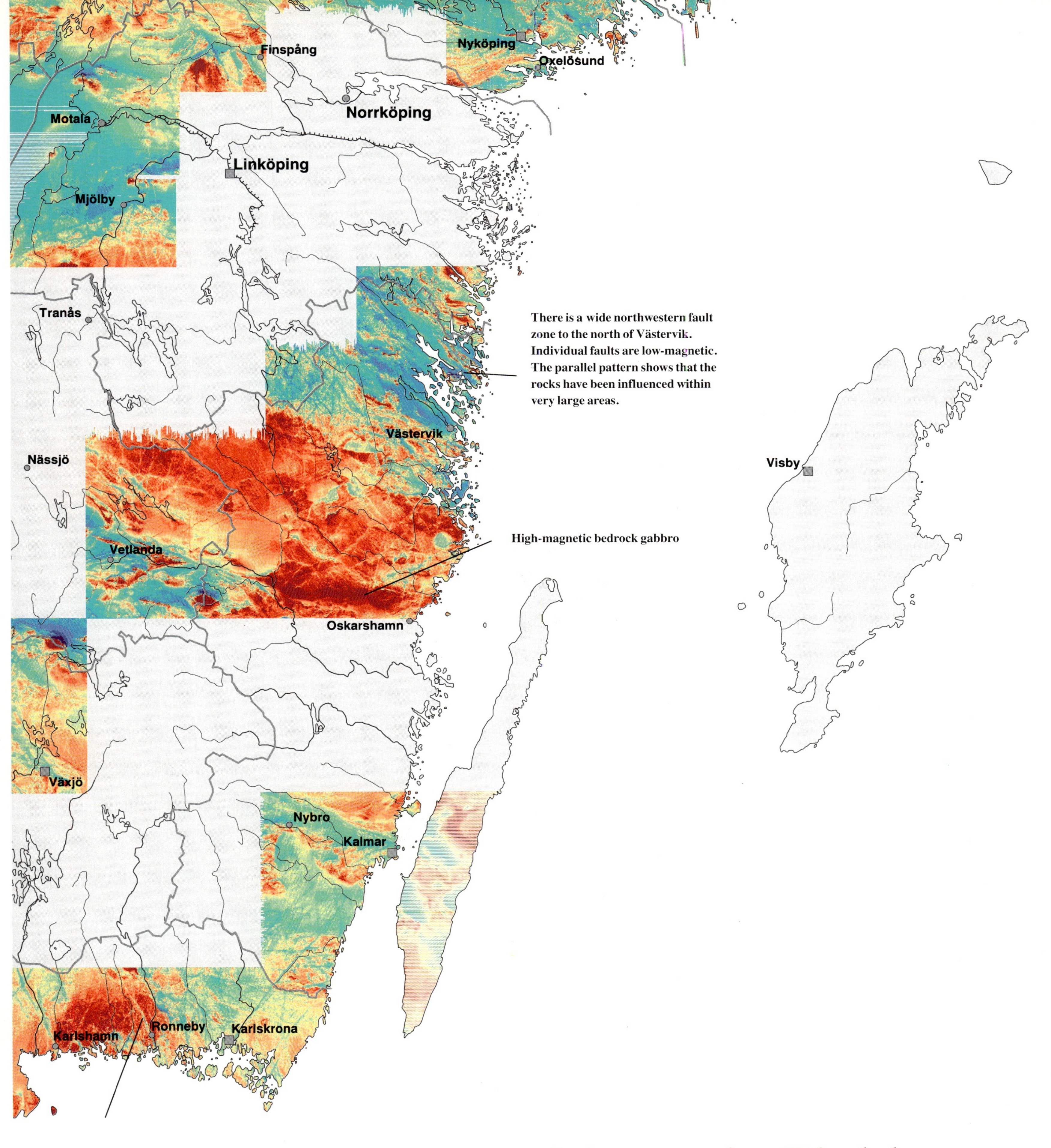

Magnetic granite in Blekinge (Karlshamn granite) is transversed by low-magnetic fracture zones. The magnetite in the fissures has oxidised and formed low-magnetic hematite.

Magnetic Field of the Earth Crust

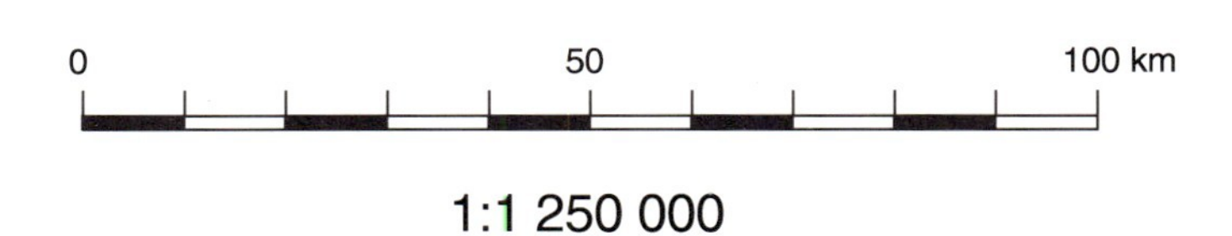

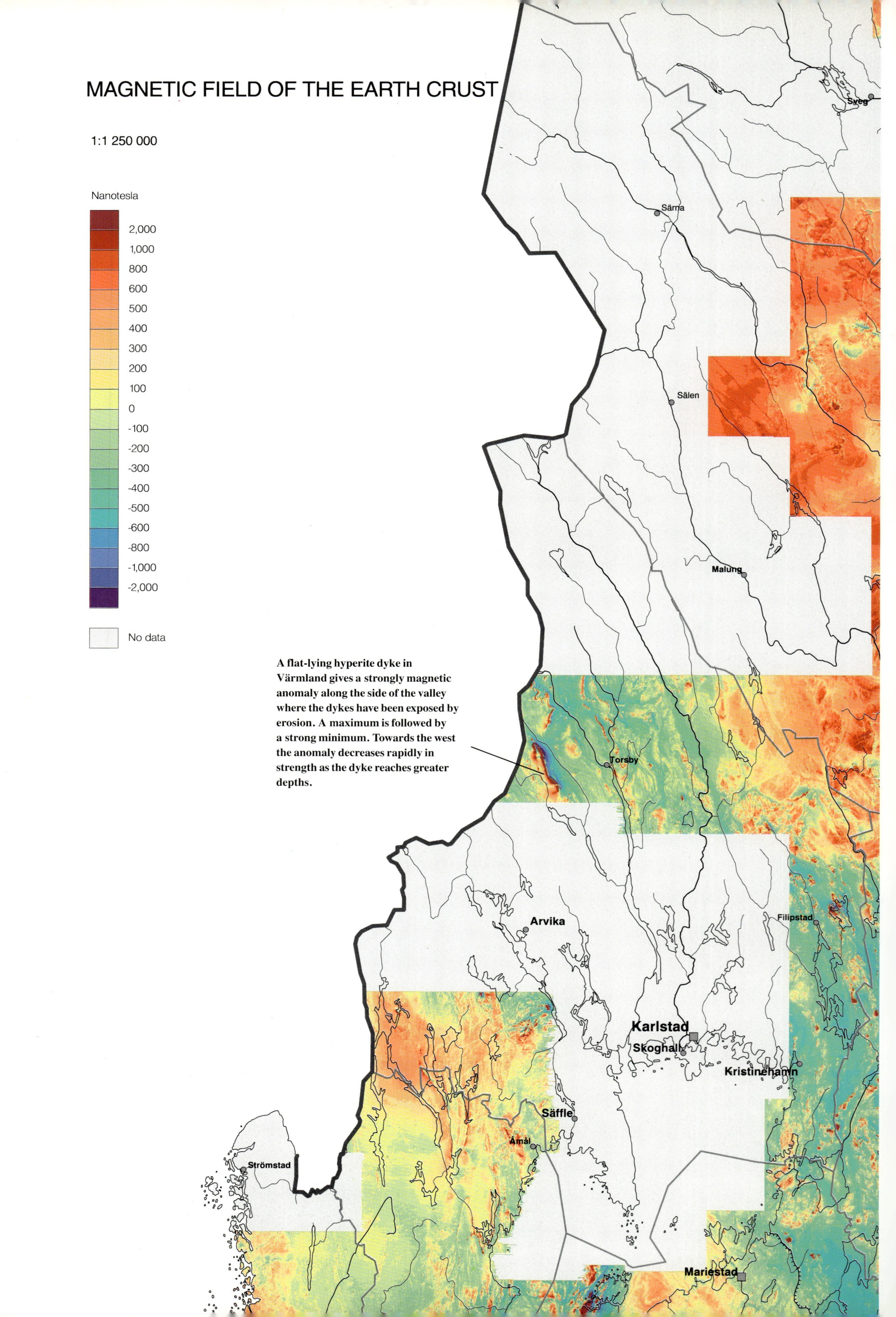

A flat-lying hyperite dyke in Värmland gives a strongly magnetic anomaly along the side of the valley where the dykes have been exposed by erosion. A maximum is followed by a strong minimum. Towards the west the anomaly decreases rapidly in strength as the dyke reaches greater depths.

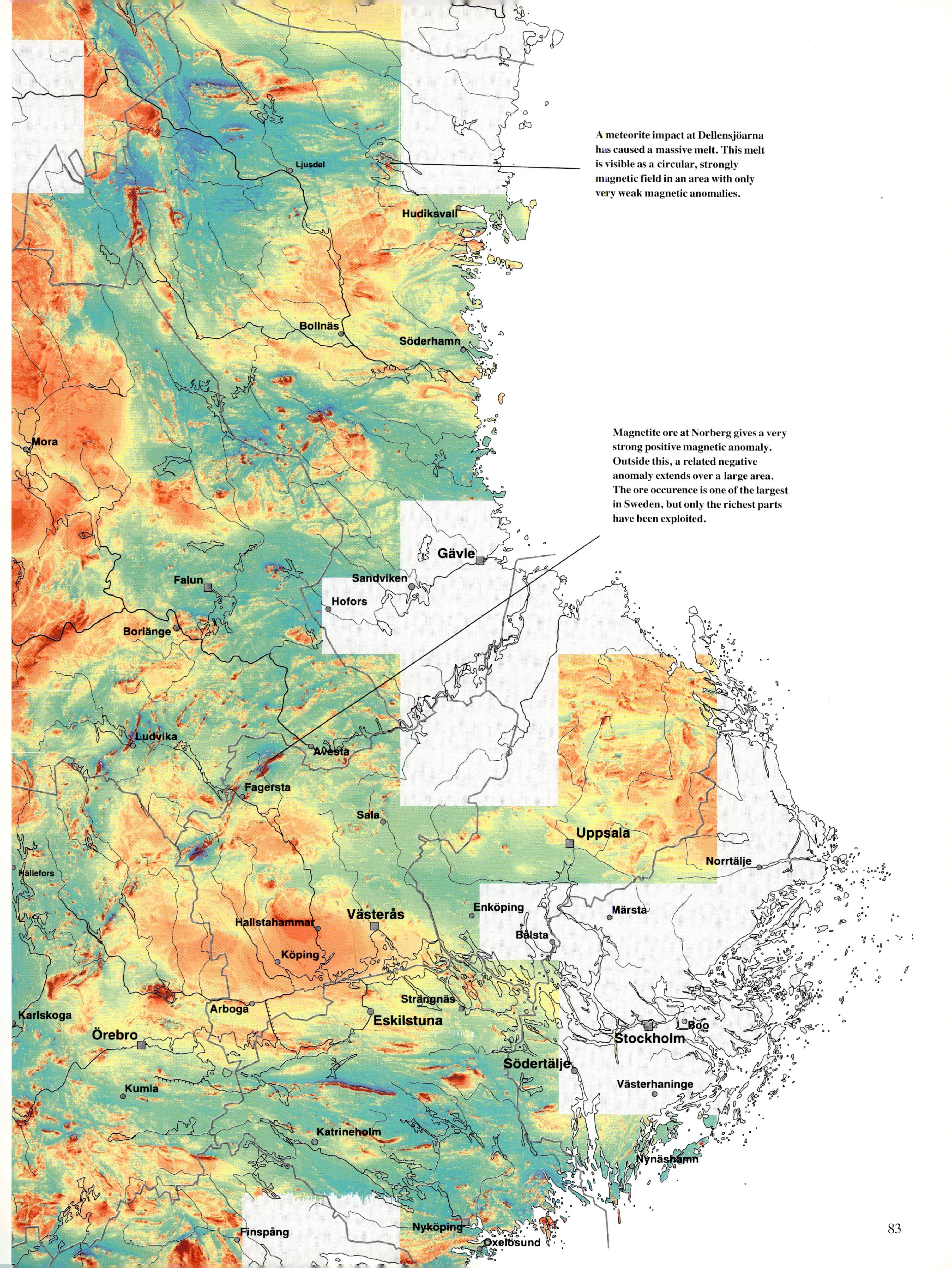

A meteorite impact at Dellensjöarna has caused a massive melt. This melt is visible as a circular, strongly magnetic field in an area with only very weak magnetic anomalies.

Magnetite ore at Norberg gives a very strong positive magnetic anomaly. Outside this, a related negative anomaly extends over a large area. The ore occurence is one of the largest in Sweden, but only the richest parts have been exploited.

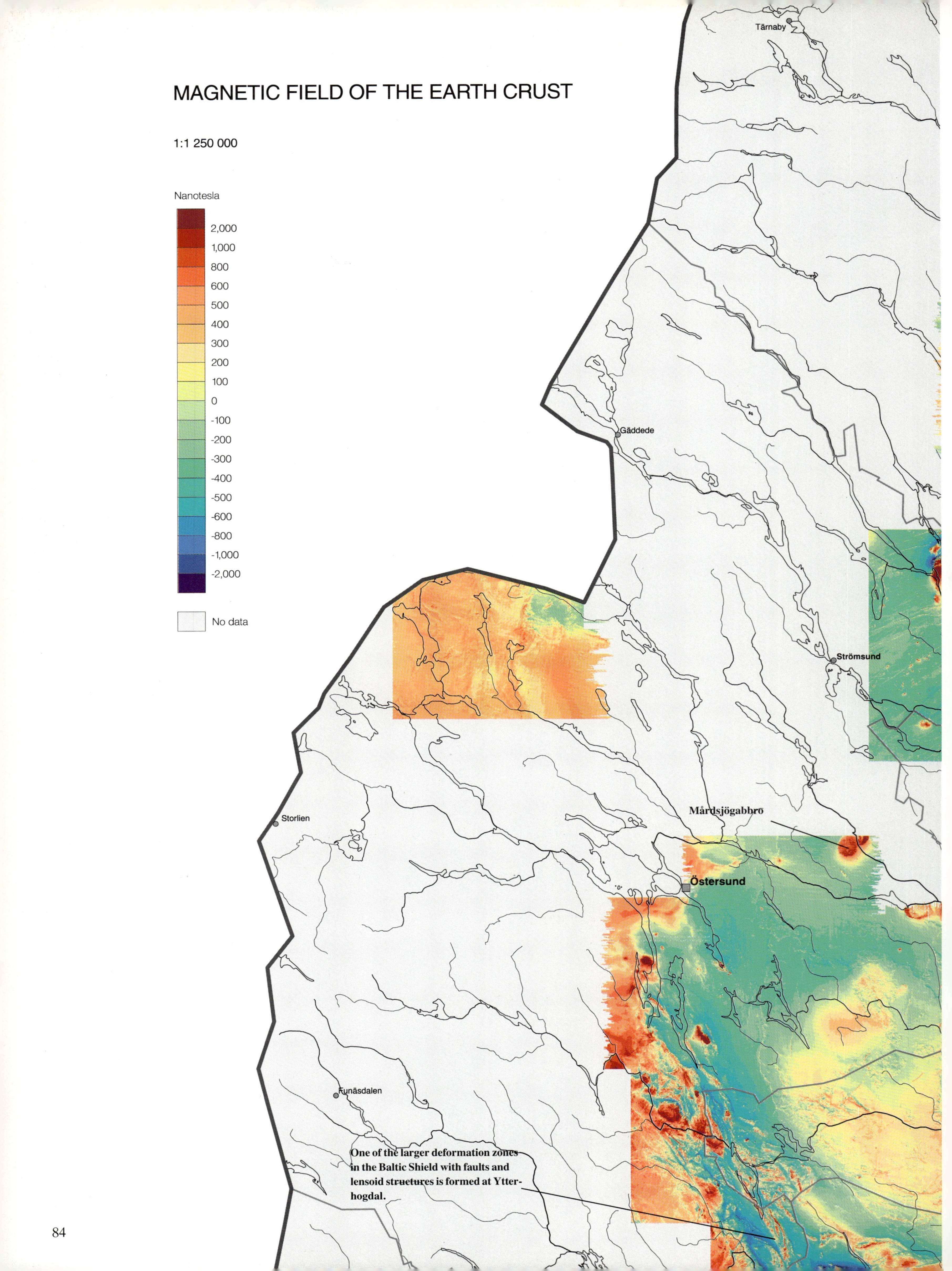

MAGNETIC FIELD OF THE EARTH CRUST
1:1 250 000
Nanotesla
2,000
1,000
800
600
500
400
300
200
100
0
-100
-200
-300
-400
-500
-600
-800
-1,000
-2,000
No data
Tärnaby
Gäddede
Strömsund
Storlien
Mårdsjögabbro
Östersund
Funäsdalen
One of the larger deformation zones in the Baltic Shield with faults and lensoid structures is formed at Ytter-hogdal.

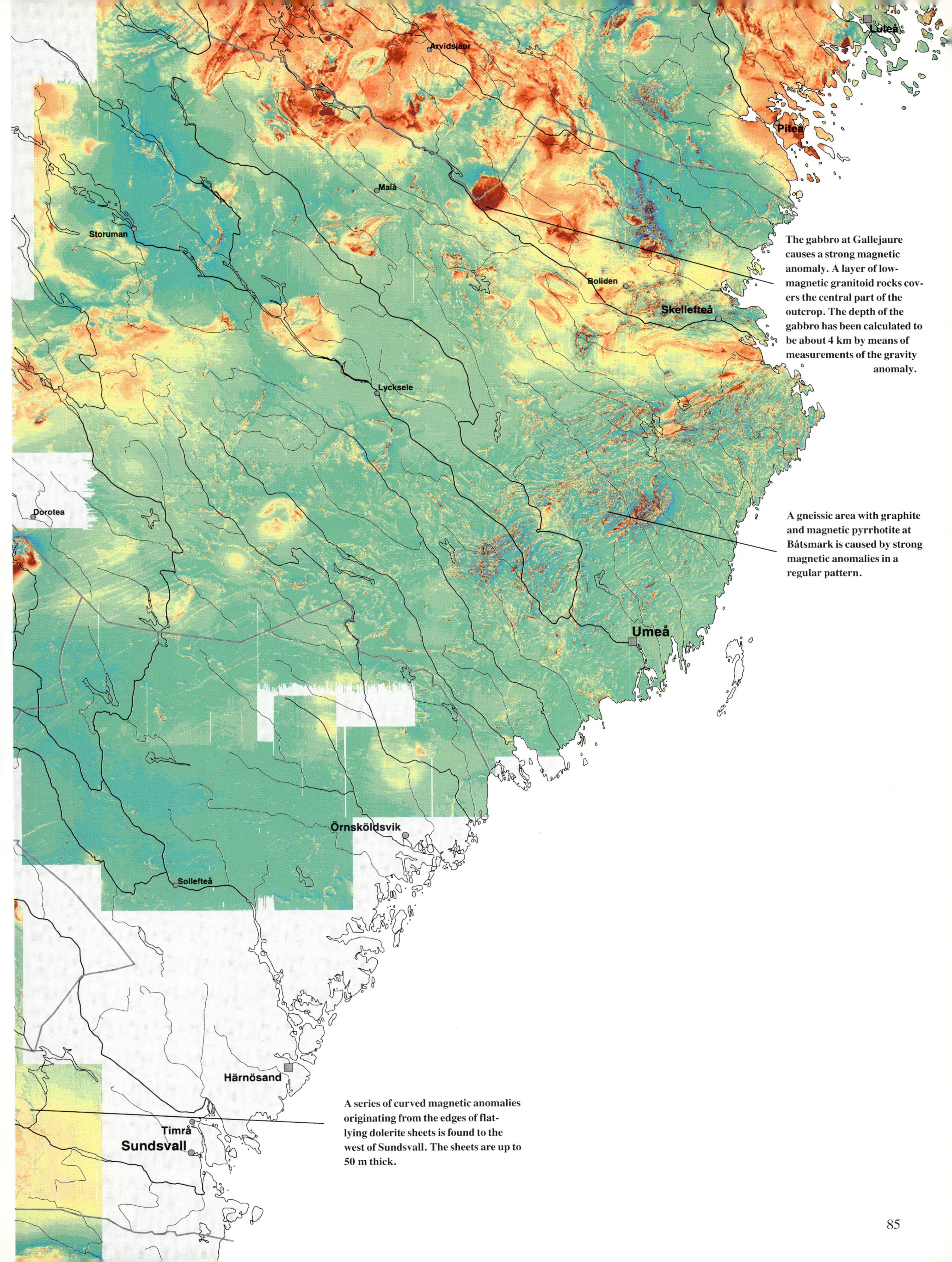

The gabbro at Gallejaure causes a strong magnetic anomaly. A layer of low-magnetic granitoid rocks covers the central part of the outcrop. The depth of the gabbro has been calculated to be about 4 km by means of measurements of the gravity anomaly.

A gneissic area with graphite and magnetic pyrrhotite at Båtsmark is caused by strong magnetic anomalies in a regular pattern.

A series of curved magnetic anomalies originating from the edges of flat-lying dolerite sheets is found to the west of Sundsvall. The sheets are up to 50 m thick.

MAGNETIC FIELD OF THE EARTH CRUST

1:1 250 000

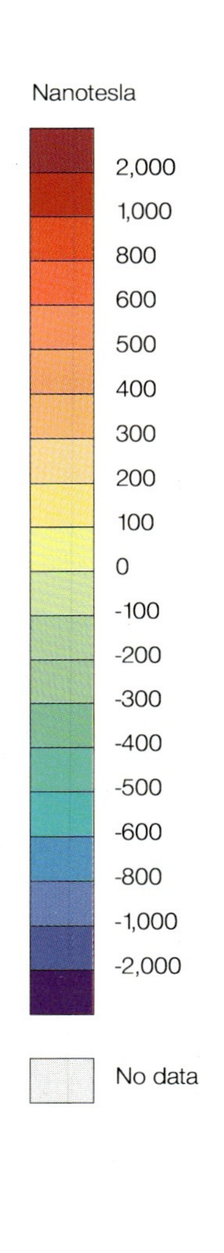

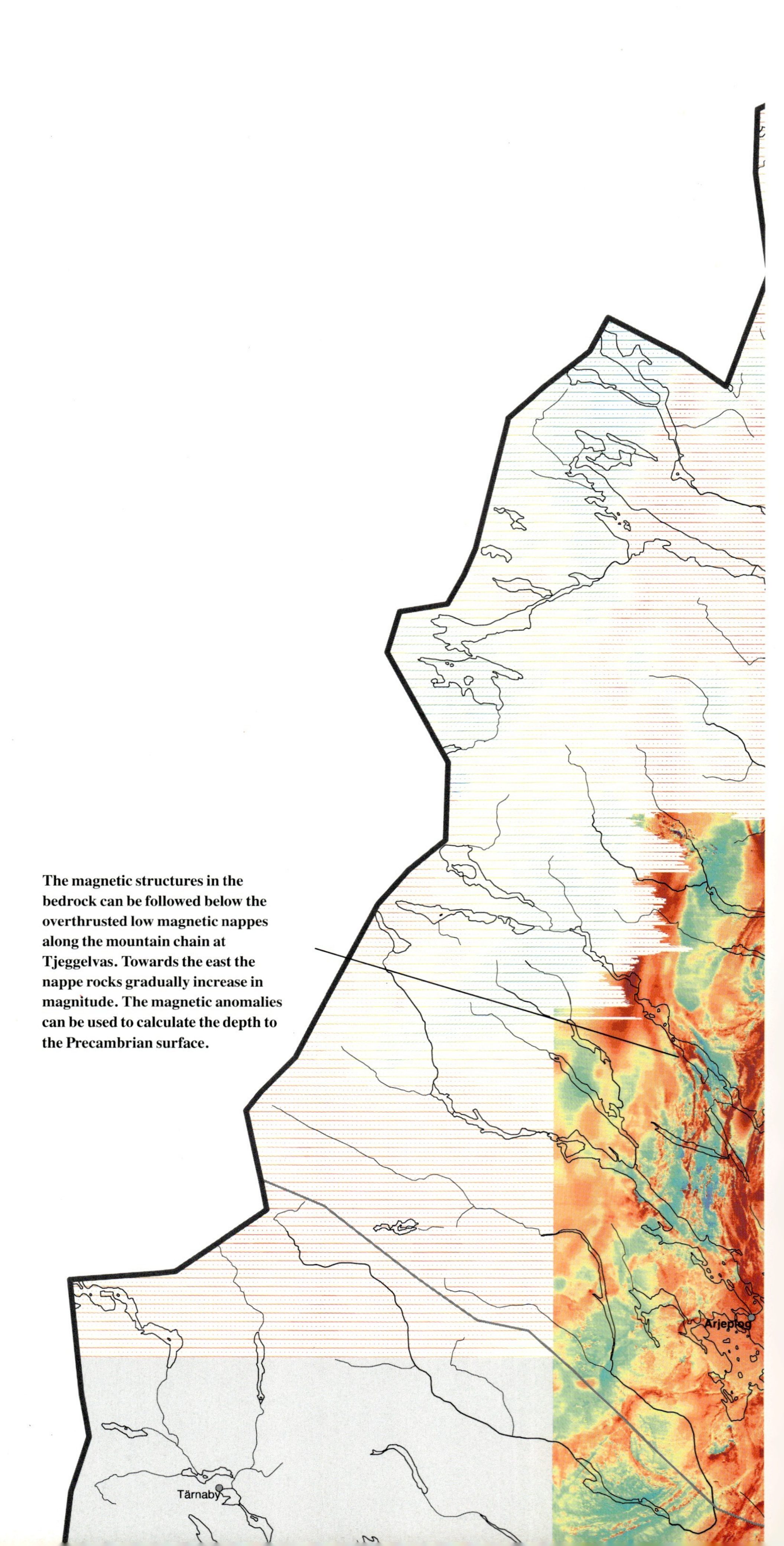

The magnetic structures in the bedrock can be followed below the overthrusted low magnetic nappes along the mountain chain at Tjeggelvas. Towards the east the nappe rocks gradually increase in magnitude. The magnetic anomalies can be used to calculate the depth to the Precambrian surface.

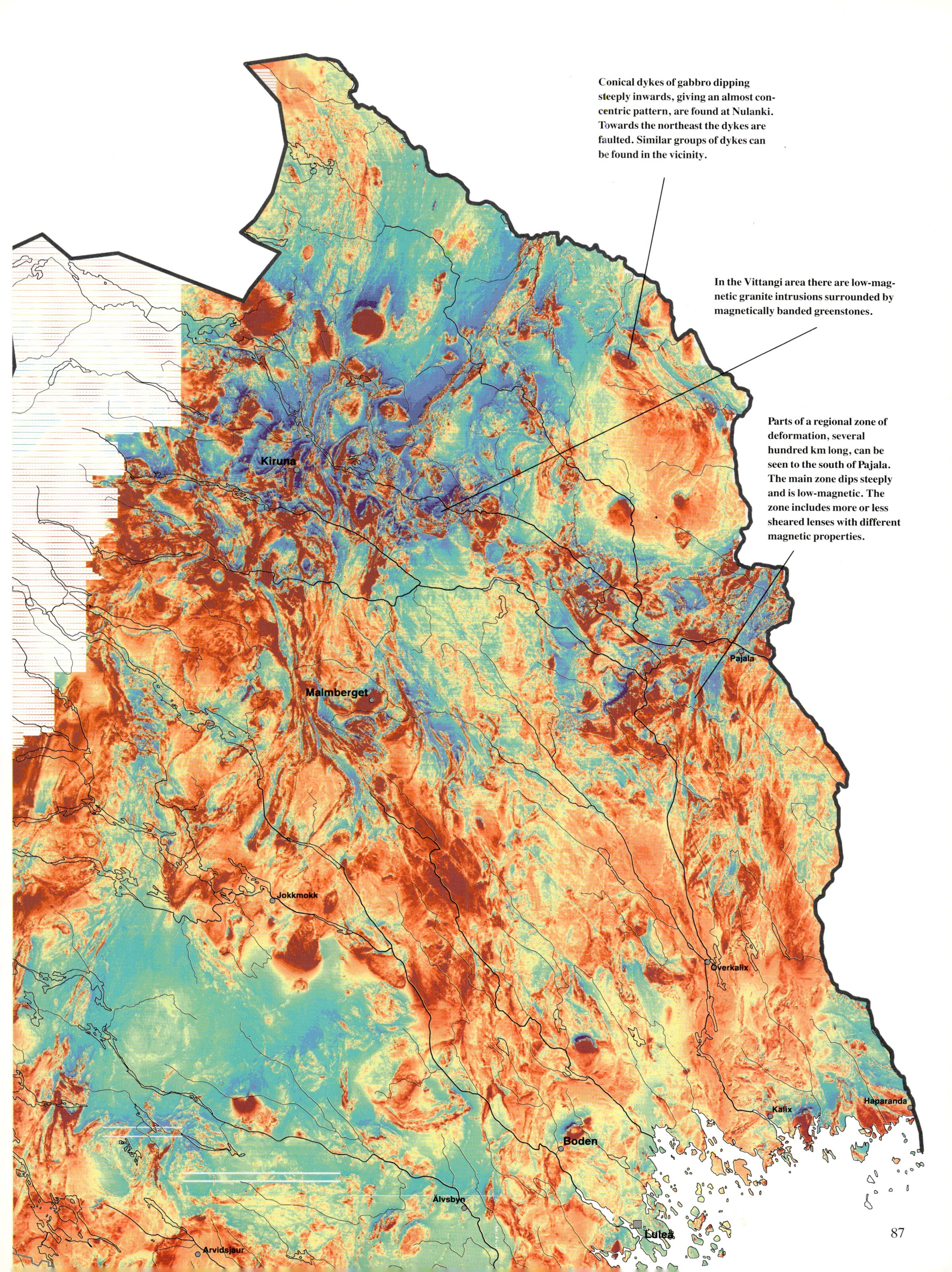

Conical dykes of gabbro dipping steeply inwards, giving an almost concentric pattern, are found at Nulanki. Towards the northeast the dykes are faulted. Similar groups of dykes can be found in the vicinity.
In the Vittangi area there are low-magnetic granite intrusions surrounded by magnetically banded greenstones.
Parts of a regional zone of deformation, several hundred km long, can be seen to the south of Pajala. The main zone dips steeply and is low-magnetic. The zone includes more or less sheared lenses with different magnetic properties.
Kiruna
Pajala
Malmberget
Jokkmokk
Överkalix
Kalix
Haparanda
Boden
Älvsbyn
Luleå
Arvidsjaur

Electrical conductivity

The *electrical conductivity* of minerals, rocks and Quaternary deposits is used when prospecting for ores and groundwater. Similar methods were already used during the 1920's to localise the sulphide ores of the Skellefte Field. The conductivity of rocks depends on their porosity and the content of dissolved salts in the pore water. When conductive minerals are present, the conductivity of the rocks considerably increase.

Common minerals with high electrical conductivity include pyrite, chalcopyrite, pyrrhotite and graphite. Sedimentary rocks have higher conductivity than crystalline rocks due to their higher porosity. Saline groundwater leads to considerable increases in conductivity both for rocks and Quaternary deposits. Normally, electrical conductivity can be determined by measuring the resistivity.

Resistivity, ohmmeter	
Graphite schist	0.01
Clay	10
Sandstone	500
Till	1,000
Granite	10,000

The height of the measuring points above sea level must be known if the gravity of the individual measuring point is to be reduced to a common reference level. In this case, measurements are made on the shore of a lake with a water surface of known elevation.

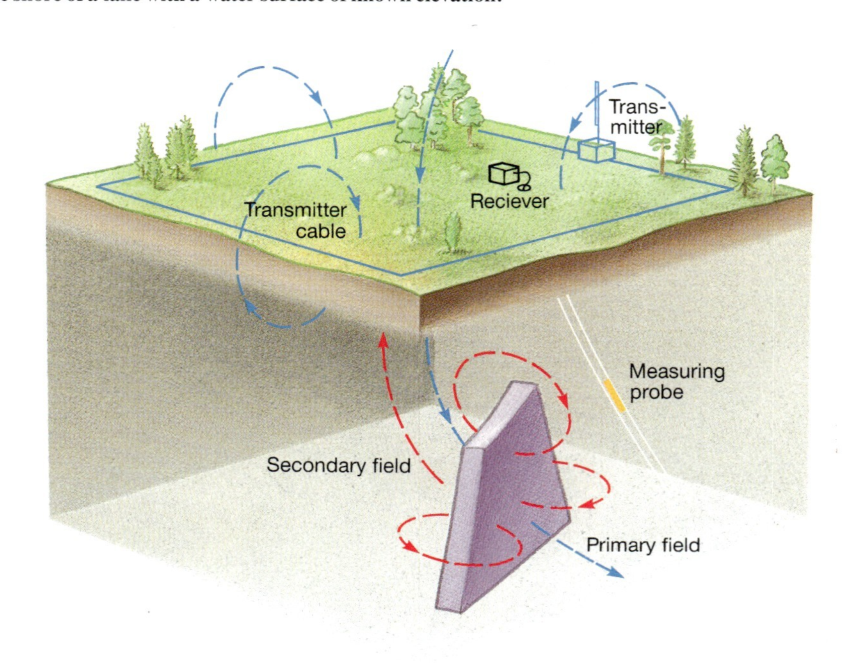

Electrical prospecting using the radio method. A variable transmitter field generates currents in electrical conductors. In turn, the currents cause secondary fields that can be registered in receivers on the ground and in drill holes.

Special control points are used to connect the gravity measurements with a reference system to the whole country. In a reference system established during the 1960's many control points were located on the steps leading into a church.

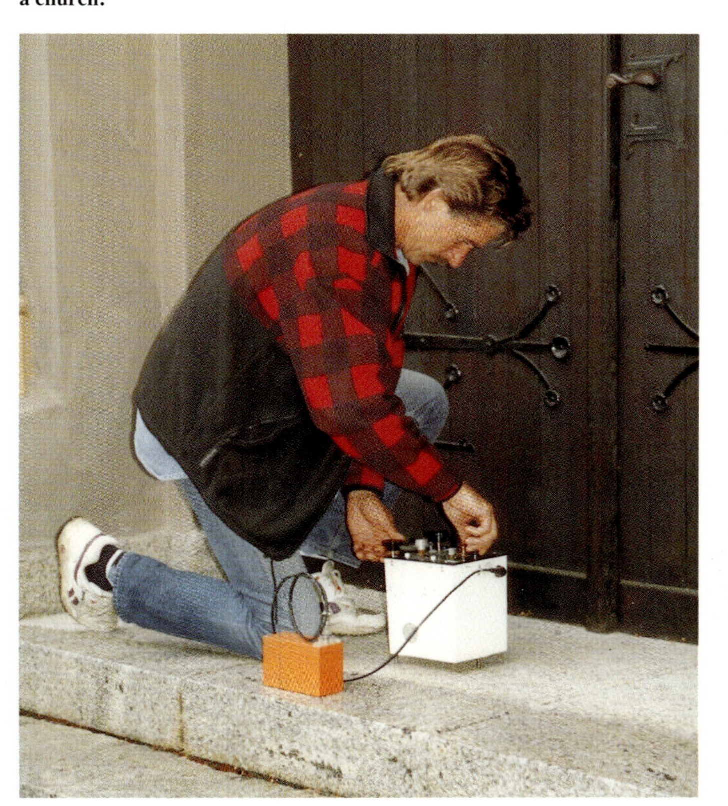

Applications

Examples of applications where measurements of the anomalies of the geopotential field are utilised are mainly found within bedrock mapping and ore prospecting. For these purposes, the force of gravity and the magnetic field are systematically measured throughout Sweden.

Measurements of gravity are made on the ground at places with known heights. The distance between measurement sites varies and is usually 1—2 km.

Magnetic measurements are made from aircraft along a system of parallel measuring lines. The measurements are made at an elevation of 30 m above the ground, and the distance between the lines is about 200 m.

At the same time as the magnetic registrations are carried out, measurements are also made of natural gamma radiation and the electromagnetic field emitted by certain radio-transmitters with low frequency and at great distances. In Sweden, a transmitter in England is normally used.

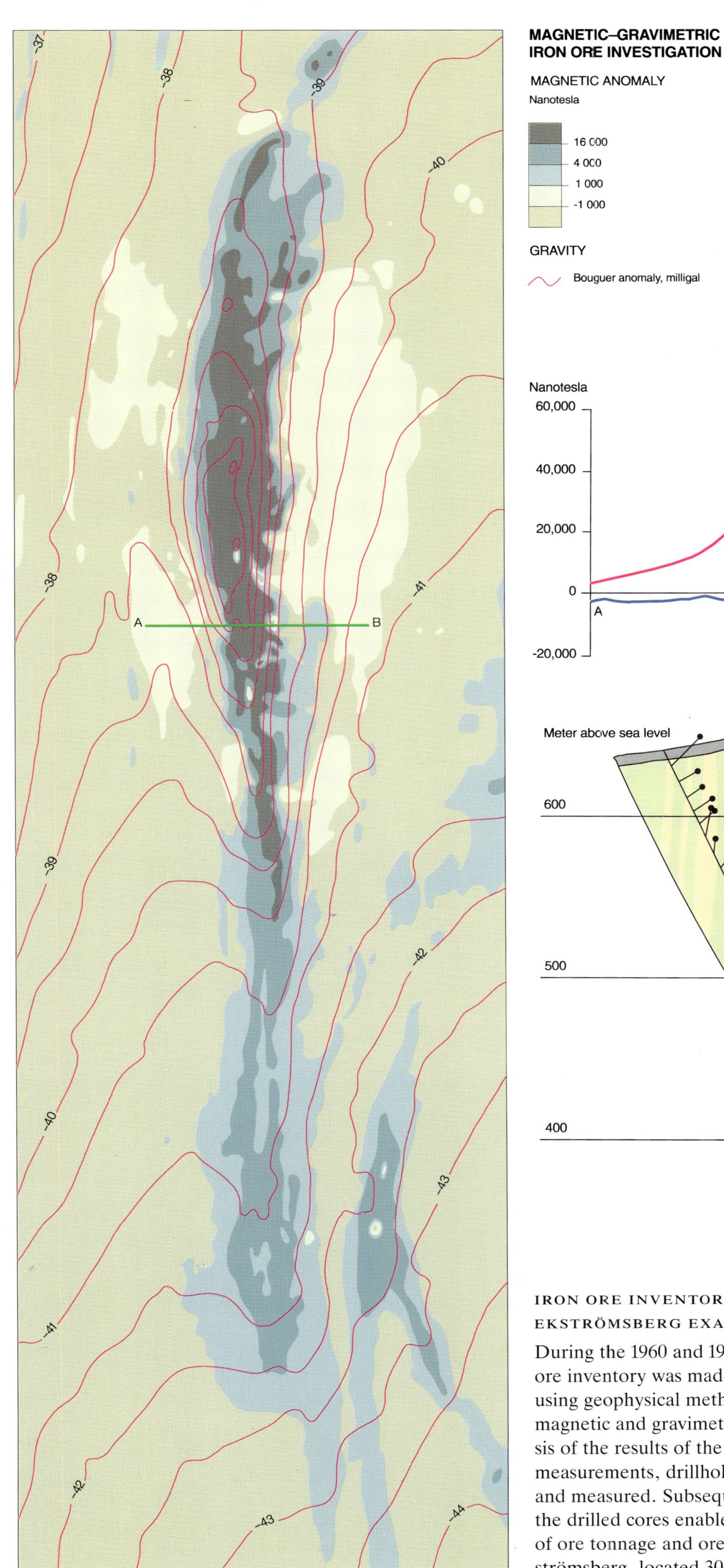

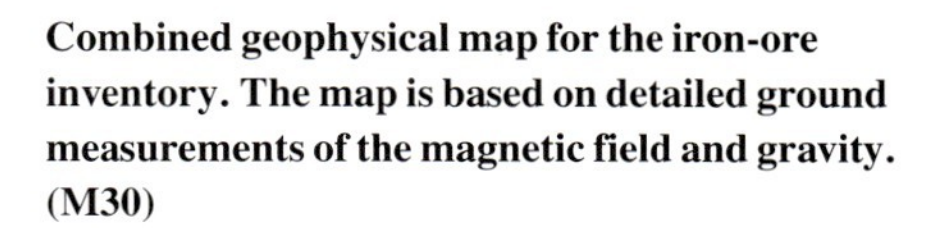

Combined geophysical map for the iron-ore inventory. The map is based on detailed ground measurements of the magnetic field and gravity. (M30)

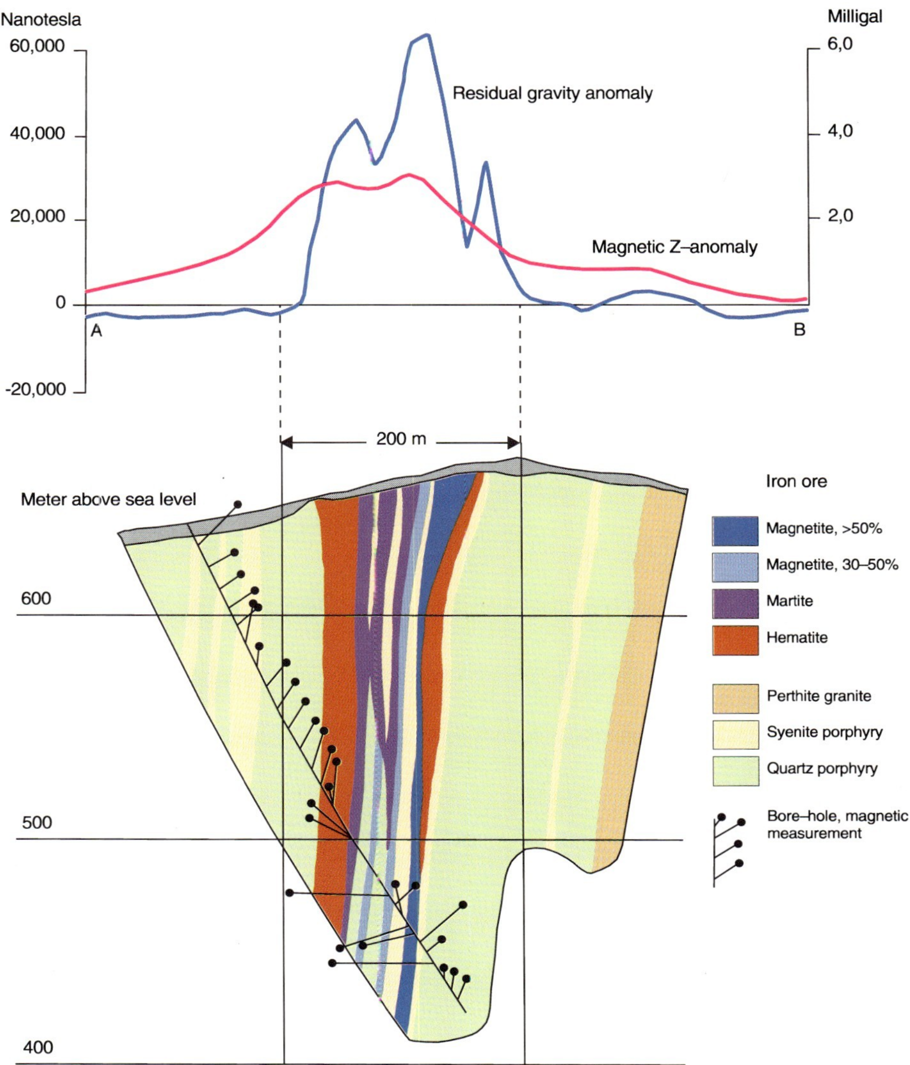

Profiles A–B with measurements of the vertical component of the magnetic field and the gravity anomaly over the ore-bearing zone. Drill-holes have already been sited on the basis of the measured anomalies. In one of the drill-holes, the size and direction of the magnetic field has been measured.

IRON ORE INVENTORY — THE EKSTRÖMSBERG EXAMPLE

During the 1960 and 1970's, an iron ore inventory was made in Norrbotten using geophysical methods, mainly magnetic and gravimetric. On the basis of the results of the geophysical measurements, drillholes were sited and measured. Subsequent analysis of the drilled cores enabled calculation of ore tonnage and ore quality. Ekströmsberg, located 30 km WSW of Kiruna, is an apatite-rich iron ore of Kiruna type. The ore body is 1,500 m long and has an area of 50,000 m^2 and extends to considerable depth. The most important ore body consists of magnetite with a width of 40–50 m. The amount of ore down to 250 m depth has been calculated to 35 million tonnes with approximately 60% iron and 1% phosphorus.

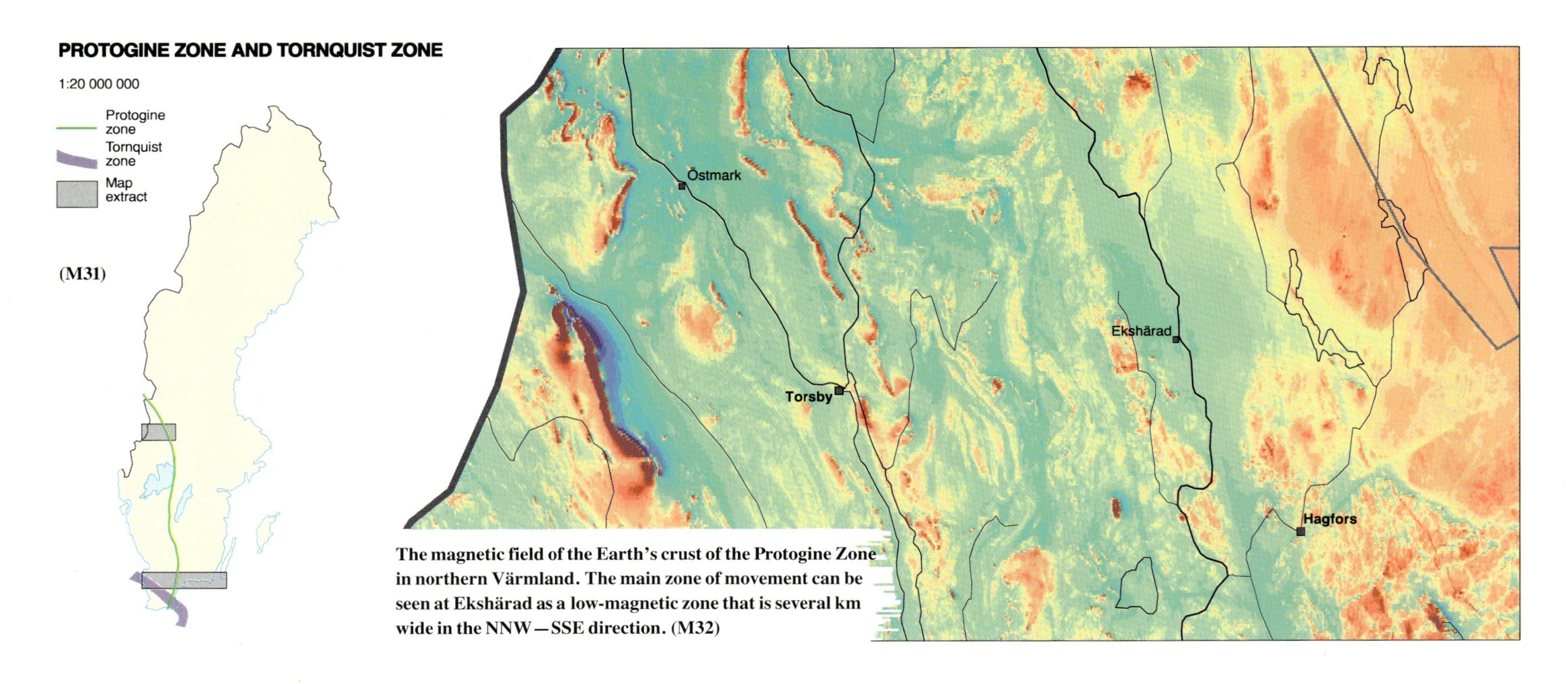

The magnetic field of the Earth's crust of the Protogine Zone in northern Värmland. The main zone of movement can be seen at Ekshärad as a low-magnetic zone that is several km wide in the NNW—SSE direction. (M32)

Nanotesla

1,600
860
460
220
80
0
–40
–140
–300
–620
–1,180

TORNQUIST ZONE

The southern boundary of the Precambrian shield passes through Skåne. Here, we can find the youngest volcanoes in Sweden, still visible in the form of vents filled with basaltic lava. These volcanoes lie in the *Tornquist Zone*, an example of the long-lived deformation zones of the continental crust. The volcanoes, active during the Jurassic and Cretaceous, mark the end of a geological phase when the Precambrian shield fractured in this area.

The numerous magnetic dolerite dykes in a WNW direction mark the start of this phase of fracturing. The swarm of dolerite dykes runs diagonally through Skåne and is about 75 km wide. The rocks in the dykes, as well as the basalts, have magnetisation directions that diverge significantly from the current magnetic field in Skåne. This demonstrates that the dolerite dykes were formed when Scandinavia was located to the south of the equator.

Apart from the dykes, the magnetic map shows numerous structures that reflect different geological processes. To the east, young magnetic granites occur and it is possible to see the extension of granite intrusions at the surface and to calculate their shape at depth. The different intrusions are almost two kilometres deep and are parts of a larger body, a *batholith*.

The magnetic granites are transsected by low-magnetic zones which were formed due to the oxidising effect of water in deeper fracture zones.

To the west, it is apparent that tectonic movements have resulted in preservation of different thickness of sedimentary cover rocks. These have extremely low magnetism, and in areas where the layers are relatively

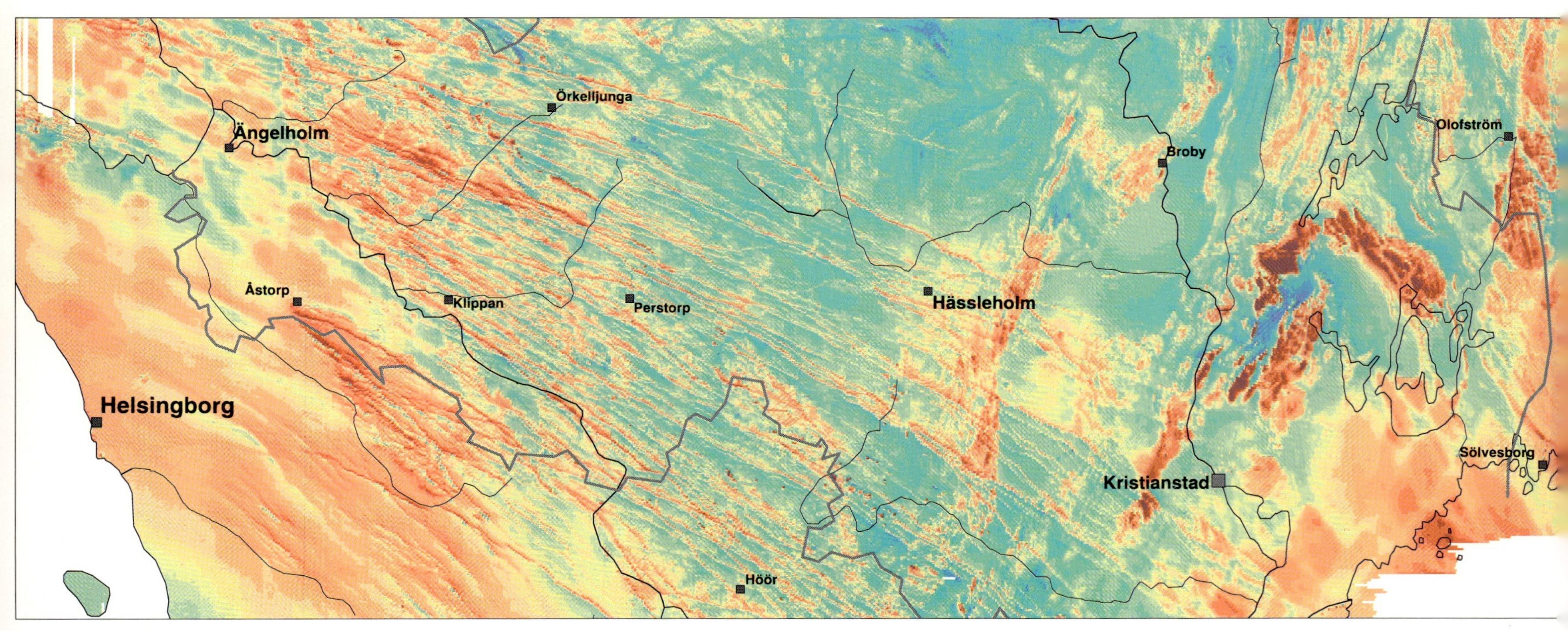

The magnetic field of the Earth's crust in a cross-section of southern Sweden from the Öresund Strait in the west to southern Öland in the east. The Tornquist Zone is dominated by strongly magnetic dolerite dykes. To some extent, these camouflage the southernmost part of the Protogine Zone—the low-magnetic area between Hässleholm and Kristianstad—that has a NNE direction. (M33)

thin, it is still possible to see the magnetic signature of the dolerite dykes. In this way, the approximate thickness of the sedimentary cover can be calculated.

THE PROTOGINE ZONE

The schistosity zone in south-western Sweden, the *Protogine Zone*, is one of the most important deformation zones in the Precambrian shield. The zone is almost 800 km long and extends from Skåne in the south to the Norwegian Caledonides in the north, where it is buried beneath the overthrust rocks of this younger orogenic belt.

The Protogine Zone in Värmland contains a low magnetic zone that dissects the magnetic pattern of the surrounding blocks. The block to the west of the zone has completely different magnetic properties and structure than that to the east.

In central Värmland, several geophysical investigations have been carried out and have allowed a three-dimensional interpretation to be made of the zone's continuation with depth. By combining information from several methods, it is possible to reduce the uncertainty in the interpretation. The magnetic measurements have been used in the profile to determine the position of the zone at the surface and its slope in the uppermost part of the Earth's crust. The deeper structure has been calculated from gravimetric measurements.

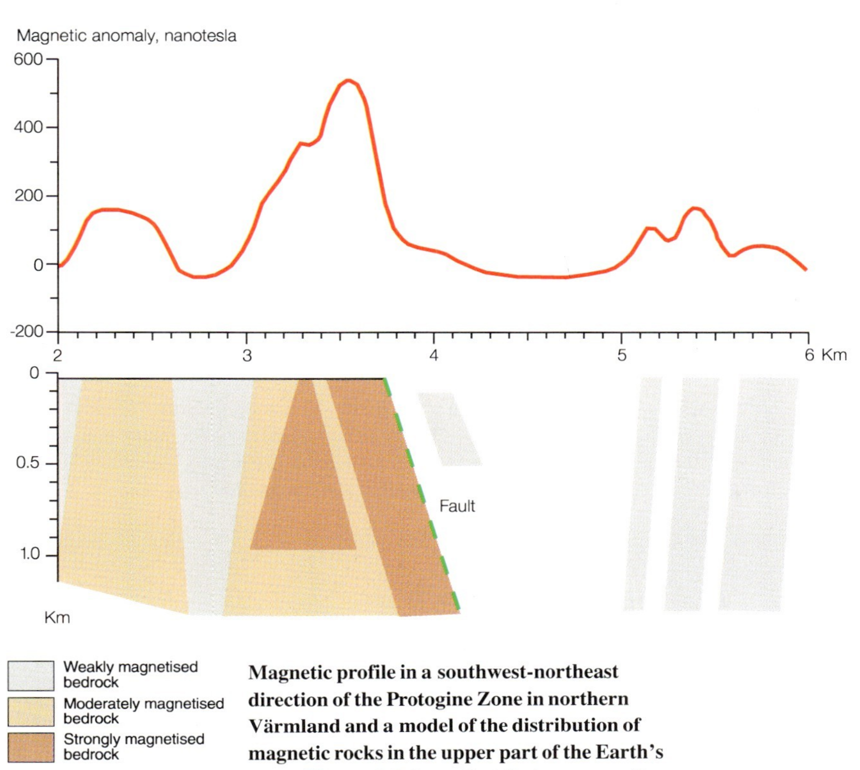

Magnetic profile in a southwest-northeast direction of the Protogine Zone in northern Värmland and a model of the distribution of magnetic rocks in the upper part of the Earth's crust. The model shows that the low-magnetic zone dips steeply towards the east.

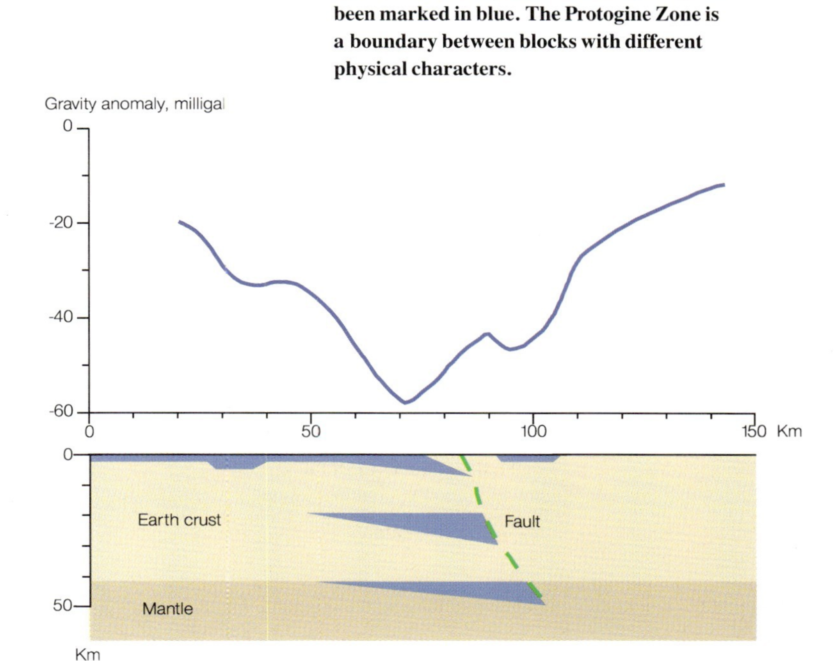

Gravity profile in a southwest-northeast direction over the Protogine Zone and a model of the structure of the Earth's crust that explains the mass deficit that has been measured. Rock that has a lower density than its surroundings has been marked in blue. The Protogine Zone is a boundary between blocks with different physical characters.

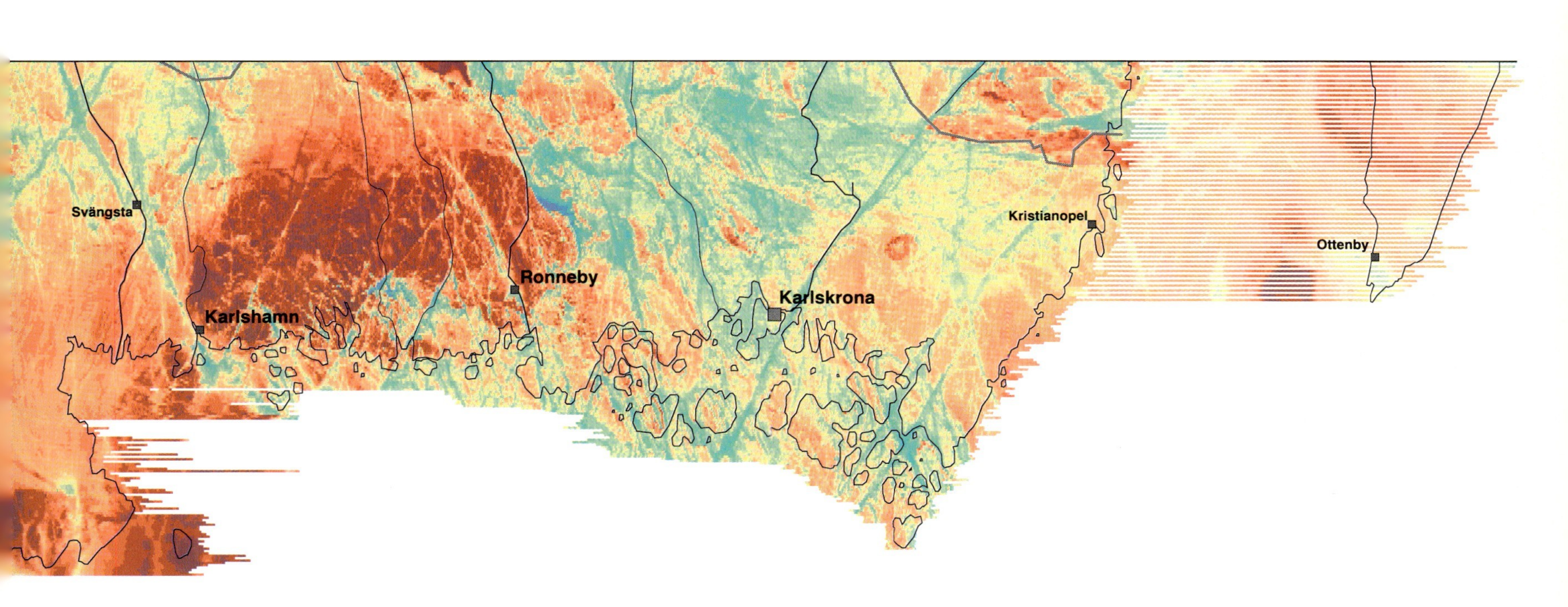

Kiruna area

The Kiruna area is one of the most spectacular areas in Sweden for ore deposits. The phosphorus-rich apatite iron ore, which is more than 4 km long and 20–200 m wide, is one of the world's largest single occurrences of iron ore.

When the airborne magnetic measurements started in 1960, the Kiruna area was chosen for study and, for the first time, a comprehensive picture of the enormous magnetic ore deposit was established. It is also easy to see the value of the magnetic map in relation to information on the general configuration of the bedrock.

The *gravimetric map* mainly shows larger structures such as the occurrence of the less dense granites with depth in comparison with the heavier rocks, e.g., greenstones. If quantitative ore calculations based on gravimetric methods are required, then detailed measurements must be made in the immediate surroundings of the ore in order to determine and separate the mass surplus of the ore in relation to the surrounding rocks.

In the same way, detailed *magnetic measurements* are made on the ground around the ore. On the basis of these maps, the geometry of the ore body can be determined with high precision.

The airborne *electromagnetic measurements* shown here are based on measurements of low-frequency radio-waves transmitted from a radio station in England. These radio-waves penetrate hundreds of metres into the ground and when electrically conductive parts of the bedrock are encountered, they cause electrical currents that influence the transmitted radio-waves in a way that can be measured.

Electrically conductive areas on the map are shown as blue lines. The method is dependent on direction in such a way that, principally, conductors lying parallel to the transmission direction will be detected. In the Kiruna area, the electrically conductive zones about 5 km to the west of Kiruna attract greatest interest. In 1973, it was possible to demonstrate that these zones were associated with the Viscaria ore that had then recently been discovered. This copper ore is located in electrically conductive graphite shale layers within basic volcanic rocks, the so-called Kiruna greenstones. In addition, electromagnetic anomalies can be caused by water-saturated fault zones, by power lines, or by railways.

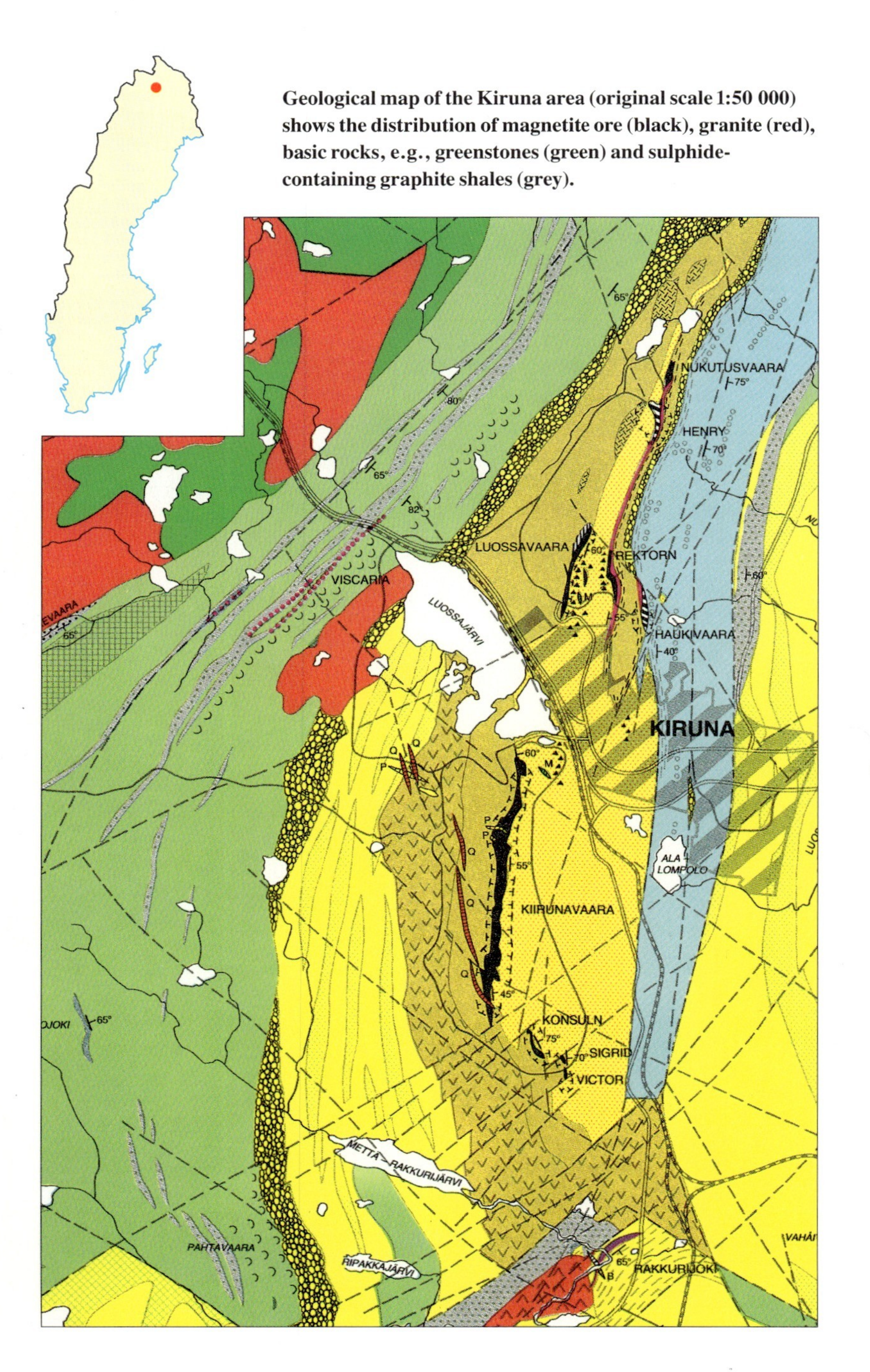

Geological map of the Kiruna area (original scale 1:50 000) shows the distribution of magnetite ore (black), granite (red), basic rocks, e.g., greenstones (green) and sulphide-containing graphite shales (grey).

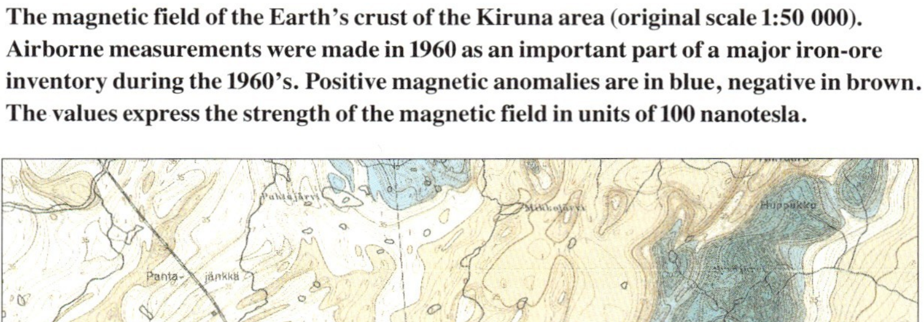

The magnetic field of the Earth's crust of the Kiruna area (original scale 1:50 000). Airborne measurements were made in 1960 as an important part of a major iron-ore inventory during the 1960's. Positive magnetic anomalies are in blue, negative in brown. The values express the strength of the magnetic field in units of 100 nanotesla.

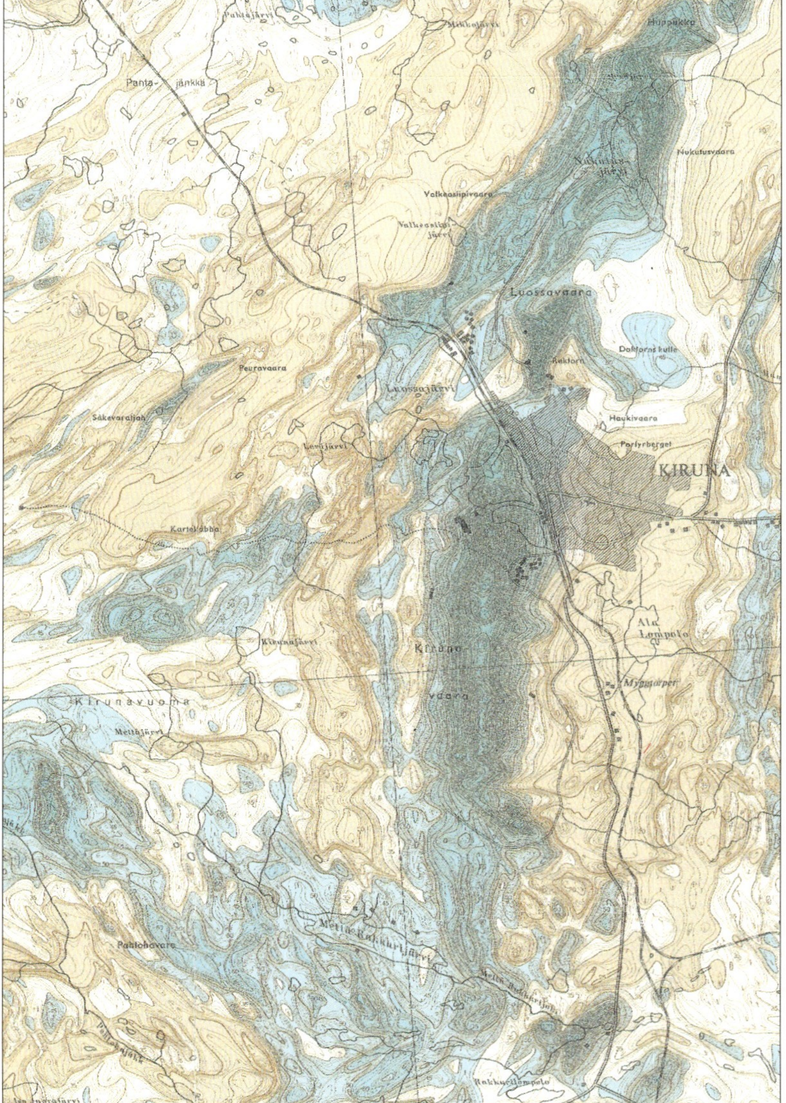

View of the Viscaria Mine, to the west of Kiruna.

The Alpine Catchfly, *Lychnis (Viscaria) alpina*, is one of the few plants that is not troubled by metal-contaminated ground. Its presence contributed to the discovery of the Viscaria deposit.

Gravity map of the Kiruna area (original scale 1:50 000). The distance between measuring points is about 1 km. Orange colour marks the mass excess owing to heavy rocks, green implies mass deficit—light rocks.

Electrical airborne measurements of the Kiruna area, original scale 1:50 000. Electric conductors in the bedrock amplify the field of radio waves from distant transmitters. Blue areas show electrically conductive zones, brown areas conduct electricity less well.

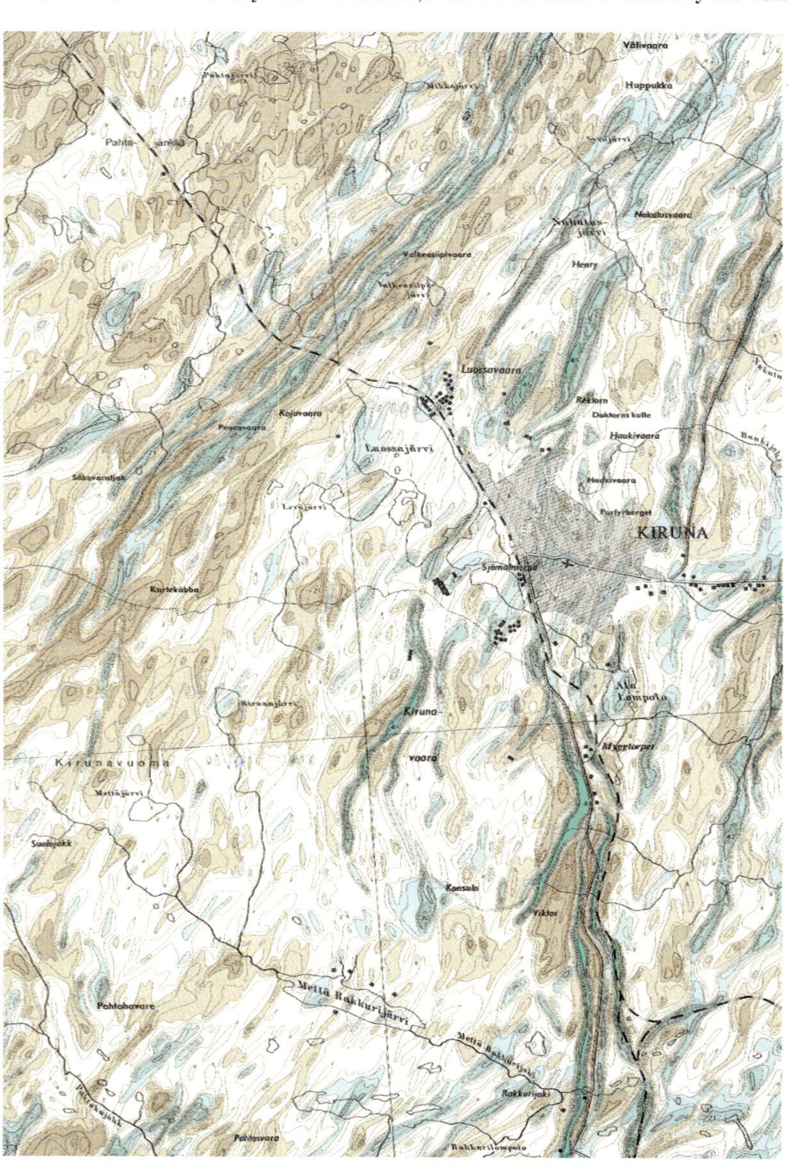

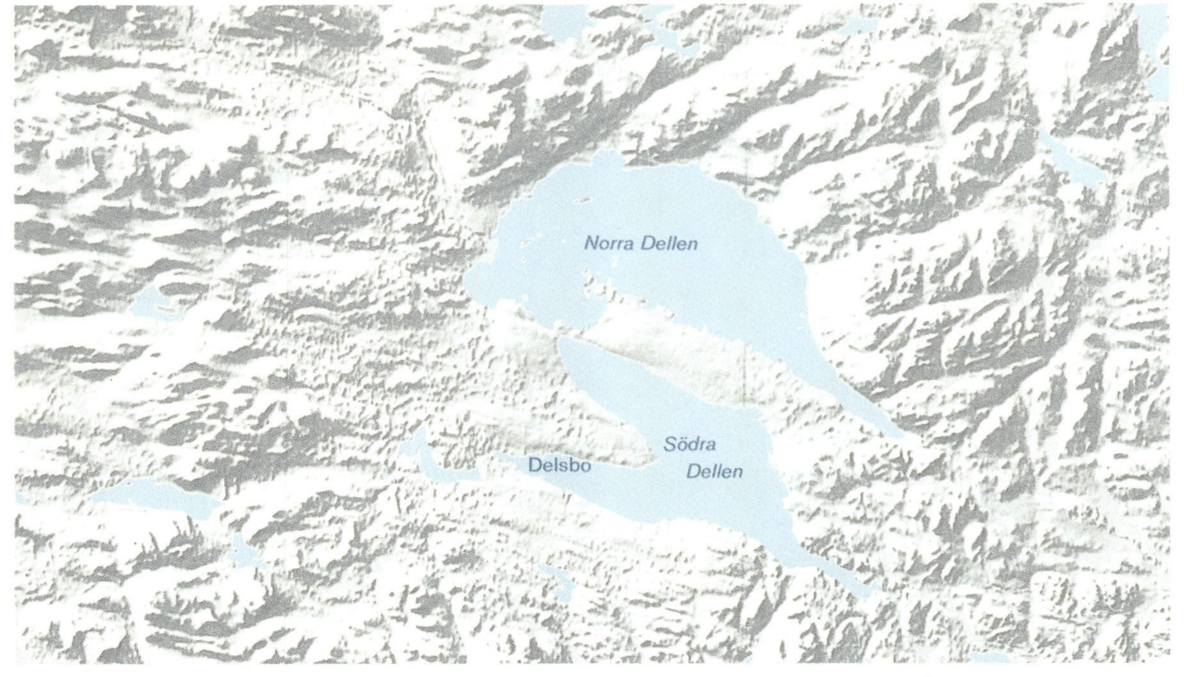

Relief map of the Dellen crater.

METEORITE IMPACT AT DELLEN

Around 90 million years ago, a meteorite impact at Dellen left a large crater in the Earth's crust that is still partly visible as the Dellensjöarna Lakes.

The impact formed a crater and the bedrock was fractured. Material from the crater dispersed in surrounding areas where it formed an impact breccia. An impact lava was formed within the crater and this is clearly visible on the aeromagnetic map as a round magnetic body with a diameter of about 9 km. This rock also has lower density than the surroundings and together with the fractured rock, gives rise to a gravity low in the area.

Both magnetic and gravity anomalies reflect the volume of rock material with different properties compared with the surroundings. The magnitude and shape of these anomalies are used to calculate the depth of the impact lava and the fractured rock. Fracturing of the bedrock gives rise to a reduction in its electrical resistivity. By carrying out ground resistivity measurements, it is possible to map the degree and spatial extent of the fracturing.

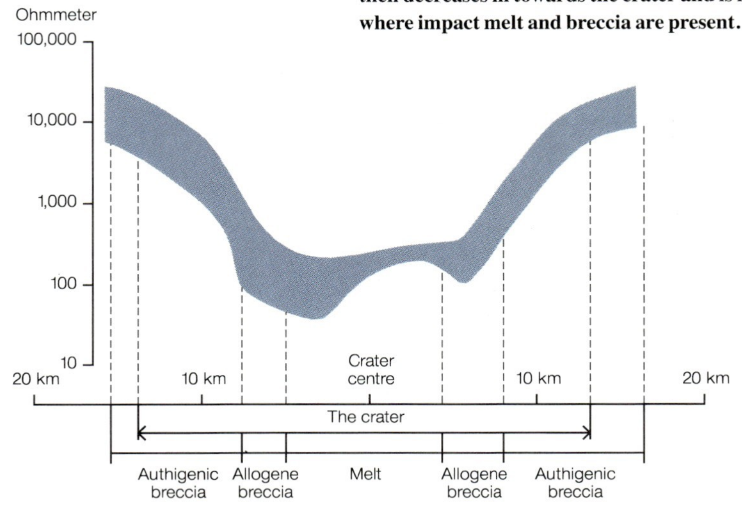

A resistivity profile over the Dellen crater. Normal values have been measured at the rim. The resistivity then decreases in towards the crater and is lowest where impact melt and breccia are present.

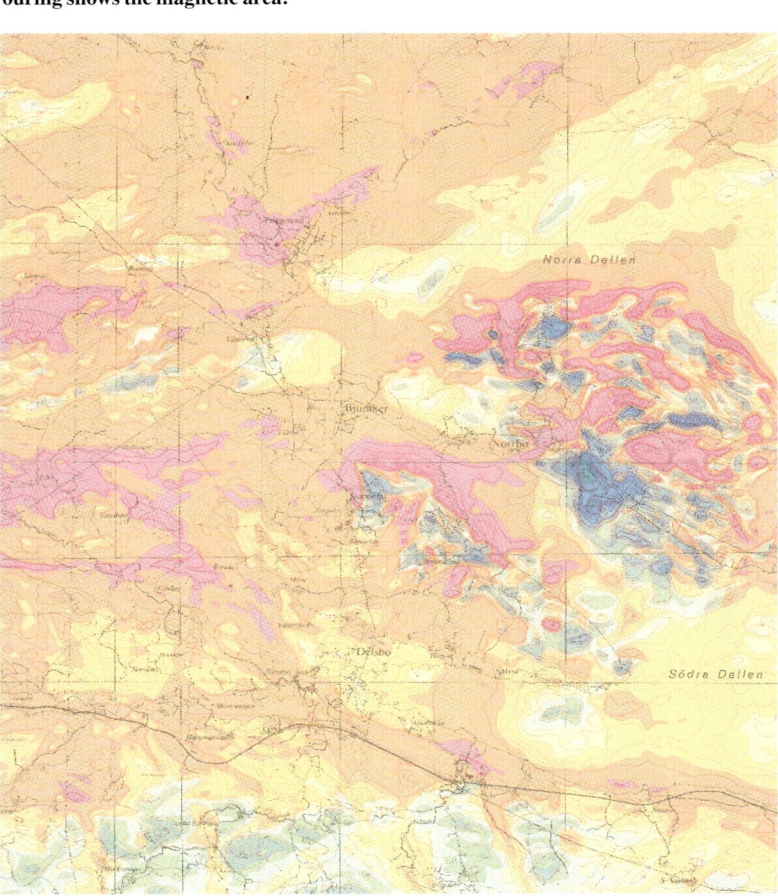

Magnetic map of the Dellen area (original scale 1:50 000). Dellenite, an impact melt, can clearly be seen as a result of its diverging magnetic properties. The blue colouring shows the magnetic area.

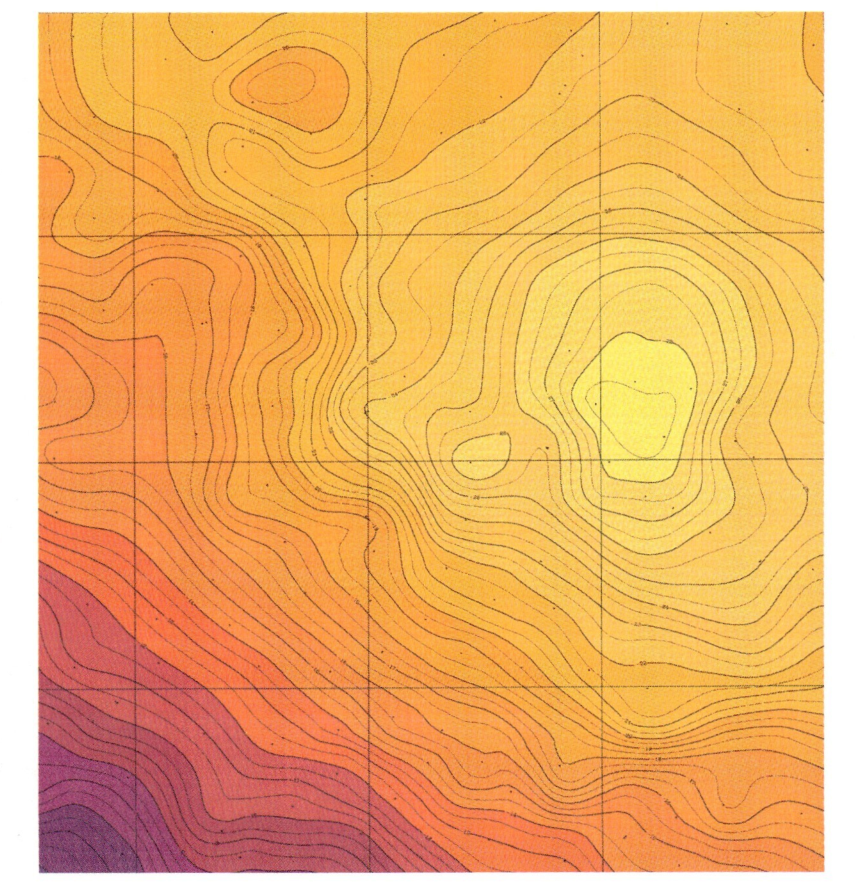

Large meteorite impacts cause mass deficiency. Here the Dellen crater can be seen as a local (yellow) gravity minimum of about 7 mGal.

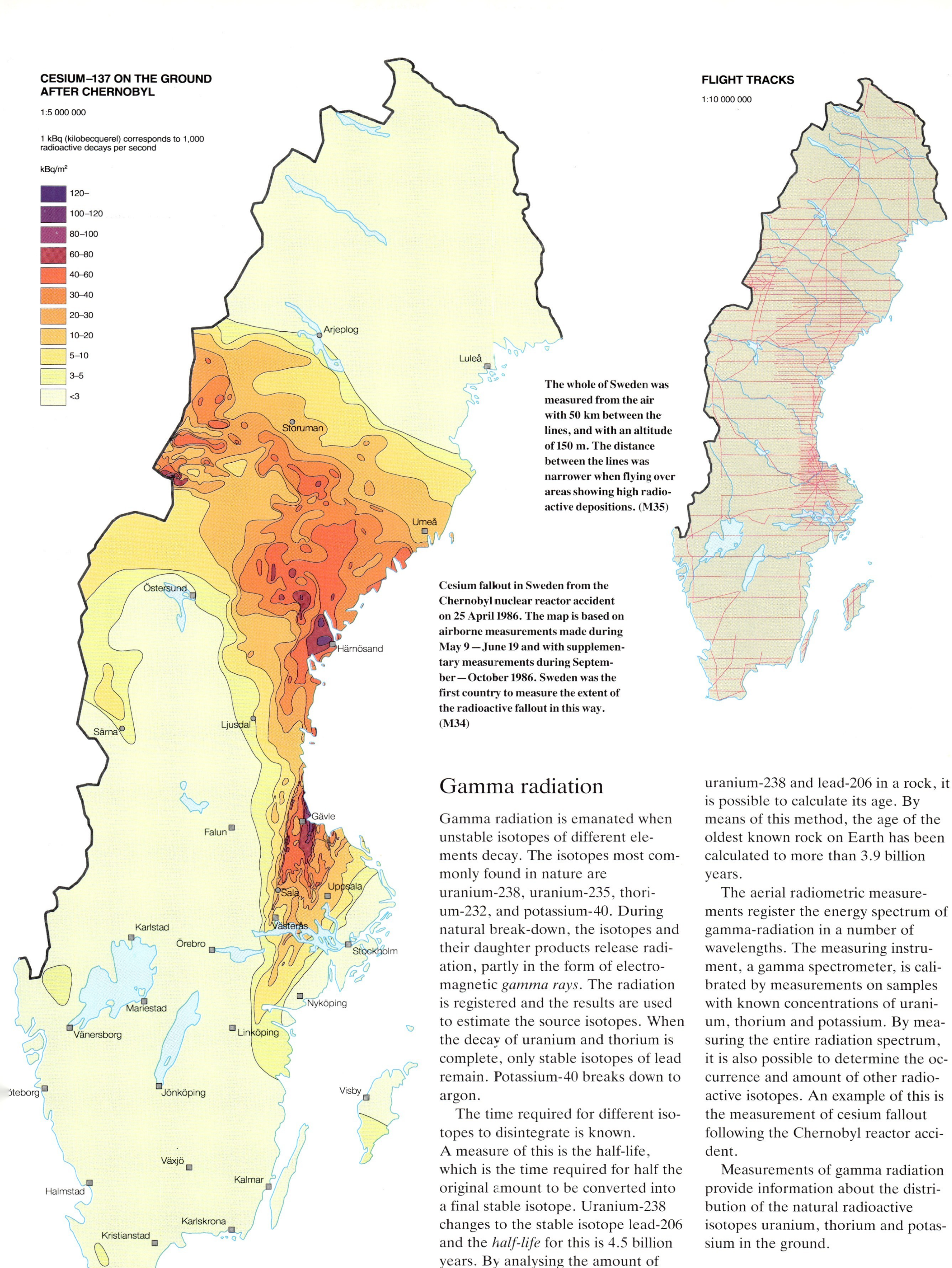

The whole of Sweden was measured from the air with 50 km between the lines, and with an altitude of 150 m. The distance between the lines was narrower when flying over areas showing high radioactive depositions. (M35)

Cesium fallout in Sweden from the Chernobyl nuclear reactor accident on 25 April 1986. The map is based on airborne measurements made during May 9 — June 19 and with supplementary measurements during September — October 1986. Sweden was the first country to measure the extent of the radioactive fallout in this way. (M34)

Gamma radiation

Gamma radiation is emanated when unstable isotopes of different elements decay. The isotopes most commonly found in nature are uranium-238, uranium-235, thorium-232, and potassium-40. During natural break-down, the isotopes and their daughter products release radiation, partly in the form of electromagnetic *gamma rays*. The radiation is registered and the results are used to estimate the source isotopes. When the decay of uranium and thorium is complete, only stable isotopes of lead remain. Potassium-40 breaks down to argon.

The time required for different isotopes to disintegrate is known. A measure of this is the half-life, which is the time required for half the original amount to be converted into a final stable isotope. Uranium-238 changes to the stable isotope lead-206 and the *half-life* for this is 4.5 billion years. By analysing the amount of uranium-238 and lead-206 in a rock, it is possible to calculate its age. By means of this method, the age of the oldest known rock on Earth has been calculated to more than 3.9 billion years.

The aerial radiometric measurements register the energy spectrum of gamma-radiation in a number of wavelengths. The measuring instrument, a gamma spectrometer, is calibrated by measurements on samples with known concentrations of uranium, thorium and potassium. By measuring the entire radiation spectrum, it is also possible to determine the occurrence and amount of other radioactive isotopes. An example of this is the measurement of cesium fallout following the Chernobyl reactor accident.

Measurements of gamma radiation provide information about the distribution of the natural radioactive isotopes uranium, thorium and potassium in the ground.

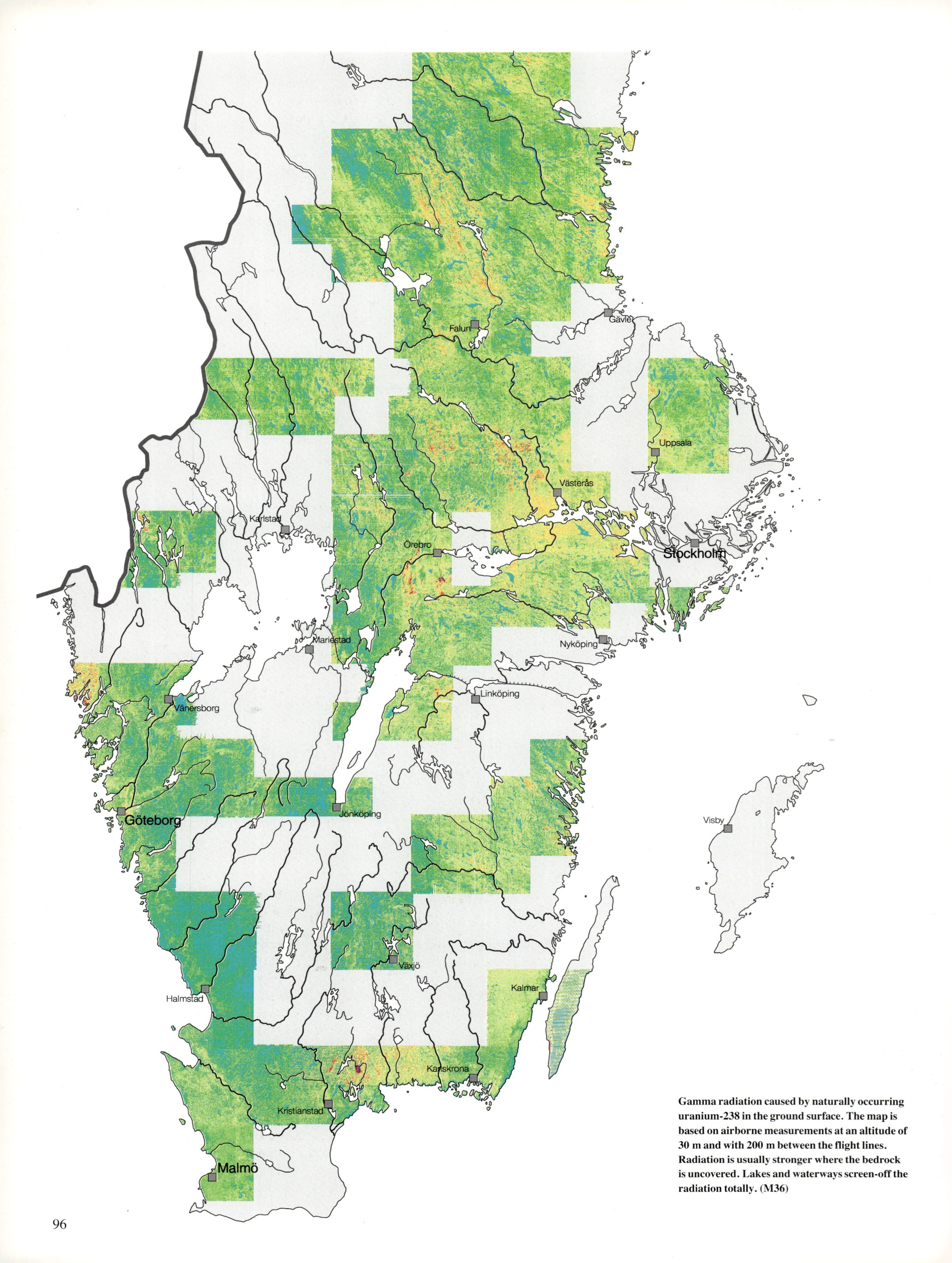

Gamma radiation caused by naturally occurring uranium-238 in the ground surface. The map is based on airborne measurements at an altitude of 30 m and with 200 m between the flight lines. Radiation is usually stronger where the bedrock is uncovered. Lakes and waterways screen-off the radiation totally. (M36)

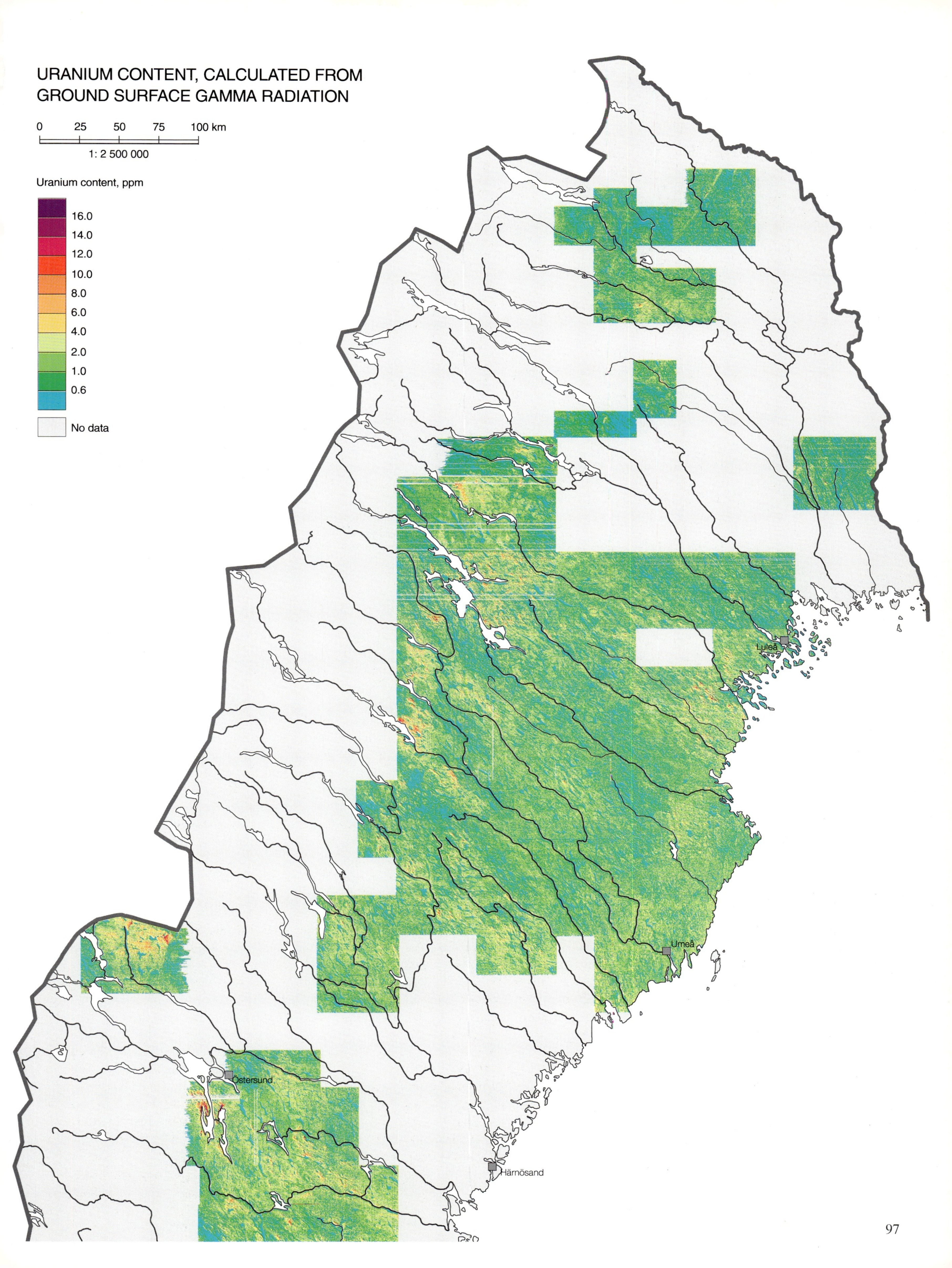
URANIUM CONTENT, CALCULATED FROM GROUND SURFACE GAMMA RADIATION
0
25
50
75
100 km
1: 2 500 000
Uranium content, ppm
16.0
14.0
12.0
10.0
8.0
6.0
4.0
2.0
1.0
0.6
No data
Luleå
Umeå
Östersund
Härnösand

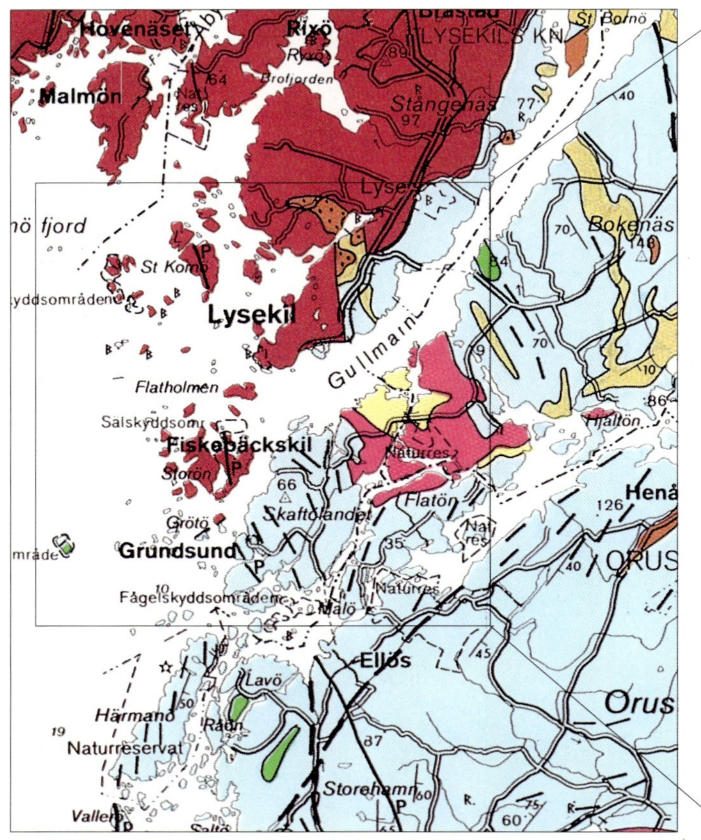

Section of a geological map showing the extent of Bohus granite (red). Scale 1:250 000.

Bohus granite has high levels of uranium and torium radiation. The map section shows the three components of gamma radiation: Uranium (red), torium (blue) and potassium (yellow).

BOHUS GRANITE

The Bohus granite is one of Sweden's youngest granites. It has been dated by the uranium-lead method to 920 million years. As with most younger granites, the Bohus granite shows high gamma activity.

As a result of its high contents of mainly uranium and thorium, the Bohus granite has almost twice the average heat production of normal granites. Consequently, the potential of extracting heat energy for neighbouring residential areas has been investigated.

As the uranium contents of the Bohus granite is much higher than the average in granitic rocks, there is a considerable risk of radon hazards. Radon is a radioactive gas formed when the radioactive isotope radium-226 decay in the decay chain from uranium-238. Radon and its breakdown daughter elements can penetrate into houses and may cause lung cancer. Since radon is an element in the decay chain of uranium, the aerial radiometric map offers an excellent opportunity to survey and systematically assess the risks caused by radon in a specific area.

Seismic methods

Sound-waves move with different speeds through different materials. The speed of sound is higher in denser rocks and Quaternary deposits. By using vibration-sensitive geophones, strategically located on the ground surface, it is possible to study sound-waves from natural earthquakes. The speed, intensity and direction of the sound-waves provide information on the structure of the Earth and the processes related to earthquakes which are caused by movements in the continental crust.

Artificial vibrations can also be used to investigate the composition of the Earth's crust. By measuring how sound-waves are reflected in different layers and horizons with differing properties, it is possible to obtain extremely detailed information down to a depth of about 50 km.

Velocity of sound-waves, m/s

Quaternary deposit	2,000
Sedimentary rock	4,000
Crystalline rock	>5,000
Earth's mantle	>8,000

EARTHQUAKES

Earthquakes occur as a result of fracturing and displacement in the brittle part of the lithosphere. They are mainly caused by the movement of various plates over the Earth's surface. A fracture in the Earth's crust will occur as a result of the successive accumulation of mechanical stresses over a long period. When the strength of the crust is exceeded, the bedrock fractures and gives rise to displacement of large volumes of crust.

The energy released causes seismic waves that spread throughout the entire globe. The magnitude of the earthquake is usually determined on a scale from 0 to 9, the *Richter Scale*, which indicates the amount of the total seismic energy released by the earthquake. Each step in the scale corresponds to an increase of 25–30 times in released energy.

In Sweden, the magnitude of earthquakes attains a maximum of about 5 on the Richter Scale. More than one million earthquakes occur annually throughout the entire globe and, in Sweden, there are about ten per year with a magnitude in excess of 2. Earthquakes are registered in a global network by means of seismographic stations.

The measuring instrument, the *seismograph*, is a pendulum that is activated by the vibrations in the ground when the seismic waves pass. From this registration, it is possible to calculate the position and magnitude of the earthquake. Most earthquakes in Sweden occur at depths of 5–20 km. It is also possible to determine how the bedrock has been displaced.

By means of global registrations of earthquakes for more than a century, it has been possible to increase our knowledge of the Earth's composition and properties. Most earthquakes occur along the plate boundaries, and particularly deep and strong earthquakes occur in the downward-moving part of the convective flow system. Here, the cold and solid downward-moving lithosphere breaks apart and is assimilated by the surrounding mantle at a depth of about 700 km. In the spreading zones, where new parts of the Earth's crust are formed, plate boundary deformation is also conspicuous. At these sites, the newly-formed lithosphere is fractured and earthquakes occur down to depths of about 60 km. Usually much weaker earthquakes are dispersed within the plates. These are related to a continuous change in the shape of the lithosphere.

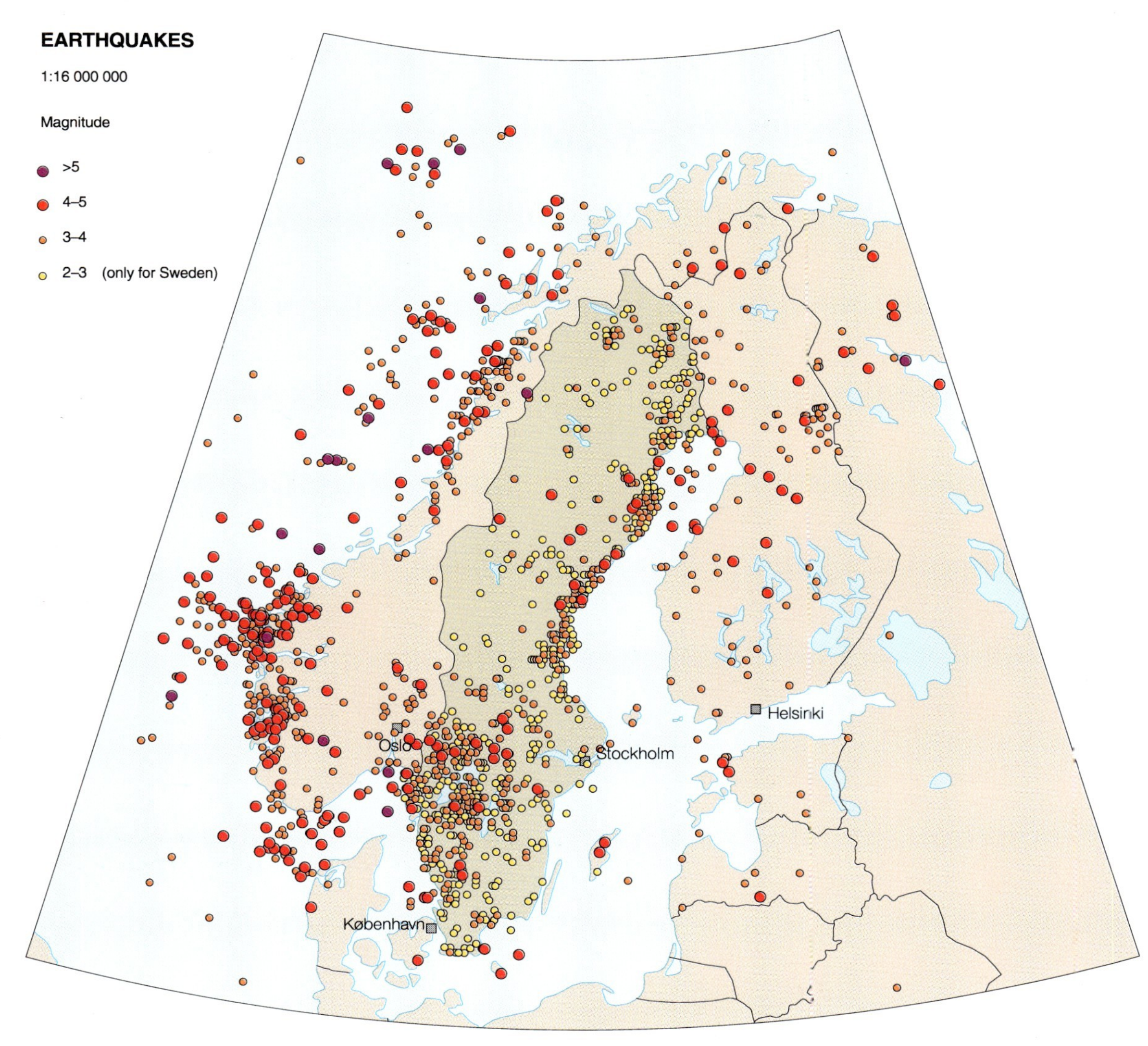

Earthquakes in Fennoscandia 1375—1992. The magnitude is given according the Richter Scale. (M37)

The Earth's crust is the part of the lithosphere where the speed of the seismic compressional wave is lower than 8 km/s. The thickness of the Earth's crust is determined by means of, for example, seismic registration of earthquakes. Calculations are made of the depth to the Moho-discontinuity where the speed of the seismic p-wave rapidly increases to more than 8 km/s when it reaches the rocks in the mantle. (M38)

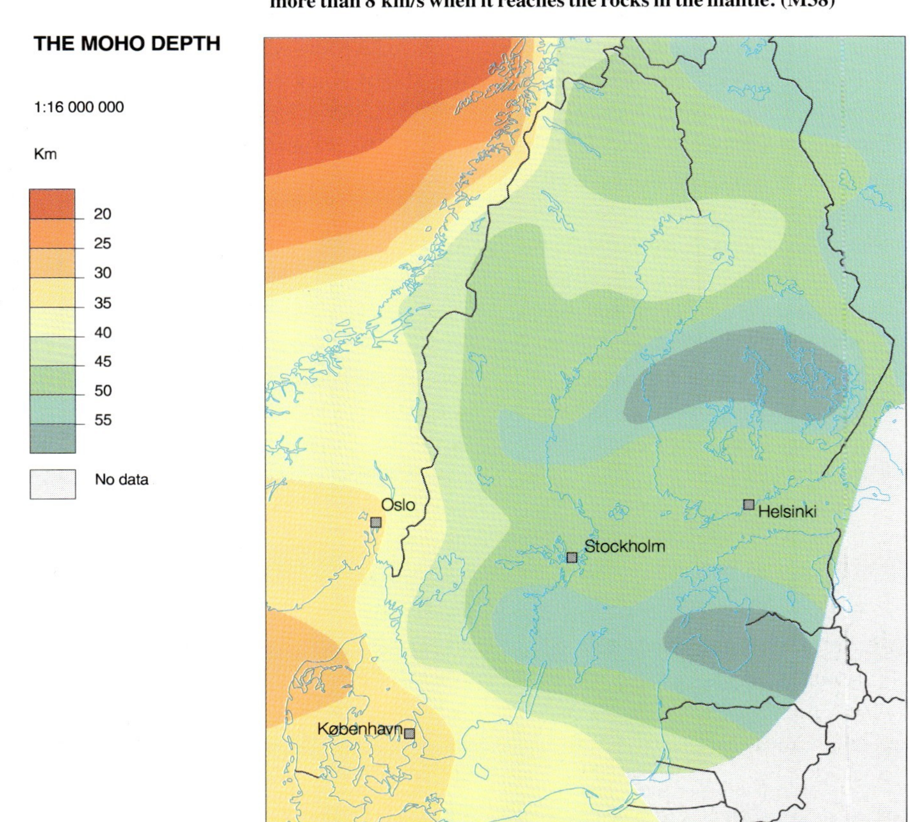

The largest earthquake so far observed in Scandinavia, with a magnitude of 5.4, took place on 23 October 1904 to the south of the Oslo Fjord. It was noticed throughout a large part of Northern Europe. (M39)

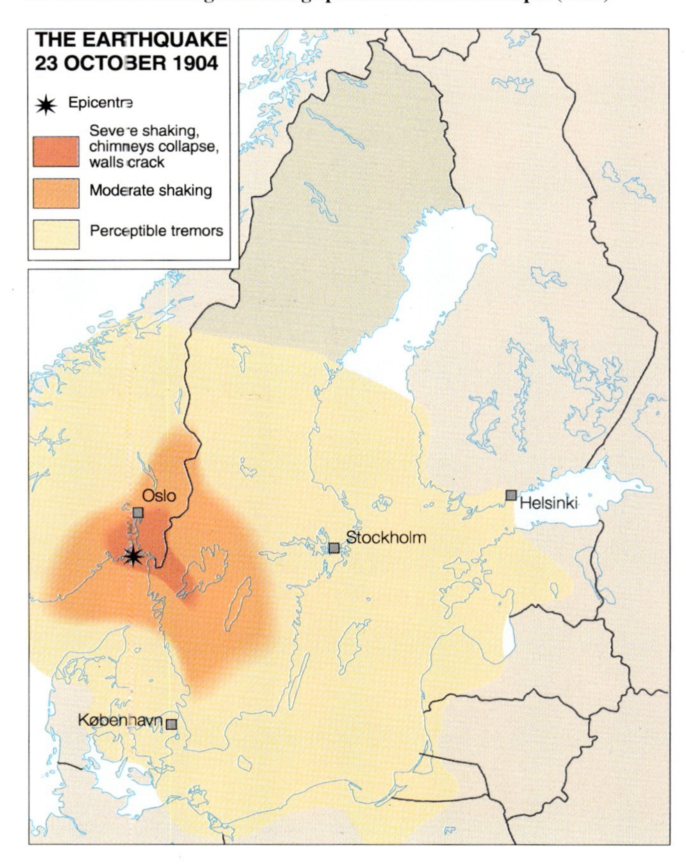

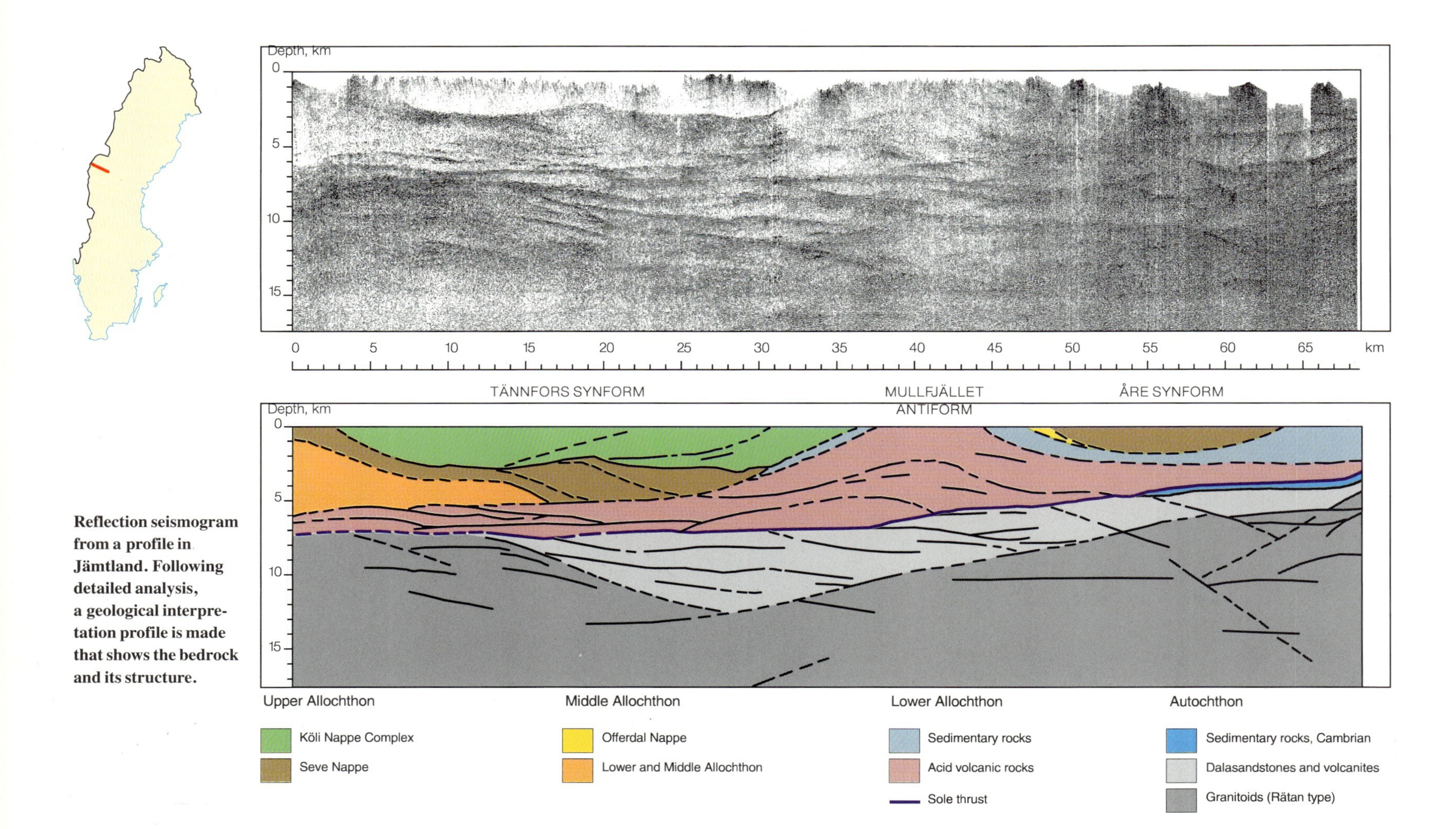

Reflection seismogram from a profile in Jämtland. Following detailed analysis, a geological interpretation profile is made that shows the bedrock and its structure.

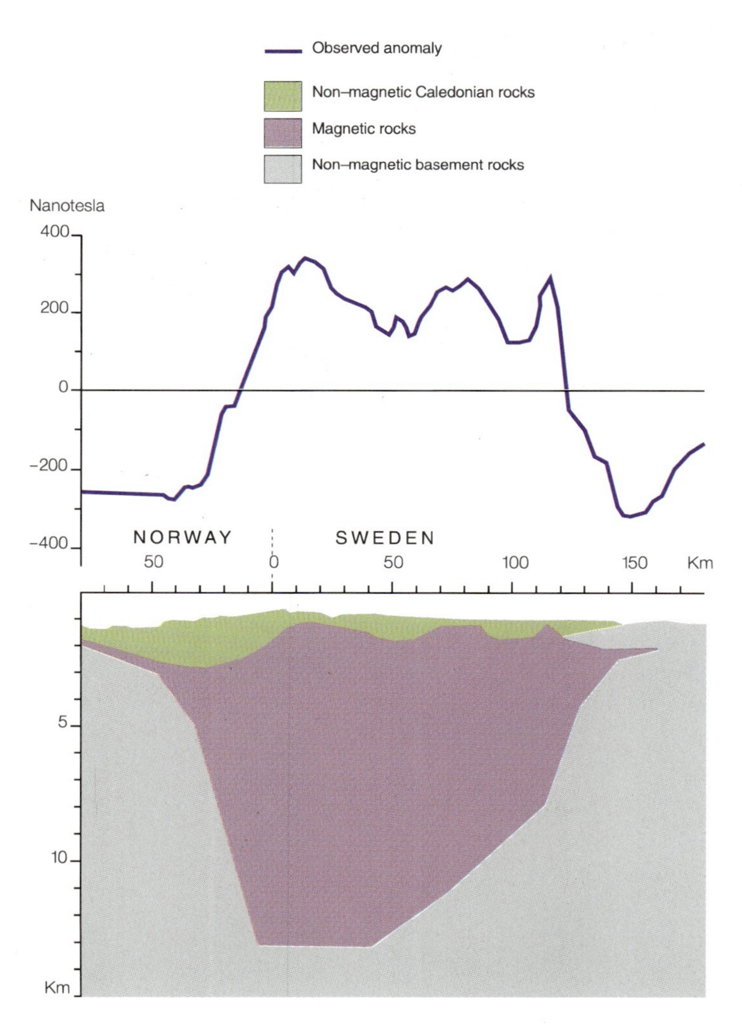

In Sweden, there are several areas with seismic activity. Lake Vänern is located in one such area and earthquakes are caused by block movements along a system of movement zones. These zones absorb most of the lithospheric shape changes. Sometimes, however, deformation of larger volumes of bedrock is required. In the case of Lake Vänern, the entire area appears to be slowly moving apart, the free surface is sinking, and a depression is being formed. In the Lake Vänern area, subsidence has proceeded for a very long period and amounts, at the present time, to several hundred metres. Deformation of the area around Lake Vänern is conspicuous in the land uplift curves. These display a rise of about 0.5 mm per year around the northern coastline of Lake Vänern.

NAPPE STRUCTURES OF THE CALEDONIAN BEDROCK

The structure of the Earth's crust can be investigated by means of reflection-seismic methods using a series of geophones located at intervals of 50 m along a line. The signals are generated with explosions and, after reflection in bedrock structures, the waves are registered along the chain of geophones. The data are stored and processed in computers and printed as seismogrammes.

A reflection-seismic profile covering about 300 km has been measured in Jämtland as part of the investigation of the structure of the Caledonian bedrock. The transition between lighter and darker parts of the seismic diagram marks the position of reflective layers where bedrock composition and structure change abruptly. Thrust zones and rock contacts can clearly be distinguished in the seismogram.

Measurements of the magnetic field enable us to calculate the depth and extent of the basement rocks beneath the thrust nappes. In the Jämtland basement, below the almost magnetically "transparent" Caledonian nappes, the northern part of a vast magnetic granite body—a *batholith*—is present which extends southwards to Dalarna. The granite, calculated to have a depth of more than 10 km, causes one of Scandinavia's largest magnetic anomalies.

Magnetic profile in Jämtland and a model calculation of a granite batholith in the basement below the nappe structures.

Land uplift

Automatic water-level meters, *mareographs*, that continuously register the sea level, were established in 1887. The series of measurements, now covering more than a century, are used to calculate land uplift along the coast.

Levellings made in inland areas and repeated at sufficiently long intervals of time, in combination with mareograph measurements, make it possible to calculate land uplift over large areas. The National Land Survey of Sweden has conducted two levellings with an interval of about 65 years. In 1980, a third precision levelling was started. Similar levellings are being carried out at the same time in neighbouring Nordic countries and it is planned to cover a measured distance of 40,000 km. The present rate of land uplift has been calculated from the results of the first and second precision levellings, together with recordings from ten mareographs. The largest land uplift rate is found in the coastal area of Västerbotten and amounts to 9.2 mm/year.

At Marstrand there is a series of measurements of the mean water level dating back to 1770. The two marks in the photograph were made in 1770 and 1847. The mean water level today is above the zone with barnacles, i.e. the distance to the uppermost line is about 60 cm. The photograph was taken in 1986.

Generalized curves for total land uplift. The greatest extent of the latest glaciation has been marked. In the northwestern corner of the area, oceanic crust is found. The area of land uplift in Northern Europe is surrounded by an area with subsidence. (M40)

Land uplift, expressed in mm per year, is based on repeated precision levellings made along roads, coastlines and railways. The values in between are interpolated. The curves are then uncertain and are shown as a dashed line. (M41)

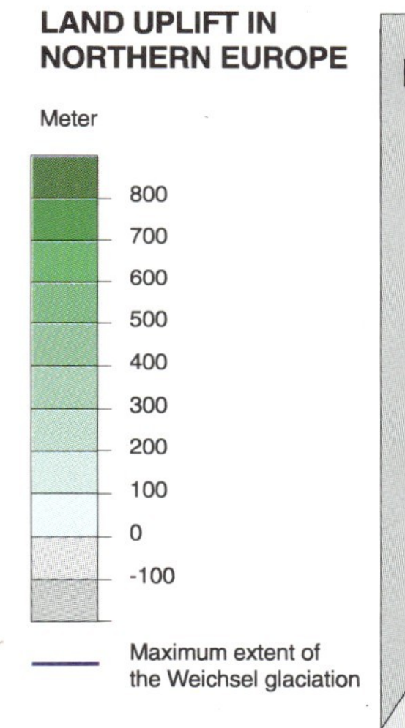

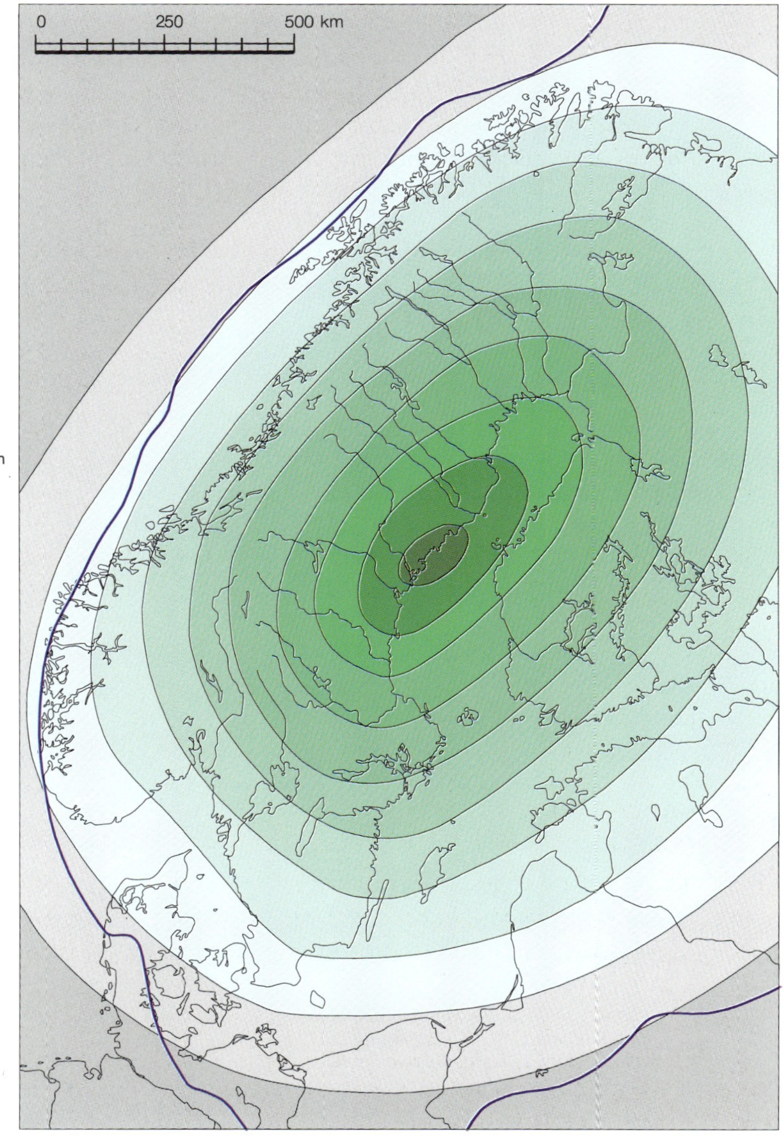

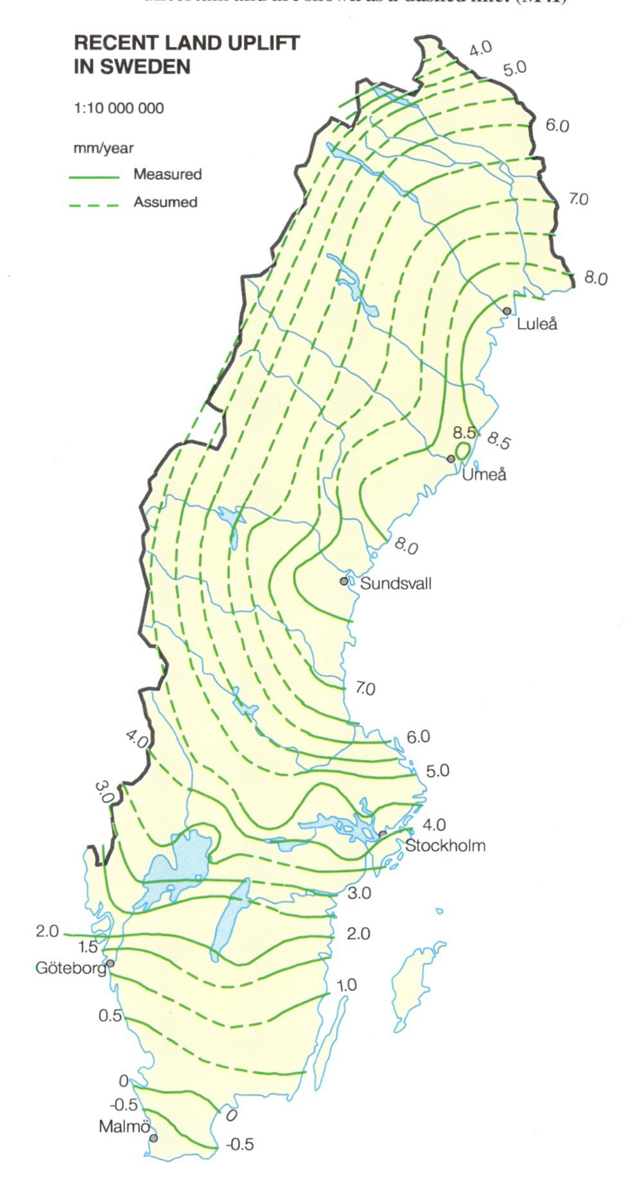

The Quaternary

The Earth's youngest period, the Quaternary, is characterised by alternating glacial and interglacial stages. The landscape is largely influenced by the latest glaciation (1–5) and its recession (6–16). Subsequently, the landscape has been reformed to a certain extent by isostatic uplift and wave-washing of the shores at that time (17–22). The wind (23) and running water (24–29) erode the landscape and create new land-forms both above and below the highest shoreline (17), peatlands expand (30–31) and anthropological activities leave clear traces (32–33). The fictitious landscape in the illustration has been formed by a continental ice sheet that has moved from left to right, the retreat of the ice margin has taken place from right to left.

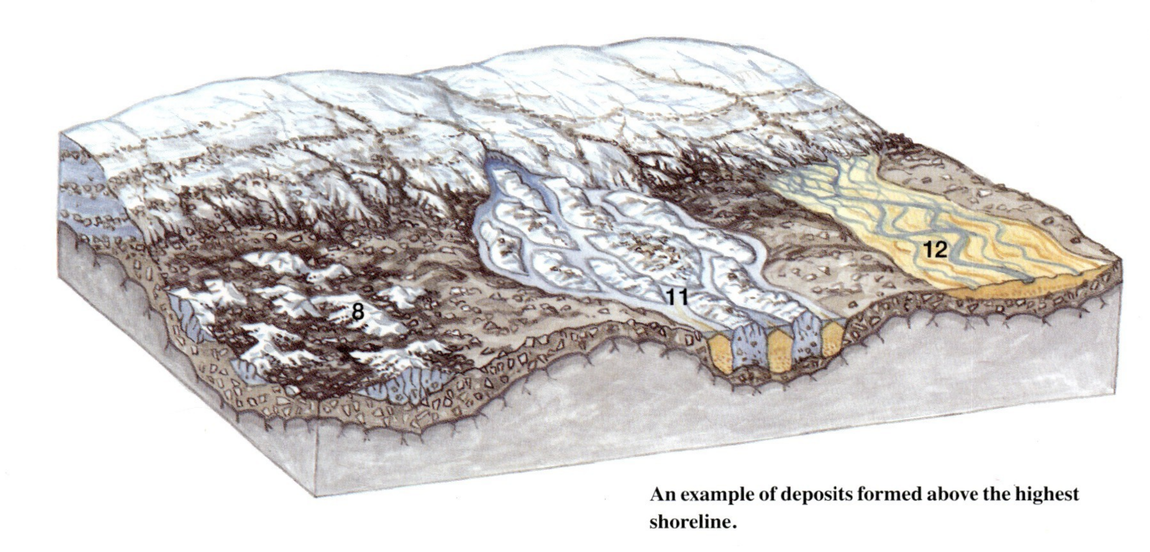

An example of deposits formed above the highest shoreline.

1. Roche moutonnée
2. Drumlin
3. Till
4. Cultivated till
5. Rocky ground
6. End moraine
7. De Geer moraine
8. Ablation moraine
9. Humpback esker
10. Esker
11. Kame
12. Glaciofluvial out. wash plain, sandur
13. Glaciofluvial delta
14. Kettles
15. Erratic
16. Clay plain
17. Highest shoreline
18. Till-capped hill
19. Shingle field (klapper)
20. Raised beaches
21. Bare-washed bedrock
22. Sand beach
23. Dunes
24. Bluffs
25. Gullies
26. Meander
27. Oxbow lake
28. Recent delta
29. Slide scar
30. Bog
31. Soligenous mire
32. Ditched fen
33. Gravel pit

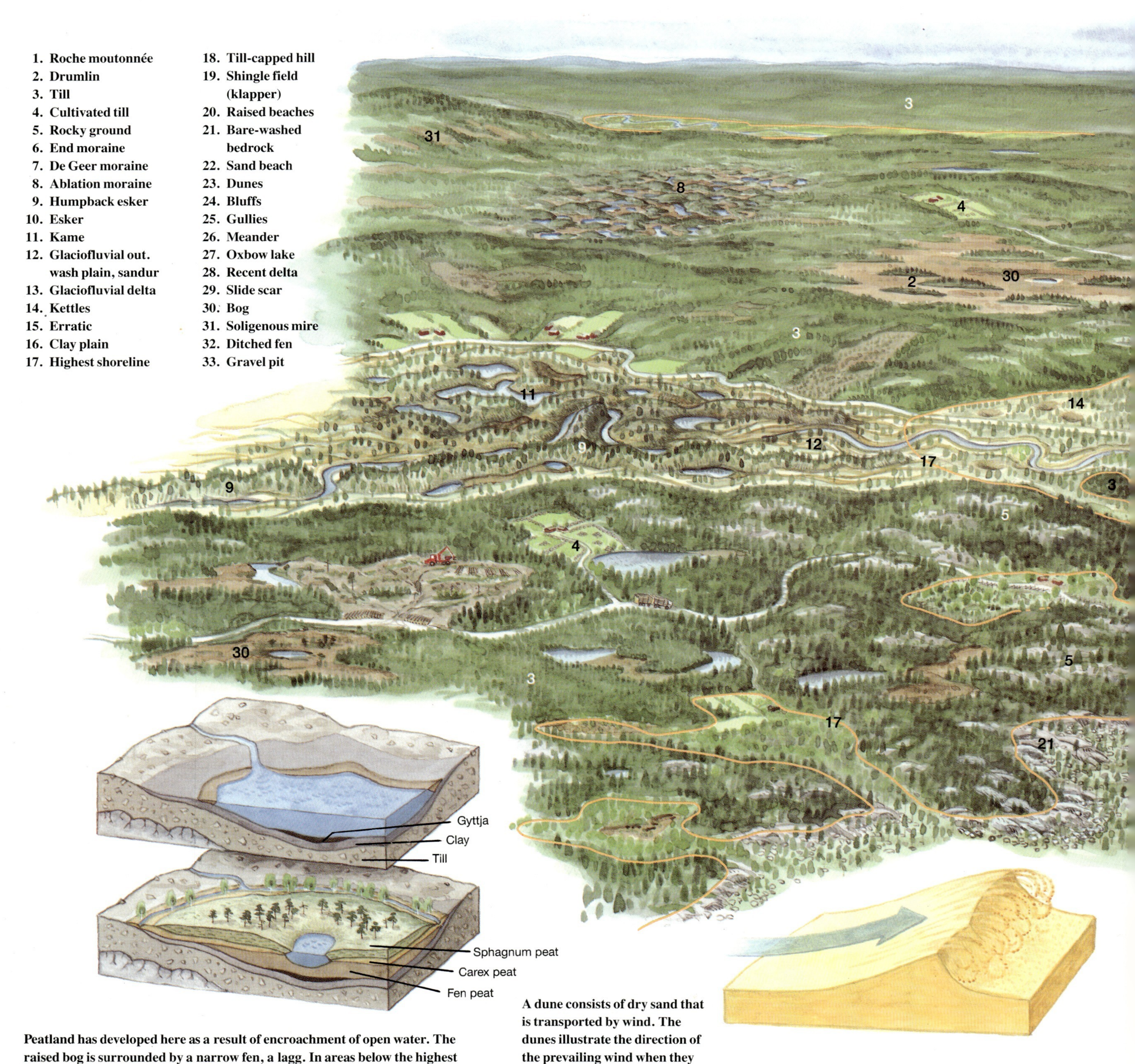

Peatland has developed here as a result of encroachment of open water. The raised bog is surrounded by a narrow fen, a lagg. In areas below the highest shoreline, clay is commonly found below the peat sequence.

A dune consists of dry sand that is transported by wind. The dunes illustrate the direction of the prevailing wind when they were formed.

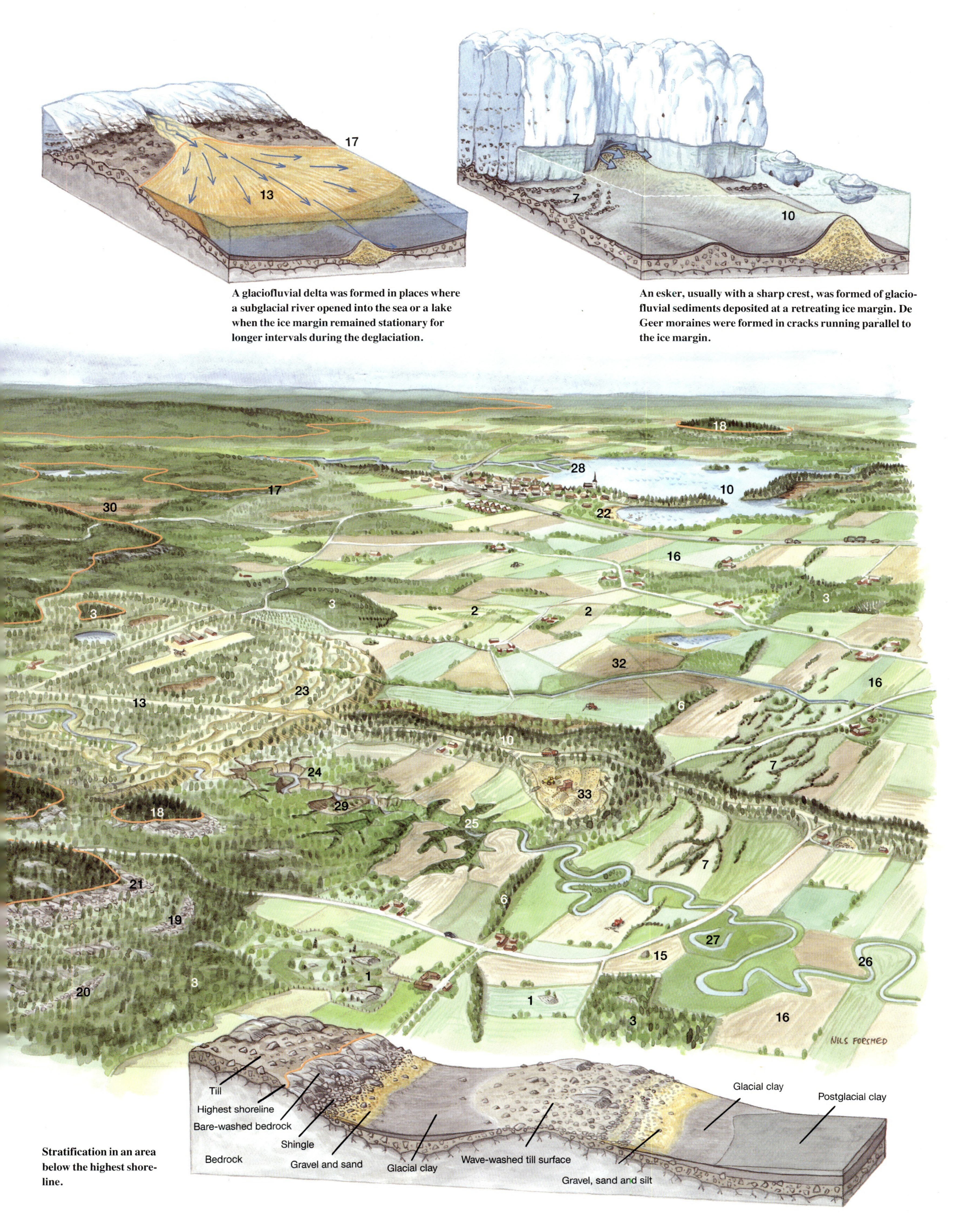

A glaciofluvial delta was formed in places where a subglacial river opened into the sea or a lake when the ice margin remained stationary for longer intervals during the deglaciation.

An esker, usually with a sharp crest, was formed of glaciofluvial sediments deposited at a retreating ice margin. De Geer moraines were formed in cracks running parallel to the ice margin.

Stratification in an area below the highest shoreline.

	0.002	0.006	0.02	0.06	0.2	0.6	2	6	20	60	200	600	2,000 mm
Clay	**Silt**			**Sand**			**Gravel**		**Cobble**		**Boulder**		
	Fine silt	Medium silt	Coarse silt	Fine sand	Medium sand	Coarse sand	Fine gravel	Coarse gravel	Pebbles	Cobbles	Fine boulders	Medium boulders	Coarse boulders

Classification and nomenclature of Quaternary deposits.

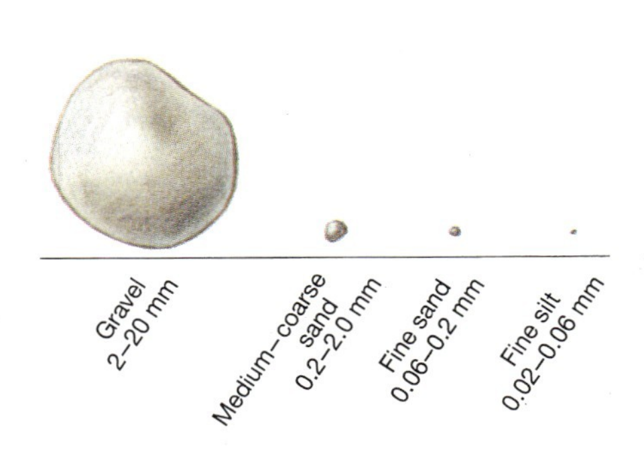

Atterberg's particle-group scale from 1903.

Quaternary deposits

The Earth's youngest period, the Quaternary, is characterized by alternating glacial and interglacial stages. Most of the Swedish Quaternary deposits were formed during and after the latest glaciation. The term *Quaternary deposits* refers to the loose overburden on the surface of the Earth, in which vegetation is or can be rooted. *Soil* refers to the part of the Earth's crust that changes under the influence of climate, vegetation and fauna, and usually has a thickness of 20–50 cm.

Quaternary deposits are grouped according to particle-size distribution and according to their genesis and the environment in which they were formed. The distribution into particle sizes is based on the particle-group scale devised by *Atterberg*; the deposits are basically named according to the dominant fraction.

Limits between particle sizes frequently coincide with important physical limits. The ability to absorb water, *capillarity*, and the ability to retain or to allow water to percolate through, *permeability*, vary foremost with particle size and degree of compaction.

Quaternary deposits made up of *fine grained particles*, for example clays, have a large capillary height of rise, but absorb water extremely slowly. Additionally, they have low permeability, implying that they are dense and consequently water can percolate through them only slowly. Fine sand absorbs most water per day. In addition, it is capable of retaining large amounts of water. Silty deposits easily become puddled when they are water-saturated, thus being susceptible to freezing and sliding. Deposits with *coarse particles*, e.g., gravel and sand, are extremely permeable; water easily flows through them, and they dry rapidly. Coarse deposits are of great importance for the supply of groundwater since they have the greatest permeability. For the same reason they are extremely sensitive to pollution.

The division of Quaternary deposits according to genesis and the environment in which they were formed consists of two main groups: *glacial* and *post-glacial*. Glacial deposits were formed by an ice sheet or its melt-water. This group includes till, glaciofluvial or glacier lake sediments, and glacial clay deposited in the sea. Post-glacial deposits were formed independently of the melting of an ice sheet.

Permeability and capillarity of Quaternary deposits in the clay—gravel fractions.

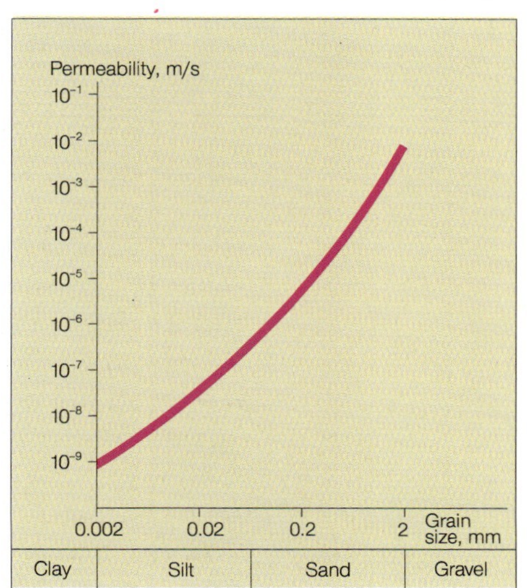

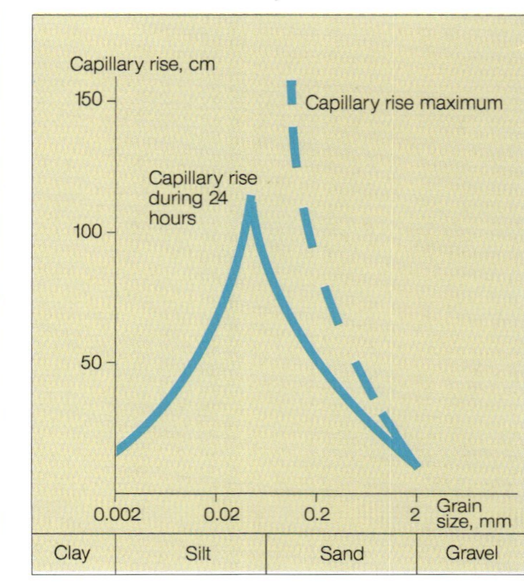

Till surfaces with high boulder frequency have very low accessibility. The boulders are usually piled upon each other. Medelpad.

Till is a mixture of all particle sizes. The illustration shows a typical basal till. Burseryd, Småland.

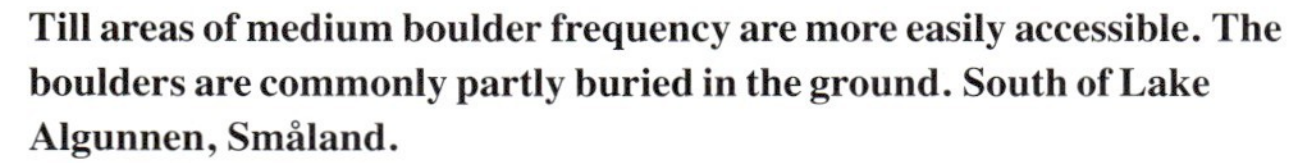

Till areas of medium boulder frequency are more easily accessible. The boulders are commonly partly buried in the ground. South of Lake Algunnen, Småland.

A radial moraine usually has abundant boulders. This radial moraine, aligned in a NNW-SSE direction, is at Strömsdal in southwestern Dalarna.

Glacial deposits

TILL

Till consists of varying amounts of boulders, stones, gravel, sand, fine sand, silt and clay. Strata of, for example, gravel, sand and fine sand occur. The particle-size composition and content of boulders in a till vary depending on the rocks included, the distance they have been transported, and the possible admixture of older sediments.

Till is divided according to the mean particle size composition of the material and the surficial boulder content. The most common type of till is sandy—fine sandy and has a moderate content of boulders.

About 75% of Sweden's land surface is covered by till. Till generally lies directly on the bedrock and largely follows its surface configuration. Till also commonly forms its own surficial configurations.

Moraines parallel to the ice direction

A *drumlin* is an oval-shaped ridge formed beneath an ice-sheet as it moved over the terrain. A drumlin frequently consists of older material that has been reworked by the ice. Drumlins are often found in groups and the ridges may extend over several kilometres. There are different types of drumlins, e.g., drumlinoid formations, made up of a nucleus of rock with accumulated till on the stoss-side and/or the lee side.

Radial moraines formed in an open crack in the margin of the ice. Their lengths vary. Surficial boulders may occur.

Moraines parallel to the ice margin

End moraines belong to this class and may either be large or small, short or long. Large and relatively continuous terminal moraines that were deposited during a longer period when the ice margin was stationary, are called *ice-marginal moraines*. Areas with large end moraines may also include glaciofluvial deltas.

A closely-related type of moraine, the *De Geer moraine*, occurs in clusters in lowland areas. These moraines are usually a couple of hundred metres long and up to 5 m high. The distance between them in the direction of ice movement is often 200–400 m. Earlier, they were considered to have been formed at the ice margin and that they were a type of end moraine. Today, however, they are considered to have been formed within the ice margin, like the radial moraines, but in cracks running parallel to the margin.

Hummocky moraines

Hummocky moraines principally occur in valleys and in flat-lying areas. Three main types may be distinguished: Rogen moraine, Veiki moraine and irregular ablation moraine. All three types have in common that they form a mosaic of lakes, tarns, peat-land and areas of minerogenic deposits. Rogen and Veiki moraines are mainly older than the last glaciation.

Rogen moraine is characterized by

The remaining snow clearly shows the pattern of the Rogen moraine with its parallel ridges. Here, at Lake Tärnasjön in the province of Västerbotten, they are aligned in a SW-NE direction.

Veiki moraine, with its characteristic plateaus separated by peatland and circular lakes, received its main features already before the latest glaciation. Ridges typical of these moraine plateaus can be seen on the clear-felled area. The districts where Veiki moraine are found are largely in the inland areas of northern Norrland, here about 25 km to the north of Gällivare.

The originally sharp morphology of this hummocky moraine landscape in Skåne to the north of Skurup has been smoothed out by cultivation.

ridges that are irregular in detail but largely at right-angles to the direction in which the ice was moving.

Veiki moraine is characterized by plateau-like moraine hills with numerous circular-formed lakes.

Ablation moraine usually forms an irregular pattern of ridges and depressions. It formed in a fractured, stationary part of the inland ice. Lakes and peat-land occur today in places where formerly there were ice residues.

GLACIOFLUVIAL DEPOSITS

Glaciofluvial sediment consists of boulders, stones, gravel, sand and fine sand that has been transported, sorted and deposited by the meltwater from the inland ice. Glaciofluvial sediment is stratified in layers with one or several particle sizes. The particles are usually rounded.

The shape of the deposits depends on the environment in which they were formed. The meltwater in the inland ice formed strongly flowing rivers in tunnels emerging at the margin of the ice. The coarser material, blocks, stones, gravel and sand, was deposited inside the tunnel or at its opening. The finer material, fine sand, silt and clay, was deposited at greater distances from the mouth of the subglacial river.

An *esker* is a long, ridge-shaped glaciofluvial deposit that was formed in a tunnel in the inland ice or in an open crack. Eskers below the *highest shoreline* are usually large and may be as much as 250 km long. Layers of clay and wave-washed sediments occur on their slopes. Above the highest shoreline, eskers are usually smaller and frequently have a sharper configuration, often called *humpback eskers*.

A *delta* is an extensive glaciofluvial deposit that has continued to develop until it has reached the surface of a lake or the sea. A glaciofluvial delta formed where a subglacial river opened into the sea or a lake, or where the ice margin remained stationary for longer periods. There may be several reasons for the stationary character of the ice margin, one of which being a result of the climate. Deltas of this kind are sometimes found in systems of ice marginal moraines. Another reason for the stationary margin is the depth of the water in relation to the thickness of the ice. Where the melting of the ice took place in the sea or in a lake, not

Outwash plain, sandur, at Ruotesjekna, Sarek, with a braided river system.

Gravel-pits are fascinating places for people interested in geology. In the pit they can identify the different phases of the deglaciation. On the slopes of this esker, the glaciofluvial sediment can be seen at the bottom, with clay and wave-washed deposits further up. Ekerö, to the west of Stockholm.

Illustration of stratification in a subaquatic esker.

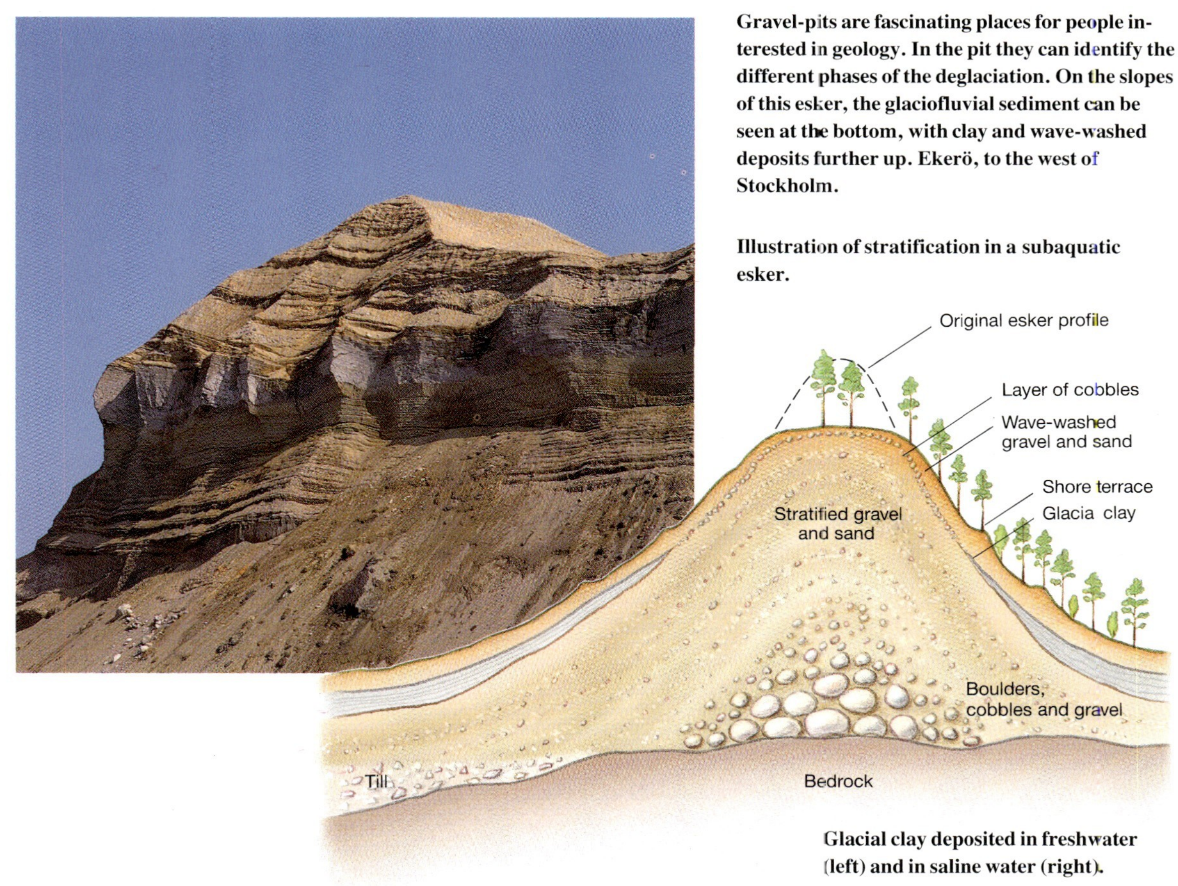

Glacial clay deposited in freshwater (left) and in saline water (right).

Varved clay in one of the clay-pits belonging to the former Vaksala brickworks, Uppsala.

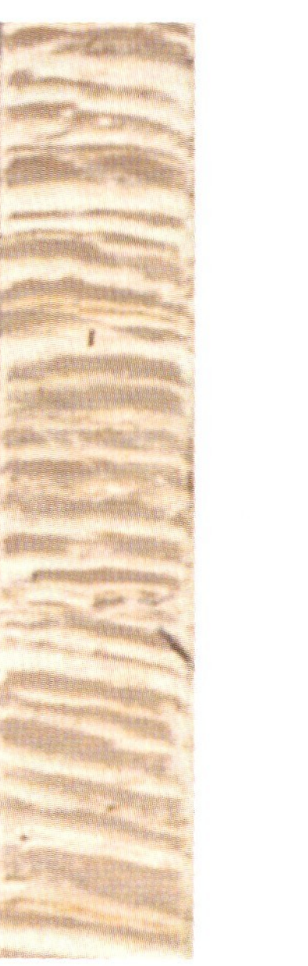

only surface melting took place but also frontal melting—*calving*. During the transition from frontal and surface melting to surface melting alone, the retreat of the ice margin may temporarily cease. The largest deltas are a result of the topography.

Kames is the name given to a hummocky area that consists mainly of glaciofluvial sediments. Depressions mark the melting of ice residues.

An outwash plain, *sandur*, is a glaciofluvial deposit formed by rivers with a network of channels that are today usually dry.

During the melting of the inland ice, the finest particles of glaciofluvial origin, clay, were dispersed in the sea and in large lakes. These particles formed clays with varying properties.

In freshwater, the particles remained suspended for long periods and sedimentation took place slowly. Depending on the seasonal changes in the melting of the ice, and thus in the flow of water, there was a regular change in the rate of sedimentation. During the spring and summer the flow of water in the glacial rivers was great and large amounts of clay and silt particles were transported. The supply of sediment during autumn and winter was, on the other hand, low. Thus, a thicker and a thinner layer together form an *annual varve*. The winter varve is usually darker in colour than the summer varve, and has a higher clay content.

In saline water, the sedimentation of clay took place faster on account of the electrolytic properties of the saline water and consequently there is no clear pattern of varves. Marine glacial clays of this kind are mainly found in southwestern Sweden.

This 13 m thick shell-bank at Bräcke near Uddevalla, was once the habitat of molluscs and crustaceans that lived about 11,000 years ago in an arctic sea. Parts of skeletons of marine vertebrates have also been found. The area is now a nature reserve.

Fossilized dune at Gualöv, Skåne. When the climate changes and the wind decreases, vegetation can migrate into the area — roots bind the sand and the dune becomes "conserved".

Delta off the mouth of the River Indalsälven. Coarse particles are deposited closest to the river channel, the finer particles being deposited further out in calmer water.

Post-glacial sediments

MARINE AND LACUSTRINE SEDIMENTS

Land uplift exposed older deposits to the influence of wave-washing and there was a more or less complete restratification. Wave-washed material was deposited along and close to shorelines as shingle, washed gravel, *wave-washed* sand and fine sand—in principle with decreasing particle size away from the shore. The restratification products with the finest particles, silt and clay particles, were deposited farthest away from the shore.

Klapper is made up of the coarsest wave-washed sediment, consisting of rounded shingles and cobbles that have accumulated in small fields, usually on a glaciofluvial deposit.

Wave-washed gravel has a variable composition. Apart from gravel, sand and stones are also included, sometimes together with boulders and fine sand.

Wave-washed sand and *fine-sand* are dominated by the sand and fine sand fraction and are sorted better than wave-washed gravel.

Shell deposits consist largely of shells and fragments of shells of molluscs and barnacles. Sometimes the waves and currents have given rise to shell banks of considerable size.

Post-glacial coarse silt, *silt* and *clay* make up the distal wave-washed sediments and are usually underlain by glacial clay.

Gyttja is deposited in open water and consists of more or less finely decomposed residues, detritus or dead organisms. The content of organic material is more than 20%. Clay gyttja contains 6–20% and gyttja clay contains 2–6% of organic material. In places where the gyttja forms surface layers, it has usually been exposed in connection with lake regulation.

ALLUVIAL AND DILUVIAL DEPOSITS

Alluvial sediments are deposited as banks along the reaches of the river or as a delta at the river-mouth. These sediments are usually well-sorted and poor in organic matter.

Diluvial sediments are deposited at high-water along the sides of the river. These sediments have a lower sorting degree and are mixed with organic material, mainly plant residues.

AEOLIAN DEPOSITS

Aeolian sediments consist of older sediments, generally glaciofluvial sediments or beach sediments.

Aeolian sand is a very well-sorted deposit consisting of medium sand and fine sand in various quantities, mainly the coarser grains of fine sand and the finer grains of medium sand. Aeolian sand forms *dunes*.

Aeolian silt consists of fine sand and coarse silt and is found usually as thin surface layers on till in the neighbourhood of areas with aeolian sand.

Peat

Peat deposits are formed when dead and incompletely decomposed plant remains accumulate at the place of growth. As a result of the high moisture in such places, the breakdown of organic matter is incomplete. This results in the accumulation of plant residues from year to year making massive layers of varying consistency and structure. Peat deposits occur both when open areas of water become overgrown and also when previously dry land becomes waterlogged.

The main types of peat deposits are *fens* and *bogs*. They are characterized by different plant communities depending on whether the supply of water is solely by means of precipitation or by inflow from the surroundings.

A raised bog separated from solid ground by a lagg. The bog is poor in nutrients and depends on nutrients in precipitation for its growth. The fen, on the other hand, is nutrient-rich due to the supply of water from the minerogenic ground. Rönningsmossen, southwest of Molkom, Värmland.

The overburden has been sorted through frost action. Boulders and stones have moved upwards to the surface and the fine material remains deeper. In this way, shallow depressions can become completely covered by boulders. A boulder depression to the north of Idre.

Patterned ground

Palsas are low mounds of peat and ice. This type of patterned ground depends on deep frost that only thaws in the surface layer during the summer. Palsa mires are typical of permafrost areas and mainly occur in the low terrain to the east of the mountains in the northernmost part of Sweden.

When stones and boulders are pushed up out of the ground as a result of the frost, the surface layer is sorted and different types of patterned ground occur. Common forms of patterned ground are *stone pits, stone rings* and *sorted rows of stones*. These are mainly found in the mountains but also on the limestone alvars on the island of Öland.

Boulder depressions and *boulder fields* occur in depressions in the terrain. They are found as far to the south as in the South Swedish Uplands, but new formations are probably not ongoing outside the mountain chain.

Solifluction lobes are formed by mass of solifluction debris flow on mountain slopes.

In conditions of severe cold, the ground cracks into a patterned system that may give rise to *square blocks*, tundra polygons. The most pronounced square blocks are found in the coldest parts of the mountain chain but fossilized examples can be seen in cuts in southern Sweden.

Tundra with permafrost in the peat, a palsa mire. A palsa of this kind may become several metres high. Tavvavuoma, northern Lappland.

Solifluction lobes on the slope of Smuolevagge near the Norwegian border in the Arjeplog Mountains. An ice-lake terrace can be seen further up the slope.

On ground with few boulders, repeated freezing and thawing leads to the formation of earth hummocks. Njarkavare, northwest of Tjeggelvas, Lappland.

Patterned ground also occurs outside the current tundra areas, such as these fossilized tundra polygons at Vessige, in the province of Halland.

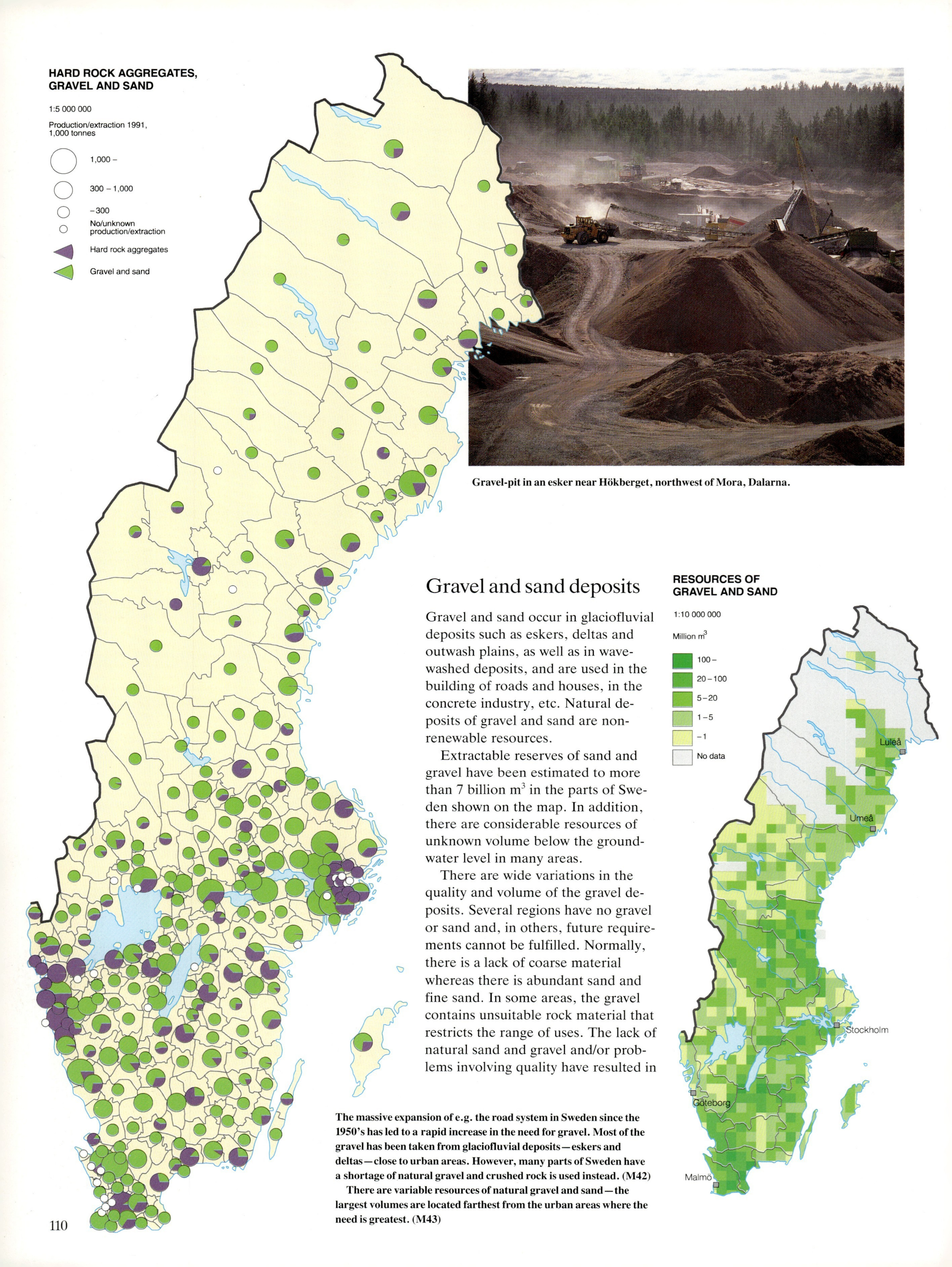

Gravel-pit in an esker near Hökberget, northwest of Mora, Dalarna.

Gravel and sand deposits

Gravel and sand occur in glaciofluvial deposits such as eskers, deltas and outwash plains, as well as in wave-washed deposits, and are used in the building of roads and houses, in the concrete industry, etc. Natural deposits of gravel and sand are non-renewable resources.

Extractable reserves of sand and gravel have been estimated to more than 7 billion m³ in the parts of Sweden shown on the map. In addition, there are considerable resources of unknown volume below the groundwater level in many areas.

There are wide variations in the quality and volume of the gravel deposits. Several regions have no gravel or sand and, in others, future requirements cannot be fulfilled. Normally, there is a lack of coarse material whereas there is abundant sand and fine sand. In some areas, the gravel contains unsuitable rock material that restricts the range of uses. The lack of natural sand and gravel and/or problems involving quality have resulted in

The massive expansion of e.g. the road system in Sweden since the 1950's has led to a rapid increase in the need for gravel. Most of the gravel has been taken from glaciofluvial deposits—eskers and deltas—close to urban areas. However, many parts of Sweden have a shortage of natural gravel and crushed rock is used instead. (M42)

There are variable resources of natural gravel and sand—the largest volumes are located farthest from the urban areas where the need is greatest. (M43)

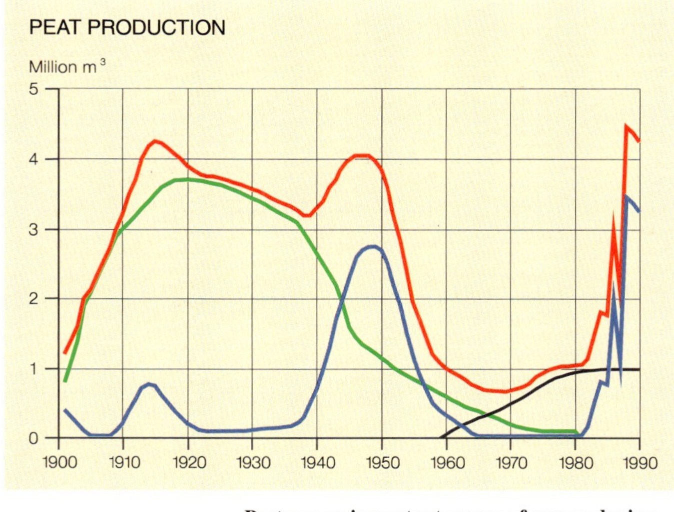

Peat was an important source of energy during both world wars. In connection with the oil replacement programme introduced during the late 1970's, there was renewed interest in obtaining energy from peat. Large amounts of peat are also used as soil conditioners, a product that is also exported.

till or crushed rock being used in many places. The availability of suitable rock is considered to be infinite.

The quarrying of hard aggregate, gravel and sand in 1991 amounted to about 92 million tonnes, which were delivered from more than 4,000 gravel pits and bedrock quarries. This corresponds to an annual consumption of about 11 tonnes/person. About 30% of the production consisted of hard aggregate.

PEAT DEPOSITS

The use of peat as fuel goes back very early in history. Torbjörn Hornklave's lay about Harald Hårfager (860–940?) mentions a man called Torv-Einar (Peat-Einar) because he used peat instead of wood as fuel. When writing about his visit to Skåne in 1749, Linnaeus tells us that *"peat is found in all dells. On the plain it is cut for burning"*. In areas without access to forests, the population had thus found an alternative to wood. Increased industrialism led to a growing need for fuel. Production of peat for burning started on a large scale during the 1890's as a result of the development in machinery. During the First and Second World Wars, the production of peat was of great importance for the fuel supply in Sweden. The energy crisis during the 1970's, with massive increases in oil prices, resulted in peat becoming an interesting alternative to oil.

The availability of raw material has in no way limited the development of the peat industry. About 15% of Sweden's land area consists of peat-land. The total peat deposits converted into energy equivalents is about 60,000 terrawatt hours (TWh). Of these, less than 10% are calculated to be extractable using conventional technology, mainly because the peat in the largest mires in the north is thin. Other limiting factors are, for instance, that the peat deposits are unevenly distributed throughout Sweden, factors related to nature preservation, chemical composition, or other quality aspects that make peat unsuitable for industrial activity.

Apart from its use as fuel, peat has been used as litter in barns and in privies, as growing substrate, soil conditioner and insulation in blind floors, for manufacturing of textiles, nappies and medical compresses, for packing and storage of food, for storage of ice, for manufacturing peat coal, activated carbon, wax, oil absorbants and filter material, etc.

It is estimated that about 15% of Sweden's land area is covered by at least 0.5 m of peat. Peat for energy purposes is processed at about 150 places (1994). Apart from peat for energy, about 50 bogs are used to produce peat for cultivation purposes, mainly in southern and central Sweden. Peat mining takes place on about 0.1% of the total peat area of Sweden. (M44, M45)

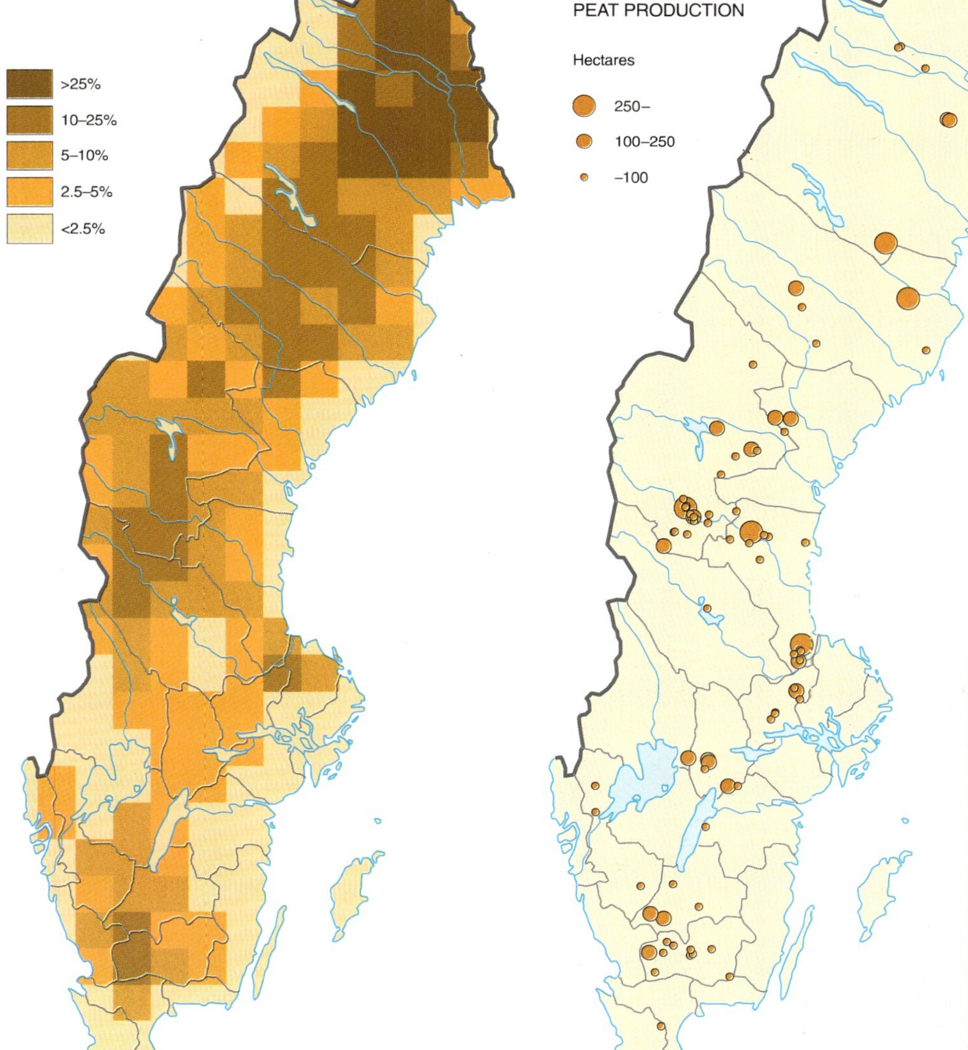

Peat slabs are today extracted mechanically on a large scale. Karinmossen, Gästrikland.

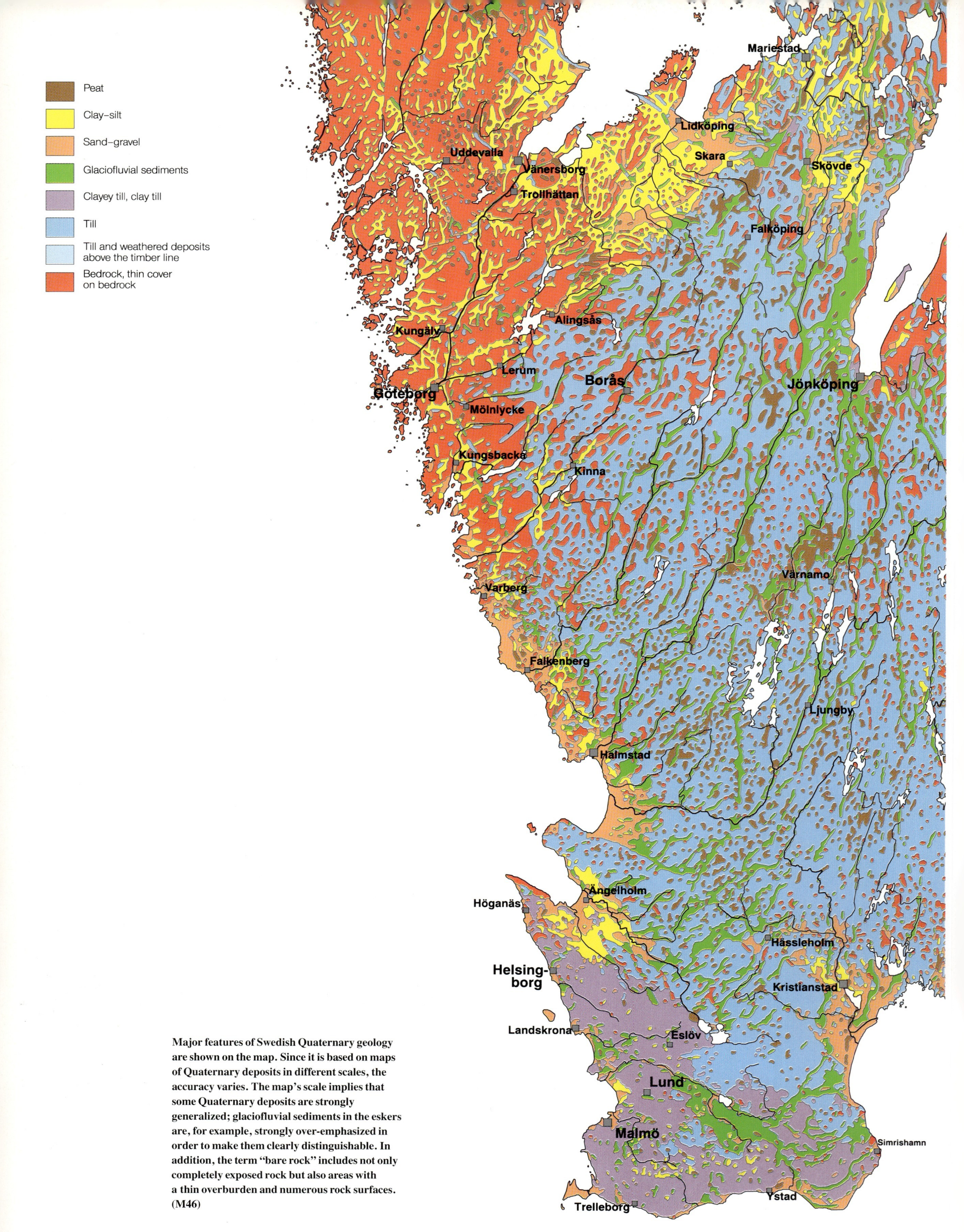

Major features of Swedish Quaternary geology are shown on the map. Since it is based on maps of Quaternary deposits in different scales, the accuracy varies. The map's scale implies that some Quaternary deposits are strongly generalized; glaciofluvial sediments in the eskers are, for example, strongly over-emphasized in order to make them clearly distinguishable. In addition, the term "bare rock" includes not only completely exposed rock but also areas with a thin overburden and numerous rock surfaces. (M46)

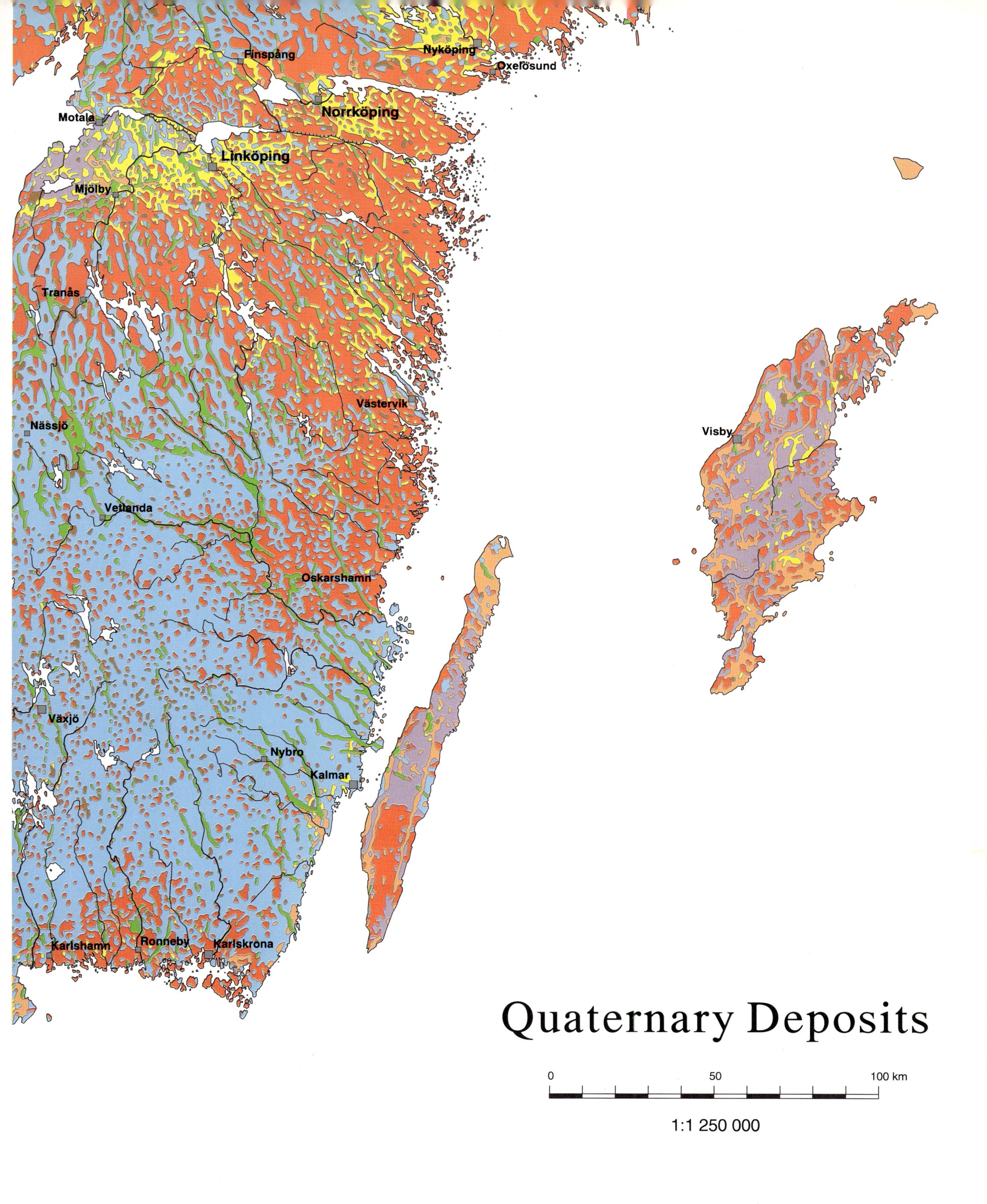

Quaternary Deposits

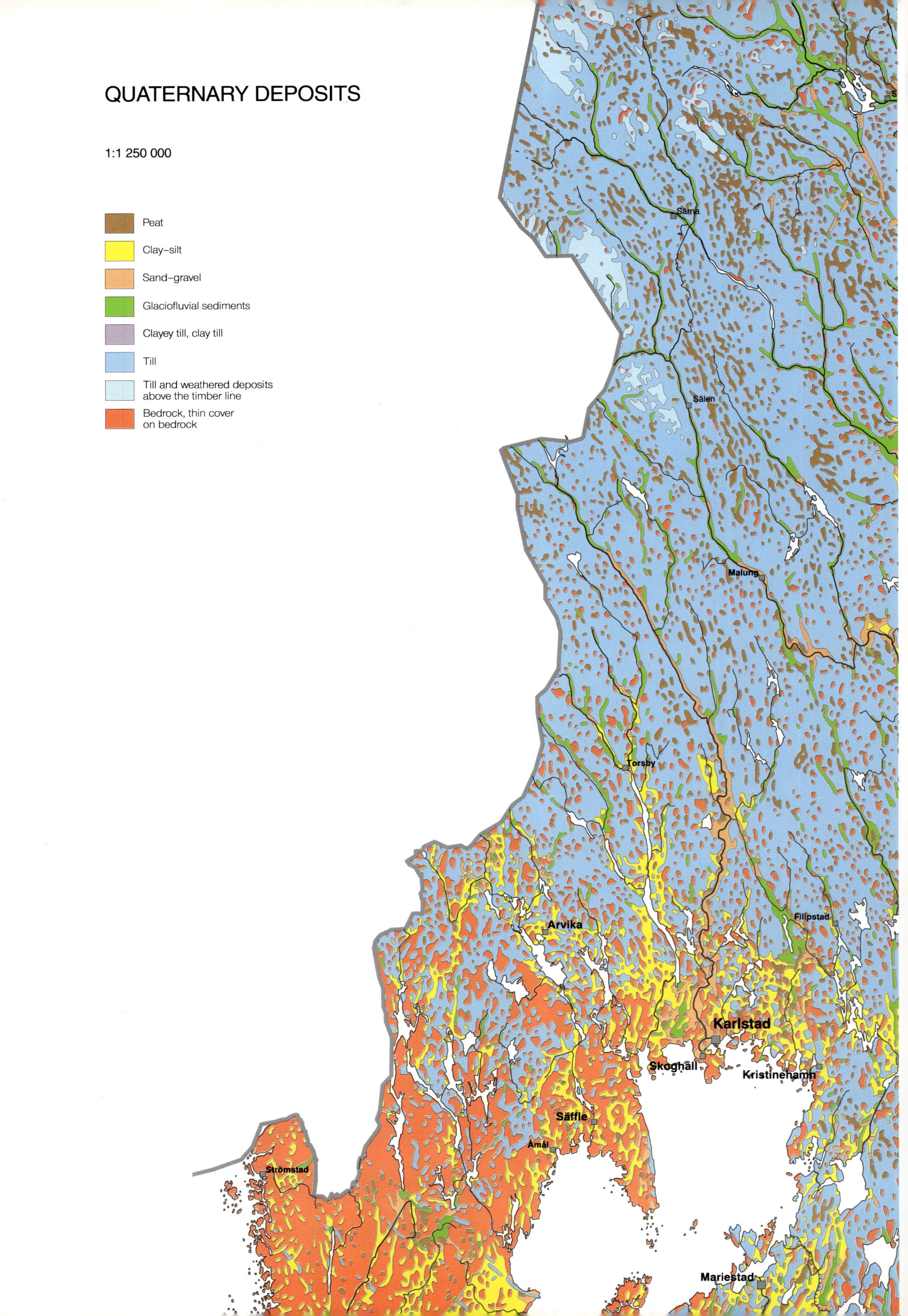
QUATERNARY DEPOSITS
1:1 250 000
Peat
Clay–silt
Sand–gravel
Glaciofluvial sediments
Clayey till, clay till
Till
Till and weathered deposits above the timber line
Bedrock, thin cover on bedrock
Särna
Sälen
Malung
Torsby
Arvika
Filipstad
Karlstad
Skoghall
Kristinehamn
Säffle
Åmål
Strömstad
Mariestad

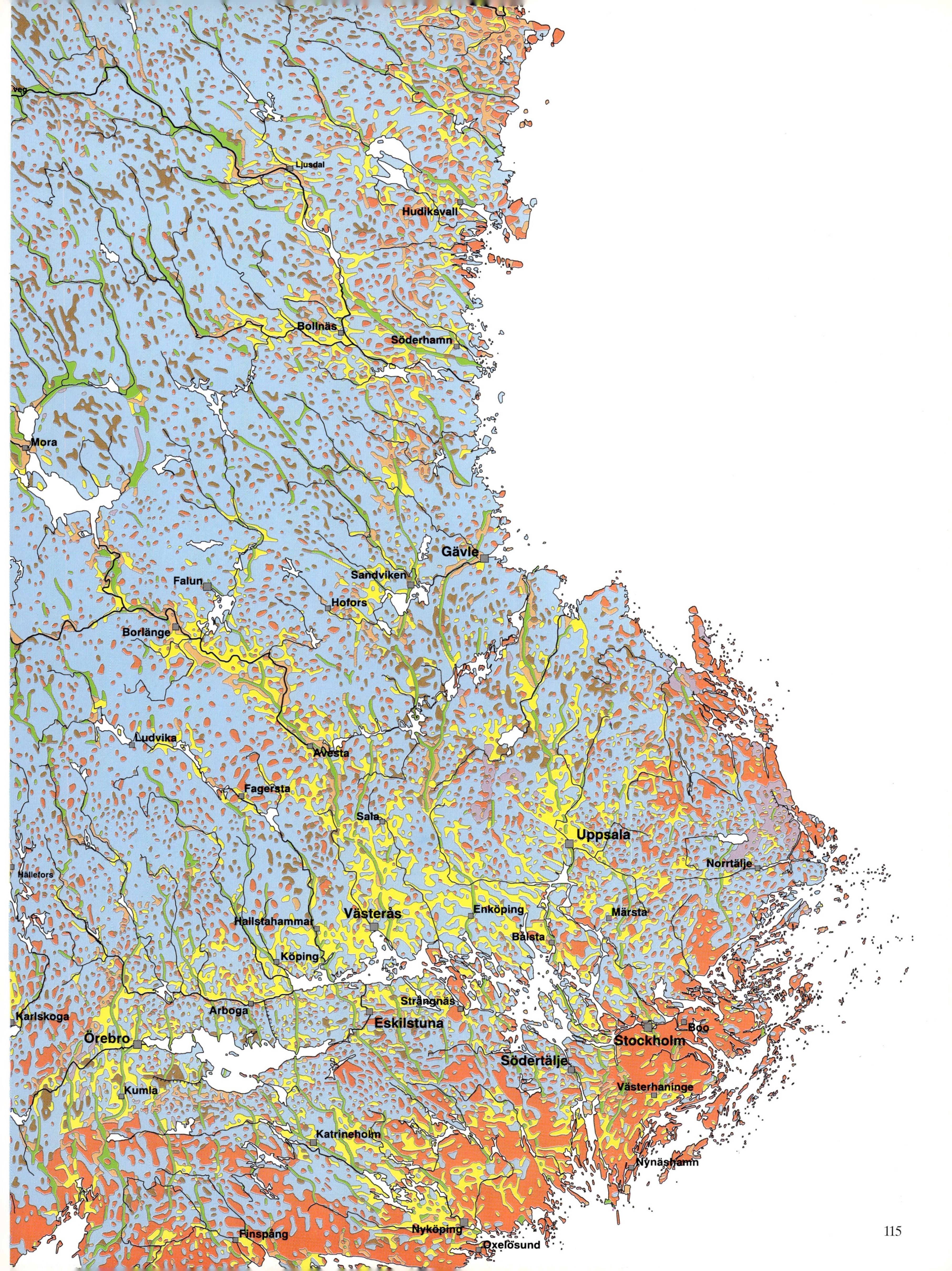
Ljusdal
Hudiksvall
Bollnäs
Söderhamn
Mora
Gävle
Sandviken
Falun
Hofors
Borlänge
Ludvika
Avesta
Fagersta
Sala
Uppsala
Norrtälje
Hällefors
Enköping
Märsta
Hallstahammar
Västerås
Bålsta
Köping
Strängnäs
Arboga
Karlskoga
Eskilstuna
Örebro
Boo
Stockholm
Södertälje
Kumla
Västerhaninge
Katrineholm
Nynäshamn
Finspång
Nyköping
Oxelösund

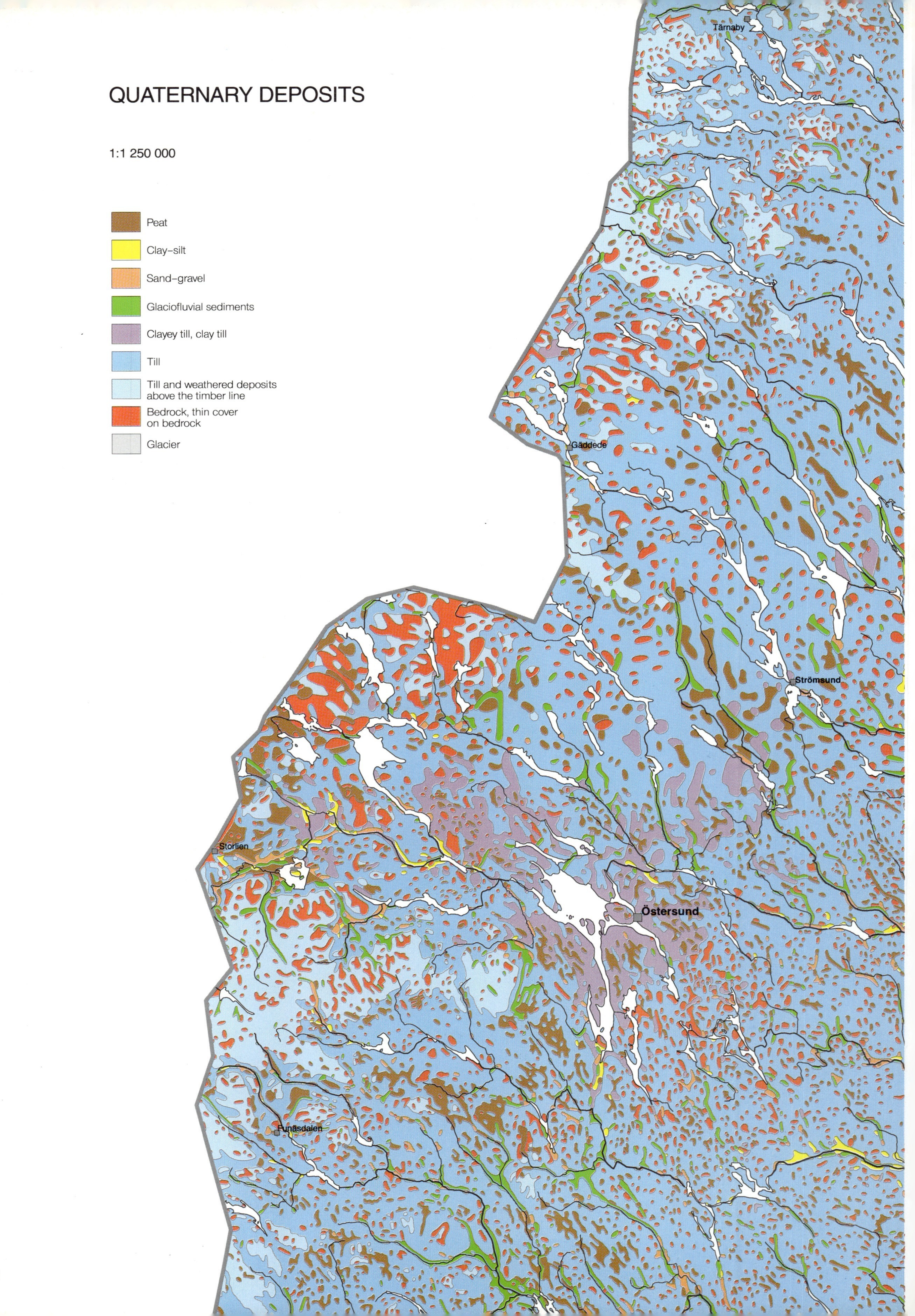
QUATERNARY DEPOSITS
1:1 250 000
Peat
Clay–silt
Sand–gravel
Glaciofluvial sediments
Clayey till, clay till
Till
Till and weathered deposits above the timber line
Bedrock, thin cover on bedrock
Glacier
Tärnaby
Gäddede
Strömsund
Storlien
Östersund
Funäsdalen

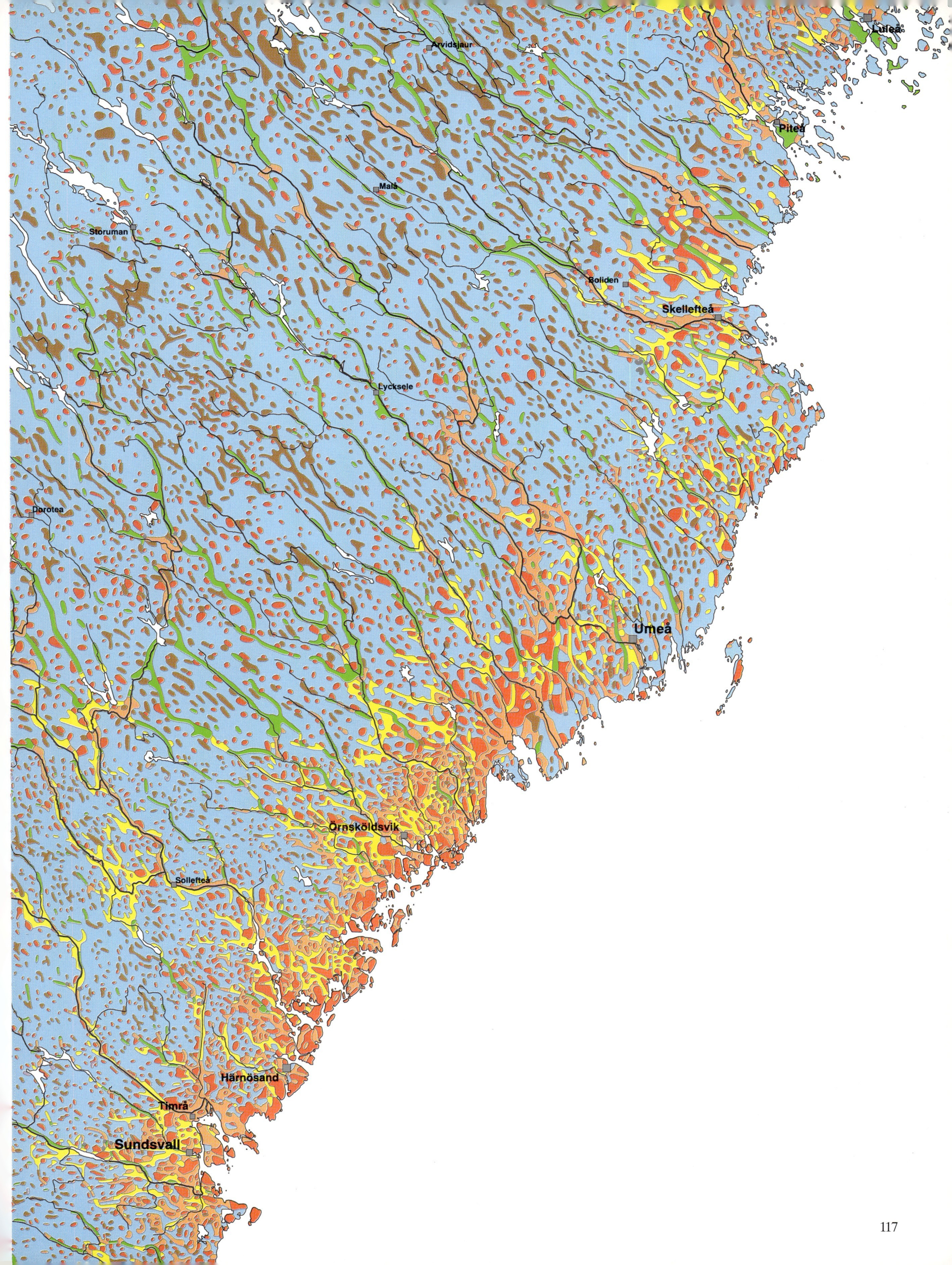
Arvidsjaur
Luleå
Piteå
Malå
Storuman
Boliden
Skellefteå
Lycksele
Dorotea
Umeå
Örnsköldsvik
Sollefteå
Härnösand
Timrå
Sundsvall

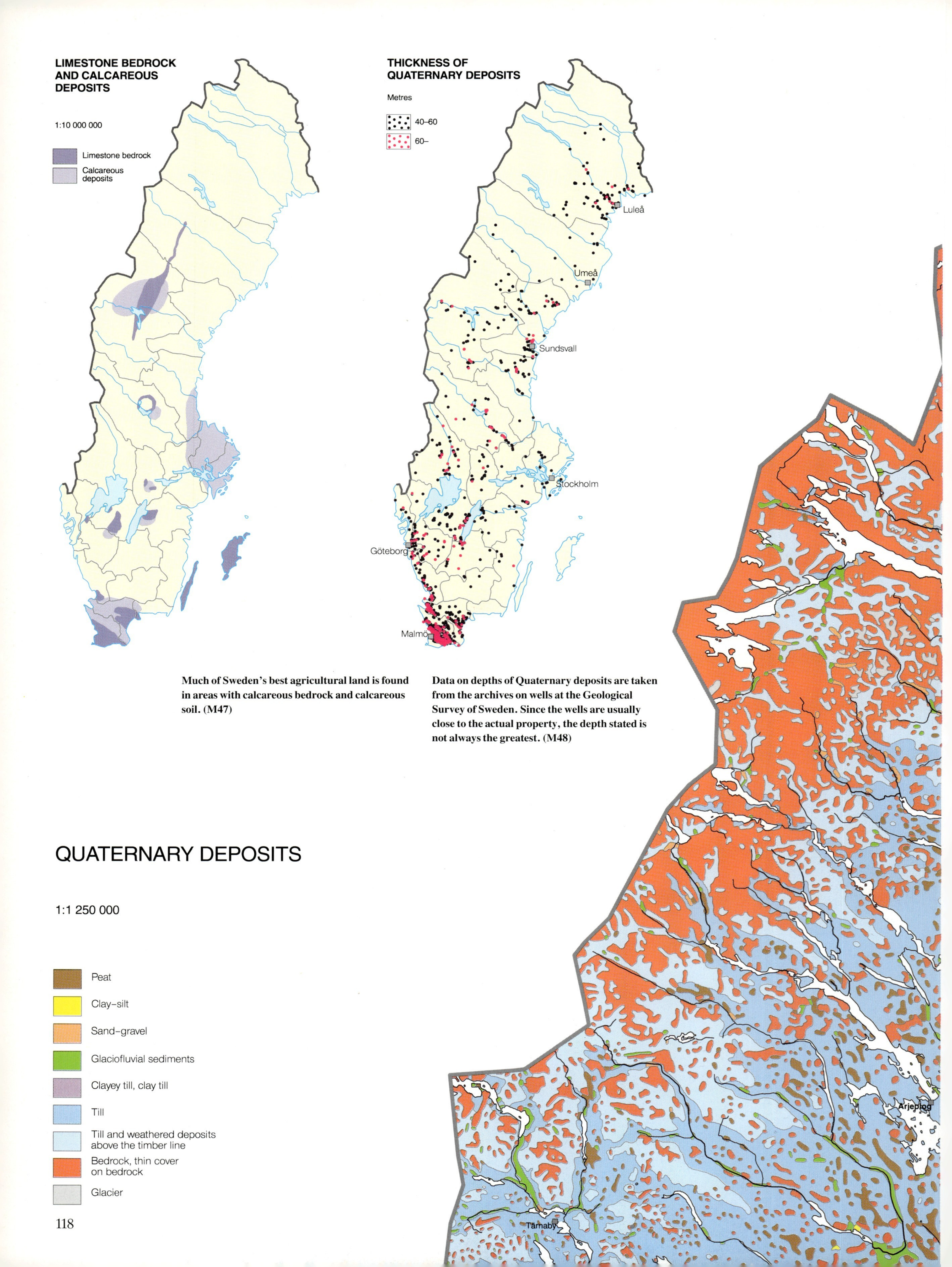

Much of Sweden's best agricultural land is found in areas with calcareous bedrock and calcareous soil. (M47)

Data on depths of Quaternary deposits are taken from the archives on wells at the Geological Survey of Sweden. Since the wells are usually close to the actual property, the depth stated is not always the greatest. (M48)

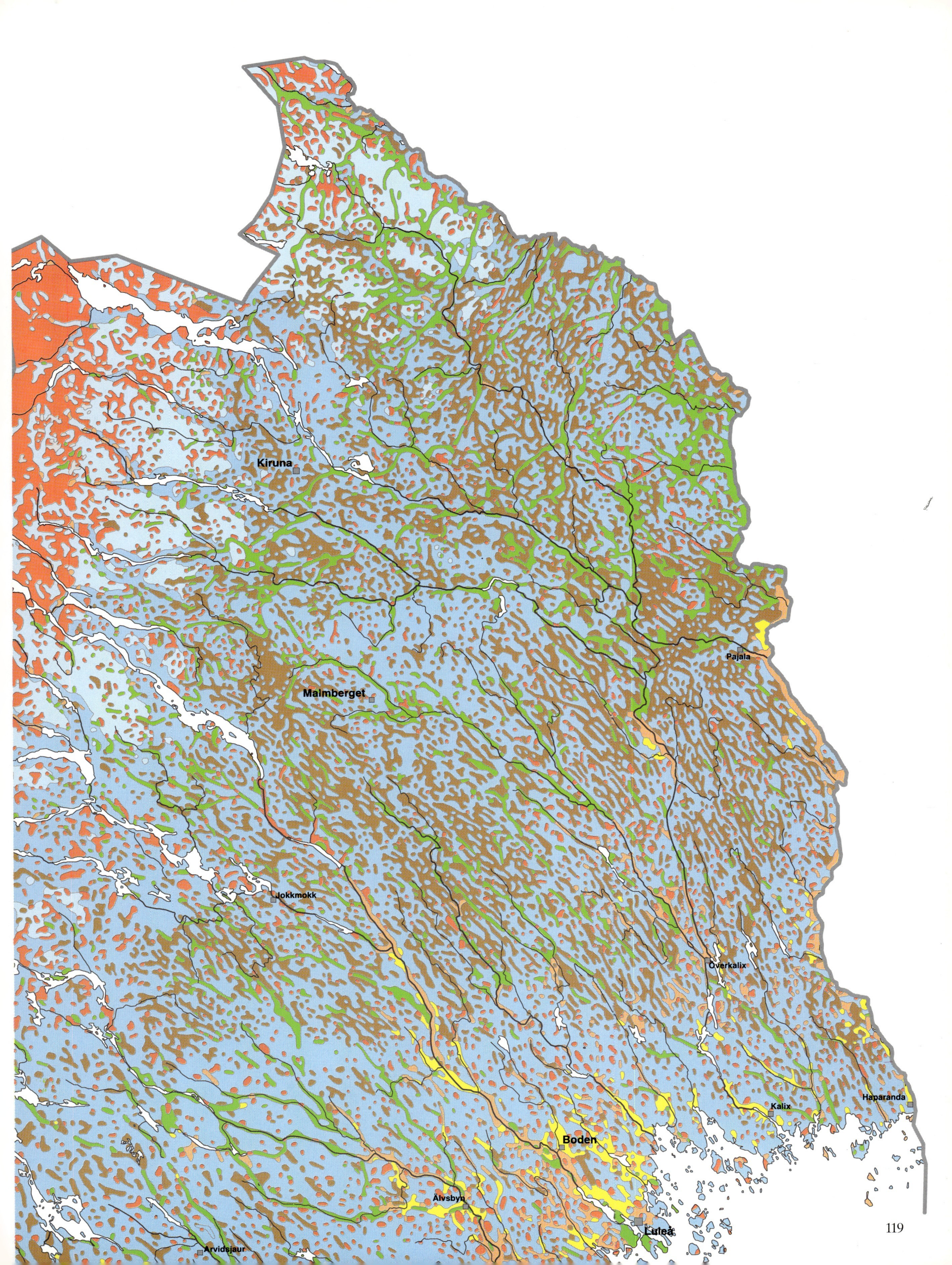
Kiruna
Malmberget
Pajala
Jokkmokk
Överkalix
Haparanda
Kalix
Boden
Älvsbyn
Luleå
Arvidsjaur

Sweden was completely covered by ice during the latest glaciation. It might have looked like this when the mountain tops started to emerge as the ice melted. Photograph from Antarctica.

Glacials and interglacials

Already during the Middle Tertiary, about 40 million years B.P., the mild climate that had prevailed until then started to become cooler. About 15–10 million years B.P. there was a marked deterioration. The temperature decreased at the same time as there were increasingly strong shifts between colder and warmer periods. This pattern intensified during the Quaternary, that started about 2.5 million years ago. Colder periods correspond to *glacials* and warmer periods to *interglacials*; the latter having an environment fairly similar to the one we experience today.

DEVELOPMENT OF THE INLAND ICE

During cold periods the glaciers expanded in the Scandinavian mountain chain. These glaciers gradually merged and extended beyond the mountain chain, leading to large ice-covered areas, the *inland ice*. The inland ice grew as a result of the precipitation falling on its upper part, the *accumulation area*, being greater than the amount that melted in the lower part and at the margin.

If the temperature is sufficiently low and the winter precipitation is sufficiently large, then ice may also be formed beyond the margin as a result of areal glaciation. Snow that falls during the winter is unable to melt during the summer and is compacted year by year until it forms glacier ice.

Depending on the way the inland ice has been formed, together with the temperature during its formation and geothermal influence, the ice may be cold-based, i.e., frozen down to the bottom, or wet-based, i.e., with a bottom that is at the pressure melting-point.

Glacial ice is plastic and flows from higher to lower levels as a result of its own weight. As it moves, material from the substratum may become frozen into the ice, e.g., at the transition from a warm to a cold zone. Since "cold" ice hinders the progress of the masses coming from behind, material that has become frozen into the ice is forced upwards and is thereby transported over long distances. During transport, carried material may scratch the rock surface below the ice, causing *glacial striation* and pressure marks. When the ice melts, the stones, gravel, sand etc., will remain as *till*, or be re-sorted by the meltwater into *glaciofluvial deposits* and other sediments.

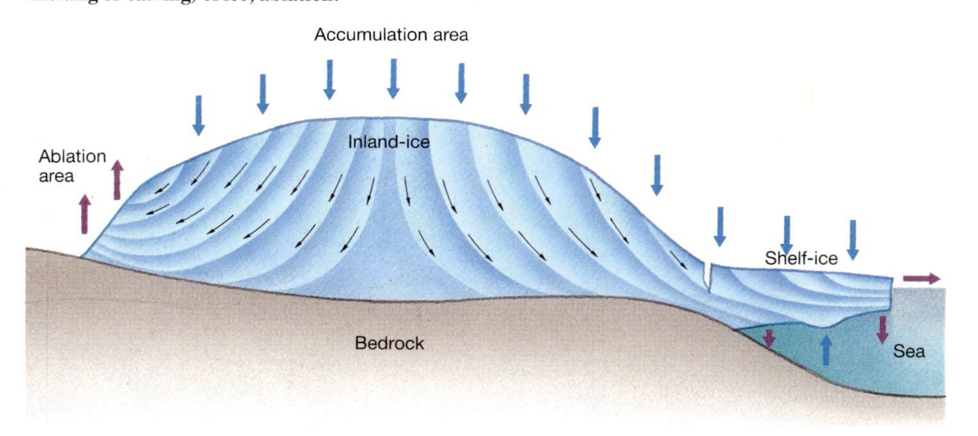

The pattern of movement in an inland ice. Blue arrows illustrate additions of snow and ice, accumulation. Red arrows illustrate removal (by melting or calving) of ice, ablation.

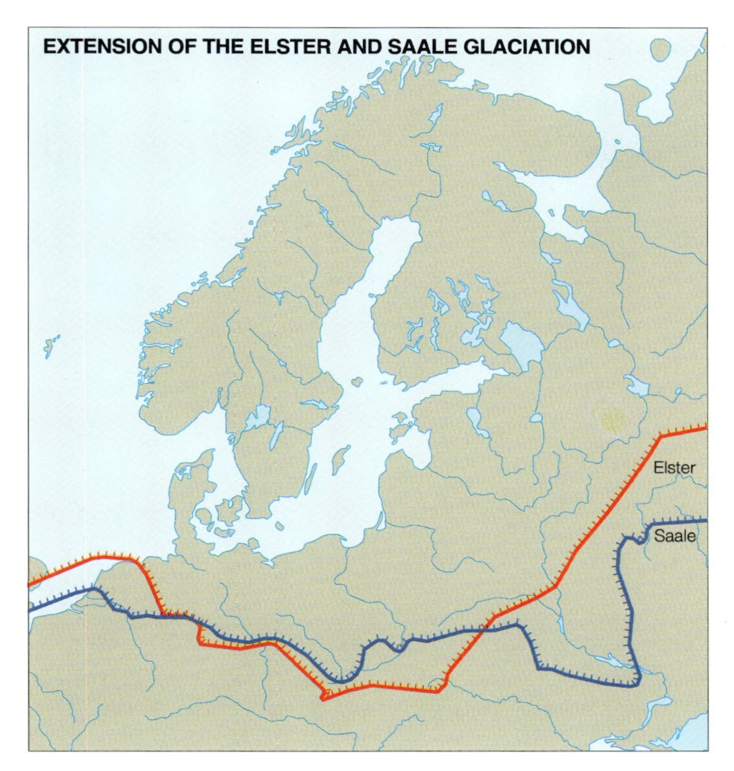

The ice sheet during, the Elsterian and Saalian glaciations, reached much farther to the south than the latest, the Weichselian.

EARLIER GLACIALS AND INTERGLACIALS

We know fairly little about the glaciations that occurred during the early Quaternary. The shapes of the mountain valleys, with deep lake basins suggest, however, that there have been several glaciations, but all traces in the form of deposits appear to have been removed.

The first indication that the inland ice extended down to the southern parts of Scandinavia is found in delta sediments in Holland. Here, there is a layer of pebbles containing rocks that must have been transported by inland ice covering the southern Baltic, before being transported west by rivers. This layer of pebbles is about 800,000 years old.

The oldest known deposits in Sweden that can be dated originate from the *Holsteinian Interglacial*, which is considered to have occurred 250,000–200,000 years ago and to have lasted about 20,000 years. Sediments from this time have been found, for example, at Öje in western Dalarna, in the Alnarp Depression in Skåne, and at Dingelvik in Dalsland. Fossils at Öje show that the forests contained coniferous trees, for example, larch, silver fir and Serbian spruce, that are not found there today. Among other trees there were numerous pines and alders but only few hardwoods such as oak, elm, lime and ash.

Podzol soils appear to have been formed at a fairly early stage of the Holsteinian Interglacial. The climate was warm and *oceanic*, i.e., moist and without particularly large differences between summer and winter. Soils formed during this period were considerably poorer than those from the two subsequent interglacials.

The till lying below the Holsteinian sediments is the oldest Quaternary formation in Sweden. It may possibly originate from the *Elsterian Glacial*, when the ice reached as far as southern Poland, but might be even older.

Left: Beech leaves from the Holsteinian Interglacial found in balls of fine sand in glaciofluvial sediment at Dingelvik, Dalsland.

Right: Interglacial gyttja and peat (the dark layer) from the Eemian Interglacial. Above this, till (light grey) from the latest glaciation has been deposited. Photograph from 1974. Stenberget, Skåne.

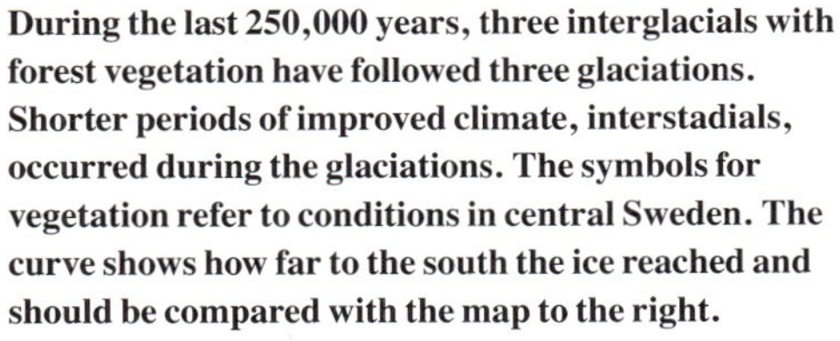

During the last 250,000 years, three interglacials with forest vegetation have followed three glaciations. Shorter periods of improved climate, interstadials, occurred during the glaciations. The symbols for vegetation refer to conditions in central Sweden. The curve shows how far to the south the ice reached and should be compared with the map to the right.

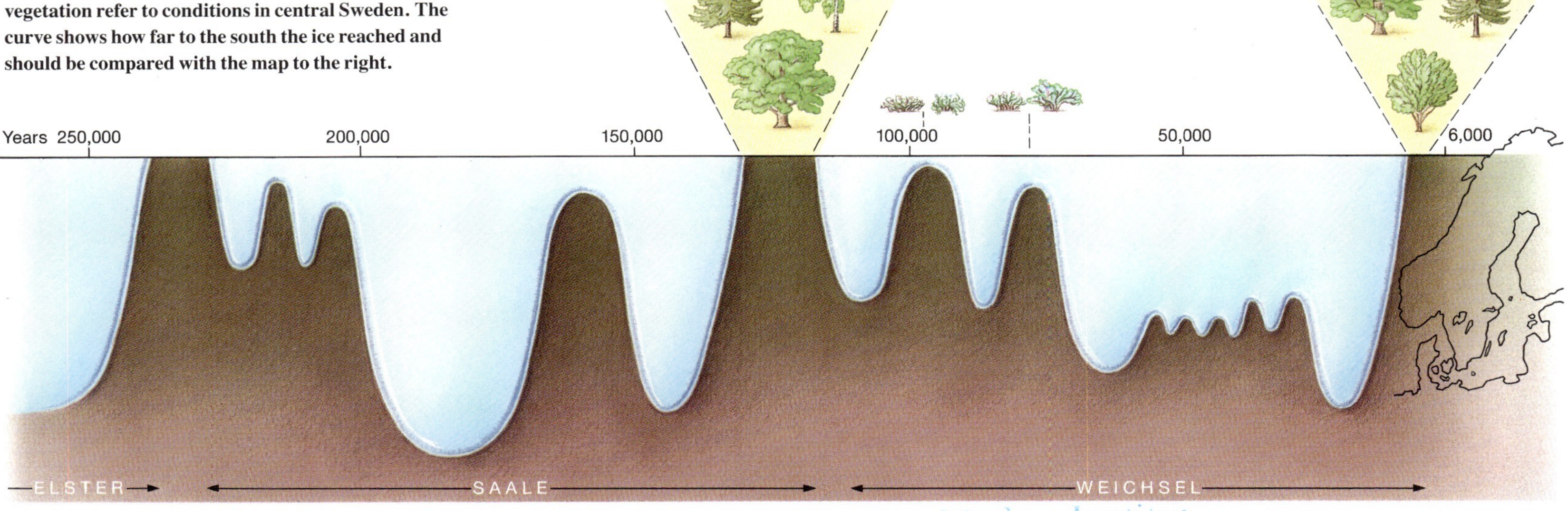

During the cold parts of the Eemian Interglacial and during the Weichselian Interstadials, the mammoth was common in the Nordic countries. This mammoth tooth was found in the lower part of a fossiliferous layer of sand at Pilgrimstad. The tooth has an impressive upper crown, 20 x 8 cm.

Interstadial strata at Pilgrimstad, Jämtland. A dark organic layer can be seen in the sand. The till lying above it is from the last phase of the glaciation and only about 0.5 m thick. Photograph from 1967.

THE EEMIAN INTERGLACIAL

The next interglacial, the *Eemian*, took place about 130,000–115,000 years ago and lasted about 11,000 years. Eemian deposits are known from several widely-spread places. Below Eemian sediments in Skåne and at Bollnäs in Hälsingland, there is till that might originate from the *Saalian* glaciation or even earlier.

The climate during the Eemian was periodically milder than it had been since the last glacial, with mean temperatures during the summers that were 1–2° higher than today's and an annual mean temperature that was probably 3–4° higher. In northern Sweden, there was a more continental climate than today, i.e., the summers were warmer and the winters colder. Peat from Svappavaara in Lappland from this period provides evidence of vegetation that has never subsequently reached so far to the north. Sweden's forests contained the same species of trees then as today, with the addition of beech. A clear zonation can be distinguished in the development of forest history. The tree species entered Sweden in the following order: Birch—pine—hardwoods (oak, lime, elm)—hazel—alder—hornbeam—spruce/larch.

At Bollnäs, the Eemian strata contain diatoms that are only found in brackish water, which suggests that the sea here was about 100 m higher than it is today. The extent of the Eemian sea can be better measured in Finland, whereas in Sweden we have only found its sediments at one single place.

THE WEICHSELIAN GLACIATION

We know more about the development in Sweden during the latest glacial, the *Weichselian*. Colder phases, *stadials*, were interrupted by milder *interstadials*, but never being as warm as an interglacial.

Several relics from these older phases can be found. The climatic development, however, is best known from strata outside the glaciated area, e.g., from the oceanic sediment. The ratio between oxygen isotopes ^{18}O and ^{16}O, determined in shells of foraminifers indicates the amount of water that was bound in glacier ice over the entire Earth. A surplus of the heavier isotope corresponds to a large volume of ice, and vice versa.

Changes in the oxygen isotope ratio in the strata of marine sediment suggest that interglacial environment prevailed 130,000–75,000 years ago. However, this period was not consistently warm at our latitudes and only the oldest part corresponds to the Eemian Interglacial.

During the interstadials, the ice margin retreated to northernmost Sweden but probably the mountain chain and neighbouring areas still remained covered by ice. The vegetation that entered deglaciated areas consisted of non-arboreal flora. Periodically, birch was the only tree species. Remnants of this vegetation have been found throughout the province of Norrland. Occurrence of

The Veiki moraine in the foreground is part of the Lainio arch and is older than the latest glaciation. Drumlins outside the Lainio arch can be seen in the background towards the east. North of Kangosfors, Norrbotten.

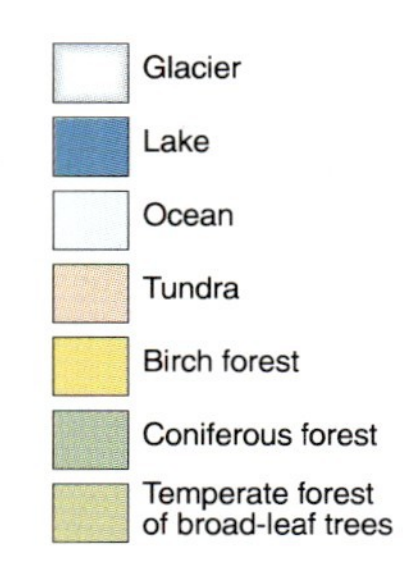

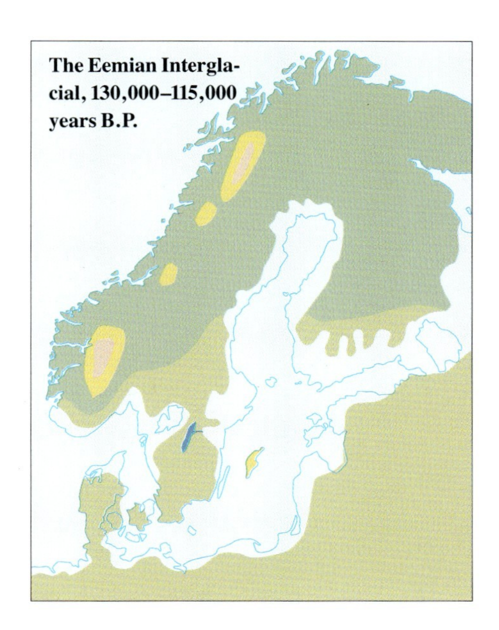

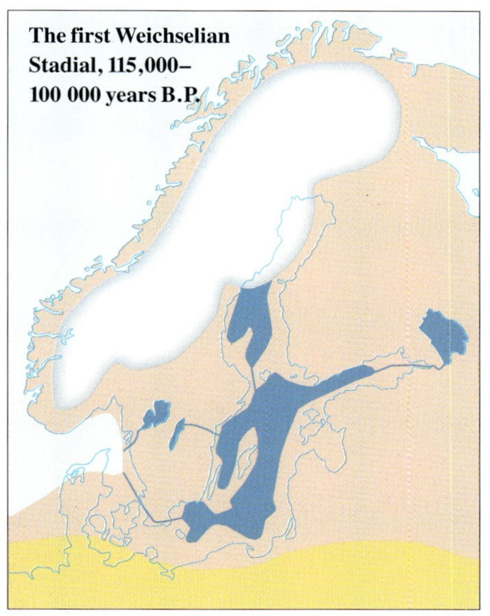

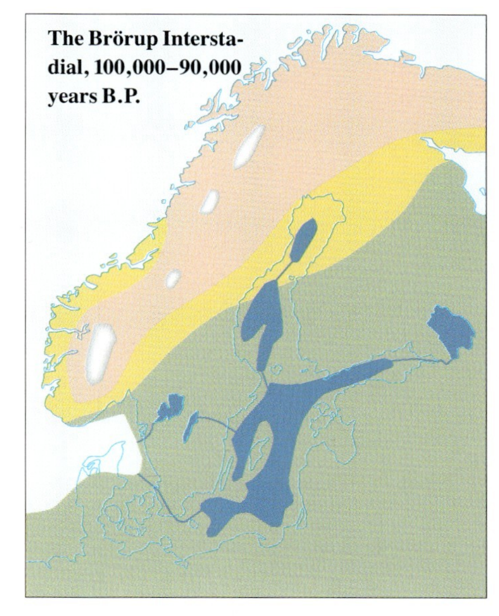

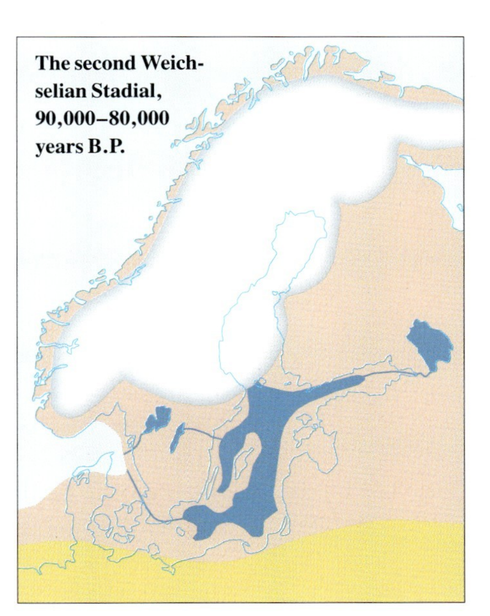

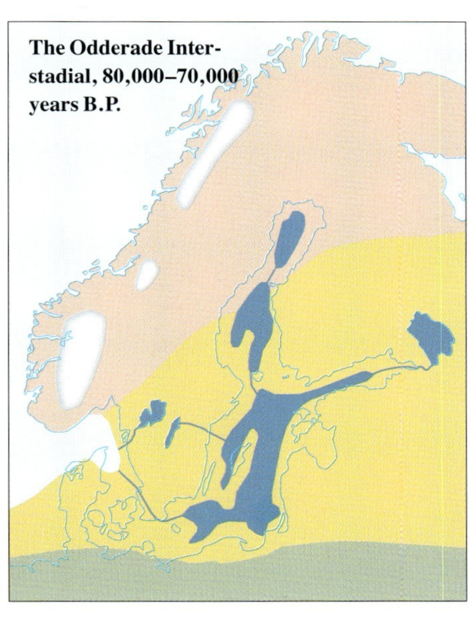

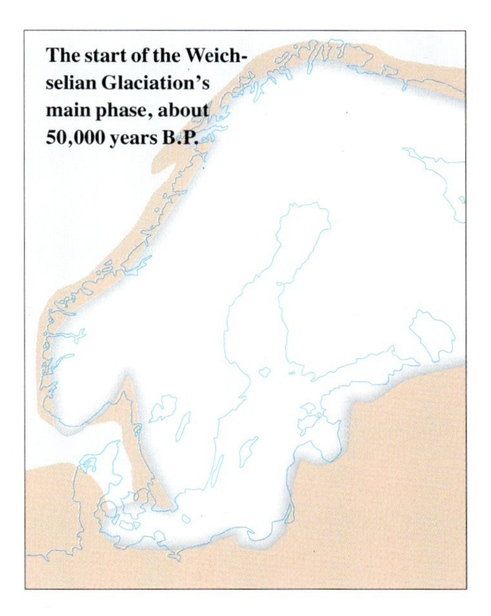

Extent of the ice cover, the sea, and vegetational zones during certain important phases since the Eemian Interglacial. The maps are partly hypothetical owing to the lack of sufficient information relating to most stages. After the situation shown on the final map, the inland ice reached its maximum position. (M49)

steppe species suggests a very dry and cold climate similar to that found today in parts of Greenland and on Svalbard.

Thus, from a global viewpoint, the entire period from 130,000–75,000 B.P. was warm. However, at Swedish latitudes the glacial phase is considered to have started already 115,000 years B.P. but was interrupted by interstadials about 100,000 and 80,000 years ago. Subsequently, there was a period during which the entire globe was colder that lasted until the end of the glacial phase about 10,000 years B.P. and that was only interrupted by moderately warmer climate about 50,000–30,000 years ago.

The oldest Weichselian Glaciation

The first Weichselian glaciation probably started in the mountain chain from where "temperate" glaciers extended out over the plains. The ice had a great effect on the surface of the land. Valleys and other major land forms had been created under comparable conditions during repeated glaciations but on this occasion, many of the types of moraine, e.g., drumlins, today found in Lappland, were formed. We do not know how far the inland ice extended and there are no reliable traces of it outside Norrland and northernmost Finland. Its margin might possibly be indicated by the thicker cover of till that characterizes the province of Norrland.

The Brörup Interstadial

The interstadial that took place about 100,000 years ago is called the *Brörup* on the continent, whereas in central Sweden we speak of the *Jämtland Interstadial* and in Lappland the Finnish name *Peräpohjola* is used. At that time the climate, when warmest, was about 3–4° cooler than today. Coniferous forest possibly extended up to southern Norrland, whereas tundra and mountain heath dominated northern Norrland. In between these two areas there was a wide zone of birch forest.

The second Weichselian Glaciation

We know very little about the glaciation that followed the Brörup Interstadial. The presence of characteristic dark clayey till suggests, hypothetically, that this ice reached the area around Lake Mälaren in the south.

The Odderade Interstadial

About 80,000 years ago the inland ice covering most of Sweden receded. This interstadial, called *Odderade* on the continent and *Tärendö* in Lappland, was very cold. Sparsely vegetated polar steppe or tundra extended far to the south in Sweden and the climate was characterized by periods of severe cold. Permafrost and solifluction mantle of arctic type were common. Outcrops and boulders were

(M50)

polished by snow blown by strong winds. This suggests that the temperature in Norrland could fall to between –40 and –50°C, when snow crystals become as hard as feldspar.

The main phase of Weichselian Glaciation

The cold climate during the Odderade Interstadial resulted in the subsequent inland ice of 70,000 B.P., being very cold. To some extent, it emerged as a result of areal glaciation. It was frozen onto the underlying surface and in some areas erosion was negligible or non-existent. Consequently, large amounts of older sediment and moraines are preserved in Norrland.

Moraines in southeastern Denmark and Poland indicate that the ice extended down to these areas soon after the Odderade Interstadial. The Baltic Basin and the whole of Scandinavia, apart from parts of Skåne, appear to have been covered by the inland ice until the end of the glacial period. Interstadial deposits in Skåne and along the west coast of Sweden and Norway demonstrate, however, that there were periodical retreats of the ice margin. However, it was much colder than during earlier interstadials and the withdrawal of the ice margin was certainly of minor, and only of local, importance.

From this position, the ice made a fairly brief advance to its southernmost position, where the margin 20,000–18,000 years B.P., was at Berlin in the south and passing through Jutland. All of Sweden was covered by ice.

Both interglacials, such as the one we are living in today, and extensive inland ices, are relatively extreme conditions. The "normal" situation in Sweden is instead the presence of minor icecaps or a glaciated mountain chain surrounded by extensive areas of tundra and birch forests.

The deglaciation

The margin of the most recent ice sheet was still at its southernmost position about 18,000 years ago. Then a marked improvement in climate took place and the ice started to melt, a process that was completed after about 10,000 years. Most of the Swedish Quaternary deposits and minor terrain configurations are a result of this deglaciation.

As long as the ice was at its largest,

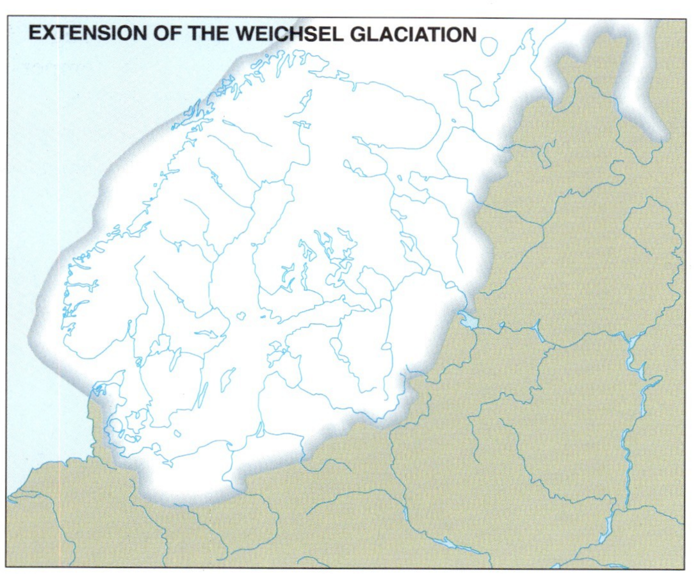

The maximum extent of the latest inland ice was around 20,000 years ago. There are different opinions as to its extent in the Arctic Ocean. (M51)

the mass moved outwards from a central area, an *ice divide* probably located in what today is the Gulf of Bothnia. As the melting of the ice sheet proceeded, the ice divide moved westward towards the mountain chain and finally broke up into several smaller divides. These different patterns of movement can be traced in the glacial striation on the outcrops and in some of the details of the tills.

MECHANICS OF INLAND ICE DOWNWASTING

The melting of the ice took place both from the surface and from the margin. The ice cover became thinner and the margin became displaced towards the final areas of the ice divides. From the Atlantic, the ice margin retreated over the mountains towards the alpine area in southern Norway and eastwards towards Sweden. From Denmark and the continent in the south, it retreated northwards through Sweden, the Baltic and Balticum–Finland. Farthest in the north, the ice margin receded southwards from Barents Sea and to the west from northern Finland.

As long as accumulation took place in the central areas, or where the ice sheet was sufficiently thick, the movement at the margin was retained. Consequently, the retreating ice margin could maintain activity. If movement ceased as a result of the lack of accumulation, the ice melted away as *dead ice*. This mainly occurred in the lee of high-altitude areas or in places where the ice was excessively loaded by morainic material.

The heavy inland ice pressed down the Earth's crust at least 800 m below its position today. Consequently, the sea was much higher than at present in relative terms, despite the fact that such large amounts of water were bound in the ice that the level of water in the ocean was at least 100 m lower than today. As soon as the pressure started to lighten, the crust started slowly to elevate. The highest situated traces of the shoreline (*the highest shoreline*) are at different altitudes throughout Sweden, depending on how far the crust had been depressed, how much the local sea surface had been elevated, and the time at which the area had become ice-free. The highest shoreline is at 285 m above sea level at Skuleskogen on the coast of Ångermanland. Furthest to the south, the highest shoreline coincides almost with the present shoreline.

In places where the ice margin was in the sea or in large lakes, it was broken by the effect of waves and the lifting power of the water at the same time as it melted. The margin was usually a tall cliff from which icebergs broke off, so-called *calving*. Residues of dead ice can be found where the water was shallow. Calving was strongest around the mouths of glacial rivers and in the lowest parts of the terrain. In such places, bays were formed in the ice margin, calving bays. Surface melting dominated in places where the ice was grounded on land. There the margin had a gently sloping shape and residues of dead ice could be formed.

The downwasting of an inland ice above the highest shoreline took place mainly from the surface. At the ice margin, glaciofluvial sediments and till are deposited. Increasing amounts of till are revealed on the surface of the ice. Vatnajökull, Iceland.

Where the ice margin of the inland ice was in the sea or in large lakes, it was usually in the form of a cliff. The recession of the ice took place by calving. The photo shows Harald Moltke Brae on Greenland. The ice margin here is similar to the parts of the ice sheet in Sweden that were below the highest shoreline.

ICE-DAMMED LAKES

In places where the margin of the ice retreated from higher to lower terrain, glacial lakes could become dammed between the ice and the open terrain. This particularly occurred in the southern and northernmost parts of the mountain chain, where the ice margin retreated from the mountains and eastwards towards the inland.

Perhaps the most important example is the Baltic Basin itself. At this time, the sea level was low and the Öresund Strait and the Danish Belts were dry. The Baltic Ice Lake covered the southern Baltic until the ice margin retreated from Billingen in Västergötland, when a link was opened up with the ocean and the ice lake was drained to sea level.

The formation of a glacial lake requires that the ice causing the damming effect is either frozen down to the bottom or is active so that a counter-pressure to the dammed water is maintained. Otherwise the water will soon find its way under and through the ice.

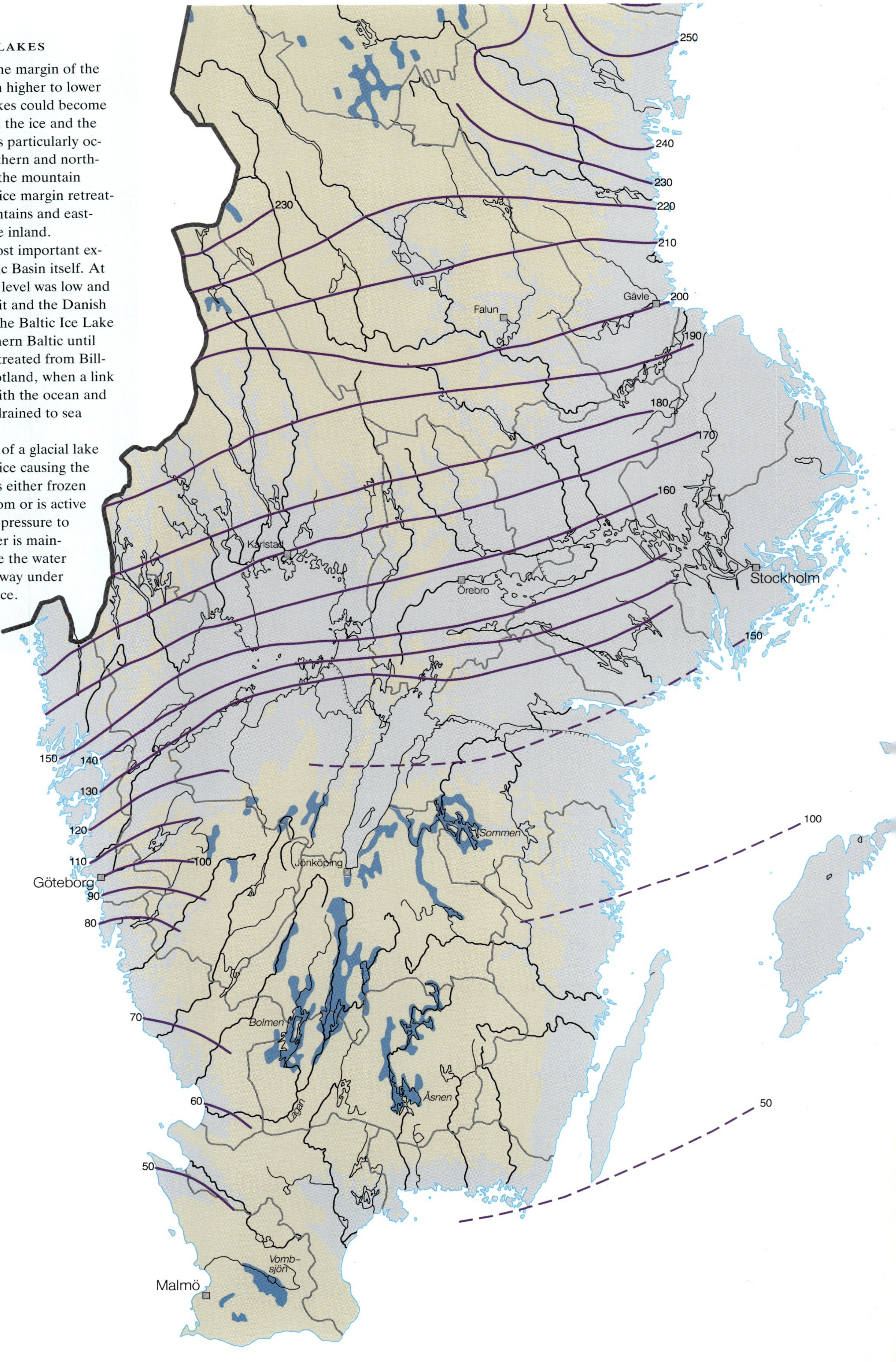

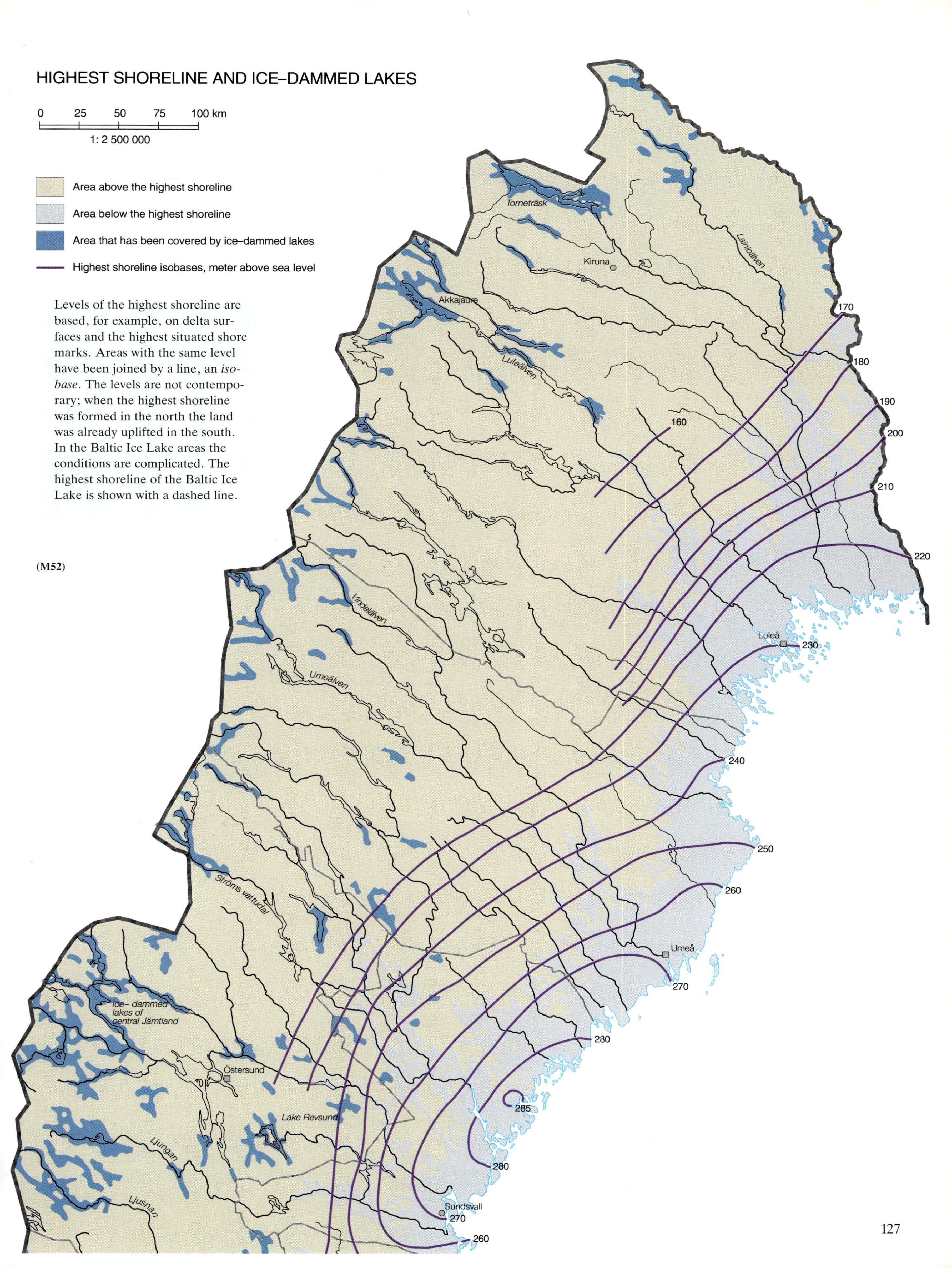

Levels of the highest shoreline are based, for example, on delta surfaces and the highest situated shore marks. Areas with the same level have been joined by a line, an *isobase*. The levels are not contemporary; when the highest shoreline was formed in the north the land was already uplifted in the south. In the Baltic Ice Lake areas the conditions are complicated. The highest shoreline of the Baltic Ice Lake is shown with a dashed line.

(M52)

Hindens Udde in Lake Vänern, is a distinct end-moraine in the Fennoscandian Ice-Marginal Zone.

De Geer moraines show a distinct contrast with the surrounding ploughed land. North of Götene, northern Västergötland.

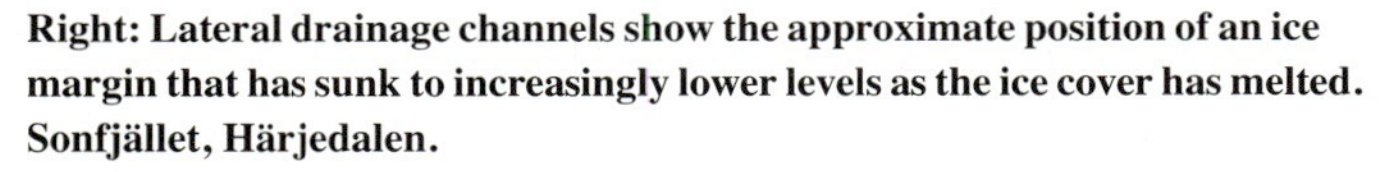

Right: Lateral drainage channels show the approximate position of an ice margin that has sunk to increasingly lower levels as the ice cover has melted. Sonfjället, Härjedalen.

Roche moutonnée with glacial striae — the ice has moved from the right towards the left.

RECONSTRUCTION OF THE DEGLACIATION

By studying different glacial geological formations, it has been possible to reconstruct the recession of the ice margin.

Endmoraines are the most reliable indication of a former ice margin but these occur in Sweden only at the west coast and in the *Fennoscandian Ice-Marginal Zone*. *De Geer moraines* are considered to have been formed within the ice margin, mainly lying parallel to it and thus provide an indication of its direction. De Geer moraines occur in an east-west zone extending from the Stockholm area across Sweden to southern Värmland. They are also found in the coastal region of Norrbotten.

Glacial striae are formed at right angles to the ice margin and the youngest striae thus provide an indication of its direction. However, there is uncertainty here since we cannot always know whether we have found the youngest system of glacial striation which might have been weathered away or even never formed.

Glacial clay with annual varves was deposited from the ice margin and outwards. Consequently, the lowermost varve must have been deposited close to the ice front at that time. By correlating series of varves from different places with each other, it is possible to identify places where the ice margin was located at the same time. In this way, we can reconstruct the positions of the margin in places that were covered by early stages of the Baltic Sea. In glacial lake areas and on the west coast of Sweden, the varves are either too irregular or too indistinct for this method to be applicable.

Varved clay and end moraines are not found above the highest shoreline. Here, however, we can reconstruct the main features of the deglaciation by means of the surface shapes of the moraines and traces of how the meltwater was drained. This particularly applies to *lateral drainage channels* that formed along the previous ice margin. Ice lakes and traces of their drainage system are also useful. On the basis of these morphological features, we can interpret areas as being free of ice, while adjacent areas were still ice-covered.

DATING OF THE DEGLACIATION

Ice margins can be dated in different ways using the methods below.

The *clay-varve method*, based on

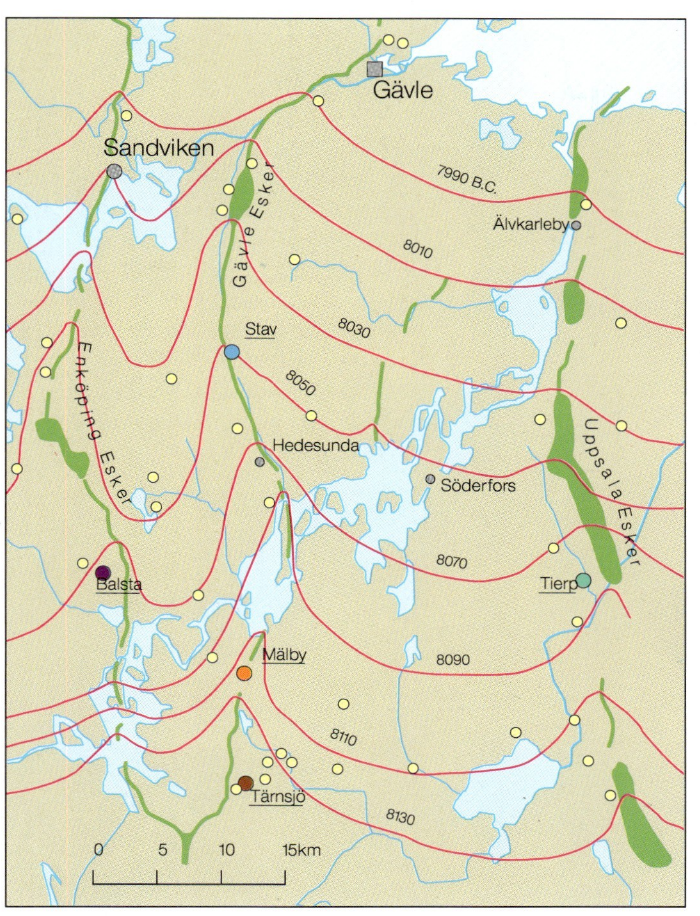

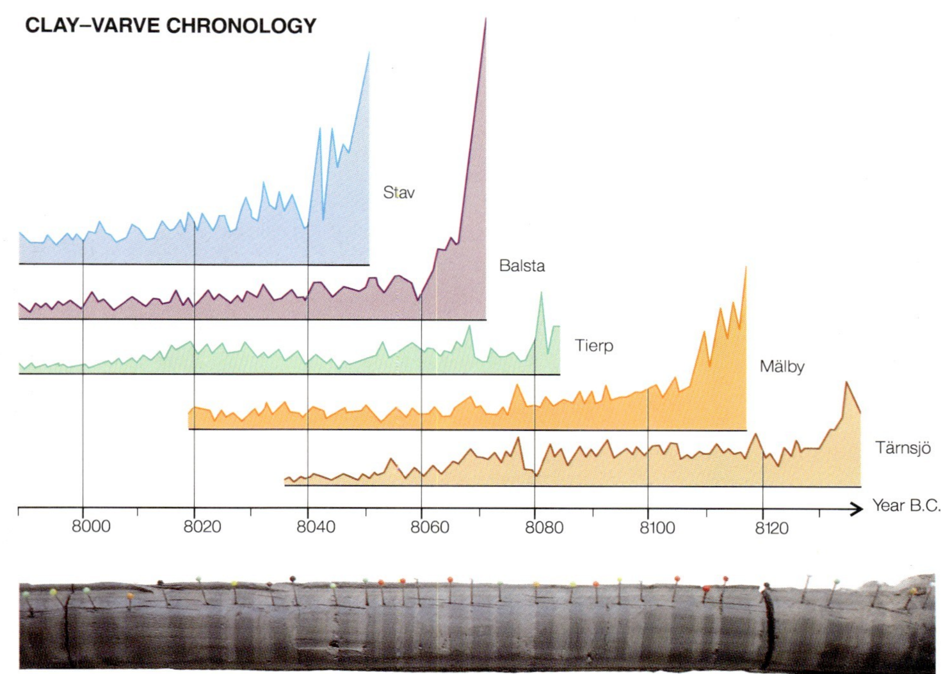

Clay-varve chronology is based on the fact that each layer, *varve,* in the clay deposited by the meltwater of the inland ice represents sedimentation during one year. Varve thicknesses illustrate characteristic changes and can be illustrated by graphs. Graphs from different places can be correlated with each other and thus it is possible to measure the time for the movement of the ice margin from one place to another.

By means of this method it is possible to establish a time scale for the deglaciation. The positions of the ice margin shown on the map are based on a large number of measuring points (coloured dots). Five of the measurements are shown in the diagramme above. The lowermost varve (on the right in each diagramme) has been deposited closest to the ice margin. (M53)

There are several methods of dating the development during the last 100,000 years, particularly the last 15,000 years. The diagramme shows the range of three of the most commonly used dating methods.

During the warm period approximately 6,000–7,000 years ago, the tree limit in the mountains was much higher than it is today. The slow decomposition in the cold mountain climate means that more than 6,000-year-old pine stumps are found today above the present tree limit.

the annual varves of the glacial clay, is in itself a dating method. Since the sediments deposited today off the mouths of the rivers in Norrland also have annual varves, the chronical time scale of the varves can be linked to the present. The entire deglaciation from Blekinge in the south to Västerbotten in the north has today been dated in calendar years. However, in the Fennoscandian Ice-Marginal Zone there are problems with the correlation between varve series.

The *radiocarbon method* is based on organic material containing a small amount of the radioactive isotope ^{14}C. As long as an organism is alive, this amount remains in equilibrium with the atmospheric content of ^{14}C. When the organism dies, it no longer receives carbon and the radioactivity declines. By measuring the amount of the isotope remaining, the age of organic material can be calculated. Organic layers that can be linked with the position of a certain ice margin, e.g., the bottom of lake sediment, can thus be used to date ice margin lines. However, it has been found that time scales based upon different methods at different times may diverge quite considerably. Thus, when looking at a period 10,000 years ago, the dating scales based on ^{14}C and clay varves diverge by several hundred years.

Fossils of different kinds, both mollusc shells and micro-fossils such as pollen and foraminifera, can be used for relative dating of the ice margin. Absolute dating requires that the levels are dated by means of other methods, e.g., ^{14}C.

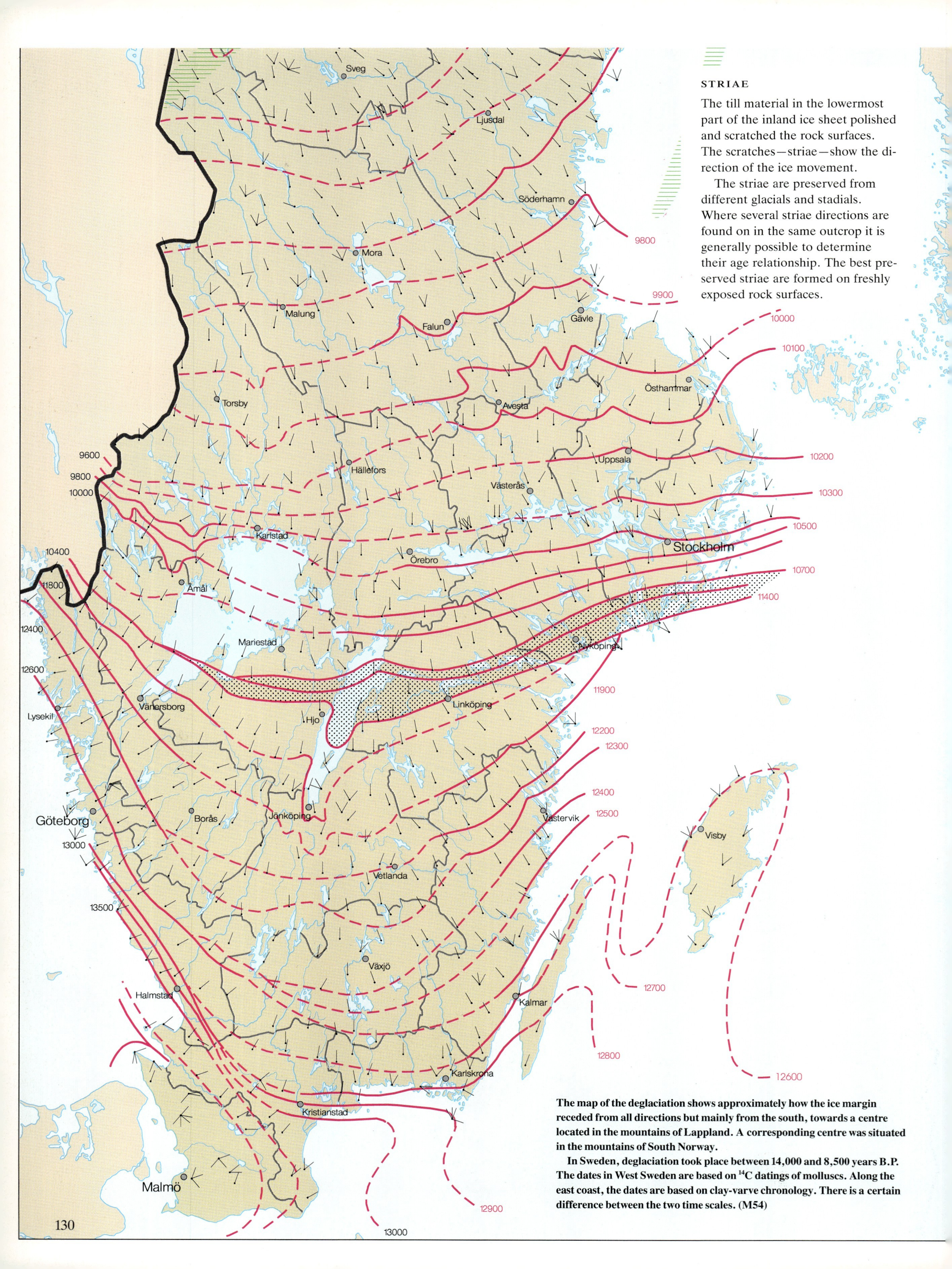

STRIAE

The till material in the lowermost part of the inland ice sheet polished and scratched the rock surfaces. The scratches—striae—show the direction of the ice movement.

The striae are preserved from different glacials and stadials. Where several striae directions are found on in the same outcrop it is generally possible to determine their age relationship. The best preserved striae are formed on freshly exposed rock surfaces.

The map of the deglaciation shows approximately how the ice margin receded from all directions but mainly from the south, towards a centre located in the mountains of Lappland. A corresponding centre was situated in the mountains of South Norway.

In Sweden, deglaciation took place between 14,000 and 8,500 years B.P. The dates in West Sweden are based on ^{14}C datings of molluscs. Along the east coast, the dates are based on clay-varve chronology. There is a certain difference between the two time scales. (M54)

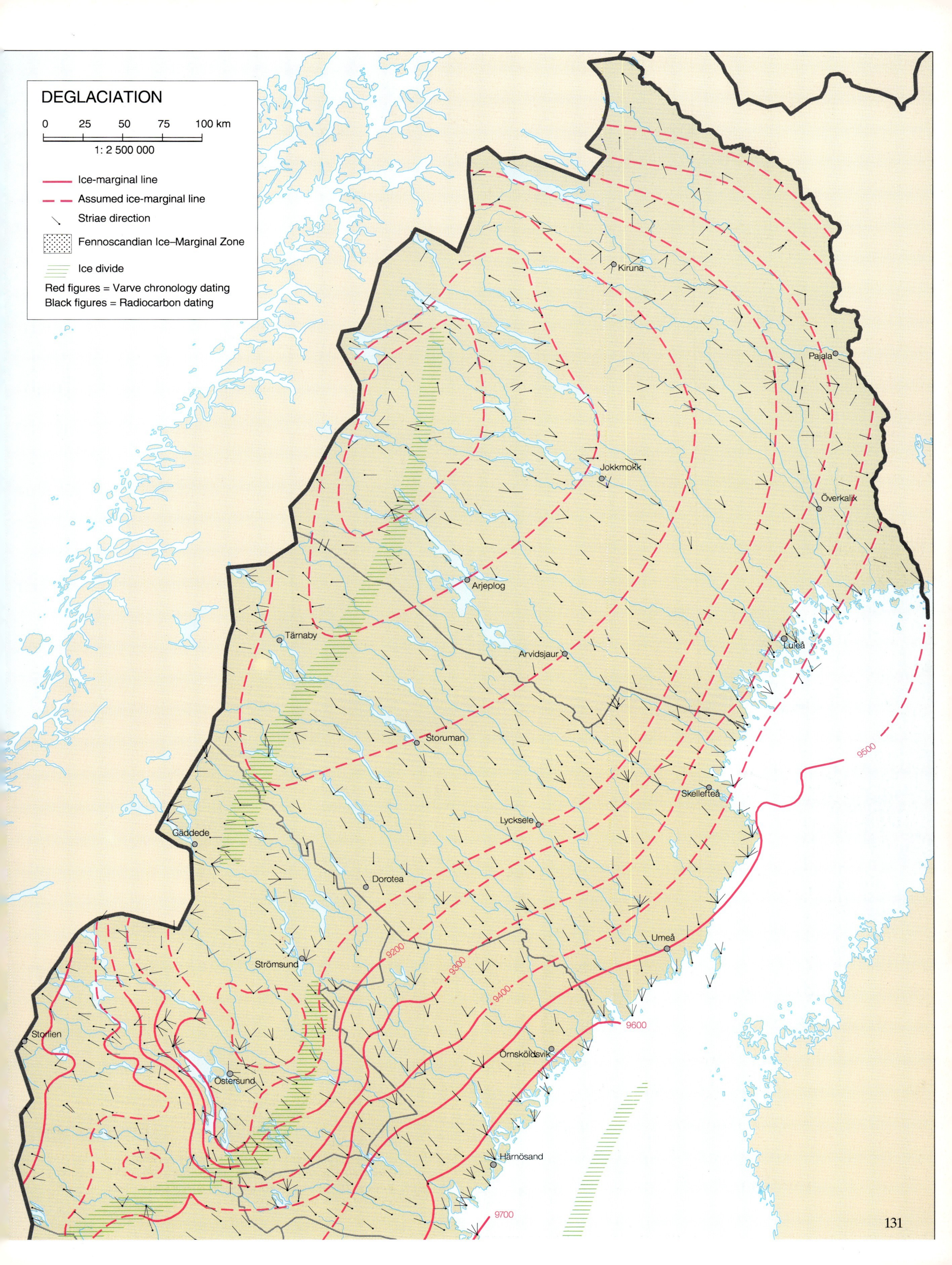
DEGLACIATION
0 25 50 75 100 km
1: 2 500 000
Ice-marginal line
Assumed ice-marginal line
Striae direction
Fennoscandian Ice–Marginal Zone
Ice divide
Red figures = Varve chronology dating
Black figures = Radiocarbon dating
Kiruna
Pajala
Jokkmokk
Överkalix
Arjeplog
Tärnaby
Arvidsjaur
Luleå
Storuman
Skellefteå
9500
Lycksele
Gäddede
Dorotea
Umeå
9200
9300
Strömsund
9400
9600
Storlien
Örnsköldsvik
Östersund
Härnösand
9700

This delta at Akkajaure was deposited in a prehistoric ice lake.

Drumlins are a type of moraine ridges running parallel to the movement of the ice. The ones illustrated, situated to the south of Tärendö, are 1–2 km long, a couple of hundred metres wide, and 10–20 m high.

The alignment of eskers is useful for approximate reconstruction of the recession of the ice sheet. An esker in the River Umeälven — only the crest of the esker emerges above the water.

Similarly, *paleomagnetism* can be used for dating sequences of sediment strata, e.g., varved clay. Measurements are then made of the magnetism preserved in mineral grains, which is determined by the geomagnetic conditions prevailing at the time of sediment deposition.

DEPOSITS FROM THE DEGLACIATION

When the ice sheet melted, the material frozen in the ice was deposited either directly as tills or as glaciofluvial deposits by the meltwater, as well as sediments in the sea and lakes. Properties of all these deposits depend on the types of rock the ice has passed, the period during which they were transported, together with the manner in which they were transported. Their landforms may provide us with information about the deglaciation.

End moraines could be deposited along the ice margin, where it was situated in water. *Drumlins* were formed beneath the ice and show the direction in which the ice was moving. Irregular *hummocky moraine* was formed when dead ice melted. *Eskers* were formed in meltwater channels beneath the ice. *Deltas* and *outwash plains* were formed close to the ice in places where the transporting ability of the meltwater decreased.

NEW OPINIONS ABOUT THE ICE AGE

During recent years we have had to revise our interpretation of the importance of the formations. Many surface formations, for example, the *Veiki moraine* and other hummocky moraines in Norrland, have been found to be formed by the downwasting of earlier ice sheets. These forms have not been influenced, or only negligibly, by the most recent glaciation. The same applies to many of the drumlins in upper Norrland, particularly the larger ones. Other types of hummocky moraines, mainly the *Rogen moraine*, were earlier considered to be ablation moraines. However, the latest glacier moved above these landscapes and merely re-shaped the remnants of former glacations.

Many eskers and other glaciofluvial deposits in Norrland also originate from the downwasting of older ice sheets. Remarkably fragile formations such as narrow hogback eskers, have been entirely unaffected by the most recent glaciation, which can only be explained by the ice being cold-based and that the movement took place higher up in the mass. A contributory reason may sometimes be that there was negligible movement down towards the bottom of the ice in the region of the ice divides.

The importance of moraines has also been re-evaluated during recent years. The presence of ablation moraine shows that movement in the ice ceased during deglaciation, but it requires that material was transported either up in, or on, the ice. In turn, this requires activity in the ice out towards the margin. Admittedly, movement had ceased by then but this had taken place successively in a fairly narrow zone at the ice margin.

Drumlin formations have been interpreted to indicate activity in the ice. However, the lack of ablation moraine that covers the drumlins may indicate that movement of the ice ceased when the drumlins emerged from the melting ice.

REGIONAL DEGLACIATION

The very first part of Sweden to become deglaciated about 14,000 years ago was the southern coast of the Bjäre Peninsula. Here, a crevasse had occurred in the ice margin between a front that was retreating towards the northeast and a tongue of ice that was melting away towards the south around the Öresund Strait. According to one theory, this tongue of ice extended from the Baltic Depression around southern Skåne and up through the Strait and the Danish Belts. This Low Baltic ice was even considered to have advanced about 13,500 years ago. Between the Low Baltic ice and the ice margin retreating to the northeast, Skåne became successively deglaciated with damming of shallow ice lakes. Another theory suggests that there was no tongue of ice and that a Low Baltic advance never took place. Instead, the ice was considered to have formed a shallow dome to the south of Skåne from where it moved outwards in all directions. The retreat across Skåne had the character of dead-ice deglaciation that proceeded from west to east. An advance took place when the ice margin was in the Simrishamn region, probably about 13,300 years ago.

The ice then rapidly broke up in the Baltic Basin. When the ice had become sufficiently thin, it started to float and disintegrated by *calving*. As a result, the inland ice remained in a curved front across southern Sweden, in the east retreating to the north and northwest. In the west, the ice margin remained stationary for long periods and formed massive end moraines. Across the South Swedish uplands, the ice retreated step-wise with rapid deglaciation and the cutting-off of dead ice during warmer periods and a slow retreat during colder periods. Differentiations have been made between the mild *Bölling* and *Alleröd* times and the *Older Dryas* time that separated them, but climatic variations were probably more complicated than that.

During the cold part of the *Younger Dryas*, 11,000–10,500 years ago, the ice margin was stationary or advanced in the west. At this time, the *Fennoscandian Ice-Marginal Zone* was formed, characterized by end moraines and large deltas, e.g., at Dals Ed and Ödskölts moar in Dalsland. Already during the Alleröd time, the ice margin had probably retreated to the north of Billingen, which enabled the *Baltic Ice Lake* to drain and become linked up with the ocean. As a result of the ice margin again advancing because of much colder climate, the Baltic Ice Lake again became dammed. When the climate became milder, the ice margin retreated and 10,300–10,200 years ago the ice lake was finally drained down to sea level.

From the Fennoscandian Ice-Marginal Zone, the margin retreated steadily across Sweden to the north. In the Bothnian Basin, it initially formed a wide dome which became increasingly separated from the ice over the land mass. When the ice in the Bothnian Sea had become sufficiently thin it rapidly broke up. The ice covering the land thus opened onto the sea, from where it melted towards the northwest. Further to the north, the margin retreated to the west from Finnish Lappland and towards the south from Barents Sea.

From the Atlantic in the west, there was a simultaneous displacement of another ice margin towards the east over the mountains. In due course, the inland ice divided into two parts, one over the south Norwegian mountains and the other across northern Sweden. The latter increasingly melted towards a centre that was probably located over the mountains around Sarek and Sulitelma. Owing to the intricate interaction between different ice margins, the deglaciation was extremely complicated, particularly in Jämtland and Härjedalen. Large residues of dead ice became isolated, for example, in the Storsjö Basin.

In many places, large *ice-dammed lakes* were formed as a result of the retreat to the east of the ice margin from the mountains, at least in the south and in the north. In the area in between, the remnants of the ice were closer to the mountains, and corresponding valleys were largely filled with ice.

Today we have little possibility to date the deglaciation in interior parts of Norrland. However, it appears to have proceeded very rapidly under the influence of an increasingly mild climate 9,000–8,000 years ago, being at least as warm as at present, and the ice sheets soon completely collapsed. It was far too mild for the formation of local glaciers, possibly with the exception of places in the highest mountains. The glaciers we have today are, thus, not residues of the inland ice, but have developed much later under the influence of a cooling climate.

The pattern of the movement of the ice over Skåne according to two models. Left: The classical picture of ice streams. Right: A model suggesting that a local cupola had formed on the inland ice, close to its margin. From this cupola the ice moved in all directions. (M55)

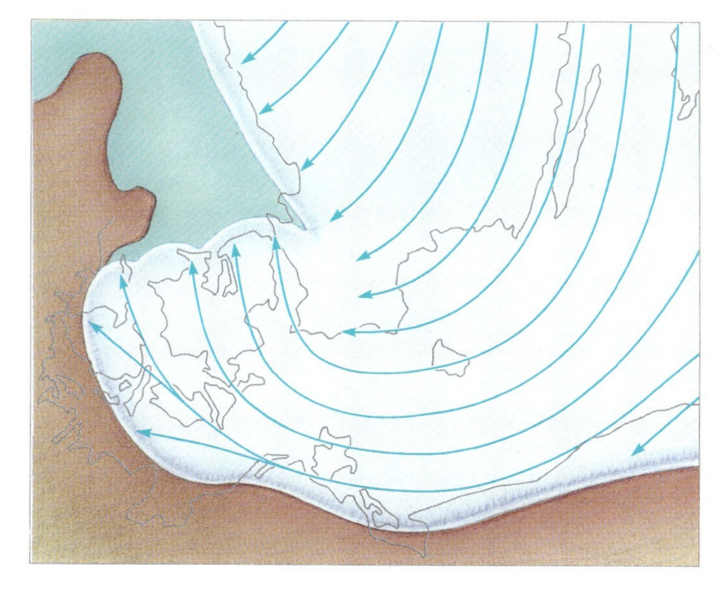

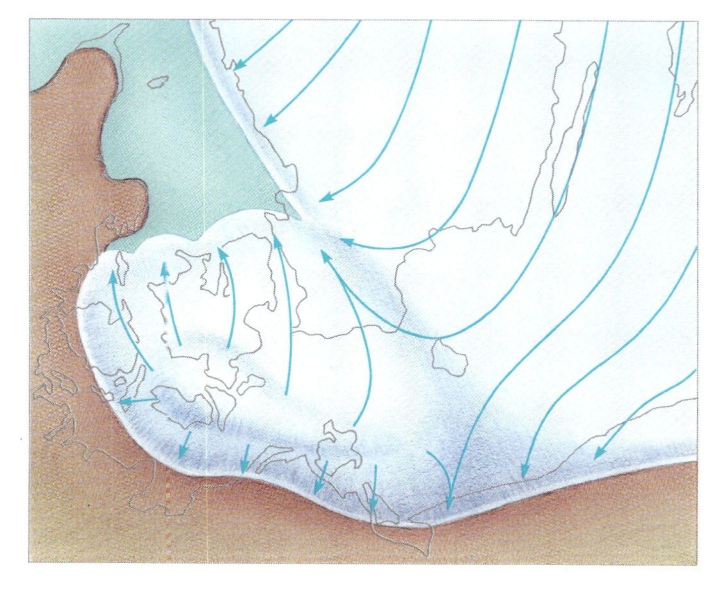

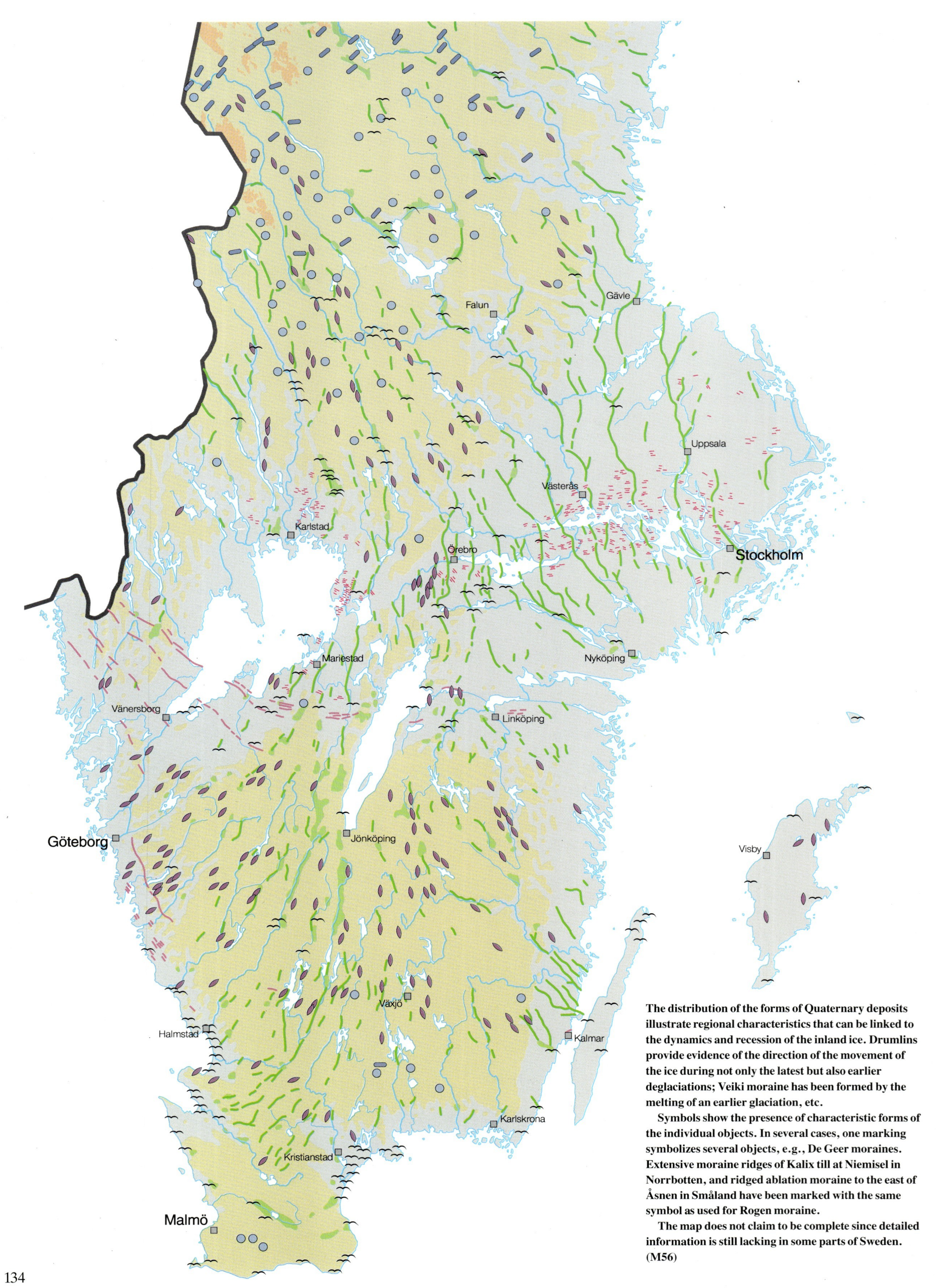

The distribution of the forms of Quaternary deposits illustrate regional characteristics that can be linked to the dynamics and recession of the inland ice. Drumlins provide evidence of the direction of the movement of the ice during not only the latest but also earlier deglaciations; Veiki moraine has been formed by the melting of an earlier glaciation, etc.

Symbols show the presence of characteristic forms of the individual objects. In several cases, one marking symbolizes several objects, e.g., De Geer moraines. Extensive moraine ridges of Kalix till at Niemisel in Norrbotten, and ridged ablation moraine to the east of Åsnen in Småland have been marked with the same symbol as used for Rogen moraine.

The map does not claim to be complete since detailed information is still lacking in some parts of Sweden. (M56)

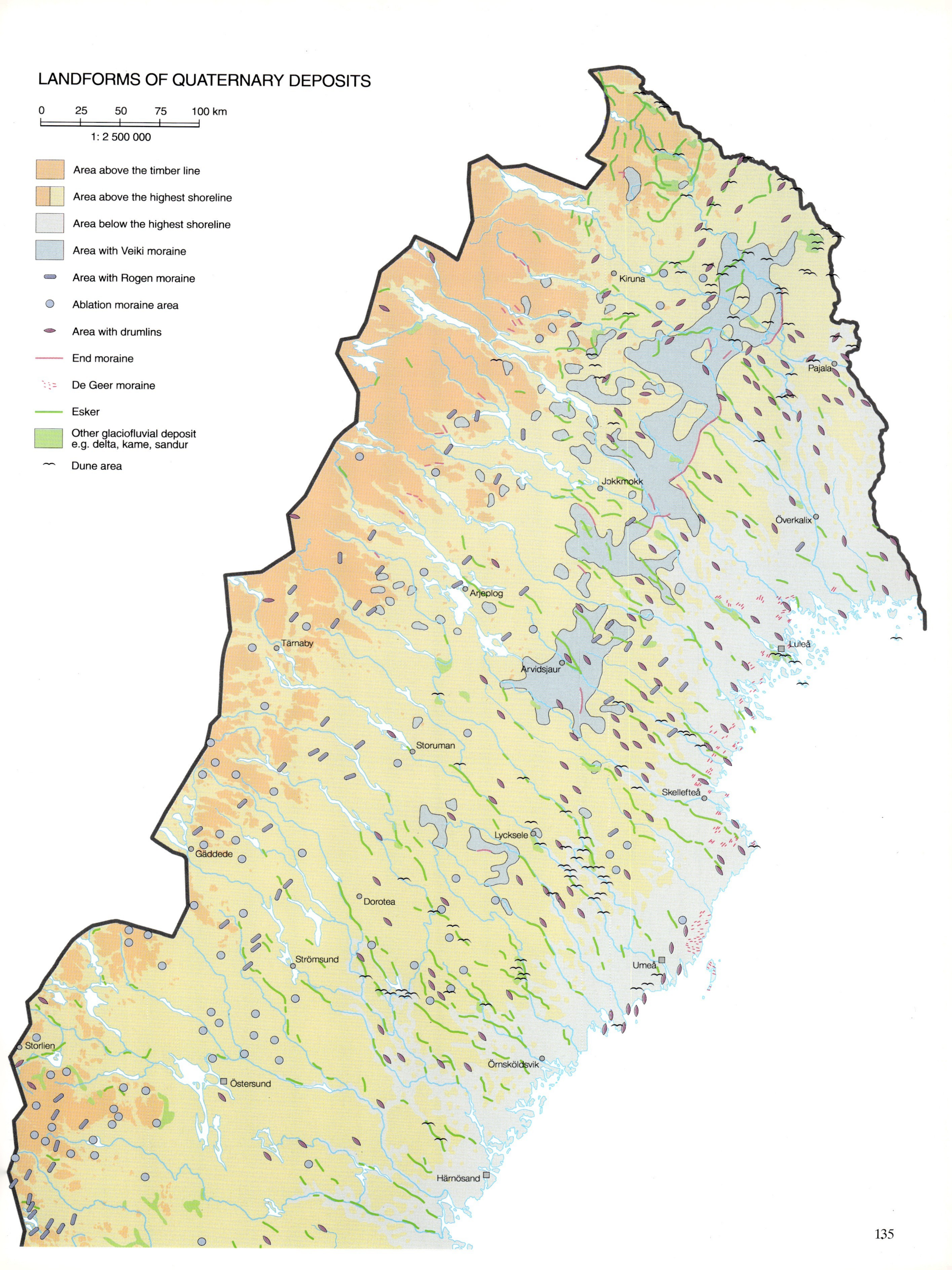
LANDFORMS OF QUATERNARY DEPOSITS
0
25
50
75
100 km
1: 2 500 000
Area above the timber line
Area above the highest shoreline
Area below the highest shoreline
Area with Veiki moraine
Area with Rogen moraine
Ablation moraine area
Area with drumlins
End moraine
De Geer moraine
Esker
Other glaciofluvial deposit
e.g. delta, kame, sandur
Dune area
Kiruna
Pajala
Jokkmokk
Överkalix
Arjeplog
Luleå
Tärnaby
Arvidsjaur
Storuman
Skellefteå
Lycksele
Gäddede
Dorotea
Strömsund
Umeå
Storlien
Örnsköldsvik
Östersund
Härnösand

Most favourable climate during present interglacial

Tundra | birch-pine forests | hazel | alder | oak | elm | lime | oak | ash

ICE-AGE | 10,000 years | 9,000 | 8,000 | 7,000 | 6,000 | 5,000

Vegetational history

Changes in climate and vegetation can be reconstructed by analyses of different components in mire and lake deposits. Variations in the frequency of microfossils—mainly pollen and spores—reflect the history of the vegetation from the deglaciation to the present.

Today, we are living in a period between two ice ages, an interglacial that has lasted about 10,000 years, ever since the latest inland ice definitely started to melt away from Sweden.

At first, where arctic conditions prevailed, there was a belt of tundra with dwarf shrubs, herbs and shrubs corresponding to the vegetation today found in the mountains. Pioneer plants spread rapidly, were sensitive to competition, were heliophytic and favoured by soils rich in minerals and lime. Examples are mountain avens, purple saxifrage, glacier buttercup, dwarf willow, dwarf birch, grasses and sedges. Juniper and sea buckthorn shrubs spread later.

Birch, aspen and pine were the first species of trees to immigrate, being followed 9,000–8,0000 years ago by hazel, alder, oak and elm. When the climate in Sweden was at its most favourable, approximately 8,000–4,000 years ago, hazel and mixed oak forests grew much further north than they do today. At that time, the vegetation had not been changed by human impact to any great extent by deforestation, cattle-breeding and cultivation. The last species of trees to immigrate to Sweden, approximately 3,000 years ago, were beech and hornbeam from the south and spruce from the east and north.

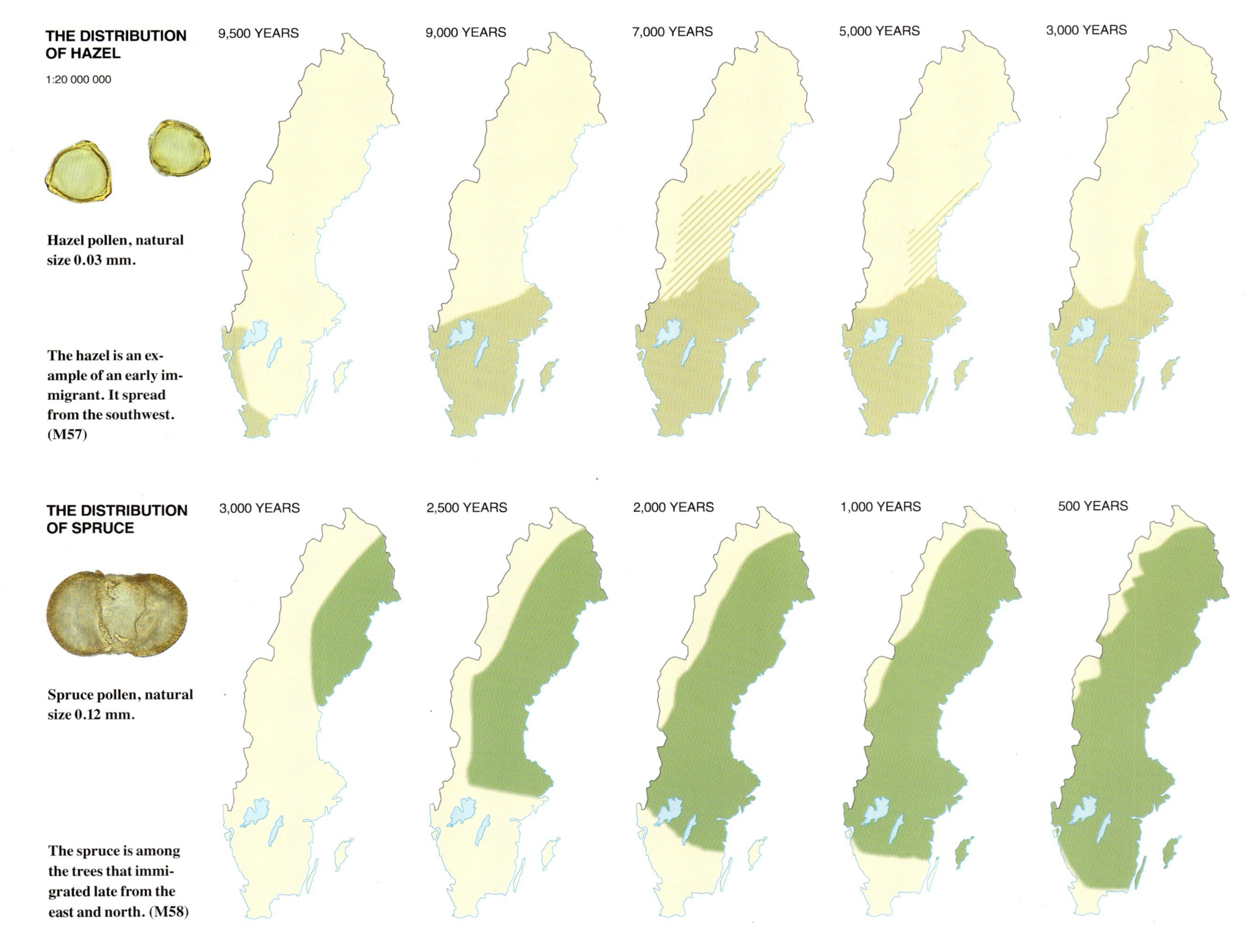

Hazel pollen, natural size 0.03 mm.

The hazel is an example of an early immigrant. It spread from the southwest. (M57)

Spruce pollen, natural size 0.12 mm.

The spruce is among the trees that immigrated late from the east and north. (M58)

broad-leaved trees spruce beech hornbeam

4,000 3,000 Birth of Christ 2,000 1,000 PRESENT NEXT ICE-AGE?

A simplified picture of forest history during the present interglacial. When the climate was at its warmest, the broad-leaved forests reached up to southern Norrland.

CLIMATE

After the deglaciation, the climate became warm and dry (*boreal*) with, for example, decreasing water levels in lakes and mires. Later, the climate changed to humid and warm (*atlantic*) with higher precipitation and higher mean temperatures than at present. A return to colder and drier conditions started approximately 4,000 years ago, with increasing precipitation and decreasing mean temperatures. The latest period with severe climate is called the "Little Ice Age" and lasted from the 16th century until the early 20th century.

Plant remains in the form of fruits, seeds, needles and pollen grains are fossil groups that are used in explaining the history of vegetation. Pollen is spread by wind, water and/or insects and becomes embedded and preserved in peat and lacustrine deposits. Analyses of pollen preserved in these deposits demonstrate how the composition of the vegetation has changed throughout the millennia. Examples of forest development following the latest glaciation in southern and central Sweden are given in the pollen diagrams below.

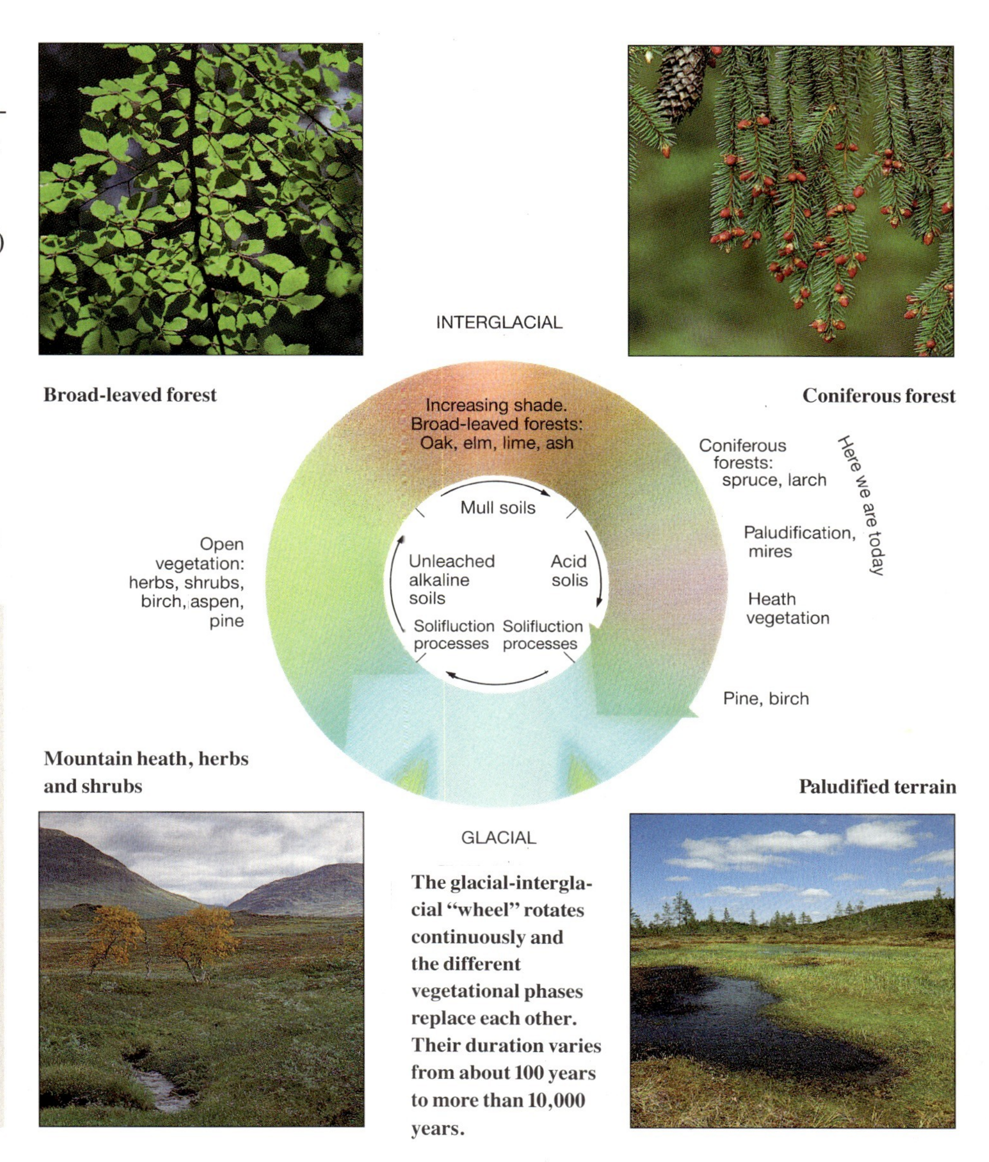

The glacial-interglacial "wheel" rotates continuously and the different vegetational phases replace each other. Their duration varies from about 100 years to more than 10,000 years.

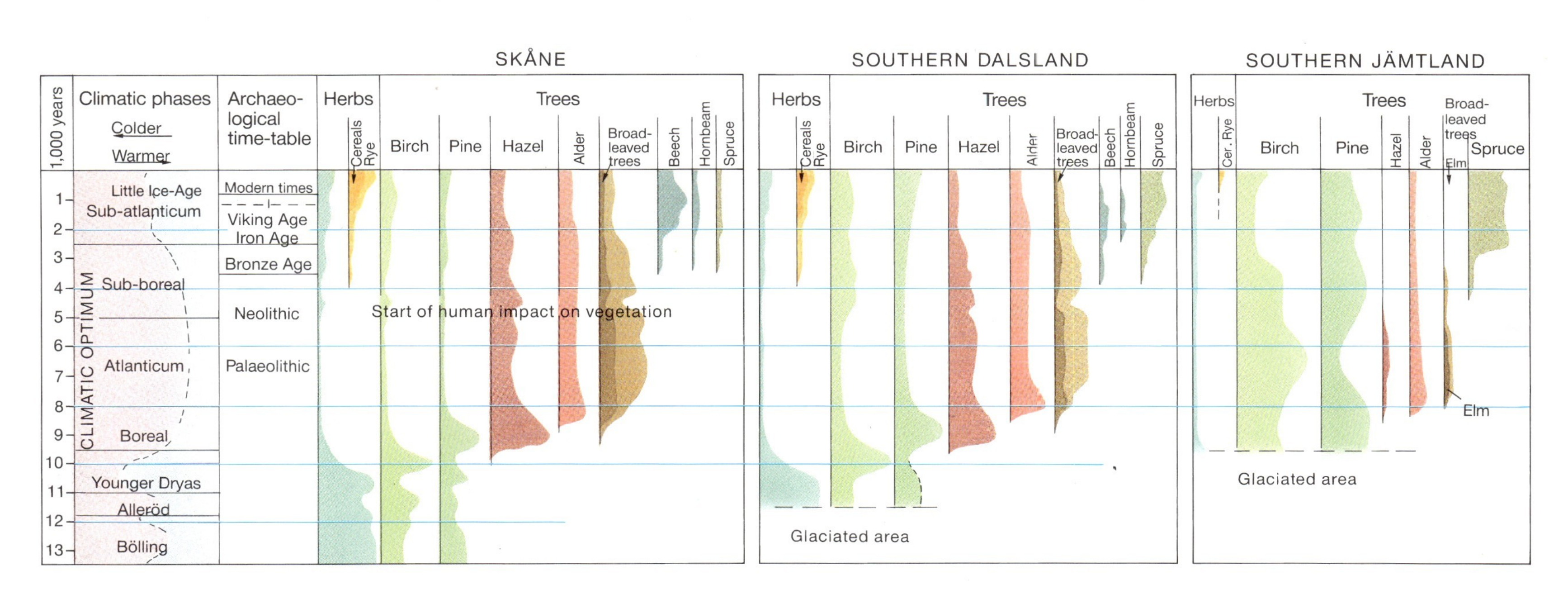

In many places on the Island of Gotland, the old raised beaches can be seen as "waves" in the landscape. They provide evidence of changes to the island's coastline during the last 11,000 years.

Development of the Baltic Sea and the Skagerrak/ Kattegat

The development of the Baltic area during the past 13,000 years has largely been known since the early 1930's. Increased knowledge and improved methods have resulted in several theories being rejected, for example, the outlet of the Ancylus Lake—the River Svea Älv at Degerfors.

Large parts of Swedish lowland areas are characterised by fertile agricultural land, mainly with fine-grained sediments deposited in marine or lacustrine environments. The distribution of these deposits is largely associated with the development of the Baltic Sea and the Skagerrak/ Kattegat.

At the time the inland ice started to melt, approximately 18,000 years ago, when the surface of the oceans was about 120 m lower than today, the level of the sea started to rise in what is called a *eustatic rise*. Approximately 13,000 years ago, the climate became warmer and large volumes of meltwater entered the oceans. At that time, the surface of the sea rose 15–20 m per 1,000 years. When the continental ice sheet melted, there was reduction in the load of the ice on the Earth's crust and, consequently, the ground slowly started to rise. This is called *isostatic uplift*.

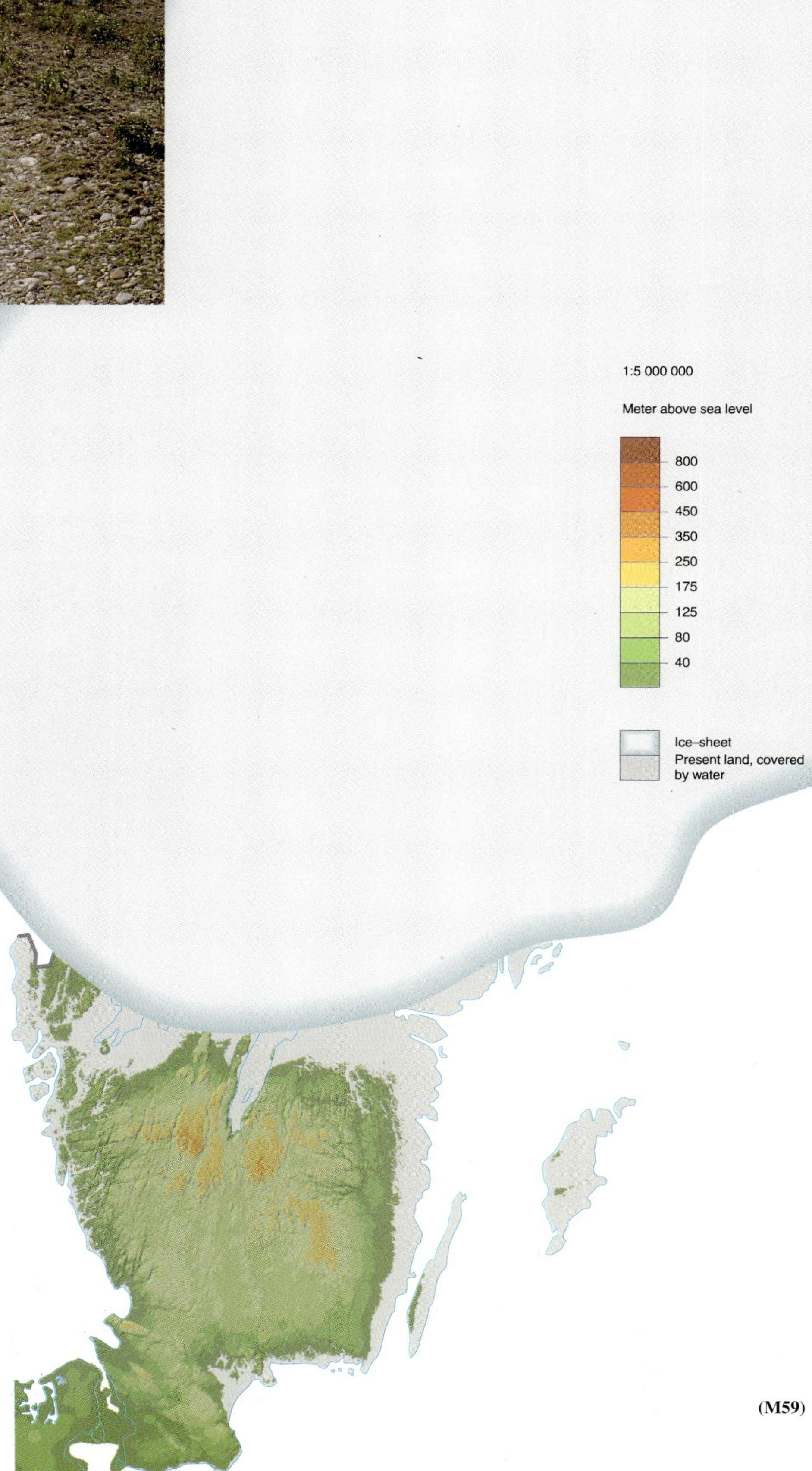

(M59)

Raised beaches on shingle field. Persön's nature reserve, southeast of Trosa, Södermanland.

The southernmost part of Sweden was deglaciated 14,000–3,000 years ago and the oldest indications of beaches—the *highest shoreline*—was formed close to the retreating ice margin. Consequently, the highest shoreline is oldest in the south and youngest in the north, with its highest point, approximately 285 metres above the sea, in Ångermanland, which became ice-free approximately 9,500 years ago.

13,000–10,300 B.P.

The initial stage of the *Baltic Ice Lake* was drained through the Öresund Strait area into the Kattegat/Skagerrak. Later, the sill of the Öresund Strait rose above the level of the sea surface and the Baltic Ice Lake became dammed-up. Approximately 11,200 years ago, the Baltic Ice Lake drained westward at Billingen hill, and the water level of this glacial lake was lowered again to the same level as the sea. A change in climate led to the growth of the inland ice, which advanced to Billingen and again dammed-up the Baltic Ice Lake. During the next 600 years, the water level of the Baltic Ice Lake rose about 25 m above sea level. Following the improvement in climate that took place approximately 10,500 years ago, and which initiated the warmer phase we are experiencing today, the margin of the inland ice retreated again northward. When the inland ice margin left the Billingen hill area about 200 years later, the water level of the Baltic Ice Lake fell by about 25 m within a few years.

10,300–9,500 B.P.

When the water level of the Baltic area again returned to the sea level, the next stage of the Baltic Sea was started, the *Yoldia Sea*, named after the arctic bivalve *Yoldia arctica* (today called *Portlandia arctica*). The massive volume of meltwater that drained to the west through the strait in northern Västergötland prevented brackish water from flowing into the Baltic Basin. This did not occur until the relatively wide Närke Strait became ice-free approximately 10,000 years ago. This brackish water stage lasted only approximately 120 years. The western part of the Vänern district retained its brackish water for another 200–300 years.

The entire Yoldia period is characterized by land uplift that was considerably faster than the rising sea level. Southern Sweden and Denmark were connected at that time; the occurrence of pine stumps at depths of 30–40 m in the Great Belt and in the Hanö Bay demonstrate that the shoreline at that time was far out from the present-day coast.

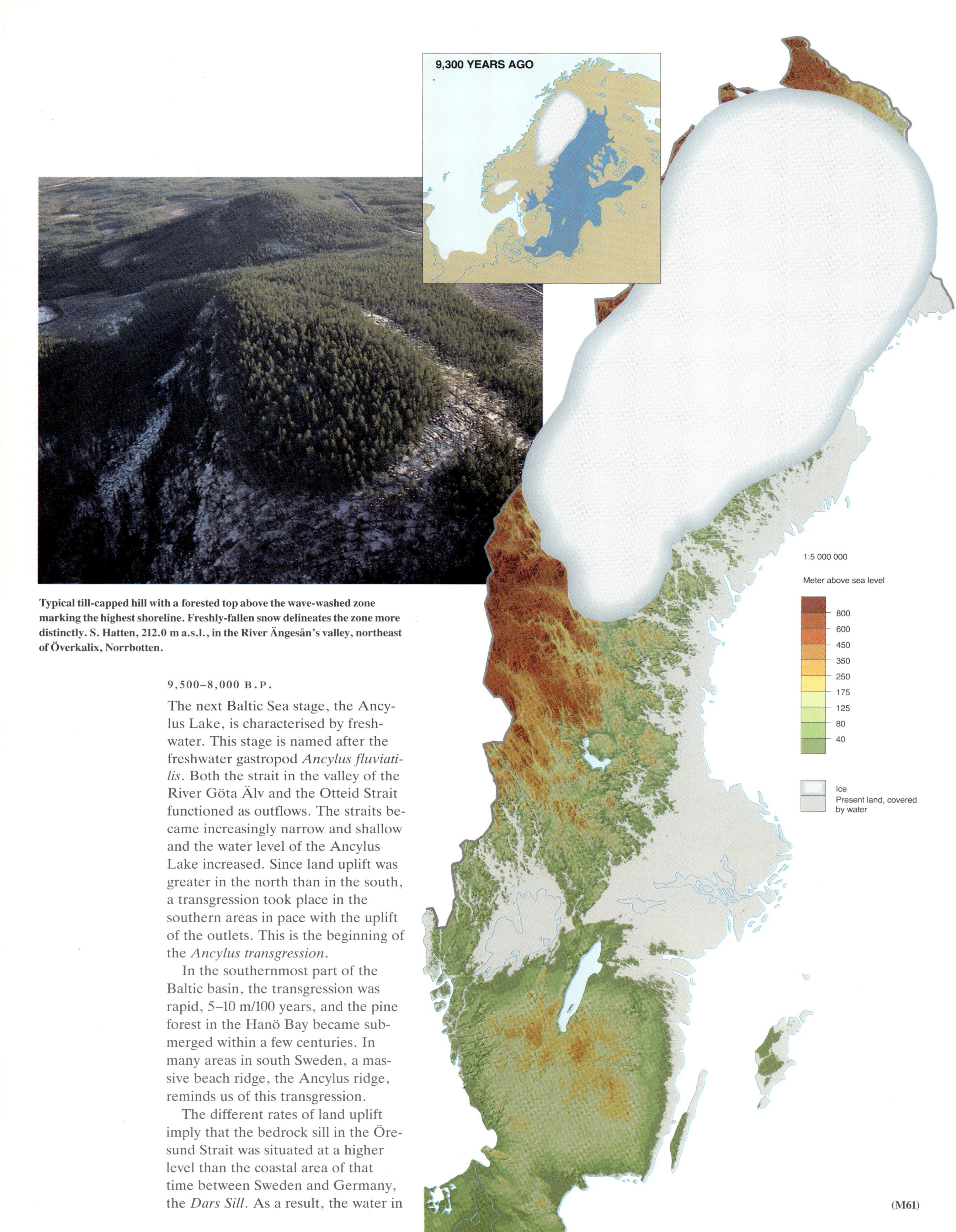

Typical till-capped hill with a forested top above the wave-washed zone marking the highest shoreline. Freshly-fallen snow delineates the zone more distinctly. S. Hatten, 212.0 m a.s.l., in the River Ängesån's valley, northeast of Överkalix, Norrbotten.

9,500–8,000 B.P.

The next Baltic Sea stage, the Ancylus Lake, is characterised by freshwater. This stage is named after the freshwater gastropod *Ancylus fluviatilis*. Both the strait in the valley of the River Göta Älv and the Otteid Strait functioned as outflows. The straits became increasingly narrow and shallow and the water level of the Ancylus Lake increased. Since land uplift was greater in the north than in the south, a transgression took place in the southern areas in pace with the uplift of the outlets. This is the beginning of the *Ancylus transgression*.

In the southernmost part of the Baltic basin, the transgression was rapid, 5–10 m/100 years, and the pine forest in the Hanö Bay became submerged within a few centuries. In many areas in south Sweden, a massive beach ridge, the Ancylus ridge, reminds us of this transgression.

The different rates of land uplift imply that the bedrock sill in the Öresund Strait was situated at a higher level than the coastal area of that time between Sweden and Germany, the *Dars Sill*. As a result, the water in

Coastal areas became submerged when the sea transgressed. Cultural strata (dark-coloured layers) of an ancient Stone Age settlement have been discovered below raised beaches in the province of Halland at Ölmanäs.

(M62)

the Ancylus Lake started to flow over Dars Sill and northwards towards the Great Belt. This phase, the culmination of the Ancylus transgression, took place approximately 9,200 years ago. During this transgression in the south, the water level in the Ancylus Lake rose 10–15 m above sea level. Since the sills in the Great Belt and Dars Sill consisted of Quaternary deposits, the constricted masses of water could relatively rapidly erode the deep valleys, that are still found in these areas, and flow out into the North Sea. At this stage, the transgression changed into a rapid regression until the level of the Baltic Sea again coincided with the sea level. During this *regression*, the water level in the Vänern basin also dropped and it became isolated from the Baltic approximately 9,000 years ago when the Närke Strait became dry. As a result, a land bridge was formed between southern and northern Sweden.

At this time, the land uplift in southern Sweden had decreased and the low-lying coastal areas became submerged by the sea, the level of which was still increasing. It took approximately another 1,000 years before the sea level had transgressed to the level of the sill in the Öresund Strait and for the water depth in the Great Belt to be sufficiently large for brackish water to enter the Baltic Basin. These events mark the end of the Ancylus Lake.

THE LAST 8,000 YEARS

In southern Sweden the eustatic rise was faster than the land uplift, which led to the start of approximately 4,000 years of transgression. About 8,000 years ago, the southern Baltic Sea started to become characterized by brackish water that gradually spread northwards to the Bothnian Sea and Gulf of Bothnia. This is the start of the Baltic's next stage, the *Littorina Sea*, named after the gastropod *Littorina littorea*.

The Littorina Limit was formed between 7,000 and 5,000 years B.P. On the West Coast of Sweden, the Tapes Limit was named after the bivalve *Tapes decussatus*. In Skåne and Blekinge, the Littorina Limit forms the highest postglacial beach ridge.

In the early 18th century the sea was known to have once extended far into Sweden. The withdrawal of the sea—"the decrease in water"—was subjected to penetrating research. It was not until the end of the 19th century that isostatic uplift could be proved to be the cause of the change in the coastline. Raised beaches at Stor-Brännberget, 15 km northeast of Kalix, Norrbotten.

STAGES OF THE BALTIC SEA, YEARS BEFORE PRESENT

Littorina Sea	8,000–
Ancylus Sea	9,500– 8,000
Yoldia Sea	10,300– 9,500
Baltic ice-lake	ca 13,000–10,300

Since uplift was greater in the north than in the south, the transgression was most noticeable in southernmost Sweden, where the Littorina and Tapes beaches were formed during a 2,000 year long complex of transgressions. Remarkable beach ridges are found along the coasts of the provinces of Halland and Skåne, and in the southern Baltic area, where it has been given the name *Littorina ridges*. In the Stockholm district, the maximum transgression took place approximately 7,000 years ago, whereas in the south it was 1,500–2,000 years later. Along the coast of Norrland, the uplift had been considerable ever since the deglaciation and here it has always been greater than the rising sea level.

Since the salinity and temperature of the sea-water were higher than they are today, there was also a partly different type of marine fauna. Oysters were found along the West Coast of Sweden and several presentday species had a more northerly and easterly distribution.

The ongoing land uplift, in combination with the slower rise in sea level rise during the past 5,000 years, has caused a regression around the Swedish coast. Salinity and temperature decreased gradually, and the conditions became increasingly similar to the ones today. At present, the greatest uplift is 9.2 mm/year in the coastal areas of the county of Västerbotten. From there land uplift decreases towards the north and south and at the coast of southern Skåne there is no land uplift at all.

COASTS AFFECTED BY TRANSGRESSION

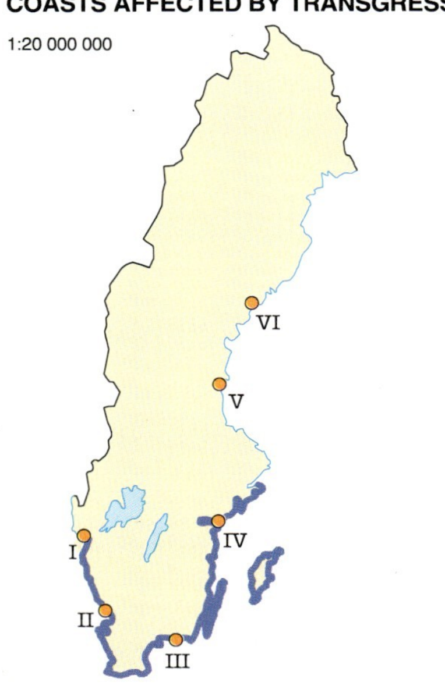

In northern Bohuslän and to the north of Lake Mälaren, the land uplift has always been greater than the eustatic rise of the sea surface. Thus, regression has been predominant along these coasts ever since these areas became deglaciated. (M63)

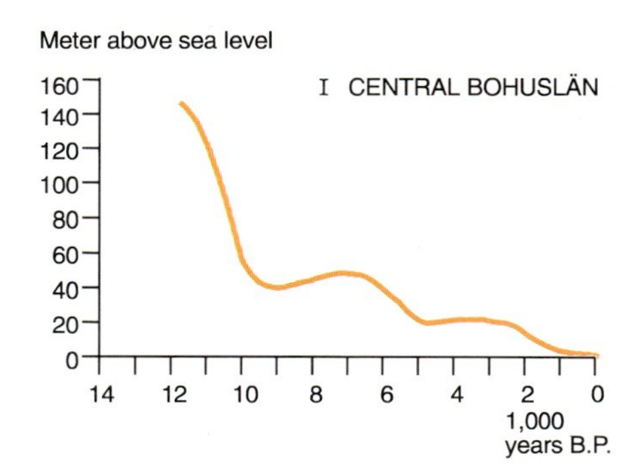

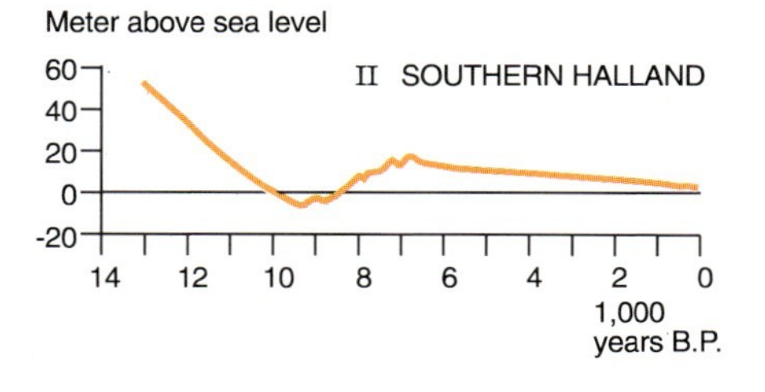

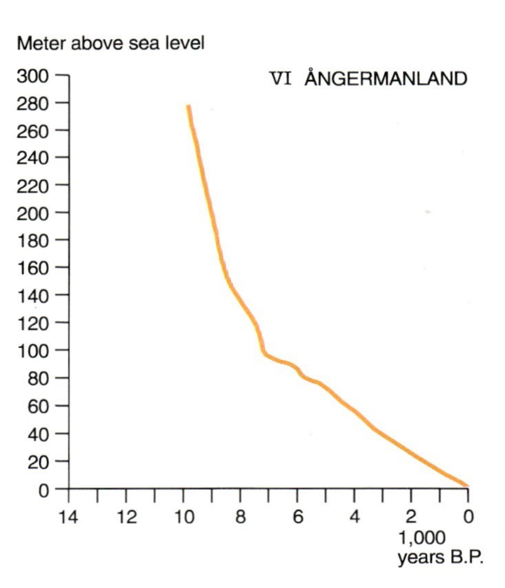

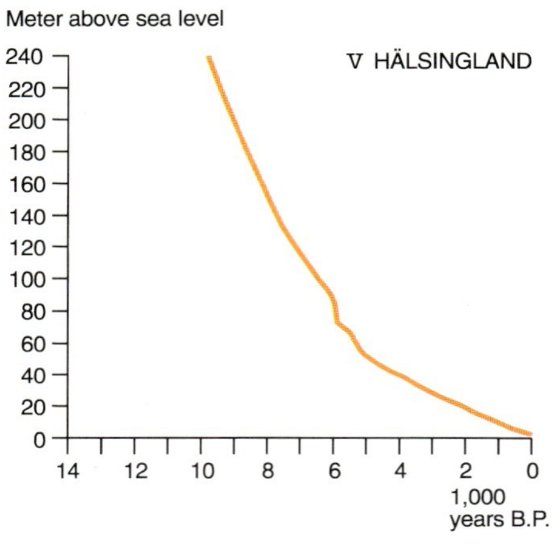

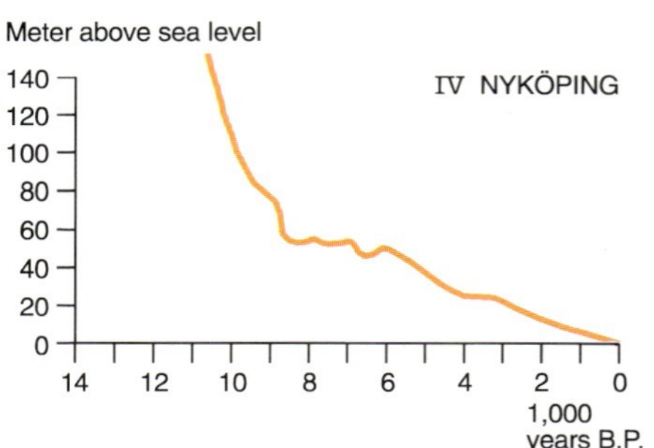

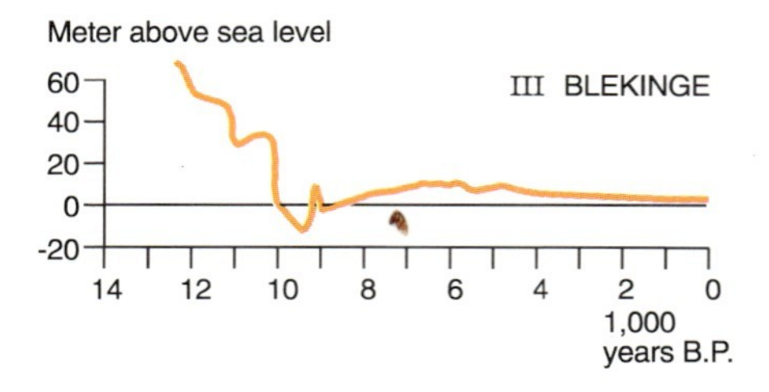

REGIONS OF QUATERNARY DEPOSITS

1:5 000 000

- Southwestern Skåne
- Till area of southern Sweden
- Cambro-silurian areas of southern Sweden
- Outcrop and clay areas of the east and west coasts
- Outcrop and clay area of the Lake Vänern region
- Outcrop, till and clay area of northeastern Götaland and eastern Svealand
- Till area of northwestern Svealand
- Till and peatland area of interior parts of southern and central Norrland
- Outcrop and sediment area of coastal parts of central Norrland
- Cambro-silurian area ot the Lake Storsjön region
- Outcrop, till and sediment area of coastal parts of northern Norrland
- Till and peatland area of northern Norrland
- Bare mountain area

(M64)

Regions of Quaternary deposits

Geology forms the landscape. Rocks and bedrock morphology, the sculpturing effect and deposits of the inland ice, post-glacial processes—all of these interact to give the scenery its basic character. Sweden can roughly be divided into a number of uniform areas. Nonetheless, the borders between them are often diffuse and within each main area it may be possible to find different and diverging sub-areas. Modern mapping of the Quaternary deposits has revealed a partly new picture of the extent and stratigraphy of the overburden types throughout large parts of Sweden.

A deposit may vary in appearance and characteristics. Within areas with crystalline bedrock, the till is mainly sandy. Limestone and shales give a fine-grained, nutrient-rich and generally limestone-rich till. Where leptites and porphyries dominate, the till is normally gravelly and poor in nutrients. Glacial clay may be varved and rich in limestone in one area but non-varved and limestone-free in another.

The thickness and surface configurations of Quaternary deposits may also vary. Along the east coast of Sweden, the clays are rarely thicker than 15 m, whereas on the west coast they may be 100 m. Glaciofluvial deposits may be in the form of eskers or as deltas. Till beds with layers of sediments between them are common in certain areas. Deposits from older glacial periods and interglacials are also found.

SOUTHWESTERN SKÅNE

Sedimentary rocks dominate, mainly mesozoic rocks. The Quaternary stratigraphy is sometimes complex, generally with considerable thickness; in the Alnarp depression, e.g., more than 120 m. Deposits from older glacials, as well as sediments from interglacials and interstadials, are found.

The area with clayey till or clay till in western Skåne has a gently undulating surface, most of which is cultivated. The stratigraphy consists of several glacial beds with sediment between them. The total thickness is often 15–40 m, sometimes more.

In some places, the hummocky moraine in southwestern Skåne is very undulating with small hills and plateaus. In the valleys, there is glacial clay on the heights, *plateau clay*, that was formed in ice-lakes.

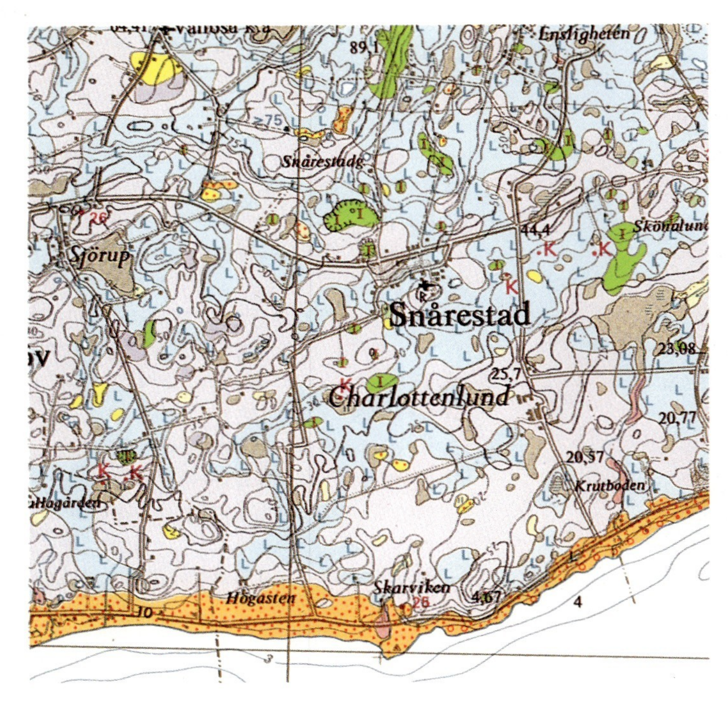

Hummocky moraine at Snårestad. The Quaternary deposits consist of boulder clay (mauve) and clayey till (L) with peat (brown) in the depressions. The glaciofluvial sediments (green) consist of intermorainic deposits (I) that are exposed in places. Wave-washed sediments (orange) are only found along the coast. Part of the Quaternary map Ystad NW, scale 1:50 000.

The gently rolling hummocky till landscape with hills of boulder clay in southern Skåne is very fertile on account of its high lime content. The terrain has been smoothed to some extent by agriculture. Snårestad Church is in the background.

The Vomb basin is dominated by sandy glaciofluvial sediments in outwash plain formations. The deposits are complex, with features of glacial clay that are sometimes as much as 60 m thick. Glacial lake sediments also cover large areas; fine sand dominates the surface but the sediments are more fine-grained with depth.

The area between Klippan and Ängelholm is a plain with glacial clay and protruding hills of till. The clay is stratified and partly rich in lime. The thickness varies and is up to 80 m. Wave-washed sediments are widespread. Sand dominates along the south coast and dunes occur.

TILL AREA OF SOUTHERN SWEDEN

The bedrock consists of granites, gneiss and porphyry. The area is mainly above the highest shoreline. Till dominates but glaciofluvial sediments and peat are also abundant.

Gravelly and sandy till dominate in the porphyry areas. Till surfaces generally have medium frequencies of boulders and are flat, sometimes with small hills. In Småland, local boulder depressions and areas with high frequency of surficial boulders occur. Drumlins with till thickness of 50 m and more can be found. Till-covered sediments, mainly gravel and sand, are found.

Glaciofluvial ridges in the form of eskers, kames and deltas are usually localised to the larger valleys. Deltas are frequently found in conjunction with the highest shoreline.

Glacial lake sediments, mainly stratified sand, is found in some valleys.

Peatland in western parts has often develloped as the result of waterlogging. These areas consist of sphagnum peat and are raised bogs. The depth of the peat is usually 3–6 m. Peatlands in eastern parts have largely been formed as a result of encroachment of open water.

Characteristic and divergent parts of this area are the coastal plain of the province of Halland with marine clays and protruding features of rock and till. In large areas, the clays are covered by wave-washed sediments. Large dunes occur, an example being the 26 m high dune southeast of Haverdal. Drumlins have been found to contain interstadial and interglacial sediments.

The Kristianstad plain is dominated by varved clay and sand with areas of sand dunes.

The coastal area of the province of Blekinge is diversified with rocks, till and clay, whereas the east coast of Småland is dominated by till and glaciofluvial sediment. Sandstone till is found close to Kalmarsund.

Cultivated basal till. Frost action has resulted in boulders moving upwards to the surface, where they have been collected in heaps. This area is hardly suitable for intensive agriculture. Kristdala, northeastern Småland.

On the alvar on the Island of Öland, the limestone is exposed in large areas or covered only by a thin layer of till.

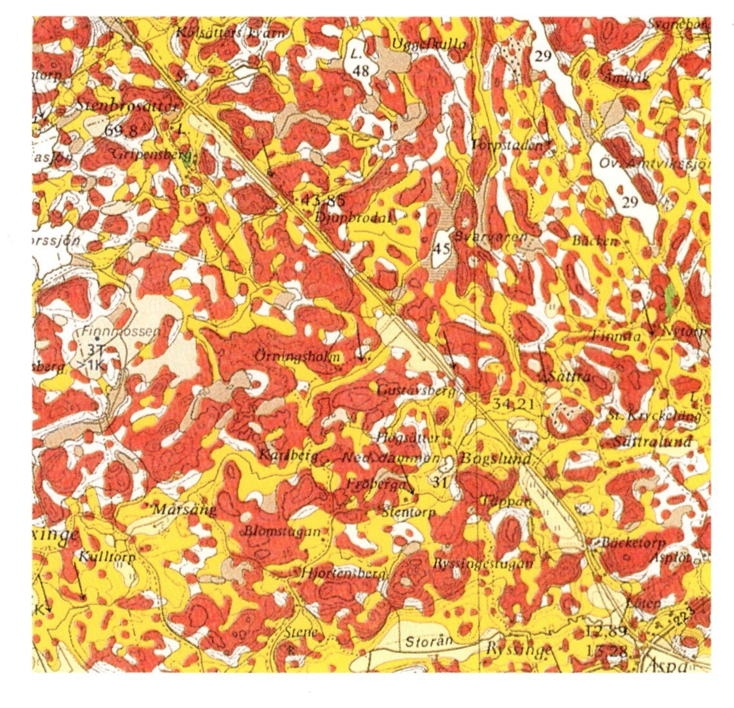

CAMBRO-SILURIAN AREAS OF SOUTHERN SWEDEN

The bedrock is dominated by Cambro-Silurian rocks that characterize the till. Most of the mesas of Västergötland are covered by a thin layer of till or peat. At Kinnekulle, the thickness of the till is greater and forms drumlins. Locally, the till has a high clay content. Large amounts of glaciofluvial sediments have been deposited along Billingen and form deltas and kames.

In the province of Östergötland, the till is clayey and sandy, as well as being thick in places. Here, there are several large glaciofluvial deposits which are part of the Fennoscandian Ice-Marginal Zone. They generally have complex stratigraphy with till and sediments. The fine-grained sediments are dominated by glacial clay, which is brown, calcareous and varved. In low-lying areas it is covered by post-glacial clays. Wave-washed sediments are found mainly in conjunction with glaciofluvial deposits.

In the province of Närke, the till is sandy, sometimes clayey with varying thickness and contents of limestone. Typical till formations are the De Geer moraines and drumlins. A number of eskers run in a north-south direction. The fine-grained sediments are dominated by red—grey clay that is negligibly calcareous and partly varved. It is rarely more than 5 m thick and is usually covered by a few metres of post-glacial clay and gyttja clay. Large areas of peat are found, for example, in the Kvismaren area.

On the islands of Öland and Gotland, the Quaternary deposits are dominated by calcareous clayey till and clay till, which generally forms a thin cover. Extensive areas of exposed rock, such as on the Great Alvar, are common. Wave-washed sediments are widespread and beach ridges from the Ancylus and Littorina trangressions are prominent. Sand dunes are found on northern Öland, on Fårö Island and on Gotska Sandön Island. The fens on Gotland are usually drained.

Landscape of rock and clay along the east coast about 5 m NW of Ludgo in Södermanland as shown on the map of Quaternary deposits. Most of this area is covered by forest. Only the larger areas of clay (yellow) in the southern part are cultivated. Bare rock is shown in red. Part of the Quaternary map Nyköping NW, scale 1:50 000.

Morphology in western Sweden is largely characterized by bare bedrock and joint aligned valleys with massive layers of clay, usually more than 30 m thick. Hisingen, near Göteborg.

OUTCROP AND CLAY AREAS OF THE EAST AND WEST COASTS

These areas are undulating and are dominated by outcrops with clay sediments in depressions and valleys.

Along the east coast, the bedrock consists mainly of veined gneiss. This area, with small hills and numerous joint-aligned valleys running in a northwest—southeast direction, is below the highest shoreline. Only small amounts of till are found here, mainly in depressions in the uplands and as narrow strips along the hillsides. Wave-washed sediments, mainly gravel and sand, are found in connection with glaciofluvial deposits. Clays on the east coast are dominated by glacial clay, that is varved and often 10 m thick or more. In lower parts of the terrain, it is covered by a couple of metres of grey post-glacial clay and gyttja clay.

Along the west coast, the bedrock is dominated by gneissic granite and Bohus granite. The area is hilly, with differences in altitude of more than 100 m. Large parts of this area are below the highest shoreline. Till is sparsely found partly as a very thin cover, only about 0.5 m thick, within some hilly areas, but also in end moraines and as stoss and lee-side moraines. Ice-marginal deposits, for example, the Göteborg moraine, consist of glaciofluvial deltas and ridges of till and glaciofluvial sediments. Stoss-side moraines are generally thick and, as at Dössebacka, may contain Quaternary deposits from older glaciations. Wave-washed sediments are found around till and glaciofluvial deposits. A characteristic Quaternary deposit on the west coast is shell deposits. Clays on the west coast are dominated

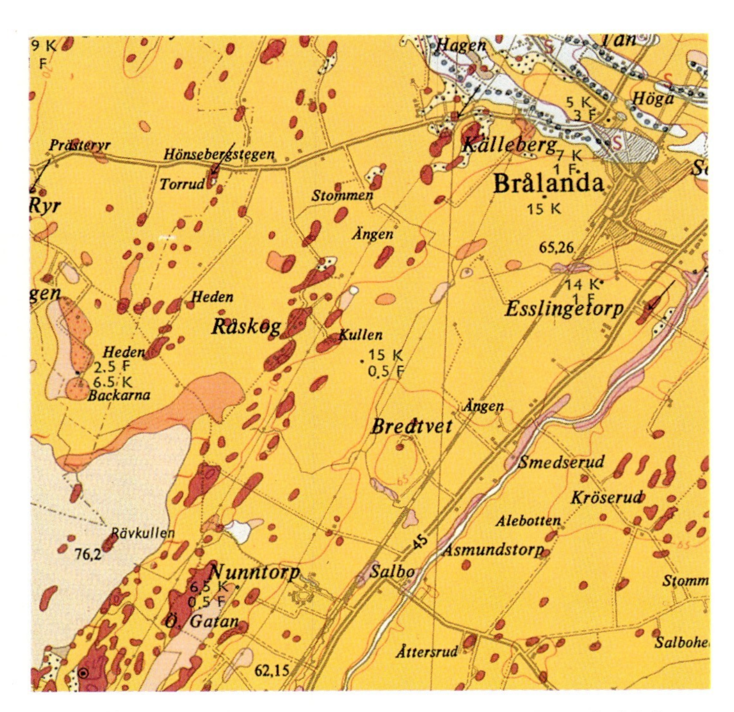

The Dalbo plain to the north of Vänersborg is very flat. Heavy glacial clay (yellow), frequently 10–15 m thick, dominates between the Archaean rocks (red). The till deposits (blue) at Brålanda are part of an end moraine zone to the south of the Fennoscandian Ice-Marginal Zone. Part of the Quaternary map Vänersborg NE, scale 1:50 000.

by glacial clay, being mainly heavy, grey and with diffuse stratification. In low-lying areas it is covered by a fairly similar post-glacial clay. The clays usually contain shells of marine molluscs. In some places, the glacial clay is of quick-clay type and may involve a risk of landslides. The total thickness of the clay is generally considerable and may be 50–100 m in wide valleys.

OUTCROP AND CLAY AREA OF THE LAKE VÄNERN BASIN

This area, which is below the highest shoreline, is hilly with outcrops and sediments. Ice-marginal gneiss and granite. Plains dominate in deposits from, for instance, the Younger Dryas, form a remarkable feature of the scenery. They consist of ridges of till and glaciofluvial sediments, locally with included, folded layers of clay.

There are numerous De Geer moraines, e.g. around Åråsviken Bay.

The fine-particled sediments are dominated by reddish-brown, generally stratified glacial clay of varying composition. The glacial clay, with a thickness of sometimes more than 20 m, is covered in the northern and western parts of the area by 2–3 m of post-glacial clays with high contents of silt and sand. The northeastern part is dominated by a clay with a very high clay content. In addition, wave-washed sediments are found and river sediments from the River Klarälven are found in the area to the north of Lake Vänern.

OUTCROP, TILL AND CLAY AREA OF NORTHEASTERN GÖTALAND AND EASTERN SVEALAND

The area is undulating with outcrops and moraine on higher parts and fine-particled sediments in depressions. The bedrock is dominated by crystalline gneiss and granites. The entire area, apart from a few upland areas, is below the highest shoreline.

The till, lying in depressions in the rock and on the slopes, is usually sandy. It is dominated by crystalline basement rocks, lime-deficient and usually 1–5 m thick.

Eskers run largely in a north-south direction through the area. In addition, there are extensive and thick deposits of glaciofluvial sediments, usually dominated by sand, e.g., in the province of Södermanland.

Wave-washed sediments, usually only a few metres thick, are found on the till slopes in exposed upland areas, for example to the south of Lake Hjälmaren and in conjunction with glaciofluvial deposits.

The glacial clay is varved, reddish-brown and locally calcareous. In connection with glaciofluvial sediments, varved clay with layers of fine sand and silt often dominates. The thickness of the fine-grained glacial sediments may be more than 15 m in large valleys. In low-lying areas, the varved clay is often covered by a few metres of grey post-glacial clay and gyttja clay.

The Lake Mälaren basin is dominated by flat areas of clay with features of outcrops and till. Boulder-rich De Geer moraines are found in clusters, for example to the southwest of Västerås. Wave-washed sediments, mainly sand and fine sand, are found along the large eskers. The glacial clay, which is calcareous in the eastern part, is generally 5–10 m thick. In basins and in lower parts of the

The undulating, diversified landscape of Södermanland, with rock (red), till (blue) and clay (yellow). The direction of the De Geer moraines (dark-blue dashes) in conjunction with the esker (green) indicates a bay in the ice front (a calving bay). Part of the Quaternary map Eskilstuna NE, scale 1:50 000.

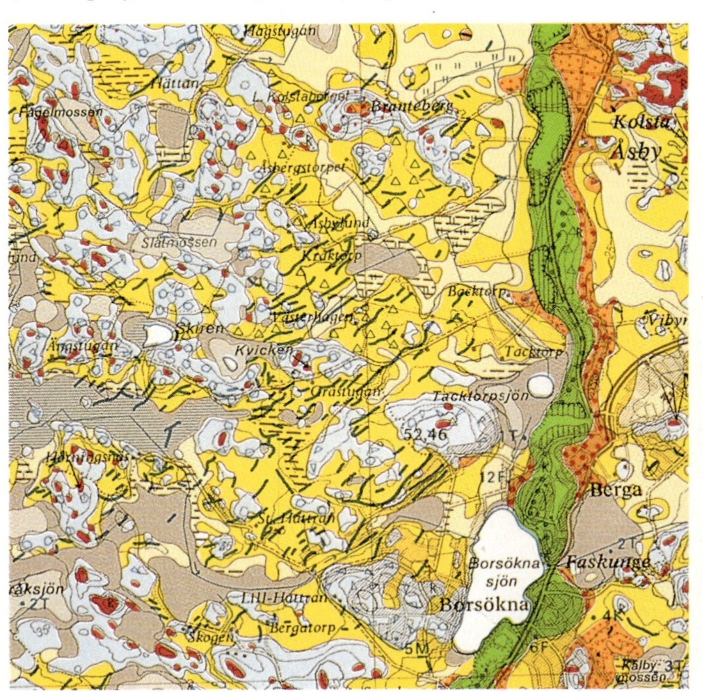

The outcrop of rock, about 50 m high, near Björnlunda in Södermanland, offers good views of the cultivated fields of glacial clay. Till is found only as narrow strips along the rocks.

valleys, it is covered by grey post-glacial clay that may be 10 m thick in some places. In the lowest parts of the basins, there is usually a thin layer of gyttja clay in the surface.

In large parts of the provinces of Uppland, southern Gästrikland and northeastern Västmanland, the bedrock surface is very flat and largely covered by till that is usually only a few metres thick. Till surfaces in some areas have abundant boulders

Large parts of eastern Värmland are dominated by till (blue) and peat (brown). Glaciofluvial deposits (green) are found along the tributary valleys of the River Klarälven, usually with a delta in the Klarälven valley. River sediments (pink) and lake sediments (orange) dominate the Klarälven valley. North of Ekshärad. Part of the map of Quaternary deposits in Värmland Province, scale 1:200 000.

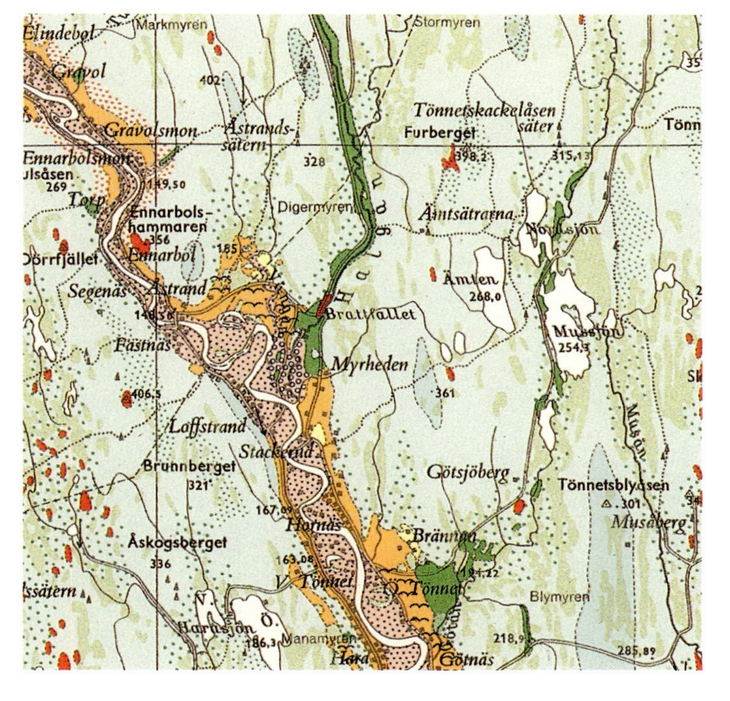

and also large boulders. In eastern Uppland, the till contains Ordovician limestone from the Bay of Gävle and is calcareous and clayey. In eastern parts of Rådmansölandet, the till has a high content of Jotnian sandstone.

Fine-grained sediments, mainly glacial clay, have a relatively minor extent and their thickness is usually less than 10 m. Wave-washed sediments occur in conjunction with glaciofluvial deposits and around exposed upland areas of till. Shingle fields occur in several places. Viksta Stentorg on the Uppsala esker is well-known, as well as the large field of shingle named Kapplasse to the east of the Lövsta Bay, that has been formed as a result of the till being wave-washed.

Peatland, the most well-known being Florarna, is widespread. Fens dominate with thicknesses of peat rarely exceeding 4 m.

THE TILL AREA OF NORTHWESTERN SVEALAND

In the southern part of the area, the bedrock consists of crystalline basement with very varying rocks. The northern part is dominated by granites, sandstones, porphyries and dolerites. The area is above the highest shoreline, with the exception of parts of Bergslagen and areas around the valley of the River Dalälven, including the Siljan basin. Tills dominate and peatland is extensive. Glaciofluvial deposits and sediments are found in the valleys.

The till in the southern parts of the area and in the province of Värmland is dominated by crystalline rocks, whereas in the province of Dalarna it is characterised by sandstone and porphyry. Porphyry usually gives rise to a till that has abundant stones. The surface is generally rich in boulders, but they are rarely large. In the south, the till is relatively thin but becomes thicker and more extensive towards the north. In the provinces of Värmland and Dalarna, the till forms a uniform cover on heights and slopes, whereas ablation moraine may be found in depressions.

Glaciofluvial sediments form continuous deposits that are generally thick in the large valleys. At the highest shoreline, they form large deltas. Fine-grained sediments are found in the valleys—silt is widespread along the River Dalälven to the south of Lake Siljan and sand and fine sand dominate along the River Klarälven.

Large dunes are found in several of the large delta deposits, such as at Morafältet. Peatland is also extensive, particularly in northeastern Värmland and western Dalarna. These large areas of mires consist of bogs, sometimes slightly domed, fens and soligenous mires.

The till area of southern Norrland is heavily forested. The undulating landscape is interspersed with long but not particularly wide valleys. Glaciofluvial sediments are found in the valleys together with, below the highest shoreline, clay and river sediments, which are cultivated. View of Ljusnan from Öjeberget, Järvsö.

TILL AND PEATLAND AREA OF INTERIOR PARTS OF SOUTHERN AND CENTRAL NORRLAND

The bedrock is dominated by different gneissic granites and granite. Porphyry is found in Dalarna. The entire area is above the highest shoreline with the exception of the coastal zones. Till and peatland cover very large areas. Glaciofluvial deposits and river sediments are found mainly in the large valleys.

Sandy till covers the largest area but in western parts of the province of Medelpad, for example, there are areas with gravelly till. In Jämtland, the till generally has a high content of fine-sand and silt. The surfaces are generally with medium frequency of boulders but in places where there is granitic bedrock there are extensive areas with abundant boulders, including large ones. In places the till is fairly thick, up to 60 m.

Till surfaces often reflect the underlying bedrock, but the till can form its own configurations such as drumlins and Rogen till. In some cases, the till formations have been found to be older than the latest glaciation.

Sequences with double tills are found. The lower till is often dark

Southern slopes with fine-grained and calcareous till are suitable for cultivation. Kaxås, northwest of Lake Storsjön in Jämtland.

grey, very hard and clayey. Till-covered sediments, both minerogenic and organic, have been found in several places. The sediments are often interstadial, for example at Pilgrimstad in the province of Jämtland.

Glaciofluvial deposits as eskers, deltas, kames and outwash plains occur. The largest outwash plains in Sweden are found in the province of Härjedalen. In the valleys, there are also different glacial and post-glacial fine-grained sediments and river sediments, which are extensive, mainly in the large river valleys.

The brown pattern of the peatlands gives a picture of an undulating till landscape (blue) with hills and ridges. The characteristic parallel ridges of the Rogen moraine can be seen in the southwestern part of the map. About 30 km northwest of Strömsund. Part of the map of Quaternary deposits in Jämtland Province, scale 1:200 000.

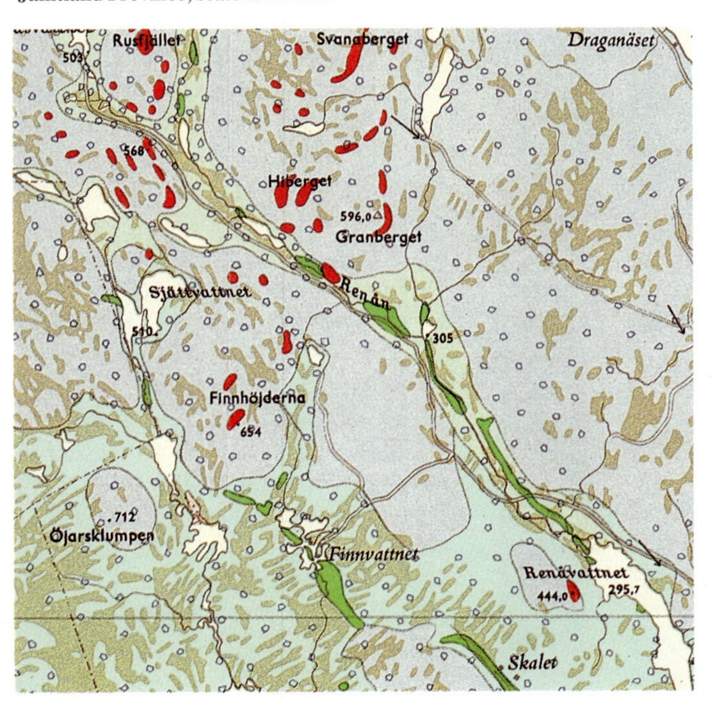

Different types of peatlands are extensive, particularly in the parts of Jämtland and Härjedalen. The peatlands consist of fens, mixed and soligenous mires and level bogs surrounded by parts of fens. The peat is usually a few metres thick.

OUTCROP AND SEDIMENT AREA OF COASTAL PARTS OF CENTRAL NORRLAND

The bedrock largely consists of different crystalline gneisses. Rapakivi granite and dolerite occur in central parts. By far the greatest part of the area is below the highest shoreline. Bare outcrops are widespread. The Quaternary deposits are dominated by wave-washed sediments and fine-grained sediments, whereas till, glaciofluvial sediments and river sediments are of a minor extent.

Till occurs mainly in conjunction with uplands and, in exposed places, is usually heavily wave-washed. Till-capped hills are characteristic—at the level of the highest shoreline they have a bare-washed zone with till above and wave-washed sediments and till below.

In the river valleys, there are glaciofluvial deposits in the form of eskers but they are generally hidden by other sediments and rarely reach the surface.

Wave-washed sediments such as shingle, gravel and sand occur in abundance, mainly on slopes in broken terrain. Their thickness varies and in extreme cases may be as much as 20 m. Shingle and gravel usually cover the surface, whereas sand dominates towards deeper levels.

The fine-grained sediments that mainly occur in depressions and valleys may be fairly thick. They are dominated by different clays with considerable features of silt. Gullies and slopes are formed in these Quaternary deposits. The clays are usually dominated by varved glacial clay overlain by post-glacial clay, sometimes containing black stripes of iron sulphide. Locally, the clays may involve a risk of landslides.

CAMBRO-SILURIAN AREA OF THE LAKE STORSJÖN DISTRICT

The bedrock consists of limestone, shales, greywackes and quartzites. The landscape has a flat and gently undulating surface, occasionally with drumlins. The completely dominating deposits are clayey sandy till and clay till. The Cambro-Silurian rocks dominate the till. Outcrops are found only to a minor extent. Despite that, the thickness of the till is often fairly shallow. The till areas, to some extent cultivated, are frequently waterlogged and there are numerous peatlands.

Till-covered sediments are found in the district around Lake Storsjön and, in some cases, these have been found to be interstadial. The sediments may consist of sand or gravel, at Vålbacken of varved clay. The varves are partly disturbed and become coarser at higher levels in the stratigraphy.

Peatlands are dominated by mixed mires with peat that is a few metres thick. In parts with fens, there are different calcicolous plants.

OUTCROP, TILL AND SEDIMENT AREA OF COASTAL PARTS OF NORTHERN NORRLAND

The bedrock consists mainly of gneiss and granite with local features of basic rock. The area is located below the highest shoreline and is dominated by till. Glaciofluvial deposits and fine-grained sediments occur in the valleys.

The till is mainly sandy and the surface usually has normal features of boulders. No particular surface configurations are usually found as the till forms a more or less uniform cover that thins out towards upland areas. Exceptions are the De Geer moraines that occur in clusters at many places along the coastland of the province of Norrbotten, for example between

Large parts of the mountains lack a Quaternary cover and in places where it is present, it is thin. In such areas the surface configuration of the overburden is not prominent. Alajaure, Padjelanta National Park.

The till and peatland area along the Caledonides. Dajkanberg, west of Storuman.

Luleå and Piteå. Till with numerous layers of sand, *Kalix till*, has been found in several places near Luleå. In some valleys in the Niemisel district, there are large transverse ridges of sedimentary till.

Glaciofluvial sediments occur in the larger valleys, generally in the form of more or less continuous eskers or as flat, domed fields dominated by sand, e.g., Pitholmsheden and Kallaxheden.

Wave-washed sediments are widespread and may have considerable thickness. Extensive fields of shingle, gravel and sand are found at Hertsön to the east of Luleå, among other places.

In the valleys, the surface layers are generally made up of river sediments that cover post-glacial and glacial clay. The clays are surficial in large flat areas, e.g., near Boden. The post-glacial clays are generally black in colour, caused by iron sulphide.

Dunes occur, for example, on Kallaxheden.

TILL AND PEATLAND AREA OF NORTHERN NORRLAND

The bedrock consists mainly of granites, gneissic granites and basic volcanites. Apart from the southeastern part, the area is above the highest shoreline and is dominated by till, glaciofluvial sediments and peatland. The thickness of the Quaternary deposits may be considerable.

The till is predominantly sandy. Coarse-grained till is found, as well as sedimentary till, dominated by sand and silt. Drumlins are common and Veiki moraine is found within a wide belt extending from Kåbdalis in the south to Lainio in the northnortheast. In several places, both till forms and thick Quaternary stratigraphy is considerably older than the final phase of the last glaciation.

Eskers, kames and other glaciofluvial deposits are common. Glaciofluvial sediments are found both from the deglaciation phase and from earlier stadials. These sediments are generally covered by a thin layer of till.

In the valleys, there are local occurrences of river sediments and aeolian deposits.

Peatlands may generally be characterized as mixed mires. The peat is generally a couple of metres thick.

MOUNTAIN AREA

The bedrock is dominated by shales, limestones, quartzites and amphibolites. Particularly in Lappland, the rock has only a thin cover and may then be eroded and cracked with scars on the slopes. Other parts are mainly covered by overburden that has been affected by frost processes to different degrees.

Till generally forms a thin cover on the slopes and on the mountain plateaus. Hummocky moraines are found in valleys and depressions. Drumlins also occur locally. The content of boulders in the till varies depending on the bedrock. Within shale areas, the till surfaces generally have few boulders while in quartzite areas, the boulders are generally numerous, e.g., in areas around the Härjång and Lunndörren Mountains in southern parts of the province of Jämtland.

Glaciofluvial deposits in the form of small and generally long eskers occur and may sometimes also form networks of eskers. The eskers may be surrounded by fine-grained sediments, sometimes deposited in glacial lakes. Outwash plains occur in the valleys.

In general, peatland has a fairly minor extent, with the exception of certain areas in the provinces of Härjedalen and Jämtland, and the peat is usually a couple of metres thick.

Characteristic features of the mountains are different secondary formations such as boulder deposits, talus, tundrapolygons, terraces and—in the northernmost mountain area—palsas. All of these are the result of frozen ground or frost disintegration. In some places they might have been formed during an interstadial.

Hard bottoms refer not only to solid rock but also to areas covered with till where fine-grained material has been washed out in the bottom surface. Remaining boulders and stones are usually covered with dense algae vegetation. Off Kullaberg, at a depth of 20 m.

On an erosion or transport bottom, the sea floor, in this case varved clay, is eroded and also partly covered by a thin layer of sand and gravel. The Bothnian Sea, to the west of Björneborg, at a depth of 49 m.

Ophiuroid and a crab on a presentday accumulation bottom where clay, silt and dead organic material are sedimenting.

The influence of waves and currents on a transport bottom moves the sand in characteristic ripples. This photograph was taken at a depth of 10 m in the Öresund Strait.

Quaternary deposits on the sea floor

The distribution of different sediments on the sea floor provides evidence of the dynamic processes that have occurred. The fine-grained sediments from areas that have had uninterrupted sedimentation and that have successively grown in thickness are particularly interesting. These fine-grained sediments provide a historical archive in which changes to the marine environment can be followed backwards in time. Variations in microfossil contents and mineralogical and chemical composition illustrate the changes that have taken place in climate, water temperature, oxygen conditions, salinity, nutrient supply and sources for the sediments.

CLAY BOTTOMS

Clay has been divided into glacial and post-glacial clay since their origins are of great importance in identifying the bottom-dynamic conditions.

Post-glacial clay and *silt* generally contain enough organic matter to be called gyttja clay or clay gyttja. These sediments have been deposited since the retreat of the latest continental ice sheet and are characteristic of bottom areas where sedimentation is still in progress, i.e., *deposition bottoms*. The gyttja clay, usually grey-green in colour, largely fills depressions in the sea floor and in this way levels out the relief. This can be found in the central parts of the Skagerrak, with its extensive deposition areas. In the Kattegat, on the other hand, the largest deposition of fine-grained sediment takes place along the western edge of the Djupa Rännan trench. Along this slope, the material deposited is mainly such that is carried by the currents around the northern tip of Jutland. Further to the south, in the winding trenches to the west of Fladen and Lilla Middelgrund, clay is deposited asymmetrically in some places. This probably depends on longitudinal bottom currents.

In the Öresund Strait and the Danish Belts, the cross-sectional area of the water masses is small and the current velocities are consequently large. In such places, fine-grained sediment is deposited only to a minor extent and only in isolated depressions. The Baltic Proper is characterized by several large, level, sedimentation basins, for example, the Arkona Basin, the Bornholm Basin and the Gotland Deep. In such places, the sedimentation rate today is between 0.5 and 1.5 mm/year in the central parts of the deep areas and decreases to zero in peripheral parts. Values of the same magnitude have also been calculated for the large sedimentation basins of the Bothnian Sea and Bothnian Bay. However, there may be considerably larger sedimentation rates locally.

The *glacial clay* is as fine-grained as the post-glacial clay but with a considerably lower content of organic matter. Outcrops of the glacial clay in the surface of the bottom indicate areas of the seabed that are exposed to erosion or transport of material as a result of the action of waves or currents. Areas with exposed glacial clay can be found in shallower waters off the Swedish West Coast. In the Baltic, the glacial clay, usually red-brown and often with distinct clear annual varves, is found on slopes down towards the larger sedimentation basins. Glacial clay is also found in exposed areas and in shallow waters along the coasts, for example, in the Hanö Bay, along the east coast of Öland, in the offshore areas of the Svealand archipelago, and in the Bothnian Sea. In the Bothnian Bay, the glacial clay becomes less common in the seabed since here it is generally covered by post-glacial sediments even far up on the slopes.

SAND BOTTOMS

Sand bottoms are usually an indication of active transport and reworking of material. Sometimes *ripples* may illustrate the prevailing direction of transport through their gently sloping stoss side and steep lee side. If sand from a shallow area is transported out over a slope and becomes accumulated at greater depths with less influence of waves and currents, then it may result in sand accumulations of considerable magnitude. Primary, glacial sand accumulations, for example eskers, are rarely found on the bottom of the sea. They are generally reworked as a result of wave and current activity, or are at greater depths, mainly covered by fine-grained sediments.

HARD BOTTOMS AND EXPOSED ROCK

Bottoms with gravel, cobbles and boulders have been included in the classification *hard bottoms*. They make up residues of till and glaciofluvial sediments that have been exposed

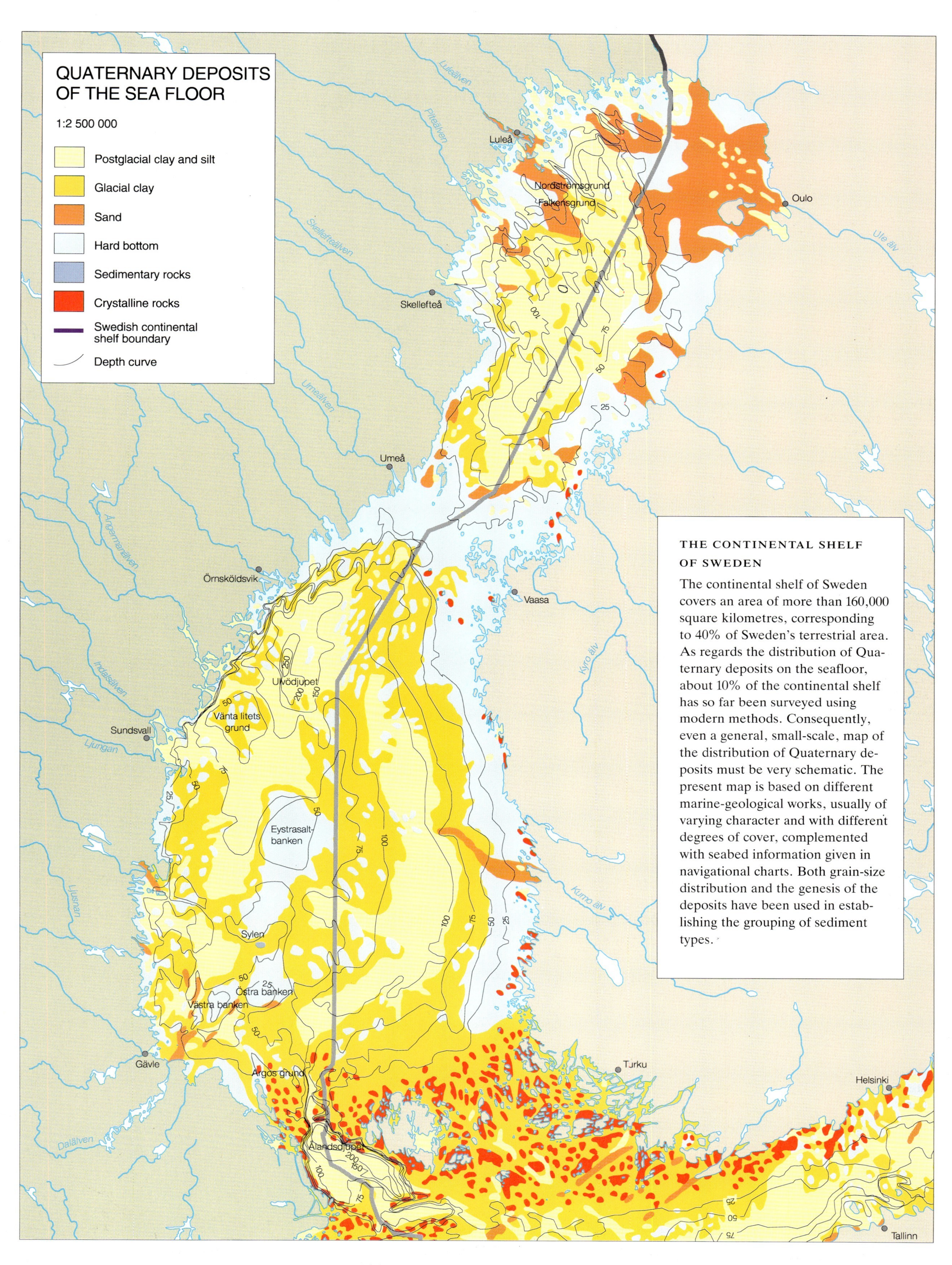

THE CONTINENTAL SHELF OF SWEDEN

The continental shelf of Sweden covers an area of more than 160,000 square kilometres, corresponding to 40% of Sweden's terrestrial area. As regards the distribution of Quaternary deposits on the seafloor, about 10% of the continental shelf has so far been surveyed using modern methods. Consequently, even a general, small-scale, map of the distribution of Quaternary deposits must be very schematic. The present map is based on different marine-geological works, usually of varying character and with different degrees of cover, complemented with seabed information given in navigational charts. Both grain-size distribution and the genesis of the deposits have been used in establishing the grouping of sediment types.

to erosion by currents or waves.

Areas with exposed rock also indicate the presence of active erosion or transportation. Surfaces with sedimentary rock are generally uniform and level because of their lower resistance to erosion. Areas of crystalline rocks, on the other hand, usually have an irregular and steep topography.

EROSION AND DEPOSITION

The influence of waves and currents causes erosion, transport and deposition of bottom material. If the hydrographical conditions change, the bottom sediments also change character.

We can distinguish between erosion bottoms, transport bottoms and deposition bottoms. The degree of influence depends mainly on water depth, current velocity, condition of the bottom material and on the sediment types in neighbouring coastal areas.

Following the deglaciation, the shoreline has been displaced both upwards and downwards depending on the interaction between changes in sea level and land uplift. This has resulted in glacial clays either being covered by sand and gravel or have been exposed to erosion in the sea floor, particularly along the coasts of Halland and Skåne.

Within the Baltic today, almost half of the sea floor is exposed to erosion or transport. This reworking of sediments has been calculated to be six times larger than the supply of sediments with the rivers. The mean value of the annual sedimentation rate within the deposition areas of the Baltic amounts to between 0.5 and 2 mm.

In contrast to the Skagerrak, the Kattegat has few deposition areas. In both the Skagerrak and the Kattegat, the amount of sediment is increasing as a result of supplies from the rivers as well as suspended material in water masses entering from the North Sea and the Baltic. Of the ca. 25 million tonnes of suspended material supplied to the southern North Sea annually, about 17 million tonnes are estimated to be taken further with the Jutland current to the Skagerrak and Kattegat. The average sedimentation rate is about 2 mm/year, but extreme rates of more than 100 mm/year have been observed in the western part of the Djupa Rännan trench.

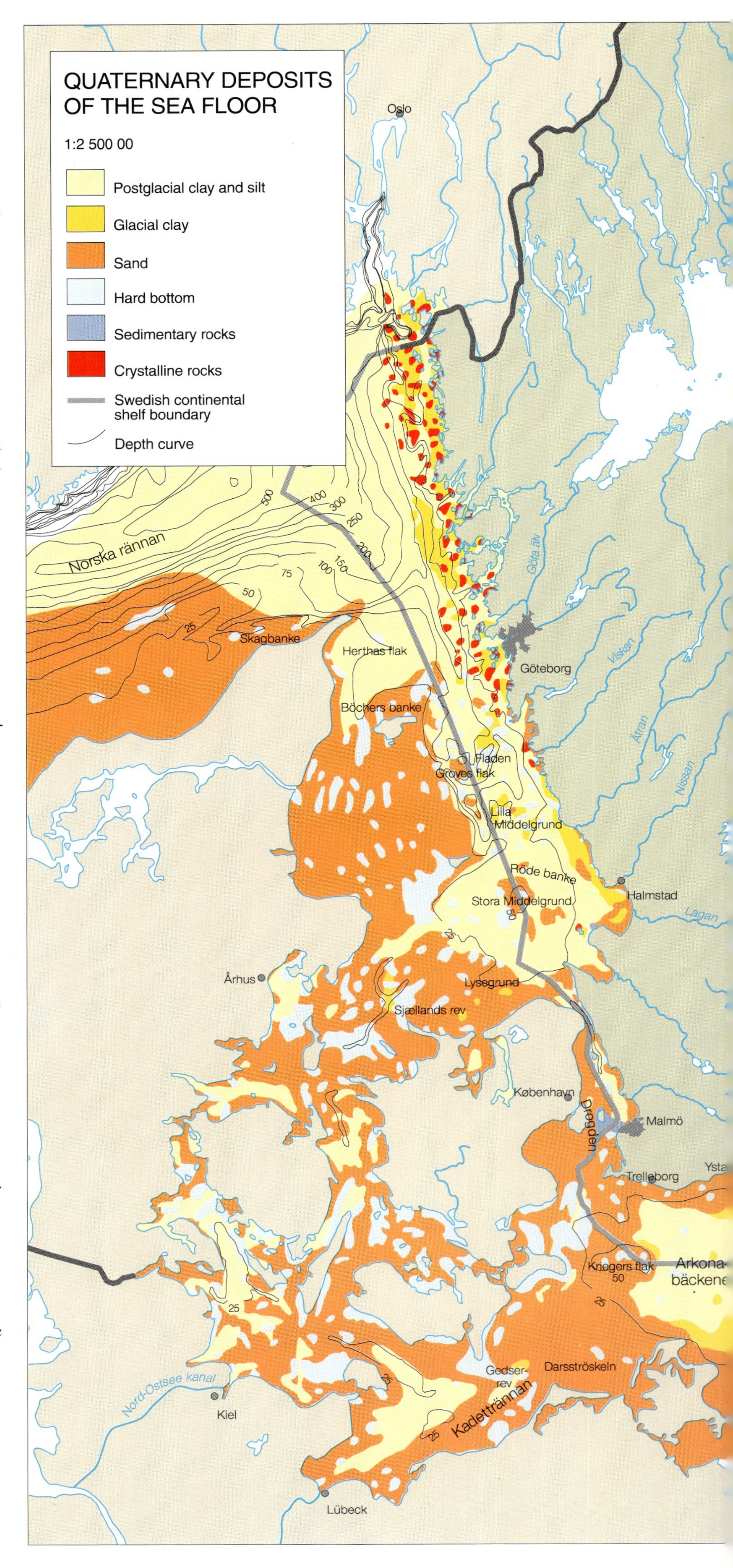

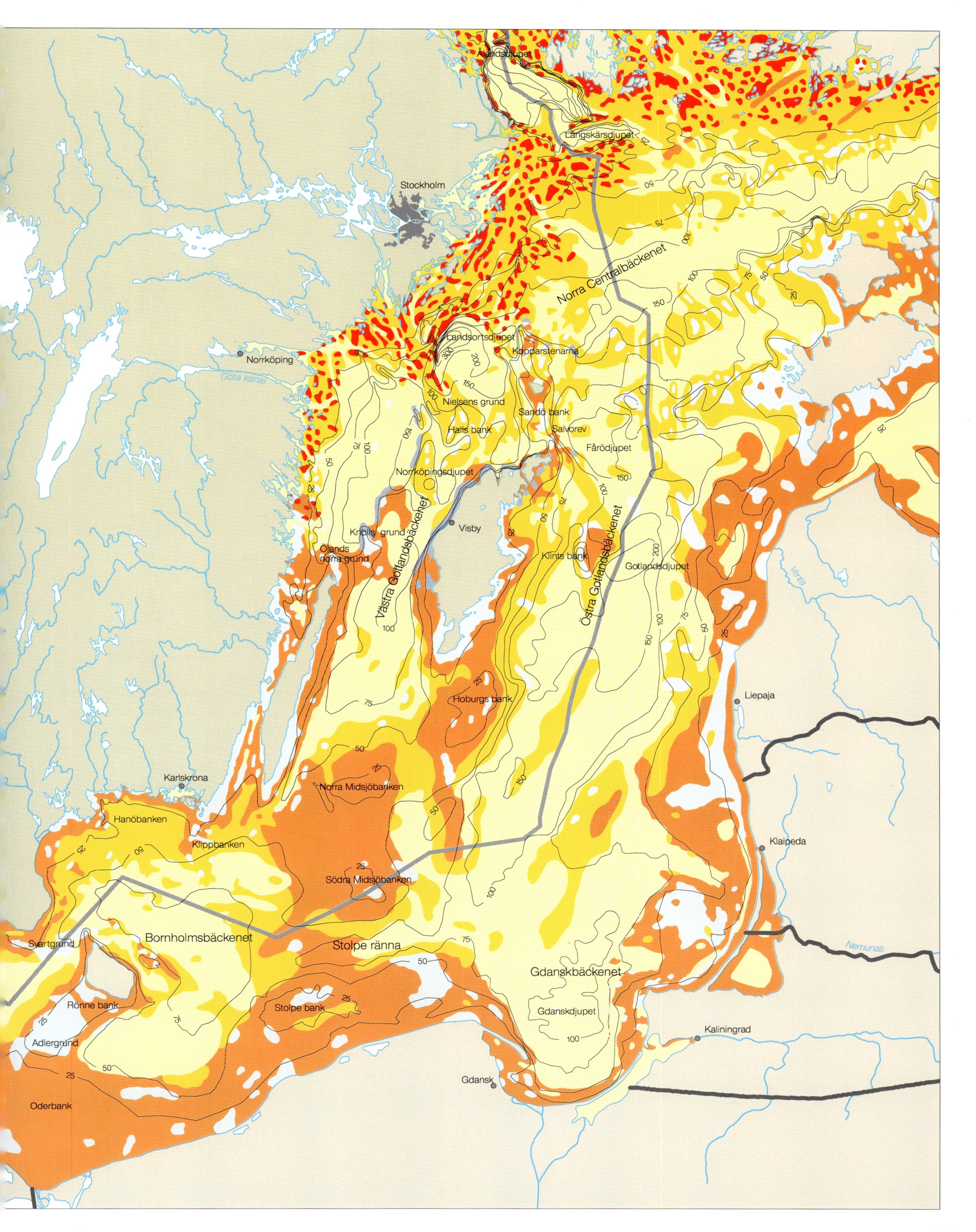

Ålandsdjupet
Långskärsdjupet
Stockholm
Norra Centralbäckenet
Norrköping
Göta kanal
Landsortsdjupet
Kopparstenarna
Nielsens grund
Sandö bank
Halls bank
Salvorev
Fårödjupet
Norrköpingsdjupet
Visby
Knolls grund
Ölands norra grund
Västra Gotlandsbäckenet
Östra Gotlandsbäckenet
Klints bank
Gotlandsdjupet
Venta
Liepaja
Hoburgs bank
Karlskrona
Norra Midsjöbanken
Hanöbanken
Klippbanken
Klaipeda
Södra Midsjöbanken
Svartgrund
Bornholmsbäckenet
Stolpe ränna
Nemunas
Gdanskbäckenet
Rönne bank
Stolpe bank
Gdanskdjupet
Kaliningrad
Adlergrund
Gdansk
Oderbank

Groundwater

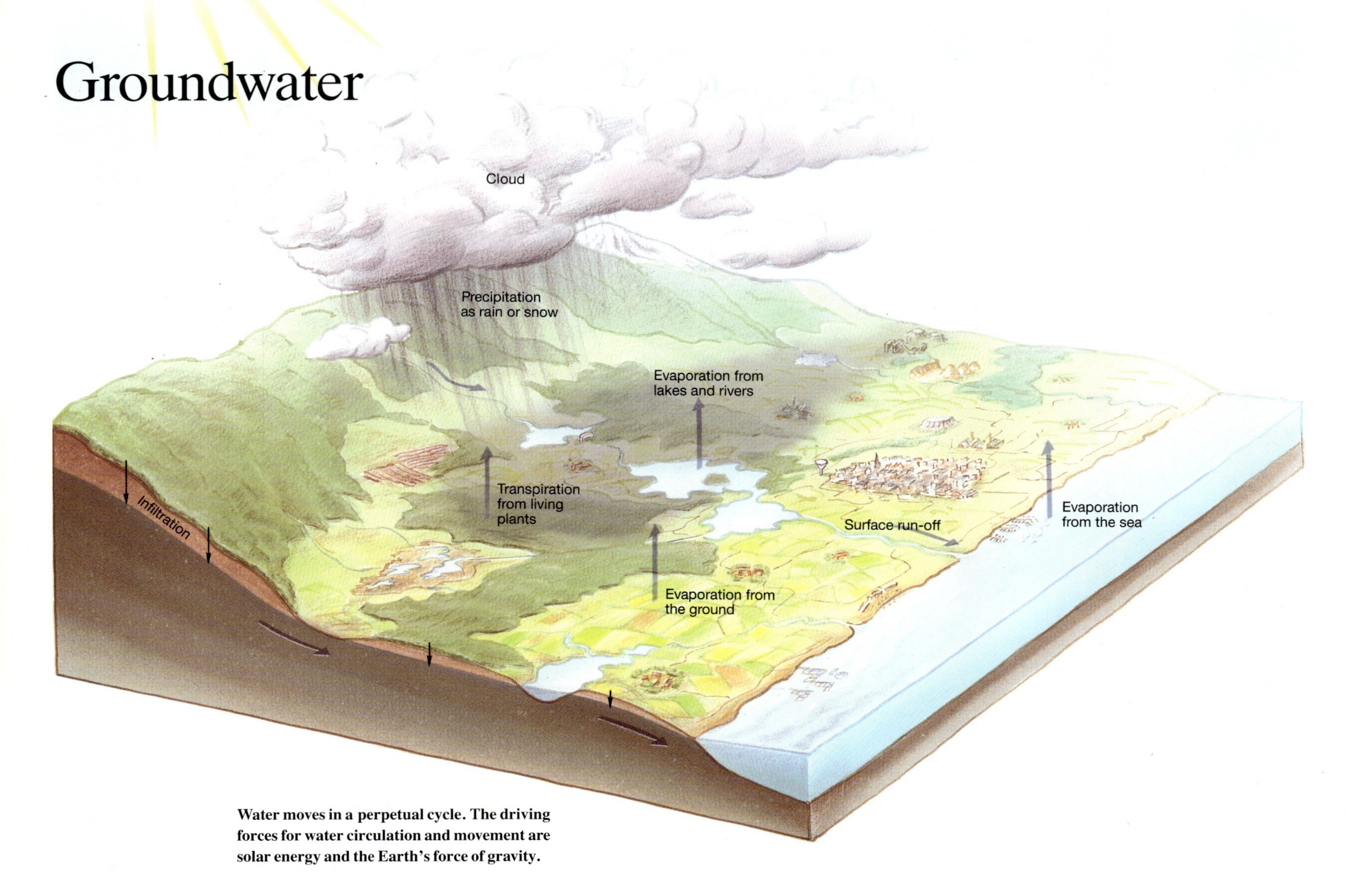

Water moves in a perpetual cycle. The driving forces for water circulation and movement are solar energy and the Earth's force of gravity.

Water that fills cavities, pores and open fissures in Quaternary deposits and in the bedrock below a certain level is called *groundwater*. Groundwater is one of our most important natural resources and is found throughout the world, but in varying amounts and at varying depths.

In the *saturated zone*, cavities are completely filled with water, whereas in the *unsaturated zone* they are mainly filled with air. Water moves down to the saturated groundwater level. However, water from the groundwater zone is capable of moving slightly upwards by means of capillary action.

Groundwater that is under sufficient pressure to rise above the surface of the ground is called *artesian water*.

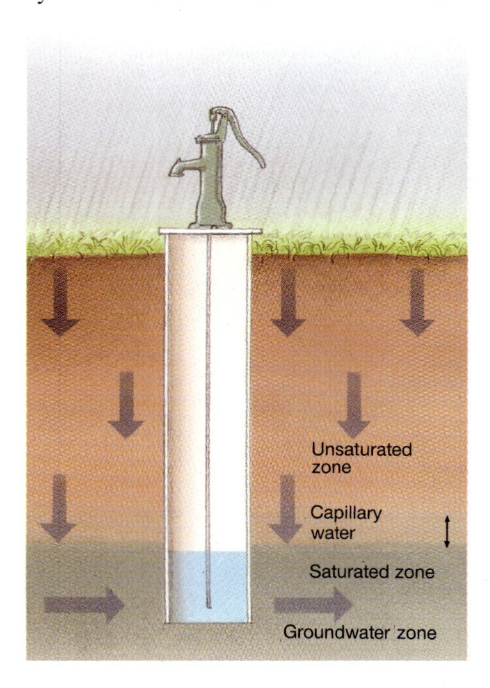

The water level in a well shows the groundwater level of the surroundings.

The hydrological cycle

The total amount of water on Earth is constant, but water moves in a continuous cycle between different areas, either as water, vapour or ice. This cycle is usually called the *hydrological cycle*. Water evaporates from the ground and areas of open water, mainly from seas, lakes and plants (transpiration), and forms clouds. Clouds release precipitation, which falls onto the ground.

Some of the precipitation evaporates and again returns to the atmosphere. A small amount flows more or less directly into lakes and watercourses, and some will penetrate (infiltrate) into the ground. Water that is not taken up by plants forms the groundwater. The amount of water in the cycle that forms groundwater mainly depends on climate, geological conditions, and vegetation.

Water in the watercourses flows fairly rapidly into the sea. However, even the groundwater moves, either directly or via surface watercourses, towards the sea. Most of the water in surface watercourses has, in fact, earlier been groundwater that has seeped through the bottoms of the watercourses and their banks.

Since the total amount of water in the environment is constant, the hydrological cycle can be described by the following formula:

Precipitation = Run-off +
Reservoir changes + Evaporation

When precipitation is greater than evaporation plus run-off, the groundwater level will rise, and vice versa. Water resources in the groundwater reservoir increase and decrease naturally as a result of variations in precipitation. In the long-term perspective, however, the amount of water in the groundwater reservoir will remain largely unaltered, provided that excessive amounts of water are not pumped out locally, or that the vegetation changes are not so radical that the natural balance is disturbed.

Groundwater resources in Sweden

An *aquifer* is a geological formation that is so permeable that water can be extracted from it in usable quantities. In basement rock, such as granite and gneiss, groundwater is mainly found in fractures in the rock—*fracture aquifers*. In glaciofluvial deposits, groundwater is found in pores between the grains of gravel and sand—*pore aquifers*. In addition, there are intermediate types such as those in sandstone, where groundwater is found both in fractures and pores.

One of the difficulties encountered when groundwater is studied is that it cannot be seen other than in places where it emerges in springs. In order to study the groundwater, we must therefore dig or drill, test-pump and measure groundwater levels. Since 1976, there is a legal obligation for well-drilling companies and consultant engineers to submit reports to the Geological Survey of Sweden, SGU, on wells they have drilled. This has greatly improved our knowledge of groundwater, and has enabled us to make better use of groundwater resources.

Where the groundwater level is higher than the ground surface, the water may emerge as a spring or seep out in a more diffuse fashion, thus giving rise to wetlands.

GROUNDWATER IN THE BEDROCK

On average, wells drilled in crystalline rock yield 600–2,000 litres per hour (l/h). In exceptional cases, wells drilled in larger fissure zones may yield 40,000–50,000 l/h.

In the northwesternmost part of Hallandsåsen on Bjärehalvön, the crystalline rocks are more extensively cracked and yield more water than most other places in Sweden. Here, wells drilled in rock usually yield about 20,000 l/h, and sometimes twice that amount.

Within the group of sedimentary rocks, there is a very large variation in groundwater resources. Cambrian sandstone in the provinces of Närke, Västergötland and Östergötland usually yields 3,000–4,000 l/h. Wells drilled in chalk in southwestern Skåne generally yield 60,000 l/h, and sometimes up to about 200,000 l/h. These wells are mainly used for irrigation in agriculture. Glauconite sandstone in the Kristianstad area is a special case. The sandstone is extremely loosely consolidated and very similar to sand. Here, for example, there are wells that yield more than 400,000 l/h. On the other hand, some clay shales give no water at all.

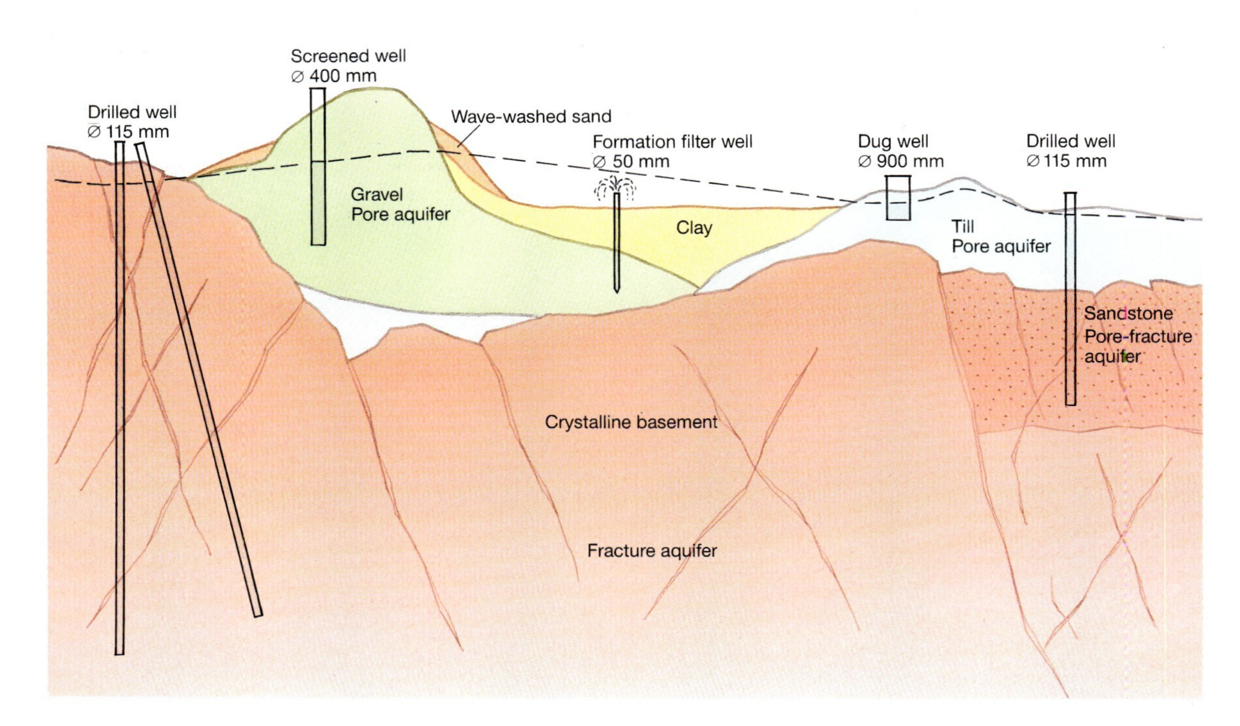

Depending on the groundwater situation, wells are constructed in different ways.

Glaciofluvial deposits, such as eskers, are the most important groundwater resources in Quaternary deposits. A small esker at Smula, to the south of Falköping, Västergötland.

GROUNDWATER IN QUATERNARY DEPOSITS

Some of the most important groundwater resources are found in glaciofluvial sand and gravel deposits, mainly in pronounced eskers and deltas.

One of Sweden's largest glaciofluvial deposits, the Badelunda Esker, yields 730,000 l/h in Borlänge. At Njurunda, to the south of Sundsvall, a continuous supply of 900,000 l/h is obtained from a group of gravel-filter wells, one of which yields 790,000 l/h. At Vivstanäs, to the north of Sundsvall, plans are being made to extend an area of wells in a sand and gravel deposit which will result in a capacity of about 1,800 000 l/h.

In these cases, the groundwater has probably not only been formed

The natural groundwater resources can be increased by allowing surface water to infiltrate in sand and gravel deposits.

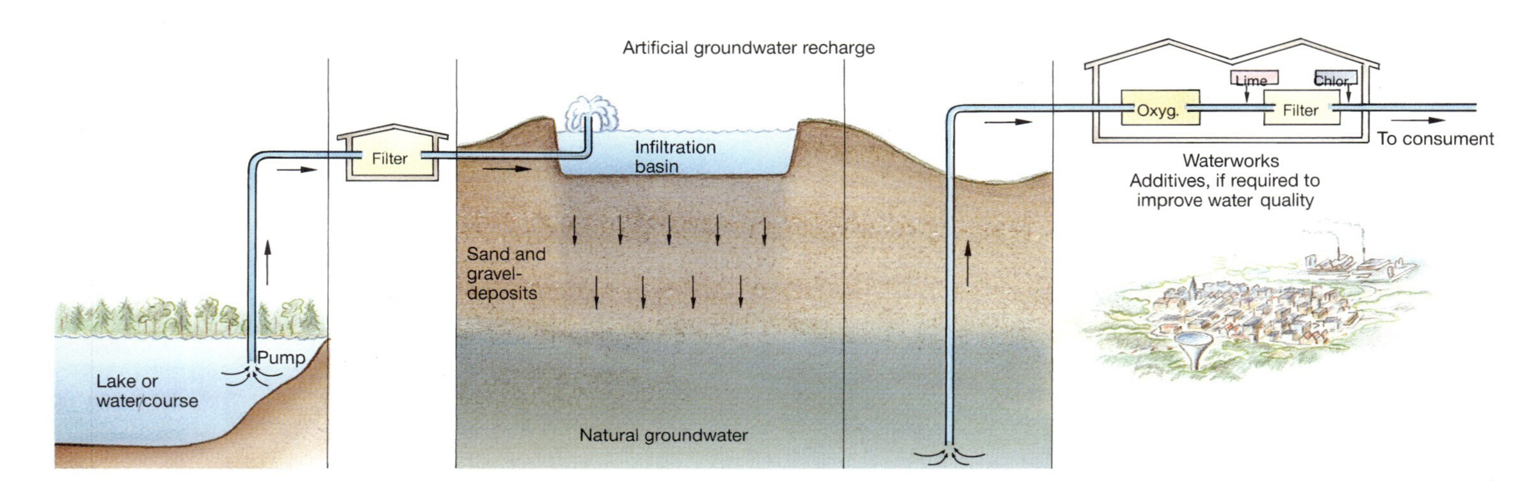

When water is filled into an infiltration basin, it is aerated; iron and manganese compounds are precipitated and form a brown sludge that becomes attached to the bottom of the basin.

The infiltration plant on the Uppsala Esker to the north of Uppsala.

Irrigation of agricultural land is being increasingly utilized, mainly in southern Sweden. In places this has led to a decrease in the groundwater level and to competition for water resources.

through infiltration of precipitation water, but also as a result of infiltration of river water, a process known as *induced infiltration*. During the filtering through sand and gravel deposits from the rivers to the wells, the water has been converted into groundwater with high and uniform quality as drinking water.

When the natural groundwater resources are insufficient, they can be improved by allowing surface water from lakes or watercourses to infiltrate through basins down into glaciofluvial deposits. One of the cities where this has been carried out is Västerås, where formerly only natural groundwater from the esker Badelundaåsen was used. By adding water from Lake Mälaren to infiltration basins on the esker, it was possible to increase the amount of water available. Water in Lake Mälaren is of relatively poor quality but after filtration through the bottom of the basin it has the same fine quality as natural groundwater. This is called *artificial recharge* and was originally used in Göteborg. Today, this technique is applied in many larger cities.

In the Alnarp depression in southern Skåne—a valley about 50 km long, 5 km wide and about 100 m deep, filled up with deposits of sand and gravel—the wells yield up to 140,000 l/h.

Wells in till are often sufficient for small water requirements such as those for individual households, even though the water quality in these wells is usually poorer.

UTILIZATION OF GROUNDWATER

About 87% of the Swedish population are connected to municipal water-works. Of these inhabitants, 38%, i.e., about 2.6 million people, are supplied with natural groundwater or with groundwater boosted by artificial recharge. Both permanent residents and recreational visitors who are not connected to the municipal waterworks generally use groundwater from their own wells.

During periods of drought, groundwater is increasingly used by agriculture for irrigation. Grassland and vegetables are irrigated during the early part of the growing season, whereas potatoes and beet fields are irrigated later.

During times of high energy prices, there has been an increased interest in using heat pumps to extract energy from groundwater. Since only heat energy is utilized, the water can either be released into a watercourse or pumped back into the groundwater reservoir. The use of closed systems allows the heat in the rock to be utilized without influencing groundwater conditions to any particular extent. Energy wells are usually drilled deeper than corresponding wells for drinking water.

Interest in using groundwater to produce bottled drinking-water has increased, often as a result of advertising where emphasis is placed on the fact that the water has been taken from wells with ancient historical traditions, mineral springs, etc.

MUNICIPAL WATER CONSUMPTION, 1989

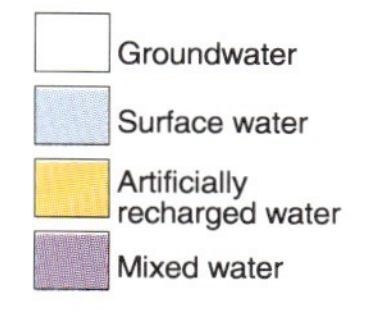

No. of waterworks

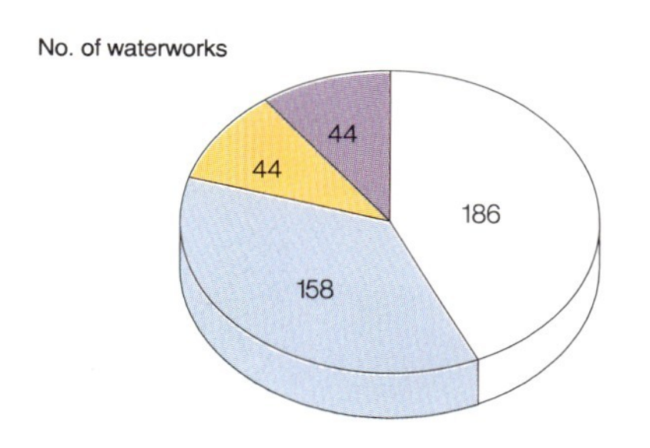

No. of persons supplied, 1,000s

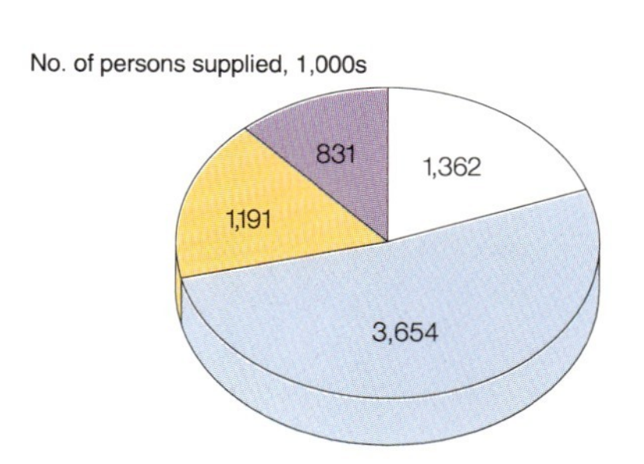

Production, million m^3

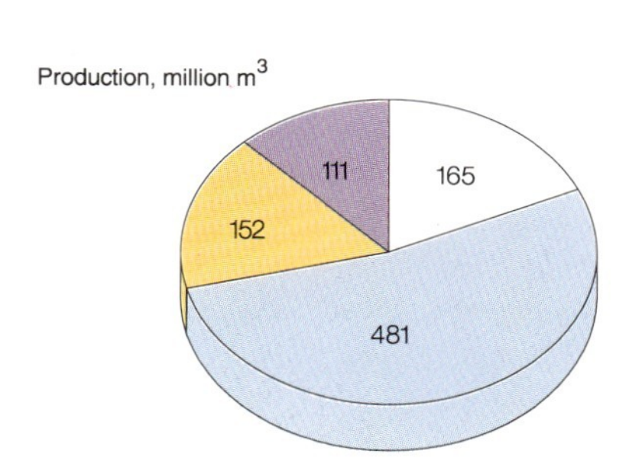

Most small and medium-sized communities base their supply of drinking-water on the groundwater.

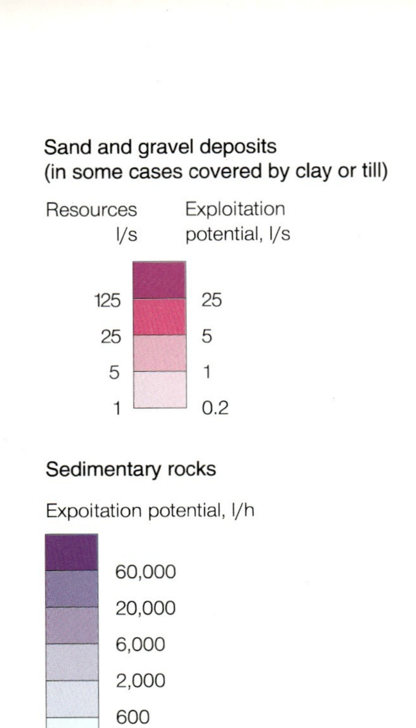

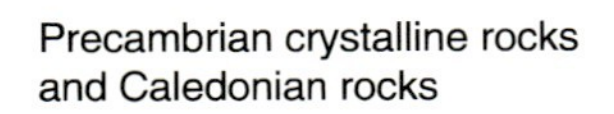

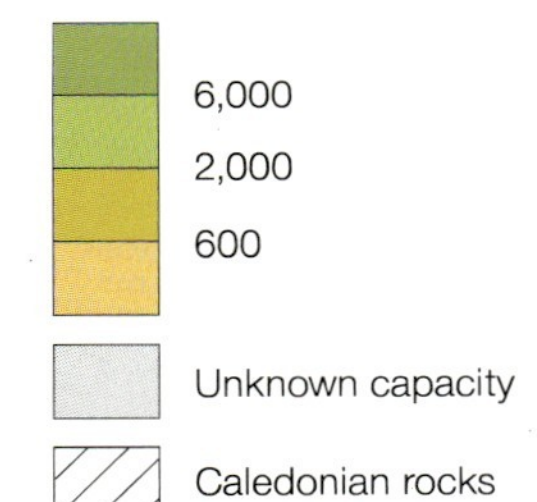

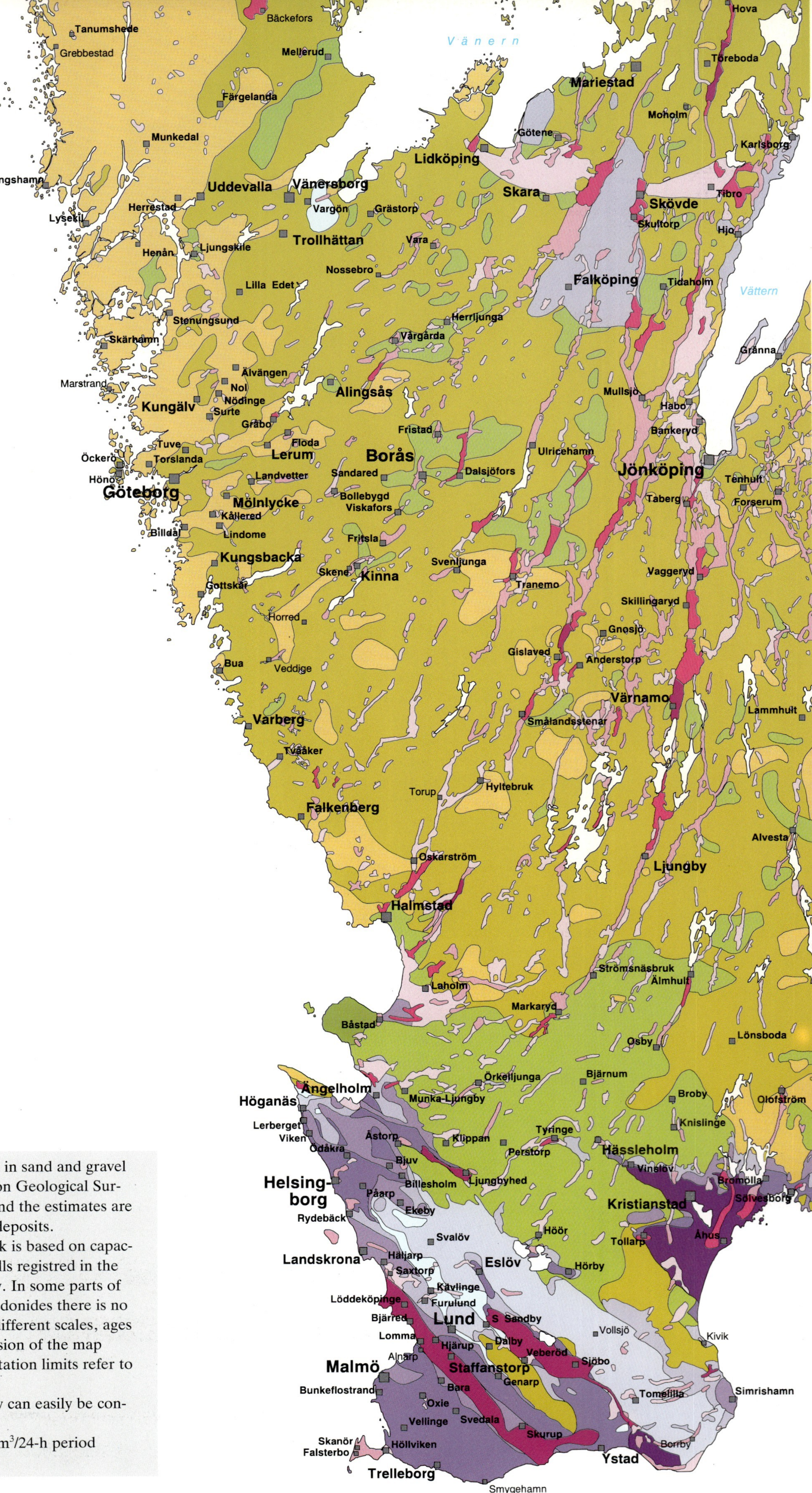

Information on groundwater resources in sand and gravel deposits in southern Sweden is based on Geological Survey's hydrogeological maps. In Norrland the estimates are mainly based on maps of Quaternary deposits.

Exploitation potential in the bedrock is based on capacity information from about 120,000 wells registred in the Well Records at the Geological Survey. In some parts of the inland of Norrland and in the Caledonides there is no information on wells. Because of the different scales, ages and quality of the data used, the precision of the map varies. Symbols used to express exploitation limits refer to minimum values.

Groundwater resources and capacity can easily be converted as follows:

1 l/s = 60 l/min. = 3,600 l/h = 86.4 m^3/24-h period

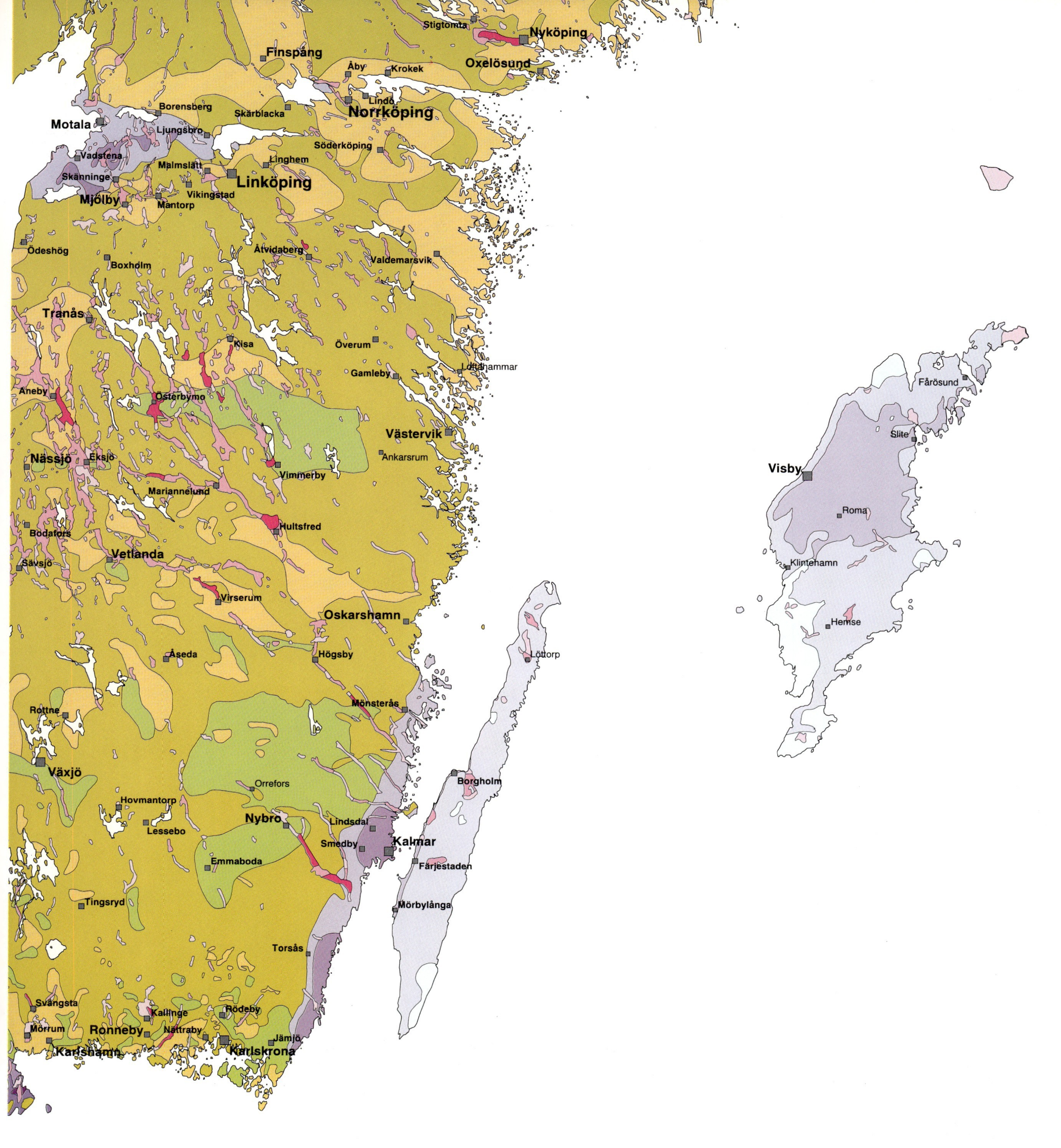

Groundwater in Bedrock and Quaternary Deposits

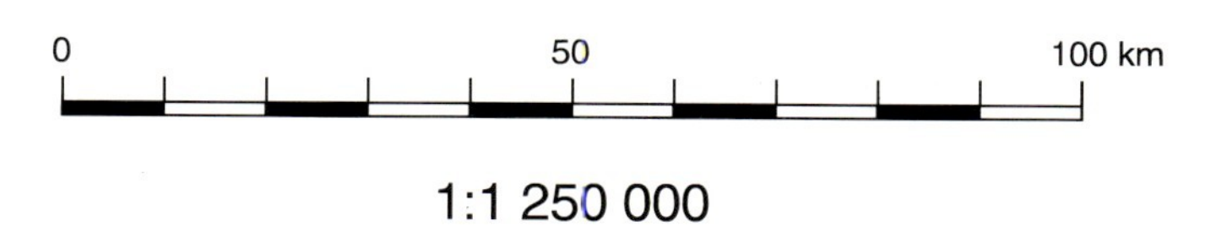

(M66)

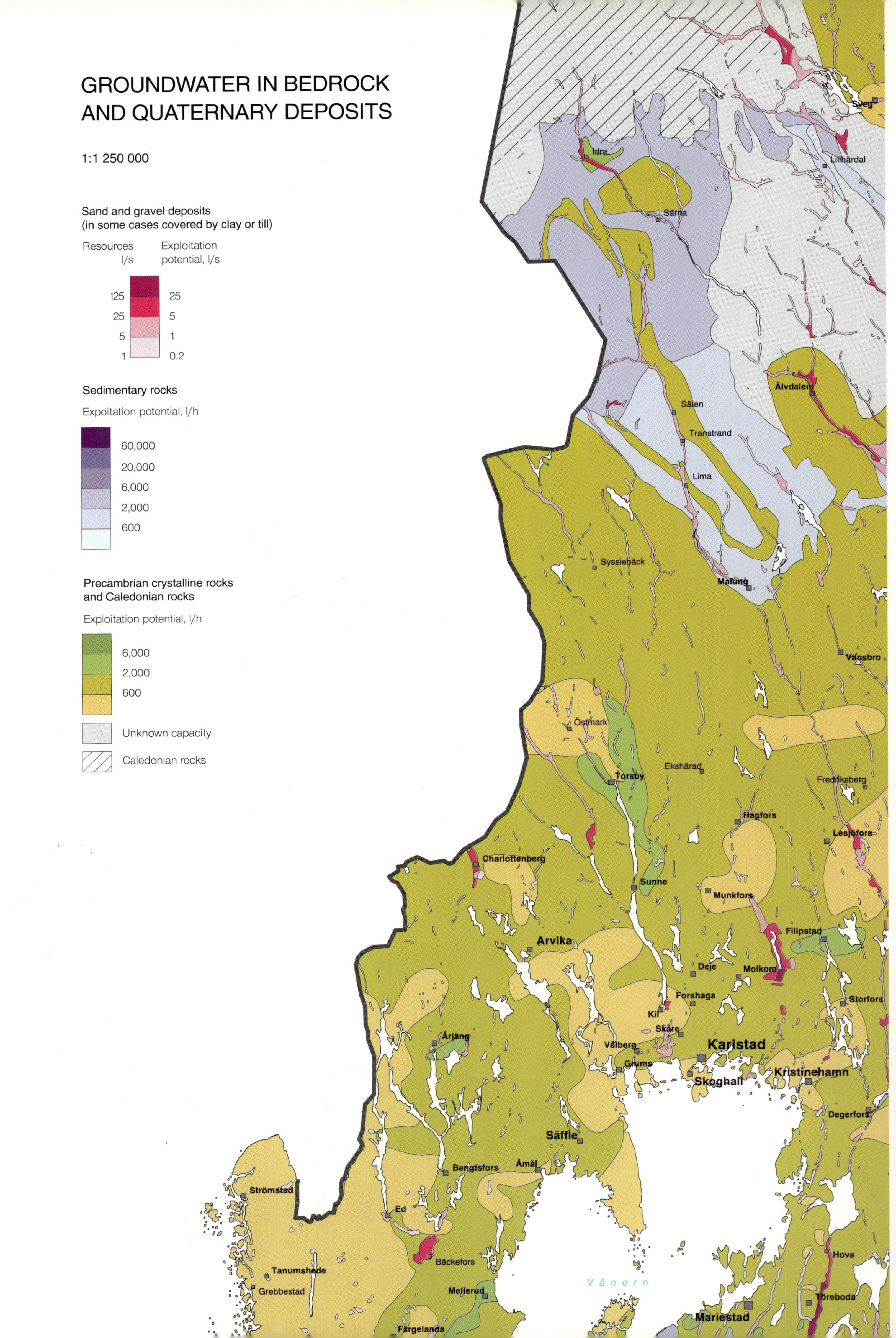
GROUNDWATER IN BEDROCK AND QUATERNARY DEPOSITS
1:1 250 000
Sand and gravel deposits
(in some cases covered by clay or till)
Resources l/s
Exploitation potential, l/s
125
25
25
5
5
1
1
0.2
Sedimentary rocks
Expoitation potential, l/h
60,000
20,000
6,000
2,000
600
Precambrian crystalline rocks and Caledonian rocks
Exploitation potential, l/h
6,000
2,000
600
Unknown capacity
Caledonian rocks
Sveg
Idre
Lillhärdal
Särna
Älvdalen
Sälen
Transtrand
Lima
Sysslebäck
Malung
Vansbro
Östmark
Torsby
Ekshärad
Fredriksberg
Hagfors
Lesjöfors
Charlottenberg
Sunne
Munkfors
Filipstad
Arvika
Deje
Molkom
Forshaga
Storfors
Kil
Skåre
Årjäng
Välberg
Karlstad
Grums
Skoghall
Kristinehamn
Degerfors
Säffle
Bengtsfors
Åmål
Strömstad
Ed
Bäckefors
Tanumshede
Grebbestad
Mellerud
Vänern
Hova
Töreboda
Mariestad
Färgelanda

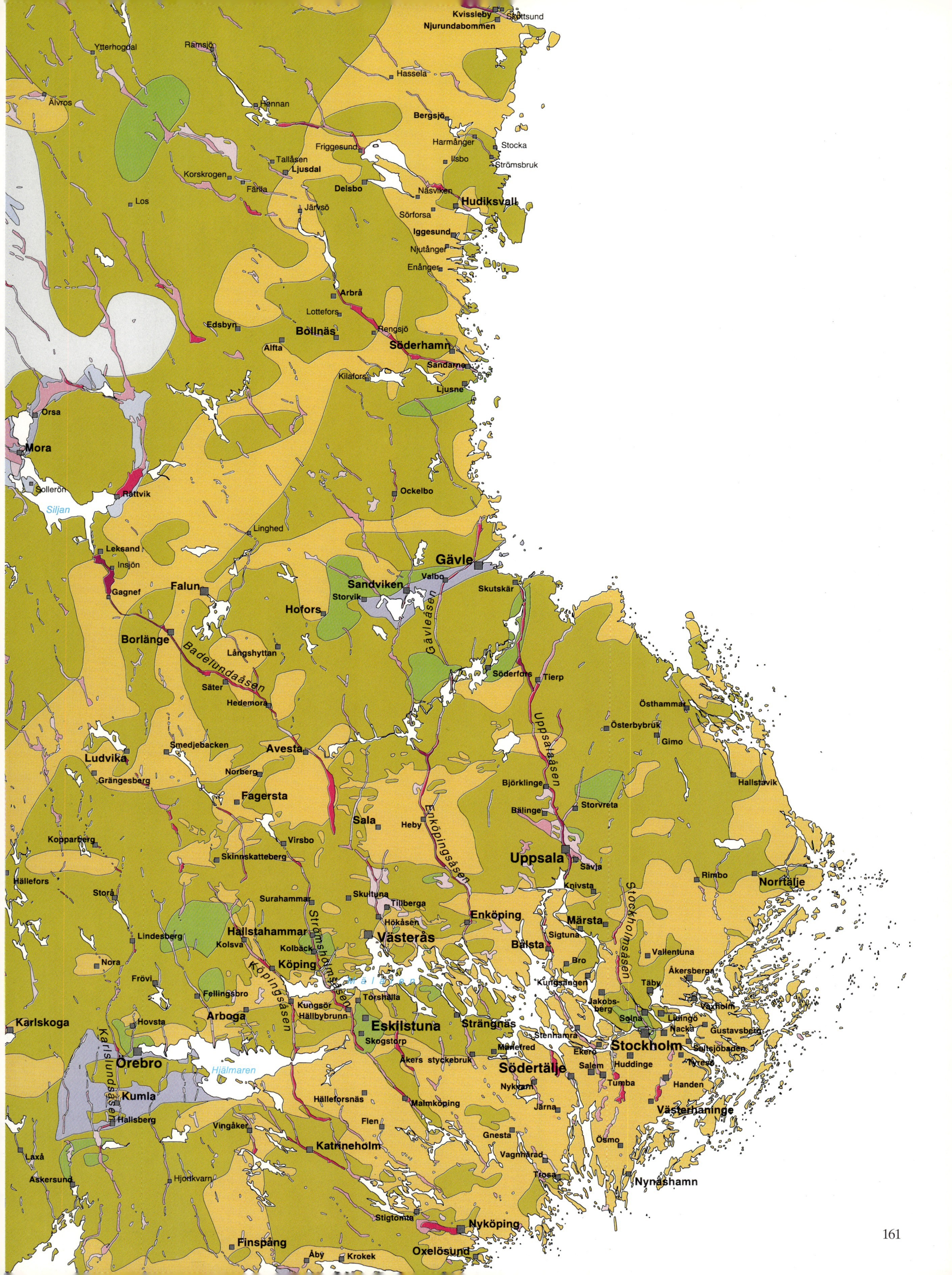

Kvissleby
Skottsund
Njurundabommen
Ytterhogdal
Ramsjö
Hassela
Älvros
Hennan
Bergsjö
Harmånger
Stocka
Friggesund
Ilsbo
Strömsbruk
Tallåsen
Ljusdal
Korskrogen
Färila
Delsbo
Näsviken
Hudiksvall
Los
Järvsö
Sörforsa
Iggesund
Njutånger
Enånger
Arbrå
Lottefors
Edsbyn
Bollnäs
Rengsjö
Alfta
Söderhamn
Sandarne
Kilafors
Ljusne
Orsa
Mora
Sollerön
Rättvik
Siljan
Ockelbo
Linghed
Leksand
Insjön
Gävle
Gagnef
Falun
Valbo
Sandviken
Storvik
Skutskär
Hofors
Borlänge
Gävleåsen
Badelundaåsen
Långshyttan
Söderfors
Tierp
Säter
Hedemora
Östhammar
Uppsalaåsen
Österbybruk
Gimo
Smedjebacken
Avesta
Ludvika
Grängesberg
Norberg
Björklinge
Hallstavik
Fagersta
Bälinge
Storvreta
Sala
Heby
Enköpingsåsen
Kopparberg
Virsbo
Uppsala
Skinnskatteberg
Sävja
Rimbo
Norrtälje
Hällefors
Knivsta
Storå
Surahammar
Skultuna
Tillberga
Stockholmsåsen
Enköping
Märsta
Hökåsen
Hallstahammar
Lindesberg
Västerås
Sigtuna
Kolsva
Strömsholmsåsen
Bålsta
Vallentuna
Kolbäck
Nora
Köping
Bro
Åkersberga
Frövi
Mälaren
Kungsängen
Täby
Köpingsåsen
Fellingsbro
Torshälla
Jakobsberg
Vaxholm
Kungsör
Solna
Lidingö
Hällbybrunn
Arboga
Nacka
Eskilstuna
Strängnäs
Gustavsberg
Karlskoga
Hovsta
Stenhamra
Karlsundaåsen
Skogstorp
Mariefred
Ekerö
Stockholm
Saltsjöbaden
Örebro
Hjälmaren
Åkers styckebruk
Södertälje
Salem
Huddinge
Tyresö
Nykvarn
Tumba
Handen
Kumla
Hälleforsnäs
Malmköping
Järna
Västerhaninge
Hallsberg
Flen
Vingåker
Gnesta
Ösmo
Katrineholm
Laxå
Vagnhärad
Askersund
Hjortkvarn
Trosa
Nynäshamn
Stigtomta
Nyköping
Finspång
Åby
Krokek
Oxelösund

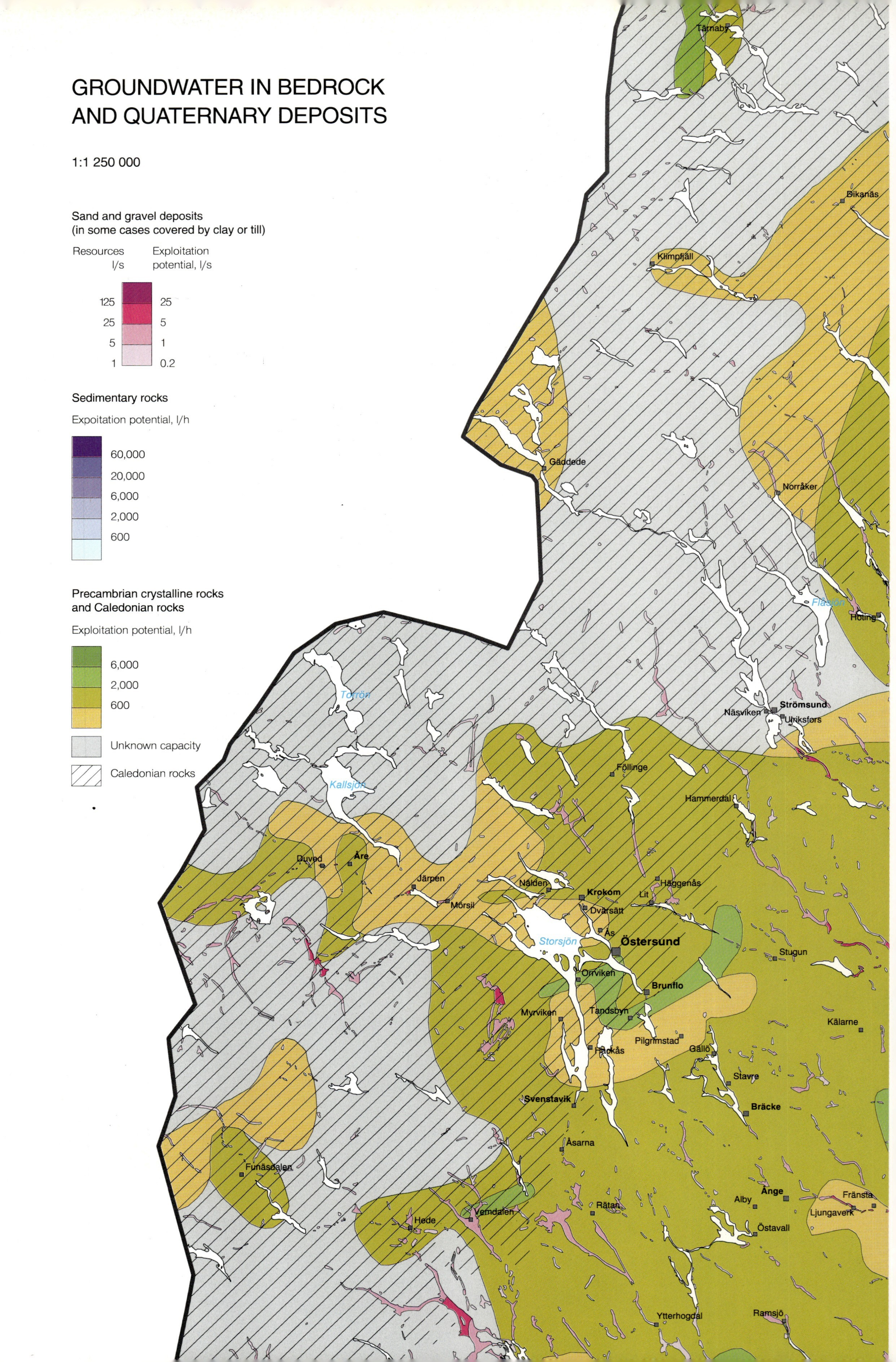

GROUNDWATER IN BEDROCK AND QUATERNARY DEPOSITS
1:1 250 000
Sand and gravel deposits (in some cases covered by clay or till)
Resources l/s
Exploitation potential, l/s
125
25
5
1
25
5
1
0.2
Sedimentary rocks
Expoitation potential, l/h
60,000
20,000
6,000
2,000
600
Precambrian crystalline rocks and Caledonian rocks
Exploitation potential, l/h
6,000
2,000
600
Unknown capacity
Caledonian rocks
Tärnaby
Dikanäs
Klimpfjäll
Gäddede
Norråker
Flåsjön
Hoting
Strömsund
Näsviken
Ulriksfors
Torrön
Kallsjön
Föllinge
Hammerdal
Duved
Åre
Järpen
Mörsil
Nälden
Krokom
Dvärsätt
Lit
Häggenås
Ås
Storsjön
Östersund
Stugun
Orrviken
Brunflo
Myrviken
Tandsbyn
Pilgrimstad
Kälarne
Hackås
Gällö
Stavre
Svenstavik
Bräcke
Åsarna
Funäsdalen
Ånge
Fränsta
Alby
Rätan
Ljungaverk
Hede
Vemdalen
Östavall
Ytterhogdal
Ramsjö

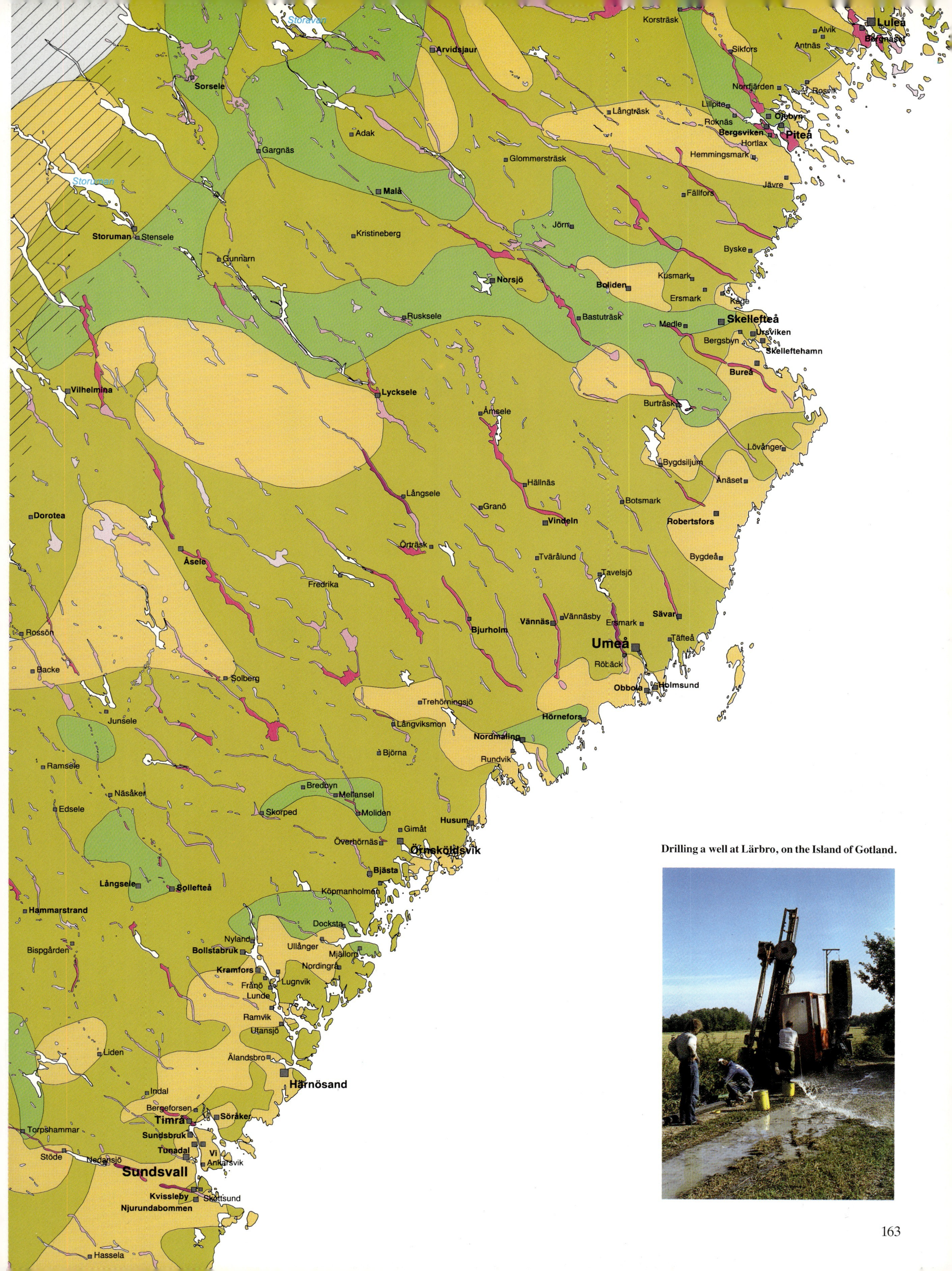

Drilling a well at Lärbro, on the Island of Gotland.

GROUNDWATER IN BEDROCK AND QUATERNARY DEPOSITS

1:1 250 000

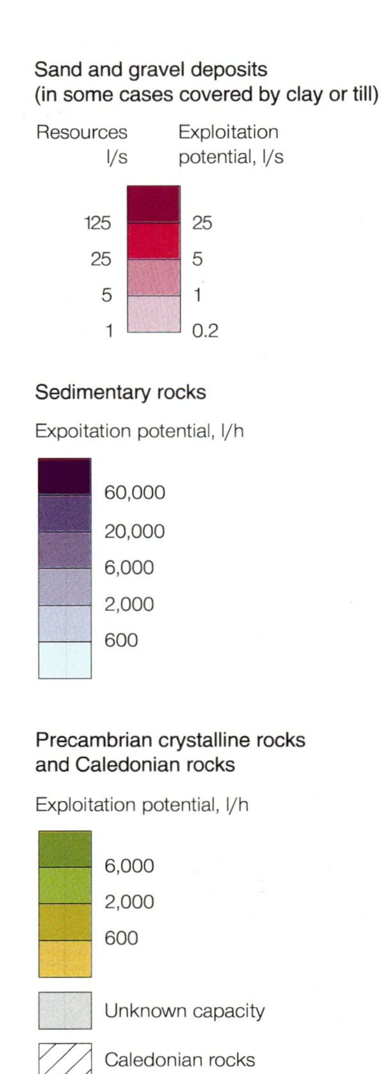

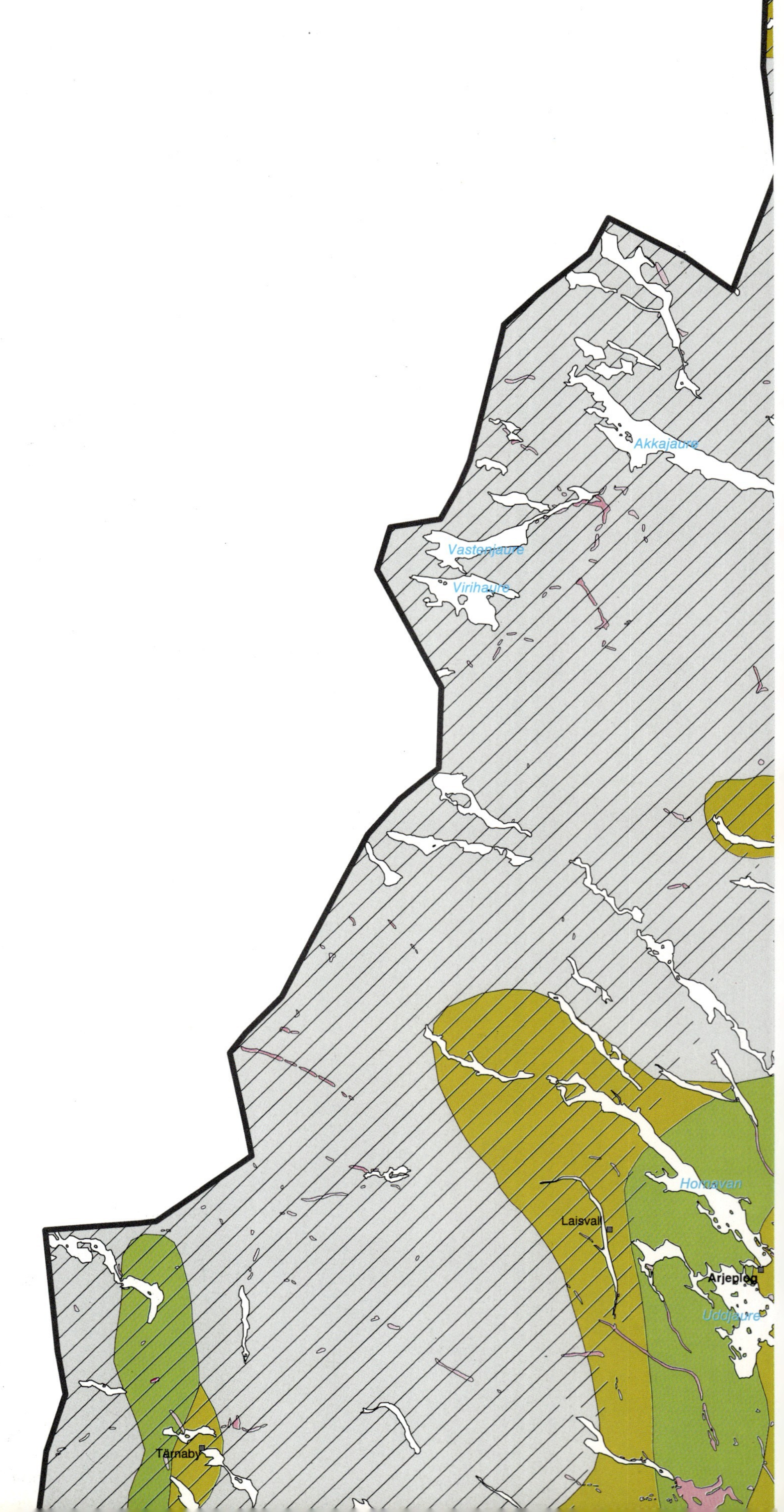

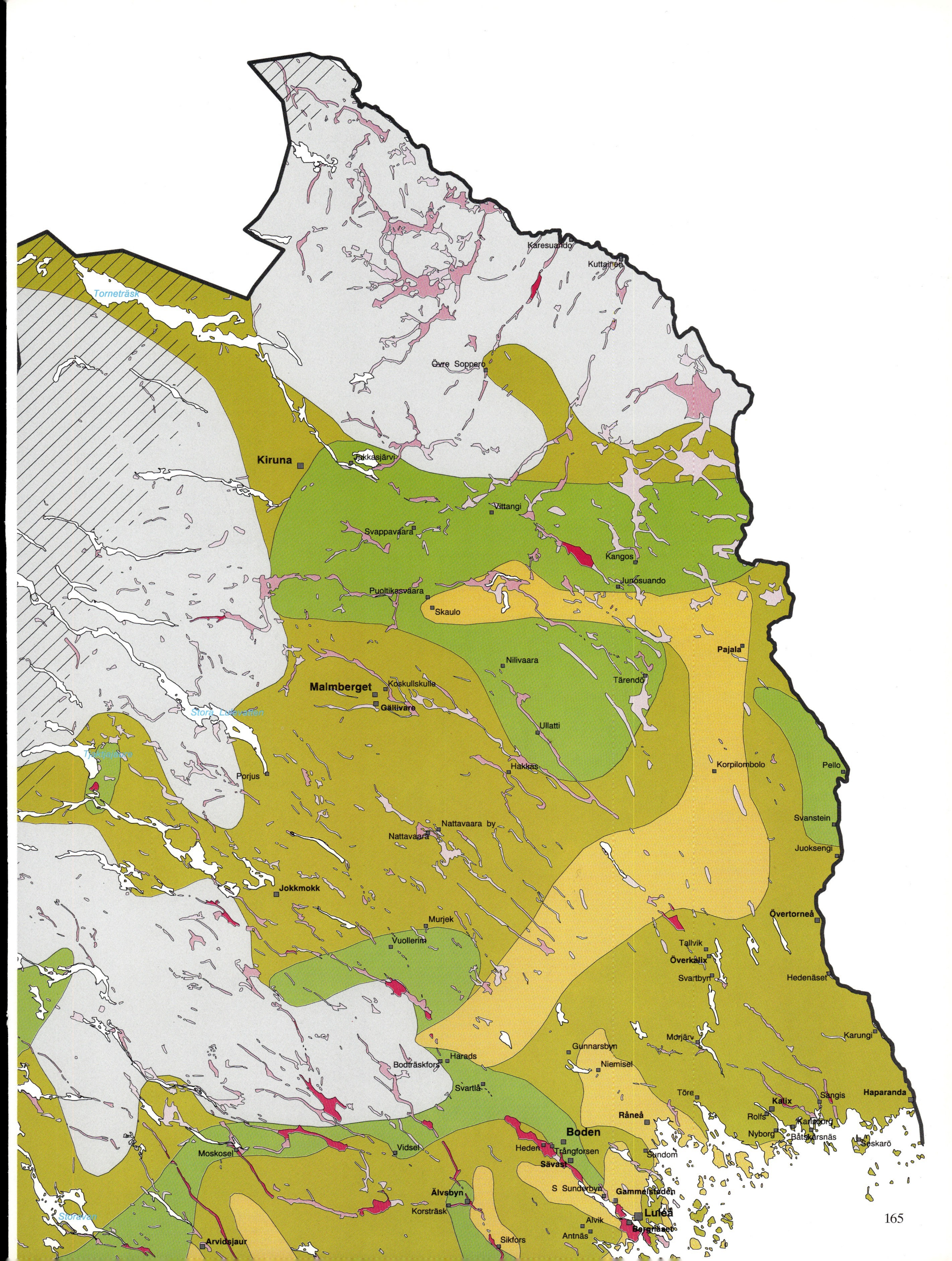
Torneträsk
Karesuando
Kuttainen
Övre Soppero
Kiruna
Tuolluvaara
Vittangi
Svappavaara
Kangos
Junosuando
Puoltikasvaara
Skaulo
Pajala
Nilivaara
Tärendö
Malmberget
Koskullskulle
Gällivare
Stora Lulevatten
Ullatti
Porjus
Hakkas
Korpilombolo
Pello
Svanstein
Juoksengi
Nattavaara by
Nattavaara
Jokkmokk
Murjek
Övertorneå
Vuollerim
Tallvik
Överkalix
Svartbyn
Hedenäset
Morjärv
Karungi
Gunnarsbyn
Niemisel
Bodträskfors
Harads
Svartlå
Töre
Kalix
Sangis
Haparanda
Råneå
Rolfs
Karlsborg
Nyborg
Båtskärsnäs
Seskarö
Boden
Heden
Trångforsen
Sävast
Sundom
Moskosel
Vidsel
S Sunderbyn
Gammelstaden
Älvsbyn
Korsträsk
Luleå
Alvik
Bergnäset
Antnäs
Storavan
Sikfors
Arvidsjaur

Groundwater composition

Groundwater is usually clear, cold, colourless and odourless, with a fresh and pleasant taste. It can usually be used as drinking-water without any purification. With regard to hygienic aspects, several substances and properties are of importance.

In poorly-constructed wells, *surface water* may easily infiltrate and, if polluted, will introduce high concentrations of bacteria and other microbiological substances, making the water unsuitable for drinking purposes. Leaking sewerage pipes will have the same result.

Water containing high concentrations of *calcium* and *magnesium* is usually called *hard*, and is found within areas with lime-rich rocks, gravel and sand. Hard water may cause problems as a result of the lime deposits formed, particularly in vessels where the water is heated, e.g., immersion heaters, washing machines, and coffee-machines. If the water is hard, then considerably more washing detergent is required to obtain the desired effect. However, from health viewpoints, hard water is considered to be more beneficial than soft water. Hard water can be softened by using an ion-exchanger.

Water with very low pH-values, i.e., *acidic groundwater*, will attack metallic piping and thus may cause leaks.

A problem that has become accentuated during recent years is the occurrence of *hydrogen sulphide*, mainly in drilled wells. If hydrogen sulphide is present, the water smells of rotten eggs. The odour usually disappears if the water is aerated.

High concentrations of *fluoride, heavy metals* and *nitrogenous compounds* may make the water unsuit-

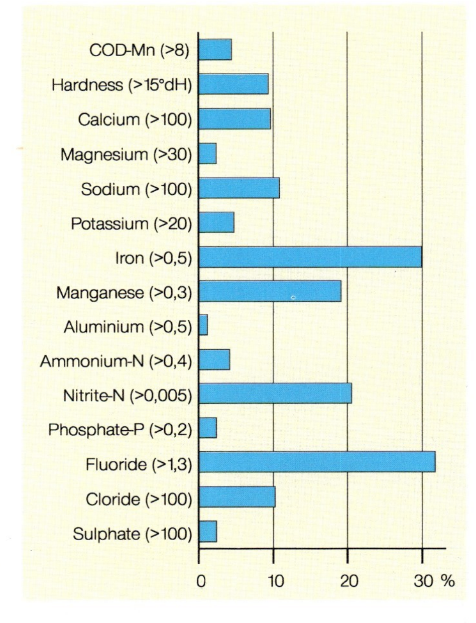

Health authorities have established a number of limit values for drinking-water quality. The diagramme shows the share of individual wells containing water that exceeds limit values for chemical substances. COD-Mn is a measure of the water's content of organic substances.

Water hardness is measured in mg calcium per litre. The hardest water is found in areas with calcareous bedrock where the ground is lime-rich. In the eastern parts of central Sweden the Quaternary deposits contain large amounts of limestone that were transported there by the inland ice from the calcareous bedrock in the Bothnian Sea. (M67, M68)

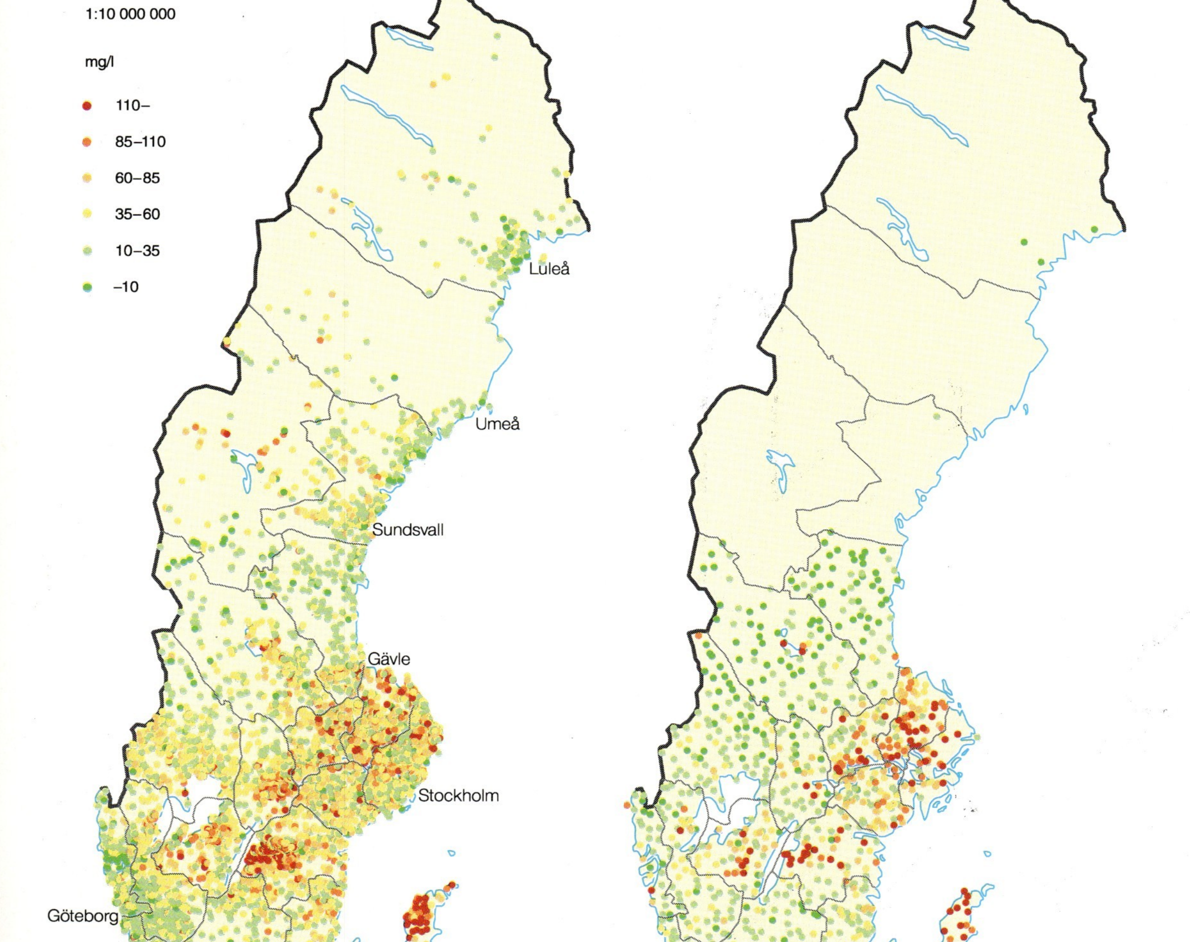

The fluoride concentration in water from wells in the bedrock is often sufficiently high to give protection against caries. However, in many cases the fluoride concentration is so high that the water is hazardous, particularly for children.

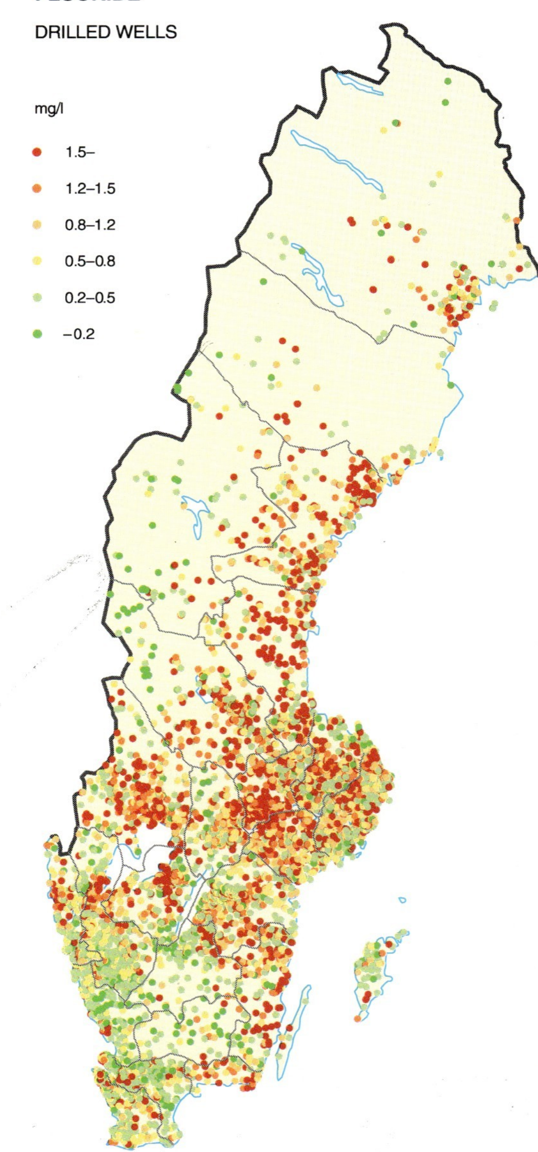

Fresh, cold and clear groundwater is a priceless renewable natural resource.

able for use. The latter are usually caused by fertilisation of fields close to the area of water supply. Low concentrations of fluoride in the water provide no protection against caries (tooth decay). Concentrations between 0.75 and 1.3 mg/l fluor on the other hand, will protect teeth, whereas higher concentrations may cause flecks on children's teeth during juvenile years. When concentrations exceed 6 mg/l, the water is considered unsuitable.

Radon gas in the groundwater, which is relatively common in Sweden, was earlier considered to be beneficial to health. Today, radon is considered by some scientists to be the most dangerous compound in groundwater. Radon in water can usually be aerated away. However, by this process it is added to other radon concentrations released from the ground and from building material and may, in cases of long-term exposure, cause lung cancer.

Almost half of all wells in Sweden have concentrations of *iron* and *manganese* that are so high that the water cannot be used to the required extent, i.e., mainly when washing clothes and dishes. Treatment of the water with filters of different kinds may, however, give considerable improvements.

Saline water is also a risk when drilling wells. The risk is greatest along the coasts and in low-lying terrain. At greater depths, the water is probably saline everywhere. Near the coasts, there is also a risk that brackish water from the sea penetrates into the wells if too much freshwater is removed. Saline water found further inland from the coastline is of greater age, originating either from sea water that has earlier covered the country or from salts that have dissolved out of the bedrock throughout time.

In most dug wells, the water has such low concentrations of fluoride that it does not give any protection against caries. (M70)

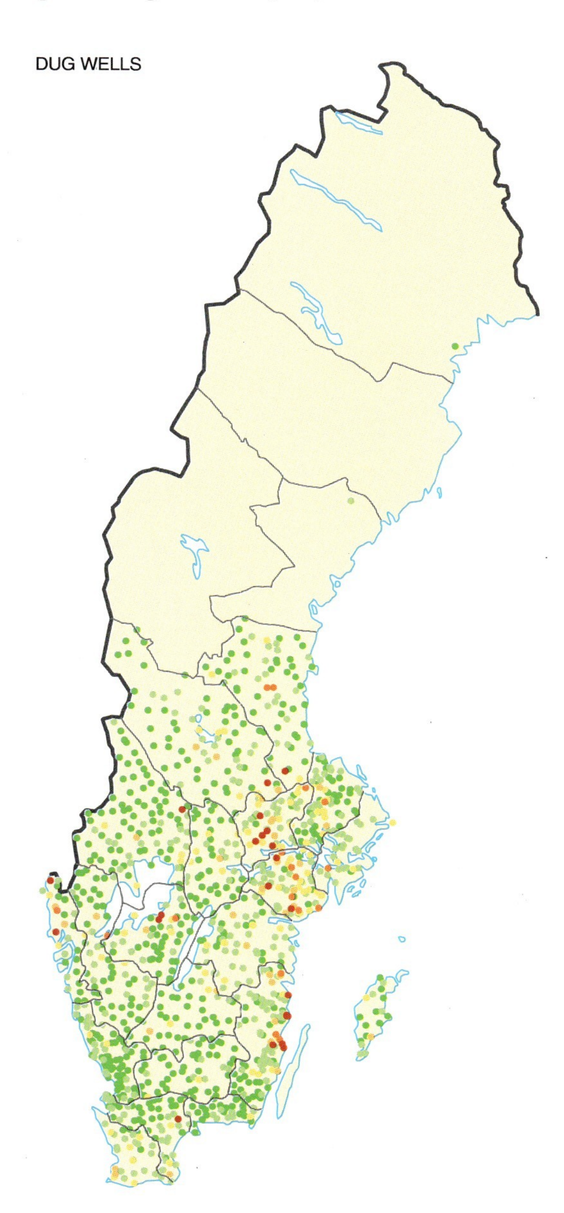

Saline groundwater, i.e., water with high chloride concentrations, is found in coastal areas and in areas that have been below sea level after the latest glaciation. (M71)

When water is pumped out of gravel and sand deposits in coastal areas, there is a very large risk that saline water will penetrate and ruin the water quality. (M72)

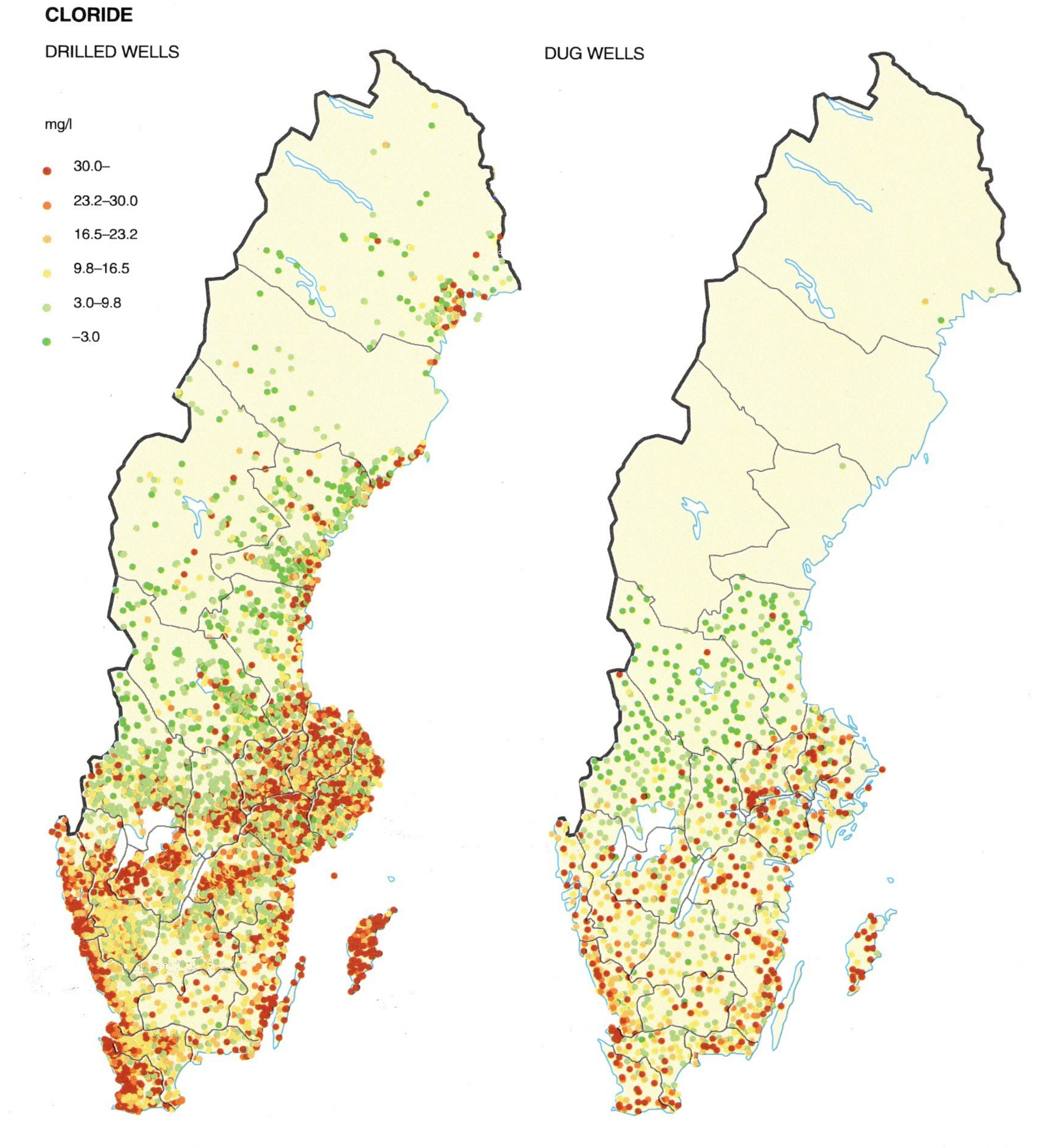

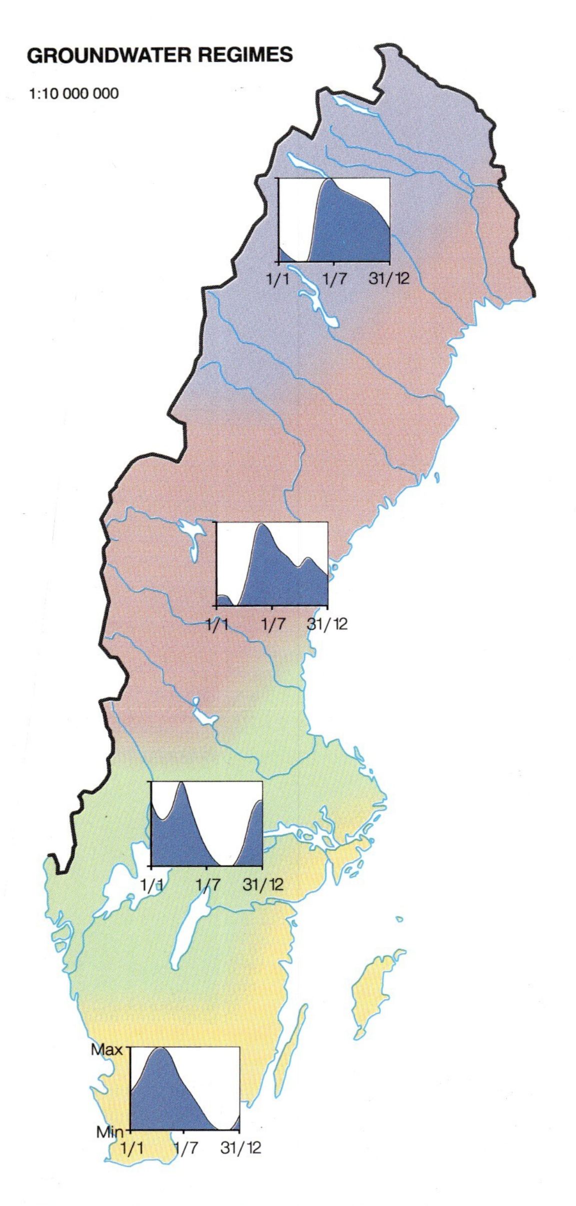

The groundwater level fluctuates with a certain regularity depending on seasonal changes in climate. The climatic differences between different parts of Sweden imply that the fluctuation patterns—the groundwater regime—have different appearances. (M73)

Variations in groundwater

The volume and composition of groundwater are highly variable. Variation in quantity, measured as changes in the groundwater level, depends on temporal variations in its recharge. The relatively regular shifts in temperature, precipitation and vegetation activity throughout the year result in seasonal variations in the groundwater level. This pattern depends on the geological setting of the groundwater reservoir, its size, position in the terrain and its geographical position. The annual amplitude depends on the pore volume available in the groundwater reservoir, including fissures.

In *crystalline bedrock* and in *till*, the volume of fissures and pores comprise a small part of the reservoir. This results in a moderate change in the amount of groundwater leading to a major change in the level of the groundwater surface.

In *fine sand, sand* and *gravel* the porosity is greater. This results, for example, in the groundwater level changing less than in till, following the supply of water with precipitation. The annual amplitude is often less than one metre in these deposits.

In large *glaciofluvial deposits*, seasonal variations may be so small that they are hardly measurable. This is because the size of the unsaturated zone above the groundwater surface has a balancing effect during time on the amounts of water that enter these aquifers.

CLIMATE INFLUENCES THE PATTERN OF VARIATION

The large climatic differences in a long and narrow country such as Sweden mean, that the variation pattern in the groundwater level differs from north to south. Sweden can be divided largely into four areas with different groundwater regimes. The different type patterns are multi-year averages of the seasonal variations in the groundwater level from small and moderately-large reservoirs.

In parts of Sweden covered by snow in winter, the main recharge of groundwater takes place in connection with the melting of the snow in the spring and early summer. In southernmost Sweden, precipitation mainly falls as rain also during the winter which permits a continuous refilling of groundwater reservoirs during the autumn and winter.

During the growing season, groundwater recharge is prevented partly by high evaporation and partly by the uptake of water from the ground by plants. Consequently, the groundwater levels become lower during the summer. They are generally lowest during the late summer or early autumn. At the end of the growing period, when temperatures start to fall and evaporation decreases, precipitation may again contribute to increasing groundwater levels. Apart from the southernmost part of Sweden, recharge of groundwater is terminated during the late autumn by frost and snow. This occurs so early in northernmost Sweden that groundwater recharge does not occur at all during the autumn and consequently the reservoir is depleted until the next period of snow-melting.

MULTI-YEAR VARIATIONS IN GROUNDWATER LEVEL

Changes in precipitation and temperature that extend over several years lead to changes in the recharge of groundwater and thus in the extent to which a groundwater reservoir is fil-

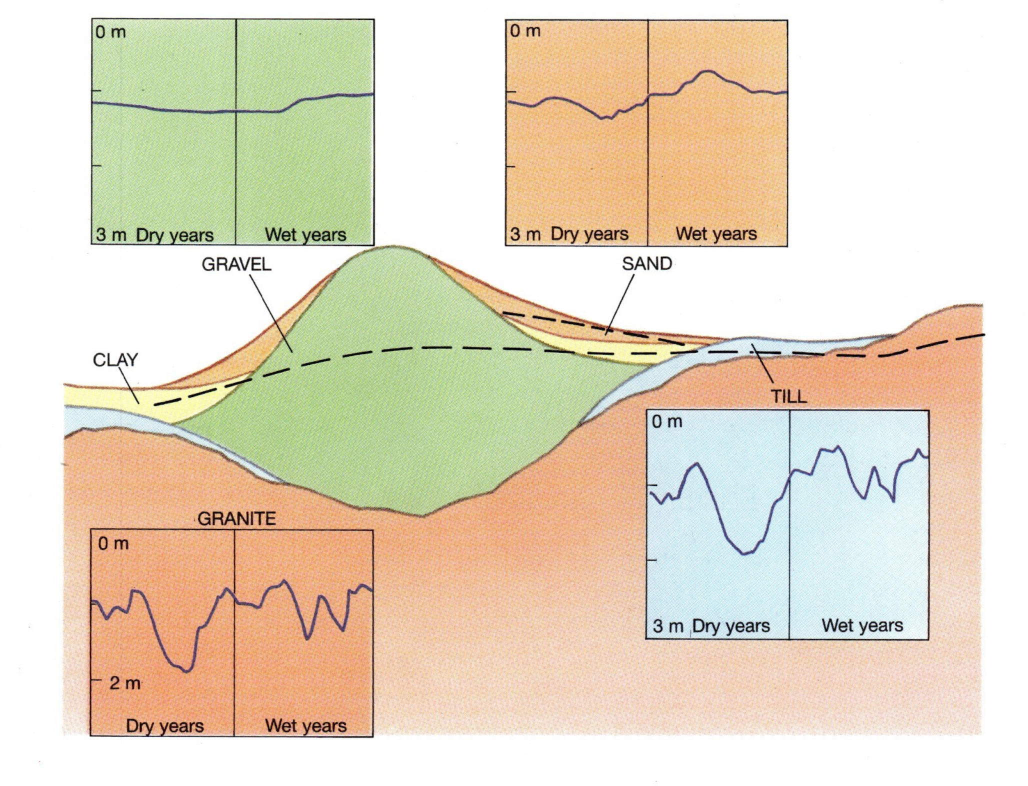

The surface of the groundwater fluctuates considerably in different types of aquifers depending on differences in pore volume and the amount of fissures.

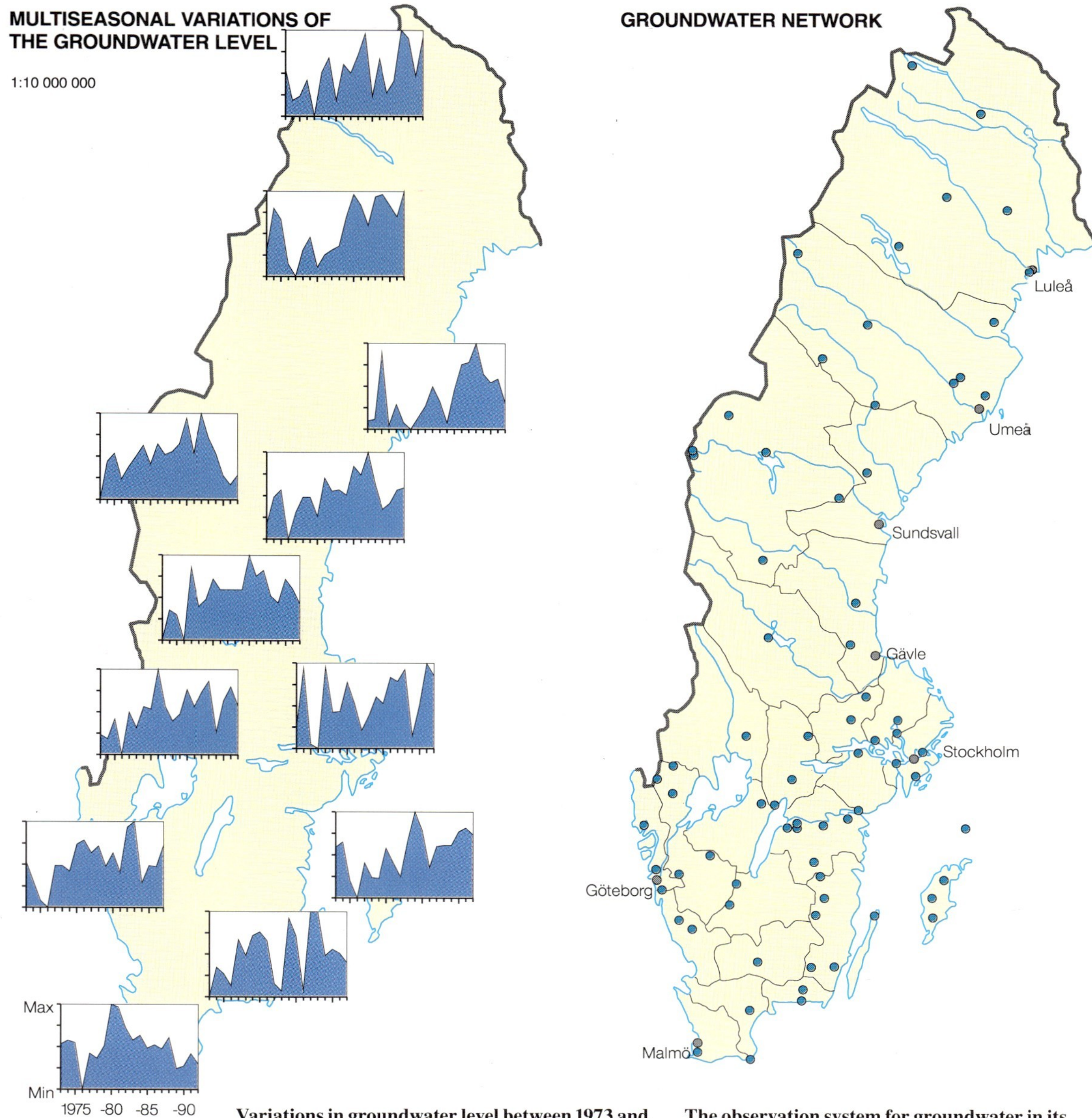

Variations in groundwater level between 1973 and 1991. The high levels during the 1980's in comparison with those in the 1970's are remarkable, as well as the three consecutive years with precipitation deficit giving extremely low groundwater levels in 1976 in large parts of Sweden. (M74)

The observation system for groundwater in its present form was initiated in 1969. Today it consists of around 700 measuring sites, distributed among about 70 areas. (M75)

led. The curves illustrating groundwater level from large reservoirs give a good picture of long-term trends since the annual variations have been largely levelled out. On the other hand, curves from small reservoirs do not give a clear picture.

Groundwater network

Data obtained from the *groundwater network* established by SGU are used in studying variations in groundwater. Information is mainly collected from areas that are disturbed as little as possible by human activities.

Collection of data on groundwater levels, groundwater temperature, frost depth, snow depth and groundwater chemistry is carried out within about 70 observation areas. Since 1978, sampling and analysis of groundwater chemistry have also been carried out within *the National Environmental Monitoring Programme* (PMK). The PMK programme aims to monitor long-term changes in the environment, collect data on environmental conditions in relatively uninfluenced areas and, together with other PMK sub-programmes, illustrate how pollution is transported not only within air, land and aquatic environments, but also between these different environments.

Threats to groundwater

Groundwater can generally be used without previous purification. However, many human activities threaten its quality. The composition of groundwater may change as a result of, for example, additions of undesirable substances, direct measures affecting flow patterns of groundwater, or biological activity within a certain area.

ACIDIFICATION

Increased combustion of fossil fuels containing sulphur is the main reason for acidic precipitation. In Sweden, there are relatively poor geological conditions for neutralization of acidic rain. Even if Quaternary deposits are capable of neutralizing the acidic precipitation, the pH-value in groundwater will not be changed. However, regardless of whether Quaternary deposits have this capacity or not, the chemical composition of the groundwater will change. This enables us to identify the place where the groundwater becomes affected by sulphuric acid precipitation of anthropological origin, i.e., originating from human activities. Groundwater may, namely, also be naturally acidic, and particularly with regard to surficial groundwater with a short turnover time.

The ratio between groundwater *alkalinity* (resistance to acidic precipitation) and total hardness is normally about 1, but decreases under the increasing influence of acidic precipitation. The maps based on chemical analyses of well-water show that groundwater in large parts of southern Sweden is severely affected. Even if the situation is considerably better regarding groundwater in the bedrock at deeper levels, it is clearly affected in areas exposed to large amounts of acidic precipitation.

The corrosive properties of acidic groundwater are a major problem. Acidic water has the ability to dissolve and transport heavy metals, both from the ground as well as from water-pipes. Corrosion necessitates the replacement of pipes. The heavy metals released during the corrosive action, mainly copper, may reach levels that imply a health hazard.

Today, acidic precipitation is the greatest threat to groundwater. Since this is a question of long-distance movements of air pollution crossing from one country to another, this problem concerns more or less all Sweden. However, there are also polluting sources within Sweden that imply a serious threat to the utilization of groundwater.

HIGH CONCENTRATIONS OF NITRATE

In some places, the use of nitrogenous fertilisers in agriculture has re-

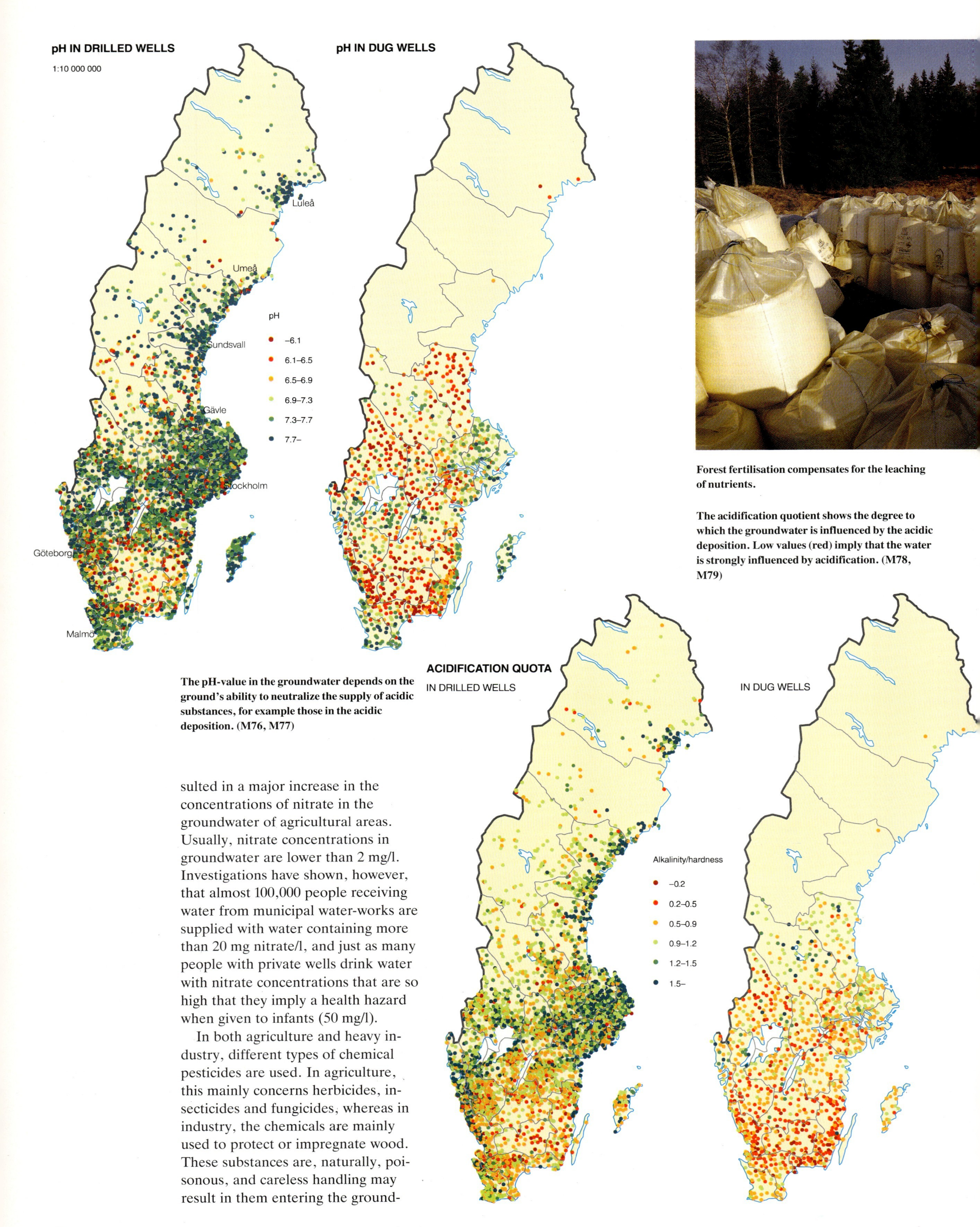

Forest fertilisation compensates for the leaching of nutrients.

The acidification quotient shows the degree to which the groundwater is influenced by the acidic deposition. Low values (red) imply that the water is strongly influenced by acidification. (M78, M79)

The pH-value in the groundwater depends on the ground's ability to neutralize the supply of acidic substances, for example those in the acidic deposition. (M76, M77)

sulted in a major increase in the concentrations of nitrate in the groundwater of agricultural areas. Usually, nitrate concentrations in groundwater are lower than 2 mg/l. Investigations have shown, however, that almost 100,000 people receiving water from municipal water-works are supplied with water containing more than 20 mg nitrate/l, and just as many people with private wells drink water with nitrate concentrations that are so high that they imply a health hazard when given to infants (50 mg/l).

In both agriculture and heavy industry, different types of chemical pesticides are used. In agriculture, this mainly concerns herbicides, insecticides and fungicides, whereas in industry, the chemicals are mainly used to protect or impregnate wood. These substances are, naturally, poisonous, and careless handling may result in them entering the ground-

Use of pesticides in agriculture may imply a threat to the groundwater.

During dry summers it may be essential in some areas to introduce a hose ban in order to ensure that there is sufficient water.

Accidents when transporting environmentally-hazardous material may lead to serious consequences for the groundwater.

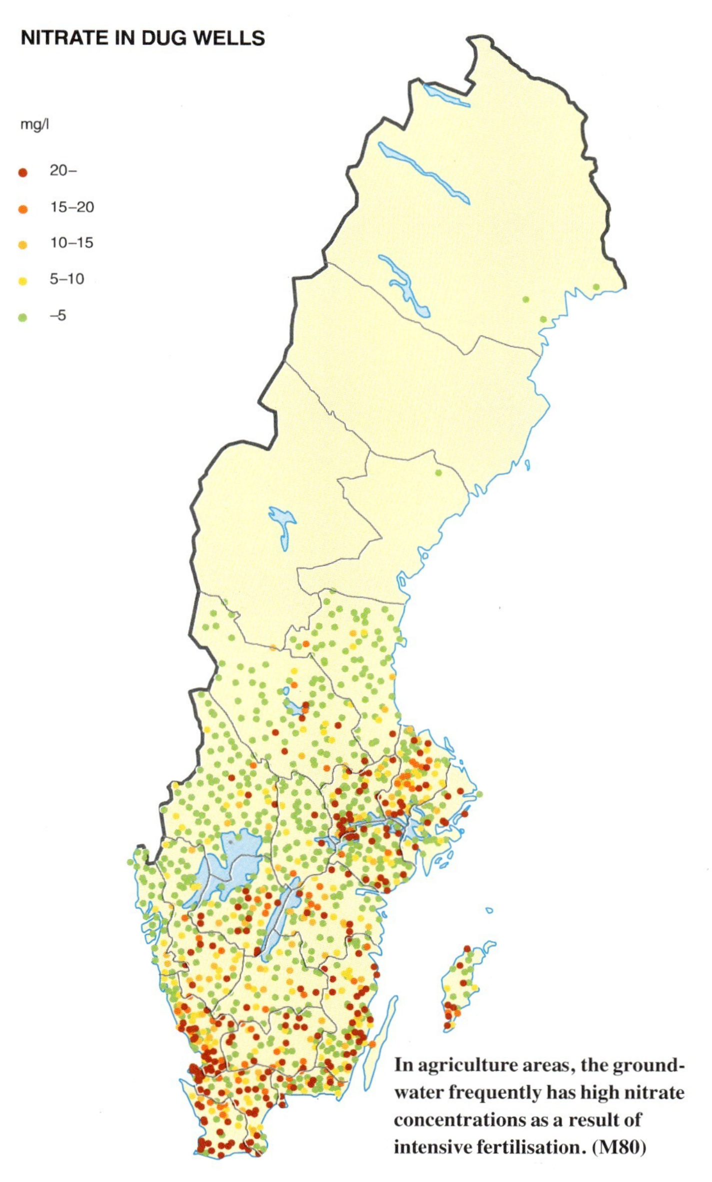

In agriculture areas, the groundwater frequently has high nitrate concentrations as a result of intensive fertilisation. (M80)

water. Analyses have identified pesticides in well-water, which has generally been explained by careless handling, such as emptying or rinsing sprayers in the neighbourhood of the wells.

WASTE WATERS

Waste deposits with unsuitable locations in the terrain may cause local damage to groundwater quality. Earlier, it was not realized that water was capable of dissolving and transporting substances that are hazardous to health and the environment. As a result, waste deposits throughout Sweden are releasing more or less hazardous substances which are then entering the groundwater. Abandoned gravel pits in eskers, potentially our most important aquifers, were used earlier as dumps for both domestic and industrial waste.

In mining districts, the leakage from by-products of the mining industry, e.g. mine tailings and sludge waste, pollute the groundwater with large quantities of heavy metals. Covering the by-products with e.g. clay or submerging them in water, will decrease the weathering and restrict the leaching of heavy metals.

HAZARDOUS TRANSPORTS

Traffic on Swedish roads poses a threat to groundwater in many ways: Accidents with environmentally hazardous transports may have serious and long-term consequences if they occur within the catchment area of water supplies, spillage of petroleum products at petrol stations and garages finds its way down to the groundwater, and the spreading of salt on roads during the winter to melt ice may result in neighbouring houses having their drinking water polluted.

OVEREXPLOITATIONS

The increasing numbers of recreational homes along Swedish coasts and in the archipelagoes often imply an excessively hard exploitation of the restricted freshwater reservoir, which may lead to salt water entering the wells. The risk for salt-water intrusion increases considerably in dry years.

Geochemistry

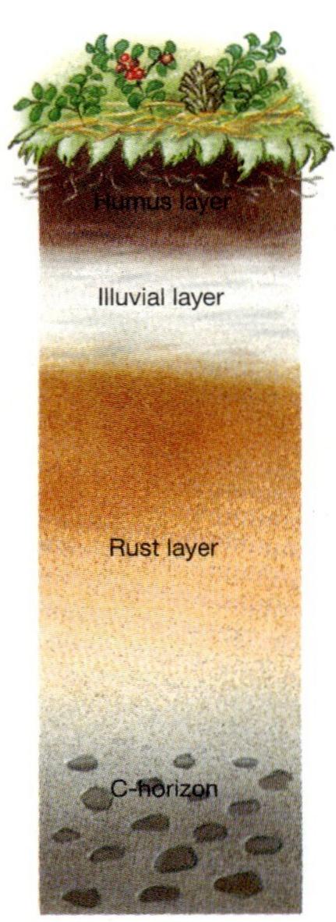

The different layers in a till podzol are usually irregular. The rust-red illuvial horizon gradually intermingles with the C-horizon.

Heavy metals in rocks and Quaternary deposits

About 75% of Sweden is composed of acidic, silica-rich rock types which contain relatively low concentrations of heavy metals. The remaining 25% consists of basic rocks enhanced with concentrations of heavy metals such as copper, chromium, nickel, cobalt and vanadium. Among the sedimentary rocks, alum shales have anomalously high contents of arsenic, lead, cadmium, cobalt, copper, molybdenum, nickel, uranium and zinc. Sandstone and limestone usually have low contents of heavy metals.

The most common overburden in Sweden is till. This consists of an unsorted mixture of all the different types of rocks that the continental ice sheet has eroded, transported and deposited. Sediments, mainly found in plains, in valleys, and in eskers, consist of Quaternary deposits that have been sorted into different particle sizes such as clay, silt, sand, gravel and cobble.

All these Quaternary deposits originated from the bedrock. Consequently, the layers have also inherited most of their metal content from this source. Since the concentrations of heavy metals differ widely in the various rock types, there are also important variations in the contents of heavy metals in the overburden from place to place. In this respect, the basic lithologies are of great importance since they may cause high concentrations of heavy metals in large areas due to glacial dispersion processes.

HEAVY METALS IN SURFACE LAYERS OF THE OVERBURDEN

Since the end of the last glaciation, weathering and other chemical and biological processes have modified the surficial layers so that the originally uniform material has been stratified into different soil horizons.

In a normal coniferous forest soil, there is a layer closest to the surface that consists only to a minor part of mineral particles. The uppermost layer is the *litter layer*, consisting of semi-decomposed plant residues. Lower down is a layer of *humus* in which the plant material is more or less completely decomposed. Heavy metals with low solubility, such as lead, mercury and selenium, have a strong tendency to accumulate in the humus layer. More easily dissolved heavy metals, such as cadmium, zinc, copper, cobalt and nickel, enter solution as a result of the acidic environment created by humic acids, and these heavy metals are transported downwards in the ground.

Below the humus layer, there is an ash-grey zone, the *bleached layer*, which has been strongly leached by the humic acids and is, therefore, poor in heavy metals and nutrients.

Beneath the bleached layer is the enriched or *illuvial horizon*, with high contents of, for example, precipitated iron, manganese, aluminium and organic compounds. Several heavy metals also tend to be precipitated in this layer since the humic acids at this level in the soil have now become neutralized. The ferrous compounds often give the enriched horizon a rust-red colour—*rust layer*.

The layers below the illuvial horizon are called parent material or the *C-horizon*. At this level, the weathering processes have only been able to proceed to a limited extent. Consequently, the heavy metal concentrations at this level reflect natural concentrations in the rock from which the parent material originated.

CHEMICAL TIME-BOMBS

Knowledge of the concentrations of heavy metals at different levels in the soil is important, particularly with regard to research on chemical time-bombs. The expression "chemical time-bombs" implies a sequence of events with increased and suddenly hazardous effects on the environment caused by different substances that have accumulated in soil, sediment or biological material. Chemical time-bombs need not necessarily have their origin in anthropological activities. They may also originate from natural occurrences of heavy metals in bedrock and Quaternary deposits.

The chemical composition of bedrock samples and the different layers of the overburden has been studied at 24 *geochemical reference stations* throughout Sweden. The results for six different heavy metals are described in the following pages. In the case of the bedrock, heavy metal concentrations have been given for the rock type that occurs locally at the site. In the different inorganic layers of the soil, the heavy metal concentration in the fraction smaller than 0.1 mm has been given. It should be observed that this particle fraction is mainly dominated by material from the local bedrock, although other bedrock material transported from greater distances by the inland ice might be included. The fine-grained deposits tend to have enhanced concentrations of many heavy metals in comparison with parent rock. The reason is that the clay minerals, in particular, are capable of binding heavy metals on their surfaces.

ROCK TYPES	INCREASED CONCENTRATIONS
Granites	Molybdenum, tin, tungsten, potassium, lead
Acidic volcanic rock types, e.g. porphyries	Arsenic, copper, lead, zinc, silver cadmium, mercury, selenium
Basic rock types, e.g. gabbro, diorite, greenstone	Chromium, cobalt, nickel, copper, titanium, vanadium
Shales, e.g. black shales	Silver, arsenic, gold, cadmium, molybdenum, nickel, lead, zinc, cobalt, uranium, copper
Sandstones and limestones	No general relationships with heavy metals of importance for the geochemical interpretation

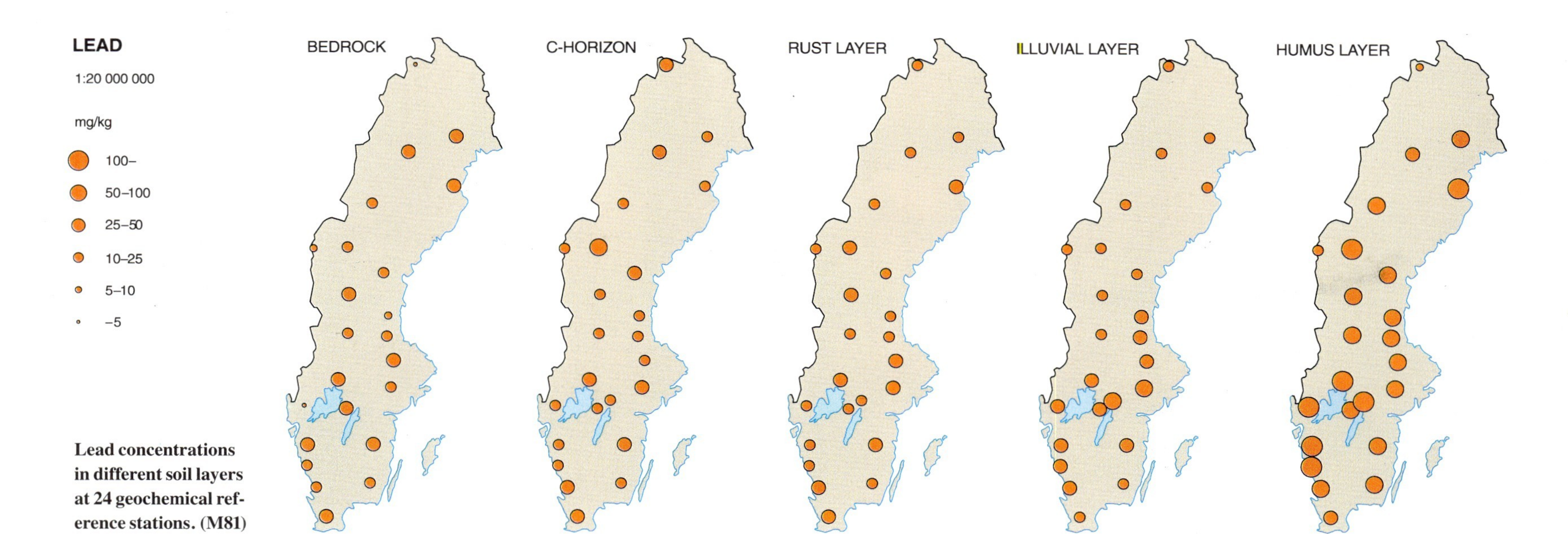

Lead concentrations in different soil layers at 24 geochemical reference stations. (M81)

LEAD

Natural lead concentrations in bedrock are highest in volcanic rocks and granite. Lead has an extremely low solubility even under acidic soil conditions and has a strong tendency to bind to humus compounds. Consequently, the humus layer acts as a trap for the lead that is supplied from the atmosphere or released by the decomposition of plants. The maps show that lead concentrations in the humus layer are highest in the southwestern parts of Sweden and decrease towards the north. An exception is found close to Skellefteå where the lead concentration is high due to the long-term discharge into the atmosphere from the Rönnskärsverken smelter. This pattern reflects the fact that a large proportion of the lead accumulated in the humus layer originates from atmospheric deposition following the massive discharges of lead in modern times from industry, combustion plants and from traffic. The high lead concentrations in the humus layer constitute a serious environmental problem, not only in Sweden but also in most industrialized countries throughout the world.

The bleached horizon has considerably lower lead concentrations than the humus layer but the average concentrations are higher than in the soil layer below. This is explained by the fact that lead to some extent has been transported downwards in the soil together with water-soluble humus and fulvic compounds that again precipitated in the bleached horizon.

In the iron-enriched horizon, the C-horizon as well as the bedrock, the lead concentrations are fairly similar and do not show any particular variation between different sites. In these layers, the lead is strongly bound as a trace-element in different silica minerals, mainly weathering-resistant feldspars.

MERCURY

Mercury is found in minor quantities in most rock types. Frequently, the concentrations are below 0.005 g/kg as can be seen on the bedrock map. Nevertheless, the presence of this metal in the environment is one of our greatest environmental problems. The reason for this is the volatility of mercury in combination with its extensive use in agriculture, the chlor-alkali industry, dentistry, in electrical objects and in measuring instruments. Thus, mercury has become widely dispersed, particularly via the atmosphere but also via watercourses. As a consequence, Sweden has more than 10,000 lakes that are black-listed owing to unacceptably high concentrations of mercury in fish.

The maps show that the humus layer contains the highest mercury contents. As for lead, mercury is bound very effectively to the humus compounds and is transported only slowly downwards in the soil. The bleached horizon as well as the iron-enriched horizon have, however, much higher mercury concentrations than the C-horizon even though the concentrations are only a fraction compared with those in the humus layer. The leaching of mercury into watercourses takes place mainly from the upper part of the soil, particularly in situations of high water levels, when large amounts of humus substances are flushed out into the drainage systems.

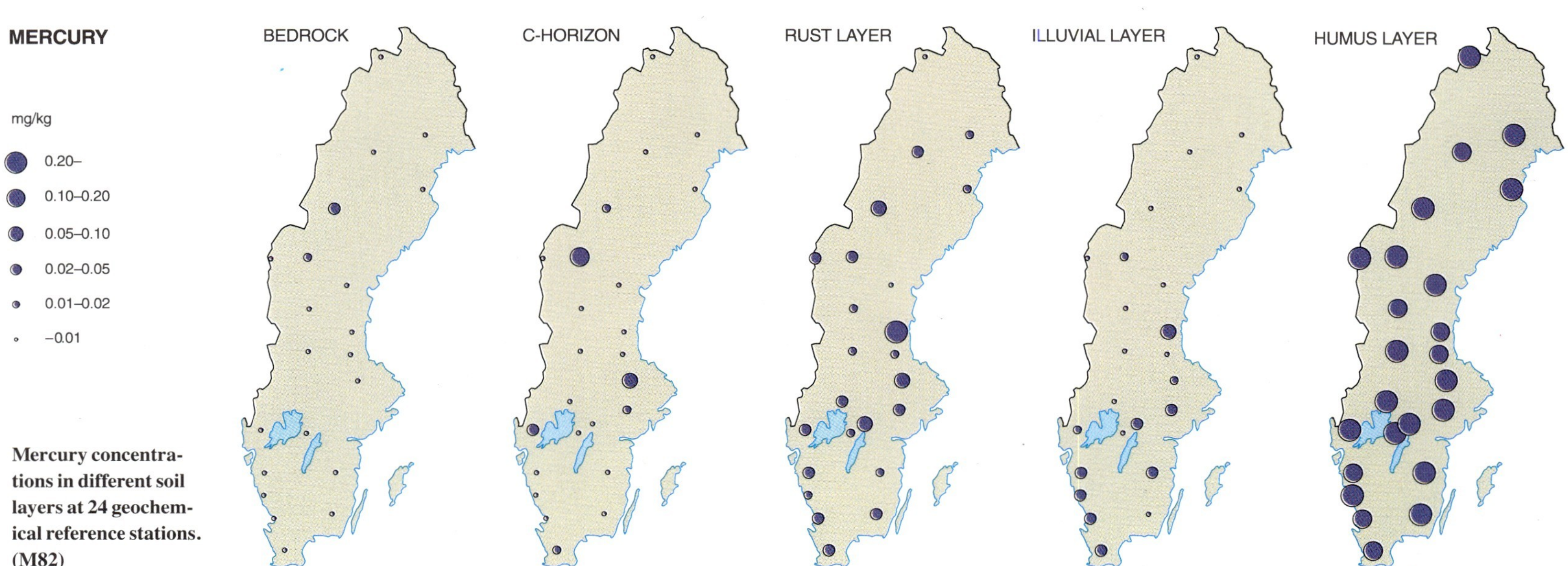

Mercury concentrations in different soil layers at 24 geochemical reference stations. (M82)

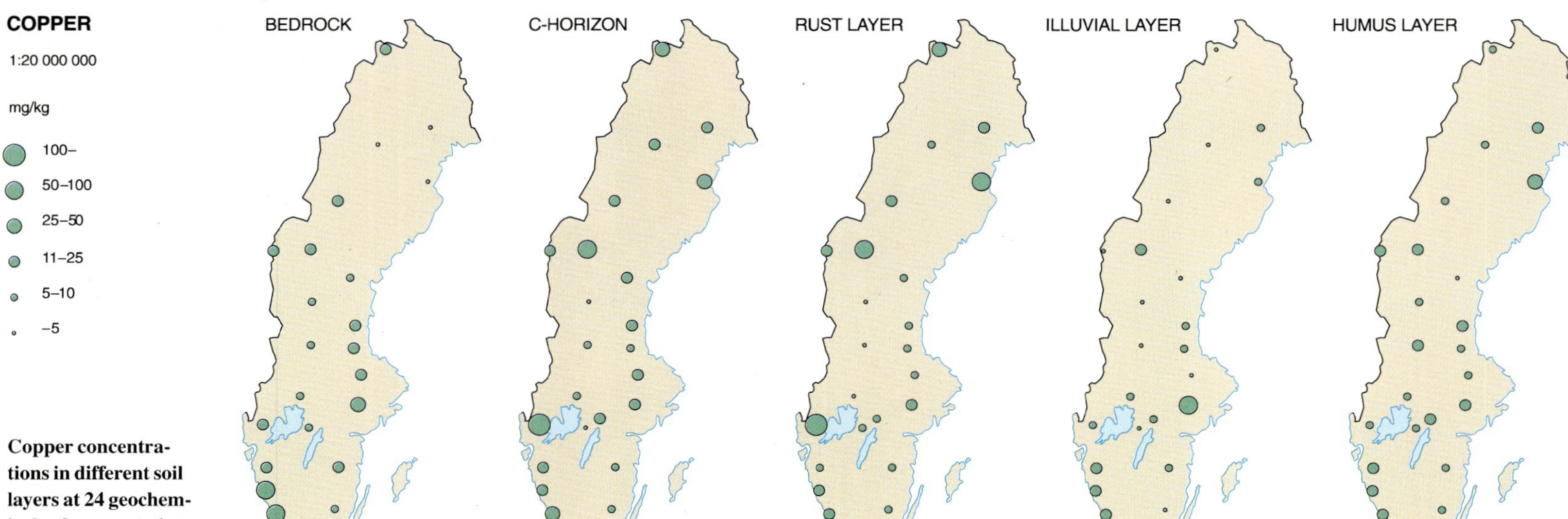

Copper concentrations in different soil layers at 24 geochemical reference stations. (M83)

COPPER

Copper is present as a trace element in most lithologies. The highest concentrations are found in basic and intermediate rock types, containing 90 mg/kg copper on average. In acidic rocks, e.g., granites, the contents are much lower, 4–30 g/kg. Sedimentary rocks usually have low concentrations, except shales where the copper concentration averages 70 mg/kg.

The bedrock map shows that copper concentrations are mainly within the concentration range of the granites. This may also be regarded as representative for the pool of copper in most Swedish soils since the C-horizon has approximately the same copper concentrations as the bedrock. However, there are sites in Dalsland and Skellefteå with higher concentrations than in the bedrock. This is due to the fact that the till at these sites contains chalcopyrite originating from copper mineralisations. At the site in Jämtland, the copper concentration in the C-horizon has been enhanced as a result of admixtures from alum shales.

The copper concentration in the upper soil layers is much lower. The bleached horizon has about one-third and the iron-enriched horizon together with the humus layer about half of the concentrations in the C-horizon. This illustrates the fact that copper is relatively easily leached from the surface layers of the soil under acidic conditions. Both in Sweden and elsewhere, there are numerous reports on copper deficiency in podzolic soils, particularly within areas where the natural leaching of copper has been speeded up by the acidic precipitation.

SELENIUM

In most Swedish rocks, selenium concentrations are low. Indeed, there are only limited areas where the concentration exceeds 0.05 mg/kg. As a consequence, Sweden has been classified as an area with selenium deficiency in an international perspective.

In the different soil layers, the highest selenium concentrations are found in the humus layer. This selenium is largely of anthropogenic origin, mainly from combustion of coal and oil but also from selenium that has evaporated from the sea. As for lead, the highest selenium concentrations are found in southwest Sweden, with decreasing concentrations towards the north. This offers strong evidence that the selenium deposition over Sweden mainly originates from other countries. In the acidic environment of Swedish podzols, selenium is converted into almost indissoluble compounds, mainly selenides. Thus, selenium present in the humus layer is hardly available to plants. The maps show that selenium to some extent penetrates to deeper levels in the soil, where it is again precipitated, mainly in the iron-enriched horizon but also in the bleached horizon.

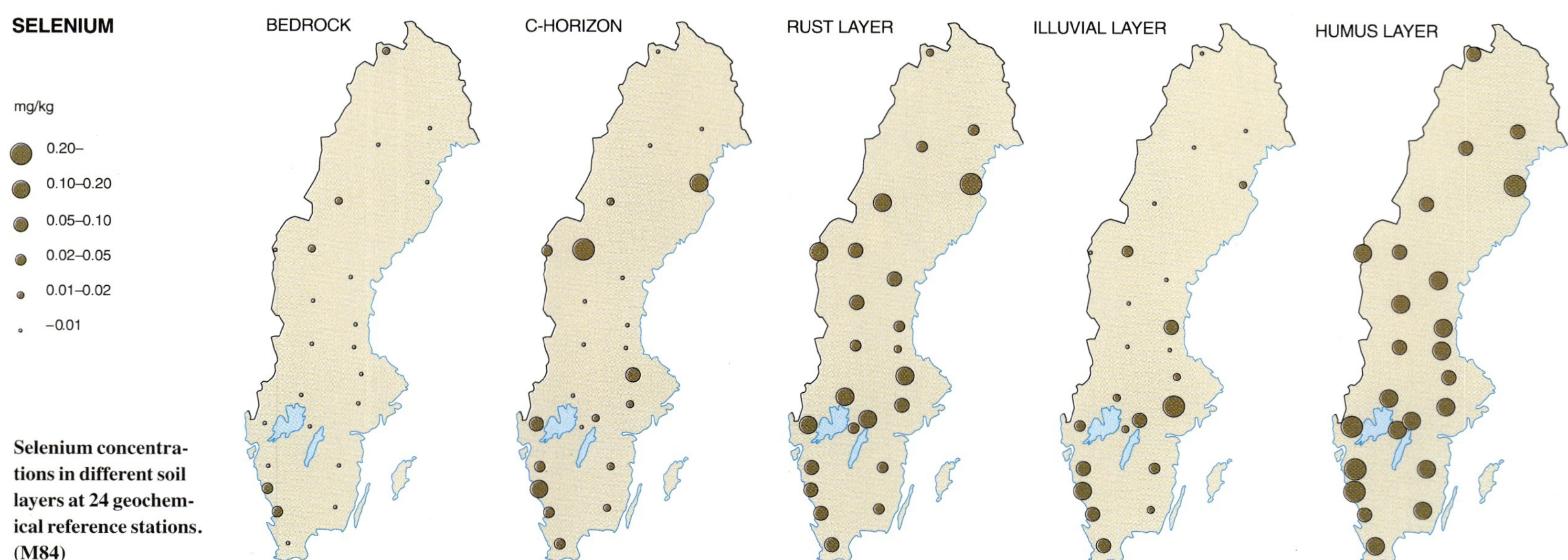

Selenium concentrations in different soil layers at 24 geochemical reference stations. (M84)

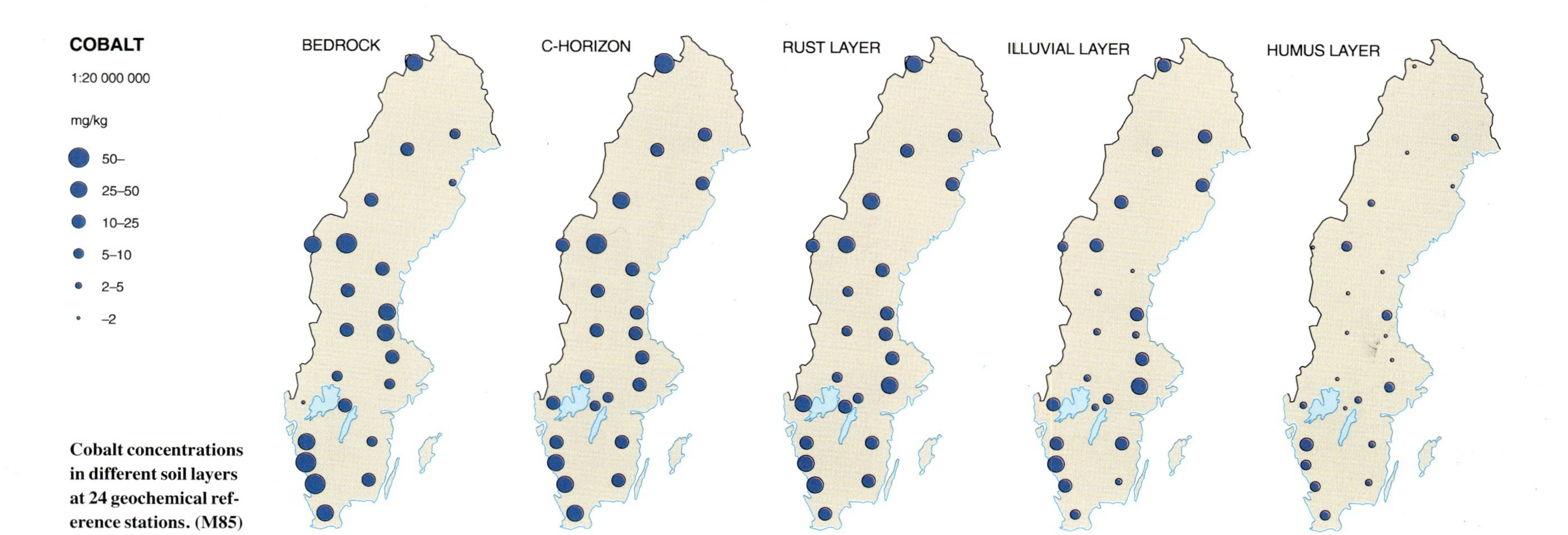

Cobalt concentrations in different soil layers at 24 geochemical reference stations. (M85)

COBALT

Cobalt is found as a trace element in most dark minerals, where it replaces iron or manganese in the crystal lattice. The metal is also found in sulphide minerals together with other heavy metals. Consequently, cobalt concentrations are highest in basic rocks such as ultrabasic rocks and diorites, and much lower in acidic rocks such as granites. In sedimentary rocks, the cobalt concentrations are highest in those containing clay minerals and organic material, such as alum shales. This is illustrated by the cobalt maps for bedrock and for the C-horizon, where enhanced cobalt concentrations can be seen in the southwest Swedish iron gneiss region, in the alum shale area of Jämtland, and in the mountain chain.

Cobalt is a relatively mobile heavy metal under the acidic and oxidizing conditions prevailing in podzolized soil. Consequently, concentrations are lowest in the humus layer and successively increase downwards in the soil profile. Acidic soils, therefore, are exposed to the risk of having such low cobalt concentrations in their surface layers that deficiencies may occur in, for example, grazing animals.

NICKEL

In the bedrock, nickel concentrations are highest in basic rocks, on average 140 mg/kg, whereas in granites they vary between 2 and 20 mg/kg. In sedimentary rocks, the concentrations are low except for black shales, where up to 1,000 mg/kg nickel may be found. The map for nickel concentrations in bedrock and for the C-horizon show patterns that are very similar to those for cobalt. This illustrates the similar behaviour of these two metals in the Earth's crust. In the upper soil layers, the lowest concentrations are in the humus layer and in the bleached horizon, where the nickel concentration is, on average, one-third of the concentration in the C-horizon. In the iron-enriched horizon, a trend can be seen for enhanced concentrations owing to the precipitation of nickel on iron and manganese hydroxides.

It still remains to be established whether nickel is an essential micro-nutrient for plants. On the other hand, nickel is an essential trace element for both animals and humans. Nonetheless, it is hardly probable that nickel deficiency could occur in Sweden since available amounts of nickel in foodstuffs are greatly in excess of the extremely small quantities required by humans and animals.

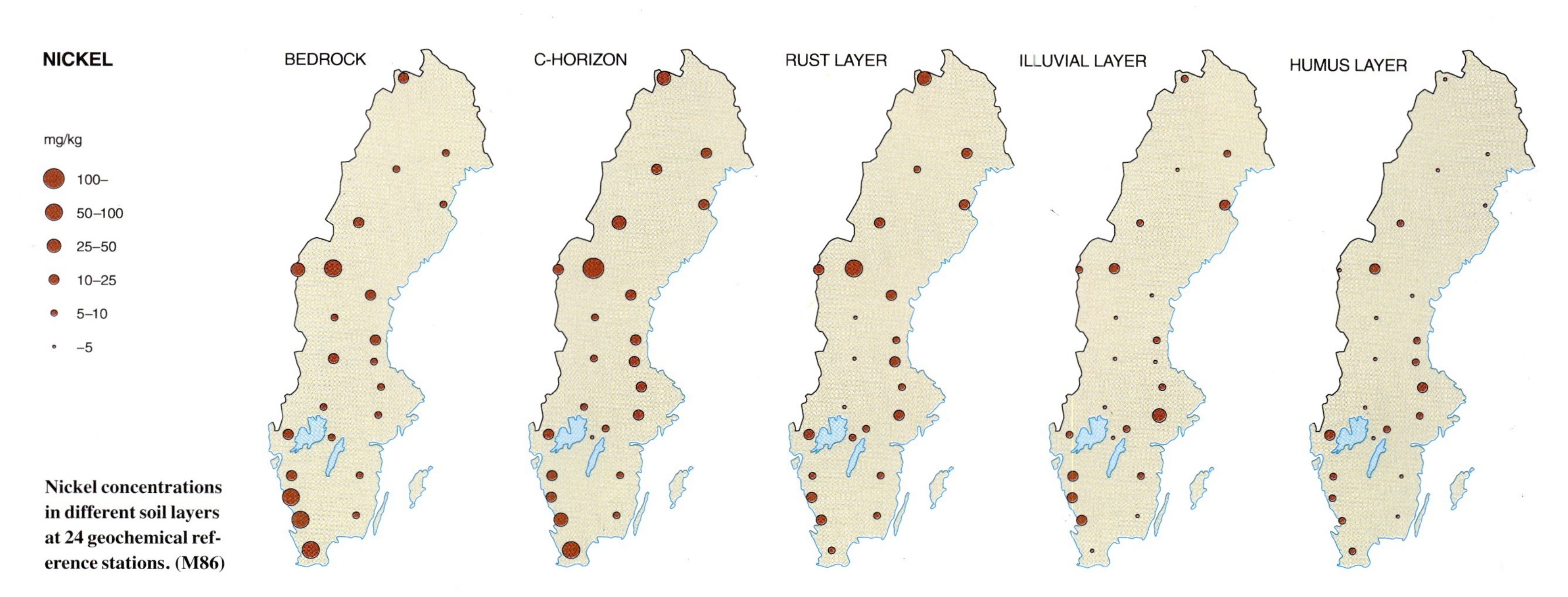

Nickel concentrations in different soil layers at 24 geochemical reference stations. (M86)

Biogeochemistry

The metals in the environment are continuously circulating through air, overburden and rock in what is known as the *geochemical cycle*. In this process, most metals are also included in the metabolism of plants and animals. All organisms, including Man, are thus affected by the geochemical cycle. The decisive factor for how organisms can absorb different metals is the *bio-availability*, i.e., how available a certain metal is to organisms.

The science dealing with the circulation of elements in the environment, how they are used and discharged by organisms, and how this knowledge is applied, is called *biogeochemistry*.

GEOLOGY AFFECTS OUR LIVING ENVIRONMENT

The mineral content of the bedrock and the overburden is of great importance for the natural occurrence of metals in the environment. When minerals in rock and overburden are exposed to weathering, metals are released and, for various reasons, start circulating in the geochemical cycle.

However, metals released when Man utilizes and extracts natural resources are also added to the geochemical cycle. Many toxic heavy metals such as lead, mercury and cadmium have, in addition, been exposed to changes in their cycles as a result of being utilized by Man. Similarly, acidification has radically changed the cycle for many metals.

WATER DISSOLVES AND TRANSPORTS METALS

One of the most important conditions for the mobility of metals in the geochemical cycle is the presence of water. Water transports metals that have been derived from weathered Quaternary deposits and bedrock. Environmentally hazardous metals are discharged from industries directly into watercourses or into the atmosphere, and subsequently deposited with rain or snow. Fertilisers, used both in agriculture and forestry, may contain trace elements and heavy metals that, together with water in the unsaturated zone, are carried to the groundwater.

The transportation of various substances in water is also influenced by acidification. Acid rain and surface water increase the leaching of metals sensitive to acidification, such as cadmium and zinc from overburden and bedrock. If the groundwater is acid, then, in addition, the metals will remain in solution more readily and can be dispersed over large areas.

TRACING METALS IN ROOTS

Metals released for various reasons into the environment will, in due course, reach the watercourses via surface- and groundwater run-off. Since most metals sooner or later enter streams during their cycles, and since streams are found throughout Sweden, they are natural and representative sampling places for metals in circulation. By collecting and analysing plants living in conjunction with these watercourses, we can find out which metals have been added to the water and their bioavailability.

Certain species of aquatic mosses and certain herb roots, for example, roots of sedge (*Carex* L.), have been found to give a good indication of the amount of metals present in the water. When plants absorb stream-water, they also absorb metals in relation to the concentration of metals in the water. Analysis of these aquatic plants will thus reveal the concentrations of the various metals in the water.

WHY USE ROOTS?

There are several reasons for analysing roots to measure the metal content in water. The metal uptake of roots is not particularly influenced by the plant's needs, and uptake and exchange of metals with the environment is a continuous process. The content of metals in roots therefore depends entirely on the supply from the surroundings. If the water contains few metals, then the roots will also contain few metals, and vice versa.

Metallic ions that have been released from soil and rock through weathering enter the watercourses. When plants along the watercourses absorb the streamwater through their roots, the metallic ions will also be absorbed. Consequently, the roots will have a metal content that reflects the chemical composition of the water and also, indirectly, the chemical composition of the bedrocks and Quaternary deposits. If the metallic concentrations in the roots are high, it may imply that an ore deposit has been found.

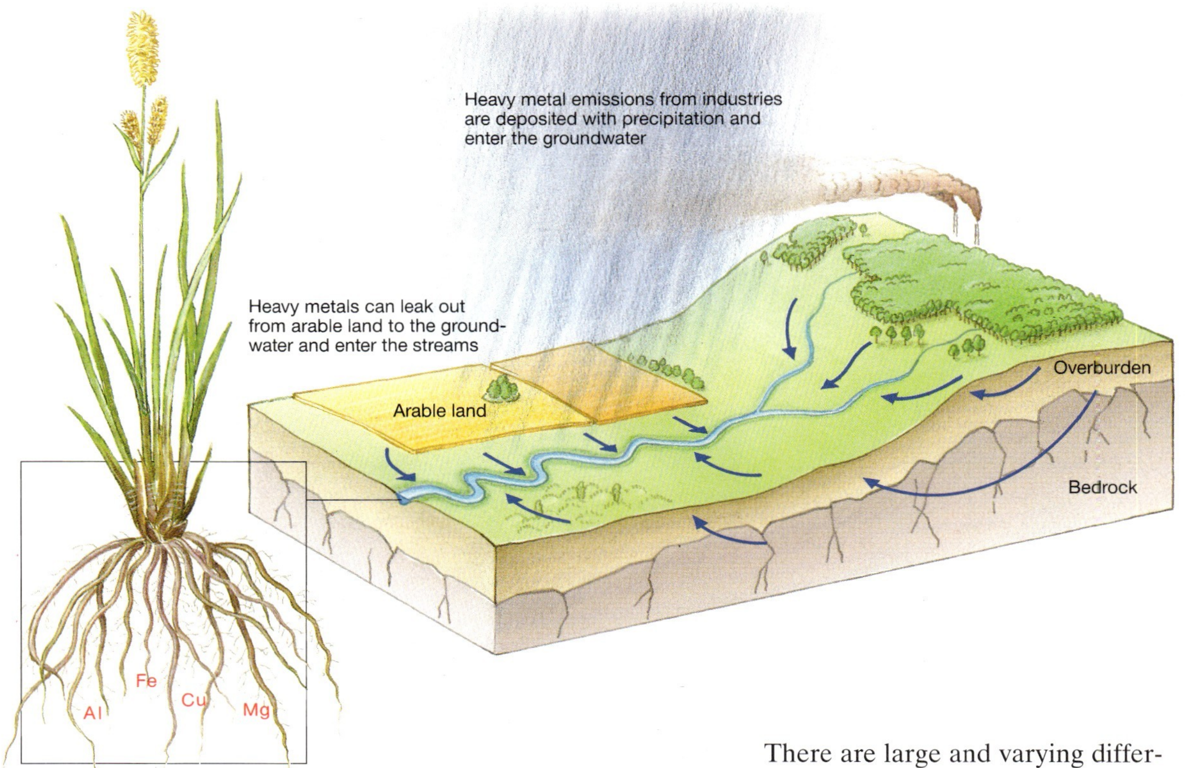
Heavy metal emissions from industries are deposited with precipitation and enter the groundwater
Heavy metals can leak out from arable land to the ground-water and enter the streams
Arable land
Overburden
Bedrock
Fe
Al
Cu
Mg
The roots from stream plants absorb heavy metals

Not only metals from Quaternary deposits and rock are made available to the environment; there are also other explanations of the high metal concentrations in watercourses. Industries, flue gases, waste handling, agriculture and forestry, etc., contribute to an increased dispersal of metals. All metals that are present for one reason or another in the circulation, will finally enter the watercourses and become available in different ways to plants and animals, including humans.

Chemical analyses of a handful of roots are sufficient to demonstrate where the high metal concentrations are present in our surroundings.

Waste dumps may leach metals that pollute watercourses. The supply of certain metals is increasing, which is shown by a high content of metals in roots of stream plants.

There are large and varying differences between contents in roots and contents in aerial parts of plants. In above-ground parts, some plants can either accumulate certain metals while others may to some extent protect themselves against excessive uptake of toxic metals. Thus, in the aerial parts there is not the same direct link between availability and uptake of a metal. Consequently, the content of metals in roots, in contrast to the content in aerial parts, will reflect the total occurrences of available metals in the water, from the lowest concentrations to the highest.

It may be considered an unnecessary complication to analyse roots instead of stream-water in order to establish the concentrations of metals in Swedish streams. In fact, water samples are more difficult to analyse and also vary in chemical composition depending on season and precipitation. In stream plants, these chemical variations in time become levelled out since the uptake and exchange of metals between the water and the plants is a slow process. The content of metals in roots of stream plants therefore provides an average measure of the chemical composition of the plant's water supply. Analyses of the chemical composition of water do not always provide information on the bio-availability of the metals present in the water, which can be found by analysing the roots.

WHAT THE MAPS SHOW

The maps show whether a metal is available in large or small amounts to the stream plants. Here, we can see where the highest concentrations in Sweden are to be found and where there is the greatest need for corrective measures. In conjunction with mineralisation or ore deposits, the natural concentrations of some metals may be so high that they can be compared with the worst form of pollution. Groundwater used as drinking water or for irrigating crops grown within such areas may therefore contain concentrations of metals which constitute a direct health hazard.

Initially, it is not always possible to establish reliability where there are high concentrations of metals within a certain area. The fact that plants contain high concentrations of metals shows, however, that the metals are, nonetheless, present and in bio-available form.

Knowledge of the amounts of metals present under natural conditions will also allow us to identify the areas where there are either too few essential metals or too many hazardous metals. At present, it is impossible to state that a certain concentration of a metal in plant roots will lead to deficiency or over-exposure to plants, animals or Man within the region in question. Nonetheless, knowledge of the variation in concentrations of metals within a region helps us to identify where further studies may be needed on how they have dispersed in the nutrient chains and, thereby, how they may be expected to affect us. In this way, the variation of the cadmium content in stream plants, as one example, has been found to show good agreement with the cadmium content in wheat. In areas where stream plants contain high levels of cadmium, we also find high concentrations of cadmium in wheat. In other areas, where availability of cadmium is lower, the opposite applies.

COPPER

Copper is of great importance as an industrial product but also as an important micro-nutrient for Man and animals. However, at high levels of intake, the metal is toxic. The highest concentrations of copper in bedrock are found in basic rocks such as greenstones. The most common copper mineral is chalcopyrite which is a sulphide mineral. Chalcopyrite is easily weathered and when copper is released from the ground and enters the watercourses, it can also be easily absorbed by plant roots. Copper concentrations in stream plants mainly show where copper-bearing sulphides are present in the bedrock.

Copper has a wide range of applications and its intensive utilization has resulted in abnormally high copper concentrations in stream plants in the neighbourhood of large cities. Industrial activities and waste- and scrap-metal handling in cities lead to large amounts of copper entering the environment and watercourses being polluted.

MERCURY

With few exceptions, concentrations of mercury in Swedish rocks are low. Almost half of all mercury present in the environment comes from pollution caused by, for example, combustion plants and metal industries. The mercury present in stream plants thus reflects only part of the natural mercury in the ground. Nonetheless, it still identifies where the highest concentrations of available mercury are to be found.

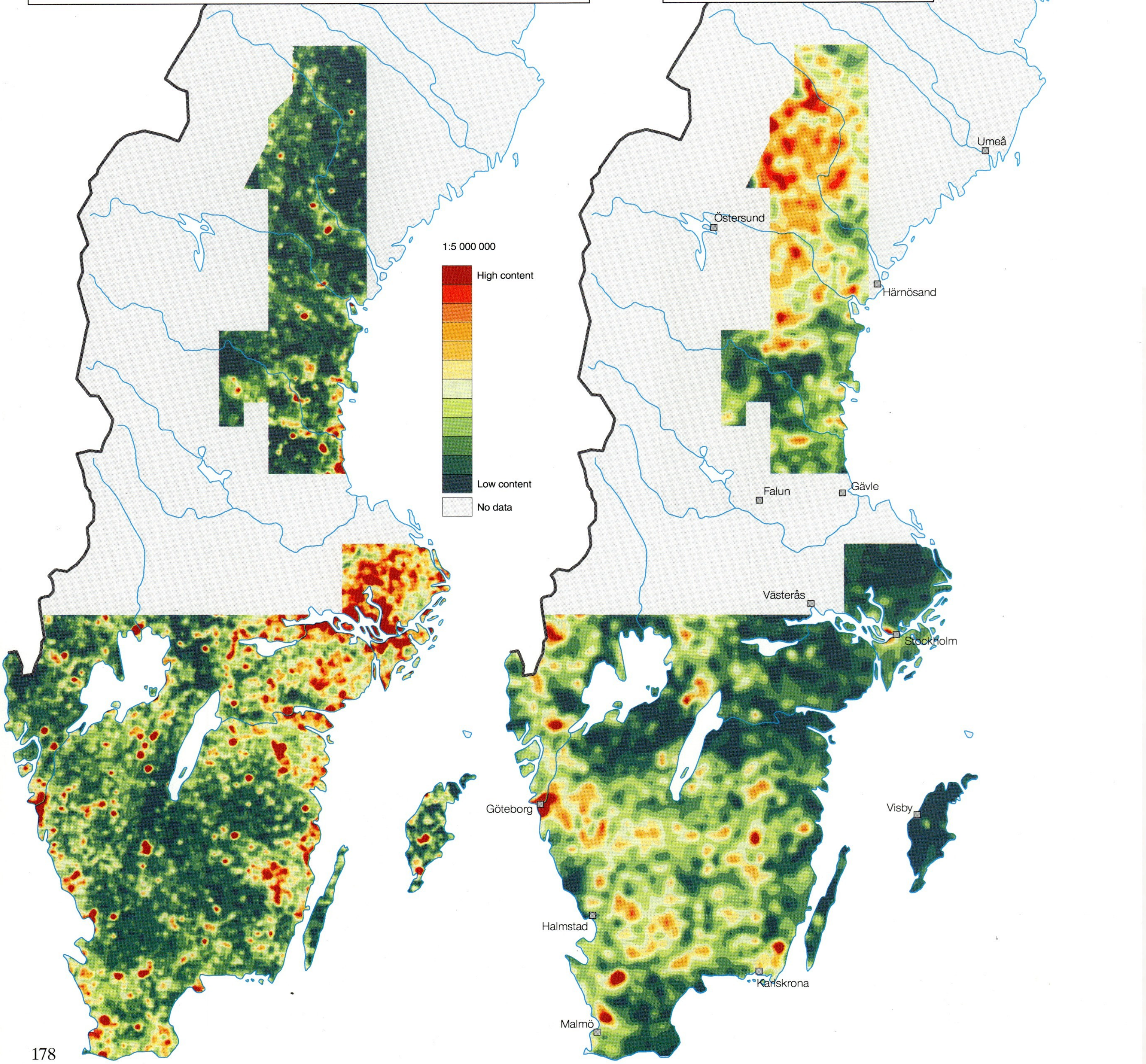

1:5 000 000
High content
Low content
No data

Luleå
Umeå
Östersund
Härnösand
Falun
Gävle
Västerås
Stockholm
Göteborg
Visby
Halmstad
Karlskrona
Malmö

SELENIUM

Selenium is an essential element that is considered to have a protective effect against the uptake of other toxic substances, for example, cadmium and lead. At the same time, selenium is toxic or even lethal in excessively high doses. This was learnt the hard way by Marco Polo when his imported pack-horses suddenly died unexpectedly during an expedition to the Chinese mountains. Later, the reason was found to be that the bedrock contained unusually high concentrations of selenium that had accumulated in certain grazing plants. The local animals had learnt to avoid these plants whereas the imported animals grazed indiscriminately and became ill or died. A similar situation of selenium toxicity in Sweden is extremely improbable since the rocks, with a few exceptions, is poor in selenium.

The relatively high concentrations of selenium in stream plants in Skåne, Västergötland and on the islands of Öland and Gotland, are caused by selenium-rich sedimentary rocks. The occurrences found within the normal selenium-poor Precambrian areas usually originate from selenium-bearing weathered sulphide mineralisations.

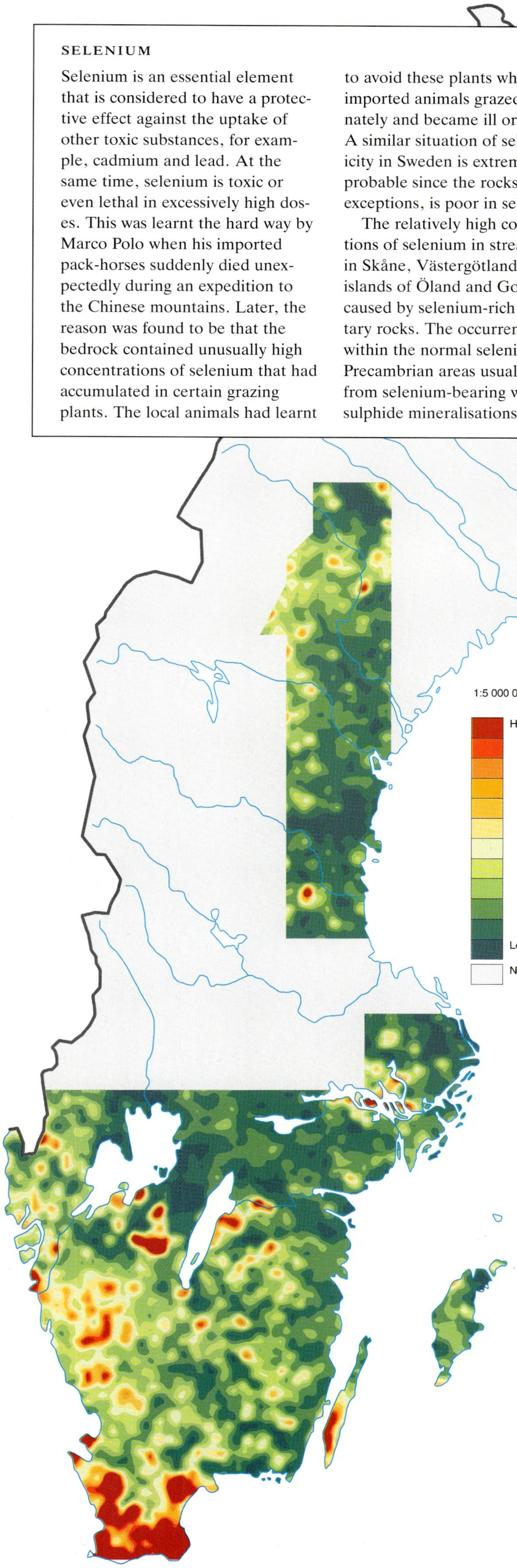

CHROMIUM

Chromium concentrations, like copper concentrations, are highest in basic rocks. Chromium is more resistant to weathering than copper and occurs in the environment in two forms: one has high solubility, the other low solubility. The more readily soluble form is the one most available to stream plants.

The highest chromium concentrations shown on the map are caused by chromium-rich ground. As for copper, the natural dispersion pattern for chromium is disturbed by the influence of industry in the vicinity of major cities.

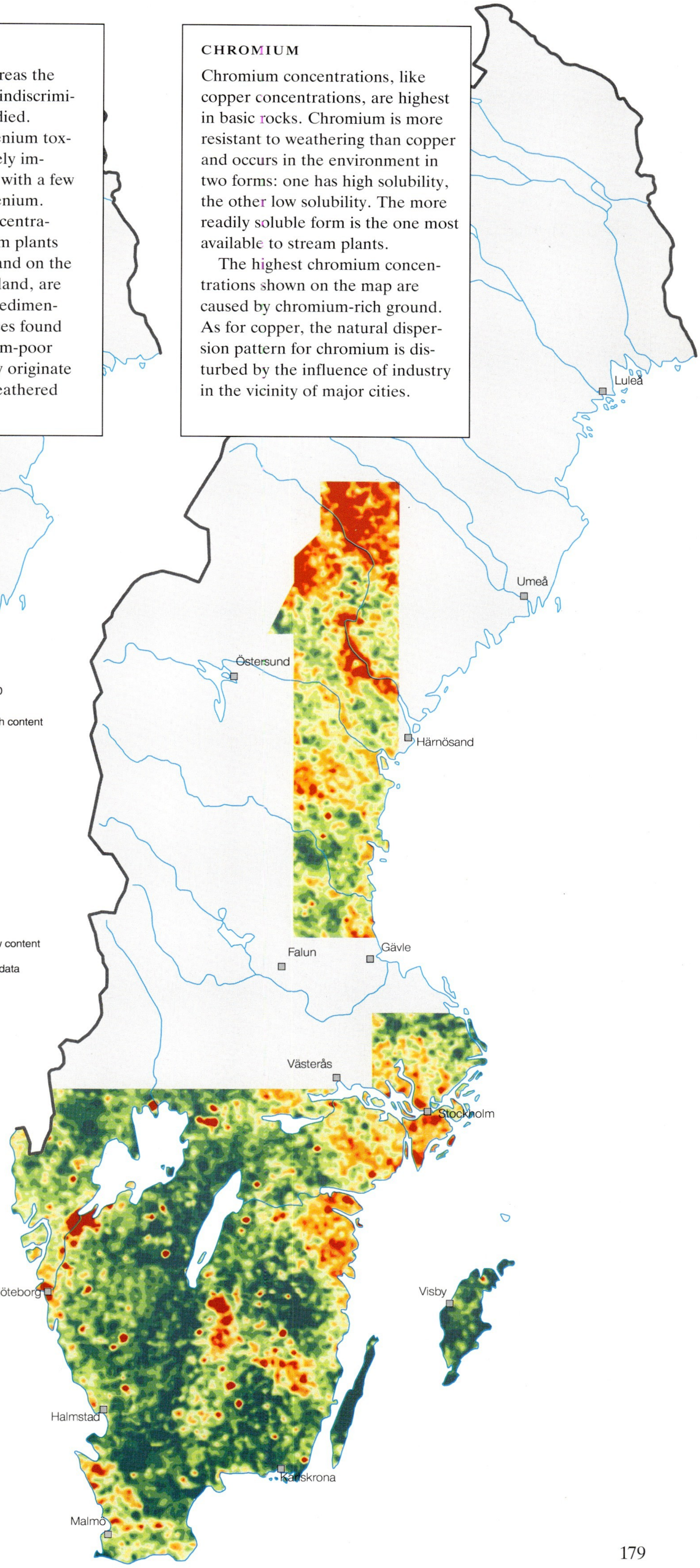

CADMIUM IN STREAM PLANTS, SOUTHERN SWEDEN

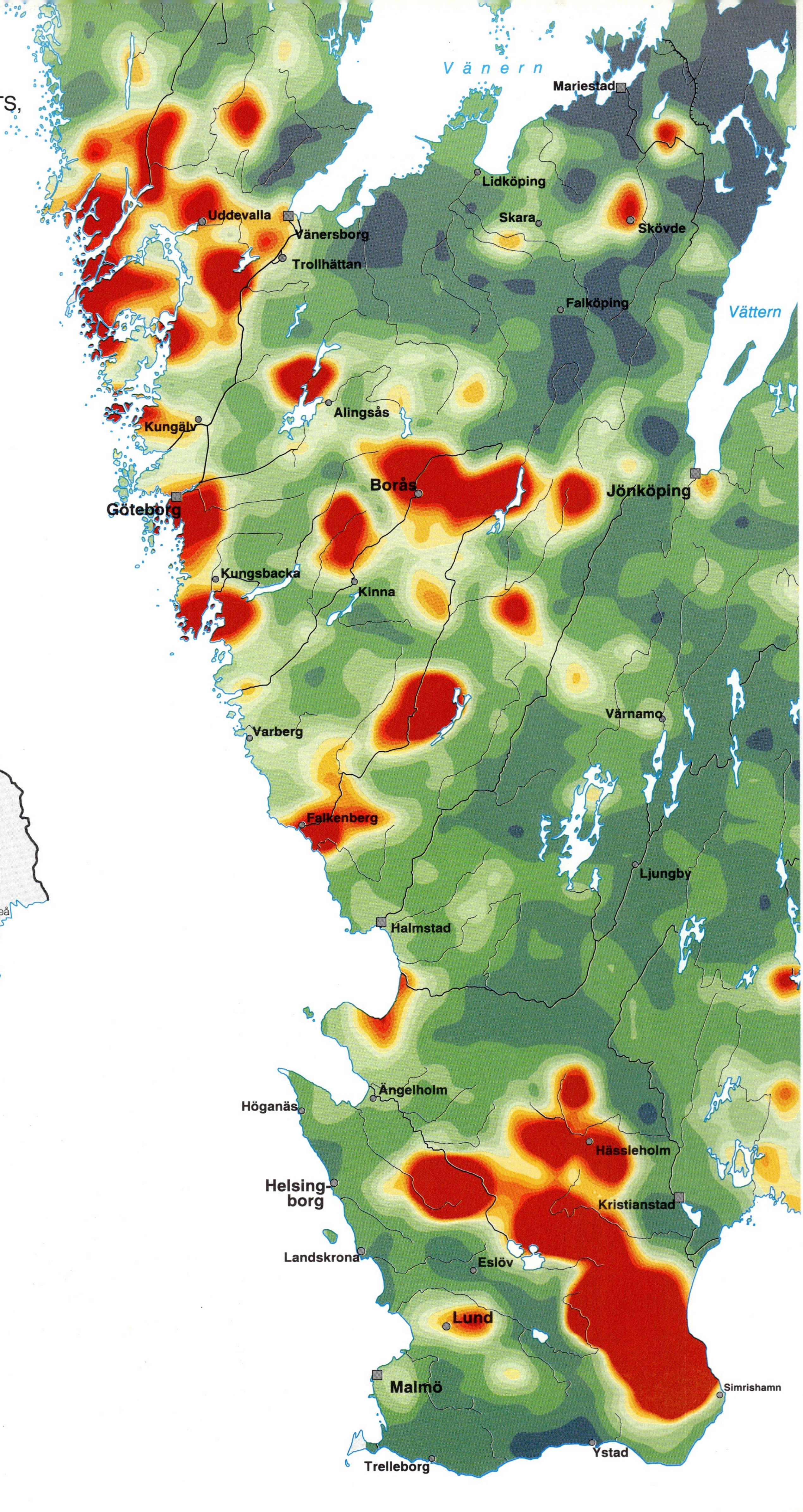

0 50 km

1:1 250 000

Mg/kg dry ashes

17.2
15.6
14.0
12.4
10.8
9.2
7.6
6.0
4.4
2.8
1.2

No data

Vänern
Mariestad
Lidköping
Skara
Skövde
Uddevalla
Vänersborg
Trollhättan
Falköping
Vättern
Alingsås
Kungälv
Borås
Jönköping
Göteborg
Kungsbacka
Kinna
Värnamo
Varberg
Falkenberg
Ljungby
Halmstad
Ängelholm
Höganäs
Hässleholm
Helsingborg
Kristianstad
Landskrona
Eslöv
Lund
Malmö
Simrishamn
Ystad
Trelleborg

CADMIUM IN STREAM PLANTS, ALL SWEDEN

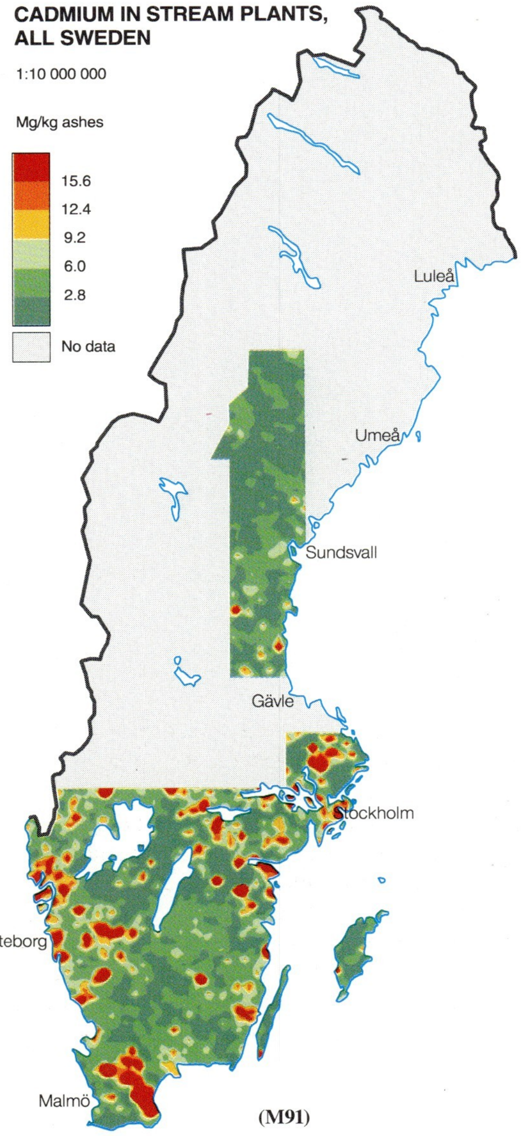

1:10 000 000

Mg/kg ashes

15.6
12.4
9.2
6.0
2.8

No data

Luleå
Umeå
Sundsvall
Gävle
Stockholm
Göteborg
Malmö

(M91)

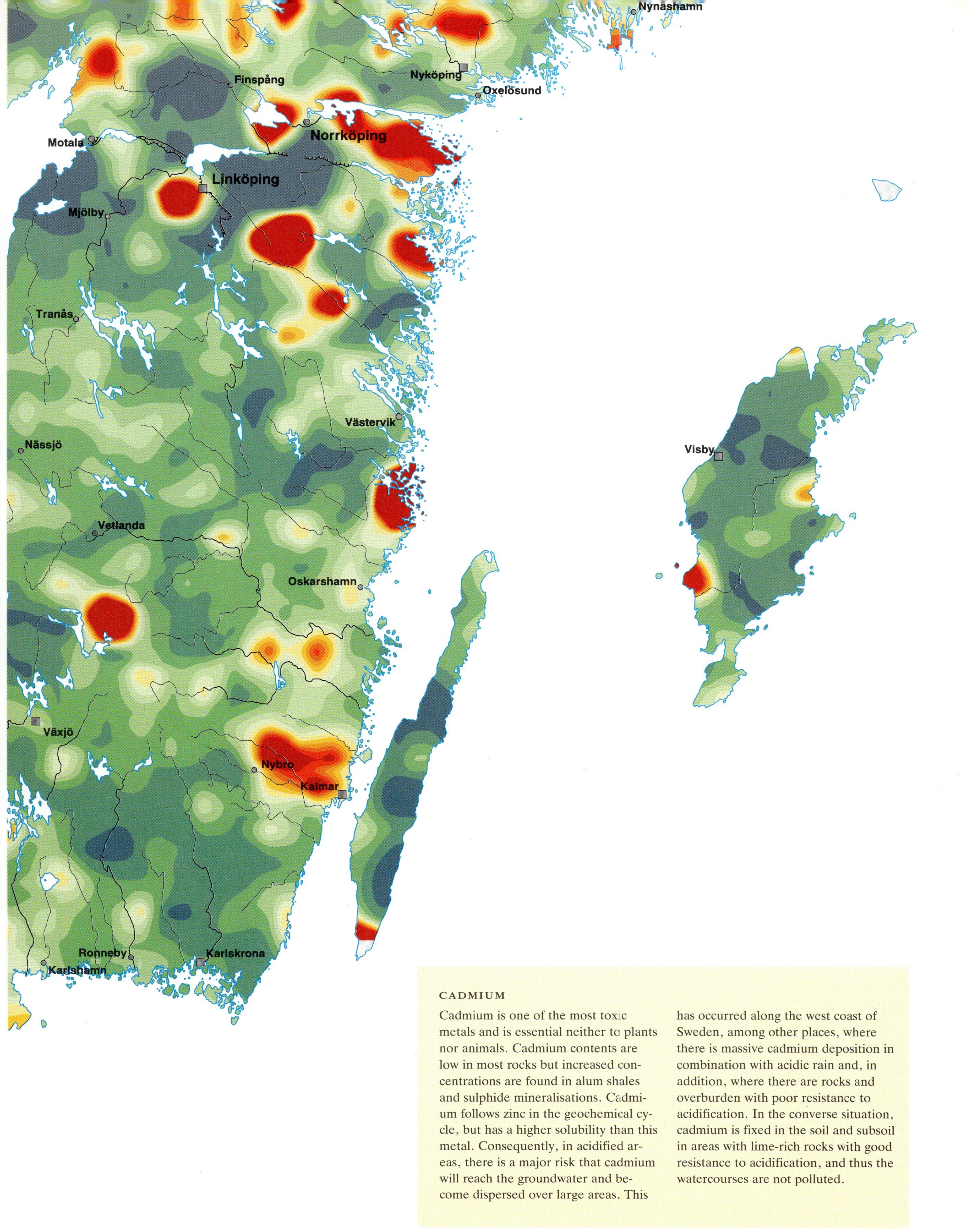

CADMIUM

Cadmium is one of the most toxic metals and is essential neither to plants nor animals. Cadmium contents are low in most rocks but increased concentrations are found in alum shales and sulphide mineralisations. Cadmium follows zinc in the geochemical cycle, but has a higher solubility than this metal. Consequently, in acidified areas, there is a major risk that cadmium will reach the groundwater and become dispersed over large areas. This has occurred along the west coast of Sweden, among other places, where there is massive cadmium deposition in combination with acidic rain and, in addition, where there are rocks and overburden with poor resistance to acidification. In the converse situation, cadmium is fixed in the soil and subsoil in areas with lime-rich rocks with good resistance to acidification, and thus the watercourses are not polluted.

URANIUM IN STREAM PLANTS, SOUTHERN SWEDEN

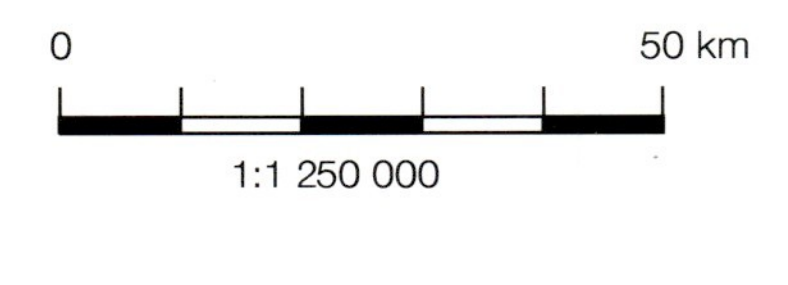

Mg/kg, ashes

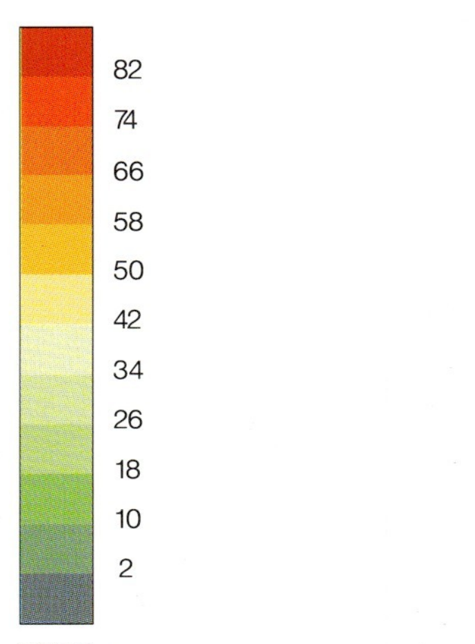

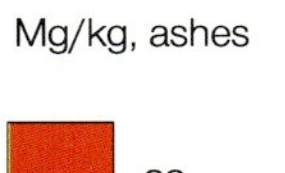

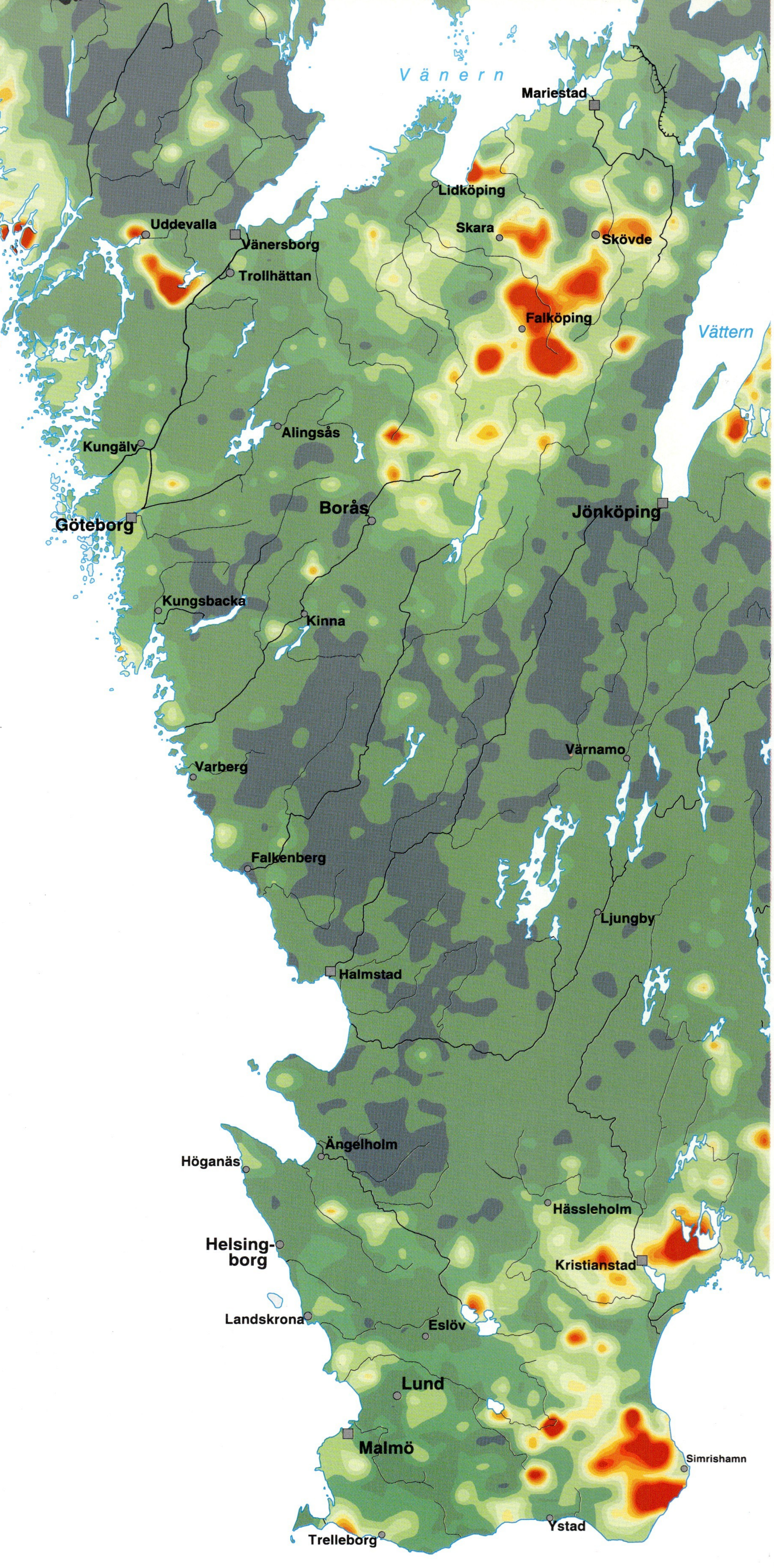

URANIUM IN STREAM PLANTS, ALL SWEDEN

1:10 000 000

Mg/kg, ashes

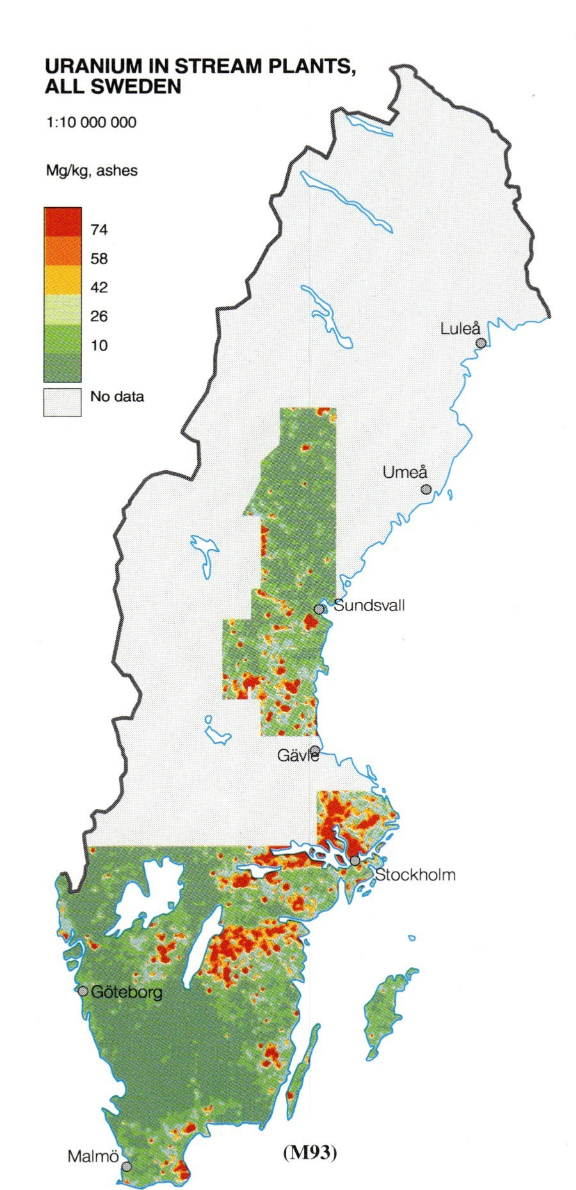

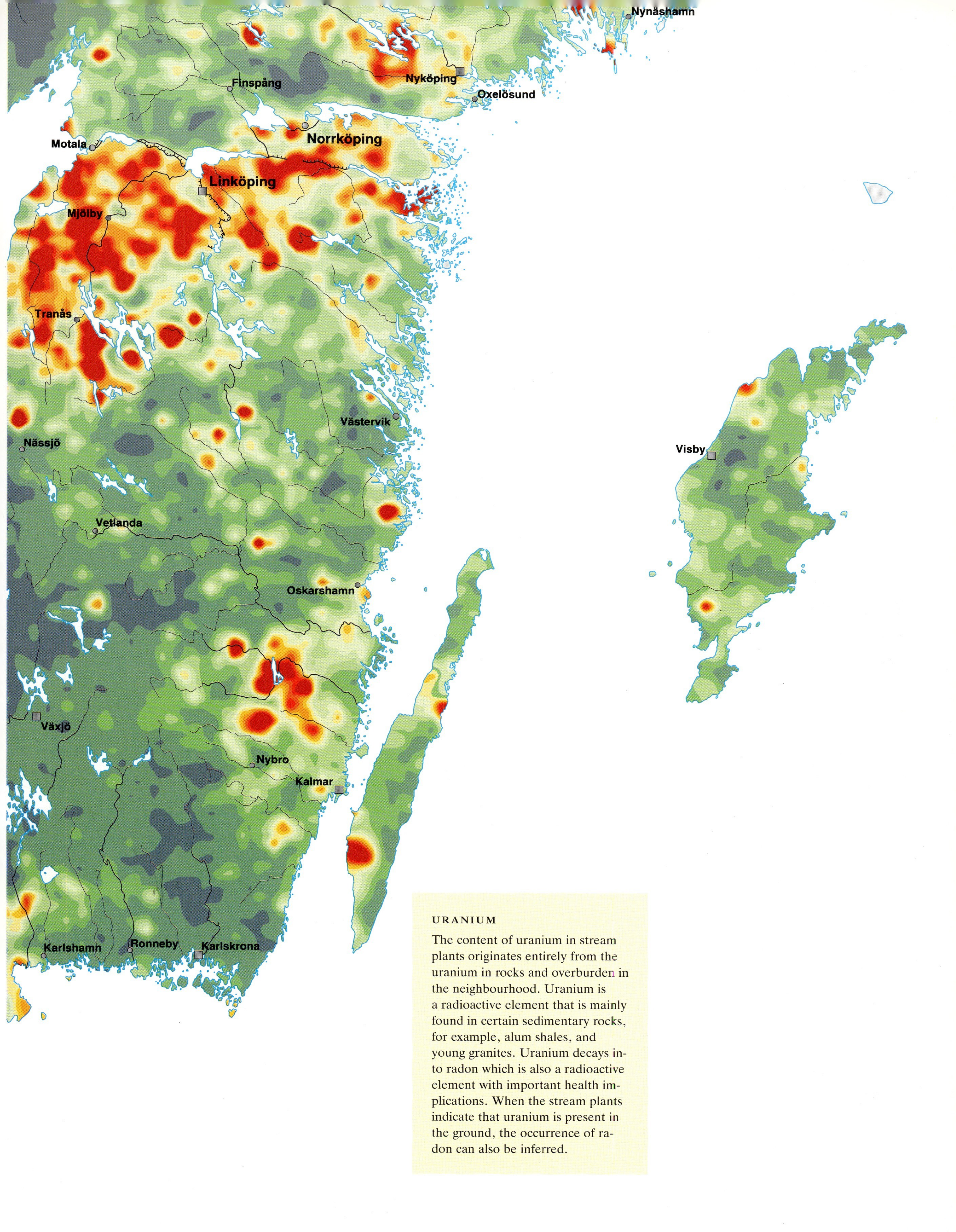

URANIUM

The content of uranium in stream plants originates entirely from the uranium in rocks and overburden in the neighbourhood. Uranium is a radioactive element that is mainly found in certain sedimentary rocks, for example, alum shales, and young granites. Uranium decays into radon which is also a radioactive element with important health implications. When the stream plants indicate that uranium is present in the ground, the occurrence of radon can also be inferred.

Geochemistry of till

MINERALS PROVIDE "FINGER-PRINTS"

Silica and aluminium are examples of elements that occur in high concentrations in almost all minerals and rocks. *Trace elements*, on the other hand, are found in much lower concentrations (measured in milligrammes per kilo), but are usually typical of certain minerals. Chemical analyses of till can therefore function as finger-prints that tell us about the rock types and minerals that characterize the till in a certain region.

GEOCHEMICAL STATUS

Our immediate ground environment—mainly soil fertility and groundwater composition—depends largely on the chemical properties of the constituents making up the overburden, i.e., its geochemistry. Different minerals may have approximately the same composition of elements. However, these may be more or less strongly tied in the different crystal lattices.

Minerals from limestones and greenstones weather relatively easily

Till is the most common Quaternary deposit in Sweden and forms large continuous areas that are frequently forested or, as shown here, emerge like islands in an archipelago of cultivated clays. Odensala, Uppland.

The mineral content in till is decisive for soil fertility and the composition of the groundwater. By analysing samples from the different layers of till, we can obtain knowledge of the supply of beneficial nutrients or whether there may be hazardous heavy metals present.

Potassium and sodium are both typical main constituents of granitic rocks but the relationship between them may vary. In the southern parts of the maps, a sodium-rich variety of granite dominates, whereas a potassium-rich type dominates in the north. Trace elements in granite are often found associated either with potassium or sodium. (M95, M96)

The cobalt map shows increased concentrations in areas where basic rocks are found in the till. Cobalt is an essential trace element. In cattle, cobalt deficiency is expressed as loss of appetite, emaciation and blood deficiency. Cobalt is fairly mobile and easily leached out of the till in acidic environments. A region with cobalt-poor till and low pH-values may, from grazing viewpoints, be considered as deficient in cobalt. (M97)

1:2 500 000

High content

Low content

No data

POTASSIUM

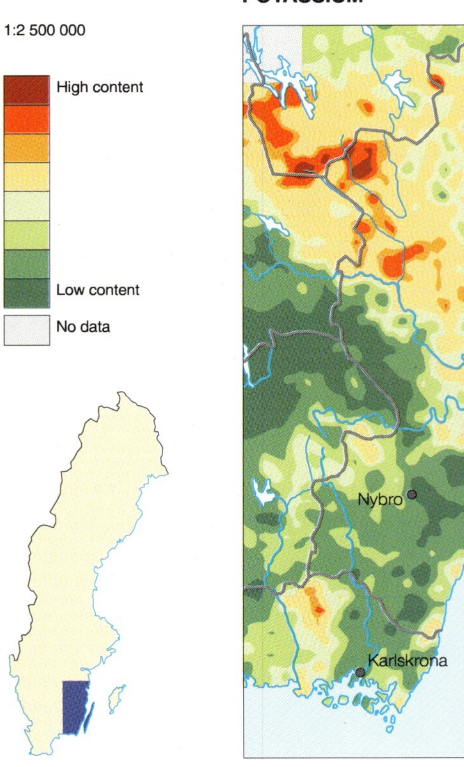

SODIUM

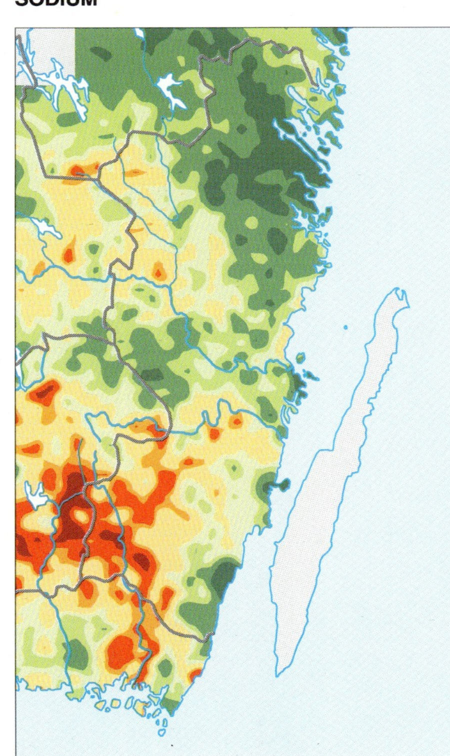

COBALT

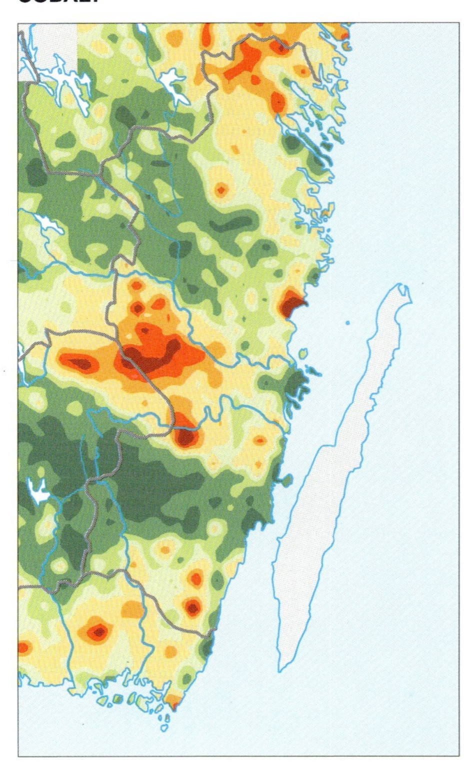

and release substances that are chemically loosely bonded. These are favourable for soil fertility and, in addition, are able to neutralize acidic precipitation. Till containing high contents of magnesium, iron, chromium and cobalt usually has a fairly high pH-value.

The opposite is found in till areas where the bedrock is dominated by acidic granites and porphyries. Such areas are characterized by relatively high contents of strongly tied potassium, sodium, lead and often tungsten. The minerals that contain these elements are fairly resistent to weathering and pH-values in the till are often low.

DIFFERENT ANALYSES GIVE DIFFERENT ANSWERS

Chemical analyses showing the total contents of different elements in till reveal only partly how strongly these elements are tied in the minerals. In order to find out how the elements can be released most easily, the weathering processes in the overburden can be speeded-up in the laboratory. Acids are used to release elements that are not tied so strongly. By combining information from a map showing an element's total dispersion with the amount dissolved in acid, it is possible to assess presumptive deficiency areas for plant nutrients. It is also possible to assess where, for example, leaching of aluminium is a risk factor.

COMPARABILITY

The elements in a soil profile vary in content from the surface downwards. Some elements have been leached out and have low concentrations in upper layers but are found further down in the unaffected strata of the till. Other elements have also been leached but have been precipitated in the iron-enriched horizon, where they have considerably higher concentrations than in the unaffected subsoil. This illustrates the importance of basing an element map on analyses of comparable samples from the *C-horizon*. Samples of this kind are not influenced by biological processes or by anthropological activity, and heavy metals and nutrients here are mainly of geological origin.

High content
Low content
No data

BARIUM

Västervik
Oskarshamn
Nybro
Kalmar
Karlskrona

PARTIALLY SOLUBLE BARIUM

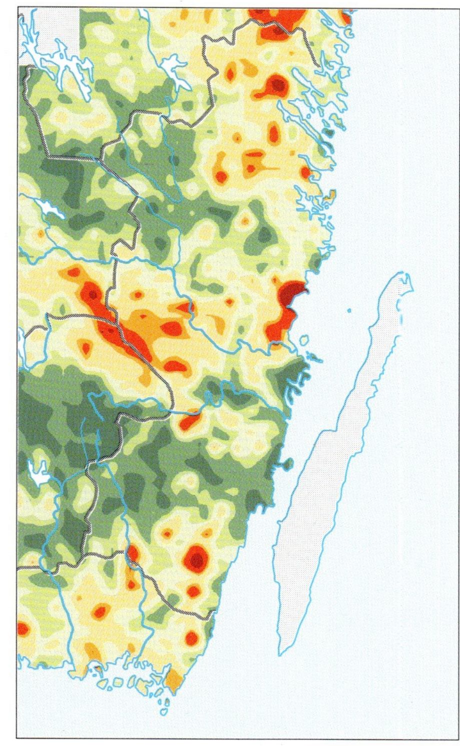

Barium is found in the mineral baryte, with low solubility, and in other more easily-weathered minerals such as the mica mineral biotite. The map on the left shows the total content of barium in till, and the map on the right shows where barium is found in a more easily-soluble form. It is not known whether barium is essential for biological metabolism but we do know that barium compounds may be hazardous. (M98, M99)

Uranium in the overburden reveals that the till contains fragments of radioactive rocks, often related to younger granites. Areas with risks for radon and radiation on levels implying health hazards are, consequently, more widespread than is indicated by the corresponding granite area on the bedrock map. (M100)

Lantanium is a rare earth metal found in different phosphate minerals. As for uranium, the presence of lantanium in younger granites can clearly be seen on the maps in, for example, the province of Blekinge. (M101)

High content
Low content
No data

URANIUM

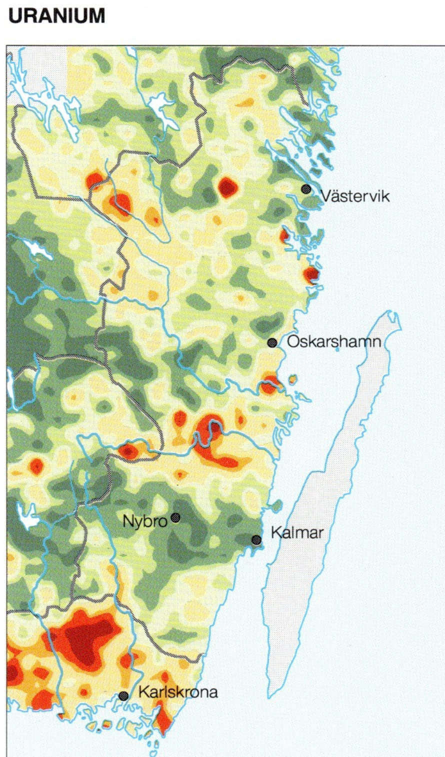

LANTANUM

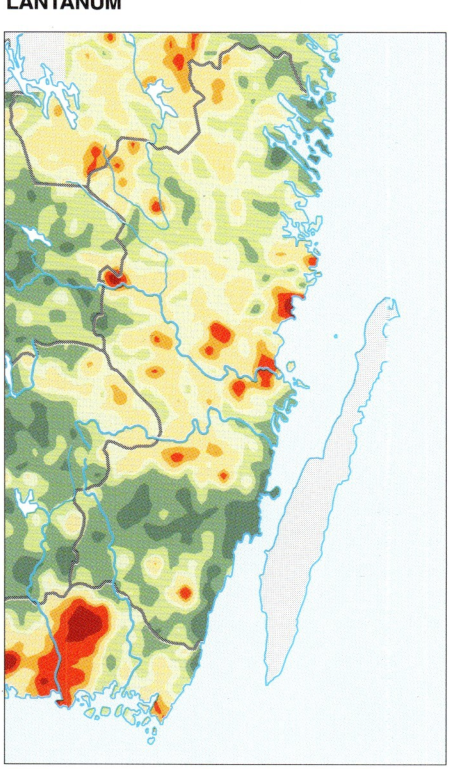

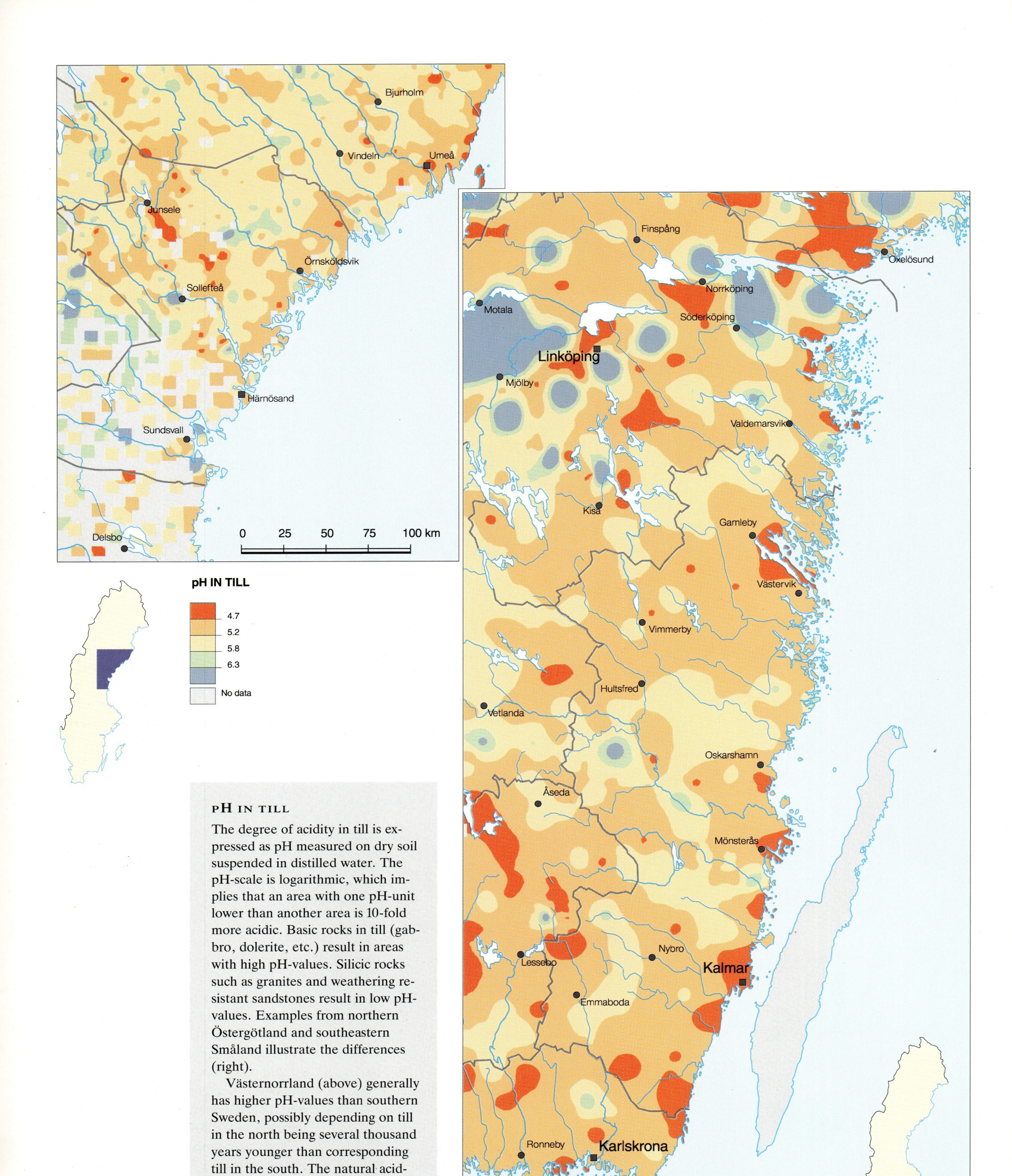

pH in till

The degree of acidity in till is expressed as pH measured on dry soil suspended in distilled water. The pH-scale is logarithmic, which implies that an area with one pH-unit lower than another area is 10-fold more acidic. Basic rocks in till (gabbro, dolerite, etc.) result in areas with high pH-values. Silicic rocks such as granites and weathering resistant sandstones result in low pH-values. Examples from northern Östergötland and southeastern Småland illustrate the differences (right).

Västernorrland (above) generally has higher pH-values than southern Sweden, possibly depending on till in the north being several thousand years younger than corresponding till in the south. The natural acidification caused by weathering has thus been able to affect the southern parts of Sweden for a longer period.

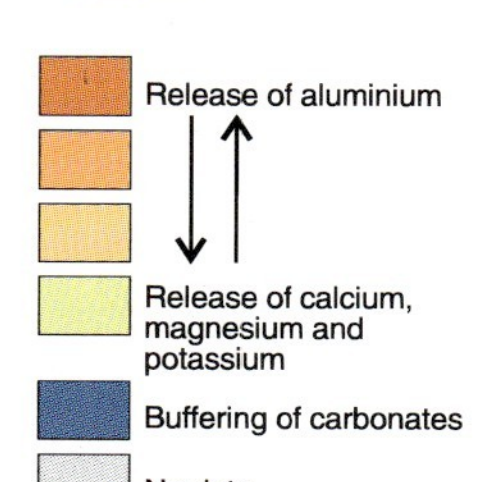

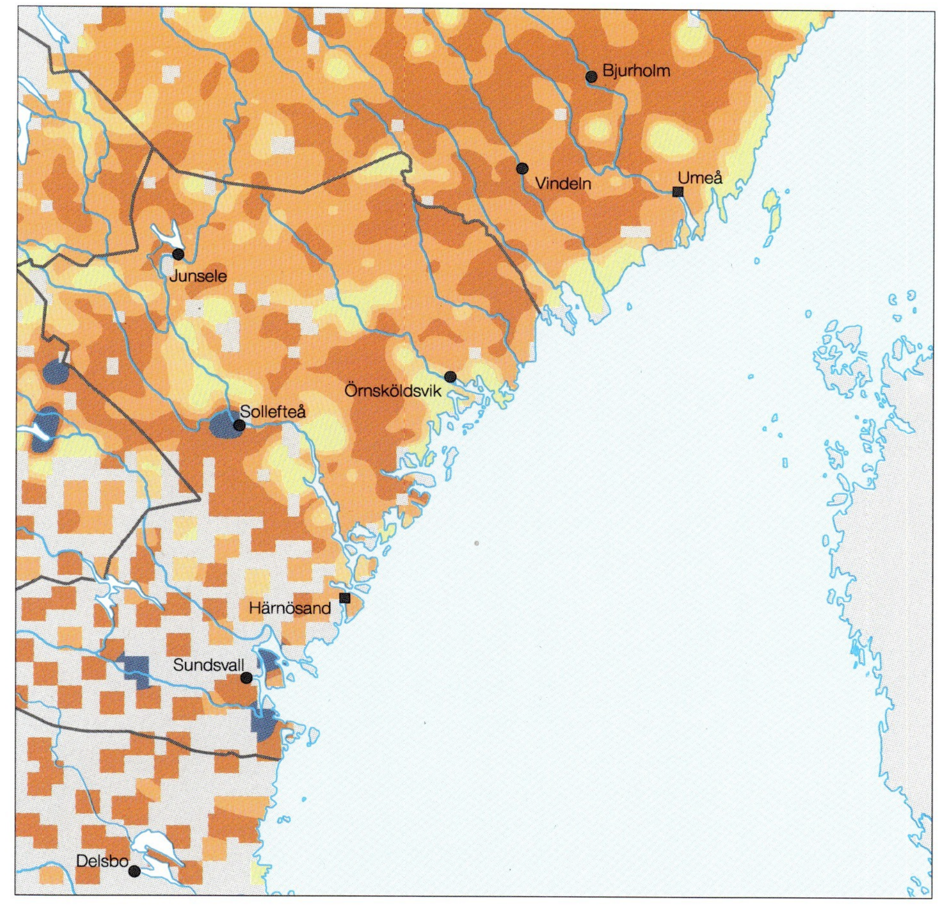

TOLERANCE TO ACIDIFICATION

The tolerance to acidification can be calculated by measuring pH in till before and after addition of a weak acid. In this way, the prevailing equilibrium is disturbed and different buffering systems are activated in order to counteract acidification. Tolerance to acidification has different effects on the ecosystem. Blue colour indicates carbonate buffering, i.e., that calcium and magnesium carbonates are present. Green and yellow show that the flows of calcium, magnesium and potassium to percolating water will increase. Orange and red show that aluminium leaching will occur. In such regions, there are risks that the ecosystem will be exposed to damaging effects.

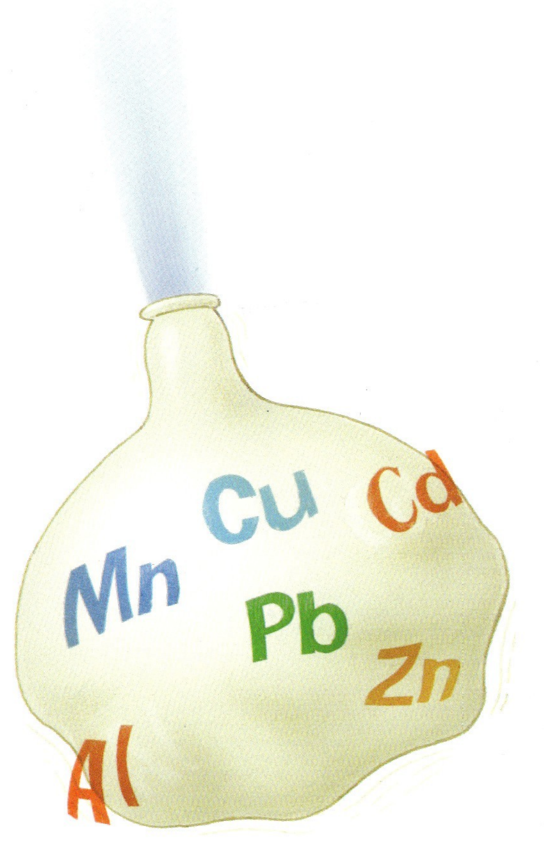

SHIFTS IN THE BALANCE

The chemical and biological processes active in the upper layers of the soil have decreased deeper in the profile. There exists, as a whole, a state of equilibrium and there is no particularly large natural influence. Whereas for thousands of years the weathering of soil minerals has released elements such as calcium and magnesium, aluminium has remained in the system and has thereby created a balance between supplied and removed nutrients. The ecosystem has adapted itself and the groundwater has a stable chemical composition.

In an increasingly acidic environment the ground's own defence against acidification—its buffering ability—becomes so impoverished that chemical limits are exceeded and the metals start to become mobile. The chemical time-bomb explodes and, like a deflating balloon, releases increasing amounts of toxic elements into the ecosystem.

However, mankind is continuously interfering with the ecosystem through changes in agricultural techniques, clear-cutting and forest ditching. For this reason, soil processes are influenced, and the chemical composition of groundwater is changed as a result of increasingly acidic precipitation. The mineral composition of the till is decisive for the ability of different buffering systems to neutralize the water, and also for the consequences of these processes.

BUFFERING SYSTEM

The acidic water percolating downwards through the ground is more or less effectively neutralized—*buffered*—by different chemical processes called the buffering system. High pH-values in soil frequently indicate the presence of lime, and in a related carbonate buffering system acidic water is neutralized very rapidly with hardly any influence on the minerals. When lime is not present and pH-values are from about 6.2 down to 4.5, the buffering takes place by means of minerals binding the acidic hydrogen ions and instead releasing beneficial basic cations. In this way, there will be an increased flow of calcium, magnesium and potassium to the surface water and groundwater.

The presence of pH-values lower than about 4.5 reveals that the reserve of easily-leached basic cations is temporarily depleted; the weathering cannot release elements at the same rate as they are consumed. Nonetheless, acidic water is neutralized relatively rapidly since different aluminium compounds dissolve and deal with the hydrogen ions. However, in this process aluminium is released and removed and in due course reaches the groundwater.

HEAVY METALS ARE MOBILE

The reduction in pH that leads to soil acidification also results in the mobility of the different metals being changed. Some achieve greater mobility, e.g., aluminium, cadmium, lead, copper, cobalt, manganese and zinc, whereas others, such as phosphorus and molybdenum, are bound more strongly in the soil.

Thus, in connection with low pH, metals normally considered to be nutrient trace elements are instead released from the soil, removed, and suddenly re-appear in high concentrations in the groundwater, i.e., in drinking water.

1:2 500 000

High

Low

No data

CALCIUM, LEACHABILITY

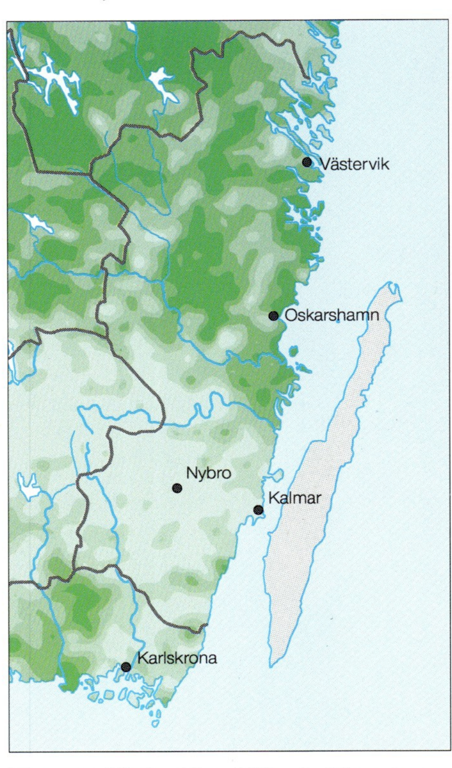

The map of the leaching ability of calcium shows where this substance is least and most soluble. In places where the leaching ability is very low, this may indicate that the mineral soil is deficient in calcium. (M105)

Vegetation takes up dissolved nutrients in the soil water. The availability of these nutrients depends on geological, chemical and biological factors.

BENEFICIAL ELEMENTS . . .

Some of the most important plant nutrients are *phosphorus, calcium* and *magnesium*. In regions with a high geochemical status, mineral soils contain sufficient nutrients for vegetation and animal life. Groundwater also has a favourable composition without hazardous heavy metals. In situations with long drought, the groundwater will supply the forest with both liquid and nutrients.

In soil regions with a low geochemical status, nutrients are tied strongly and the natural weathering that releases these elements is a slow process. If pH is so low that aluminium buffering takes place, the overburden will be a presumptive source for leaching of heavy metals. In acidic environments, weakly soluble phosphorus compounds, that are unavailable to plants, are also formed. When clear-cutting on ground of this kind, it is important to leave branches, twigs, needles and deciduous brush on the site, so that beneficial elements in them can be returned to the ground. It may also be necessary to supply suitable fertilizers if the reserve of elements in the ground is low.

The total content of magnesium in the tills of central Norrland varies between 10 and 20 grammes per kilo. High concentrations reveal features of basic rocks with easily weathered minerals that release magnesium. (M106)

1:2 500 000

High content

Low content

No data

MAGNESIUM

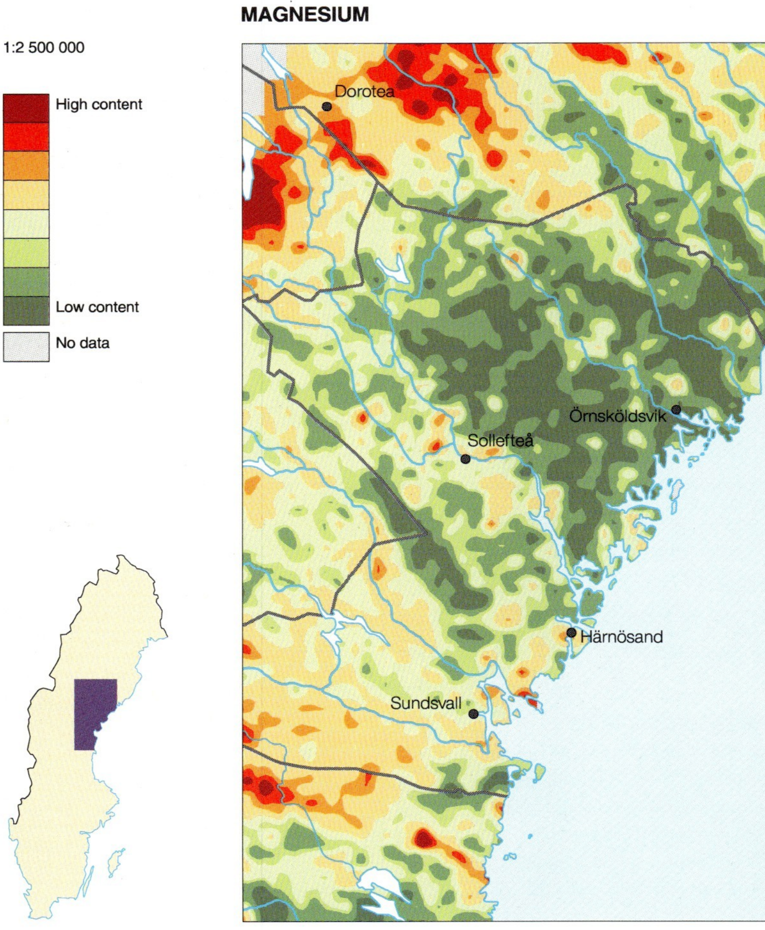

PHOSPHOROUS

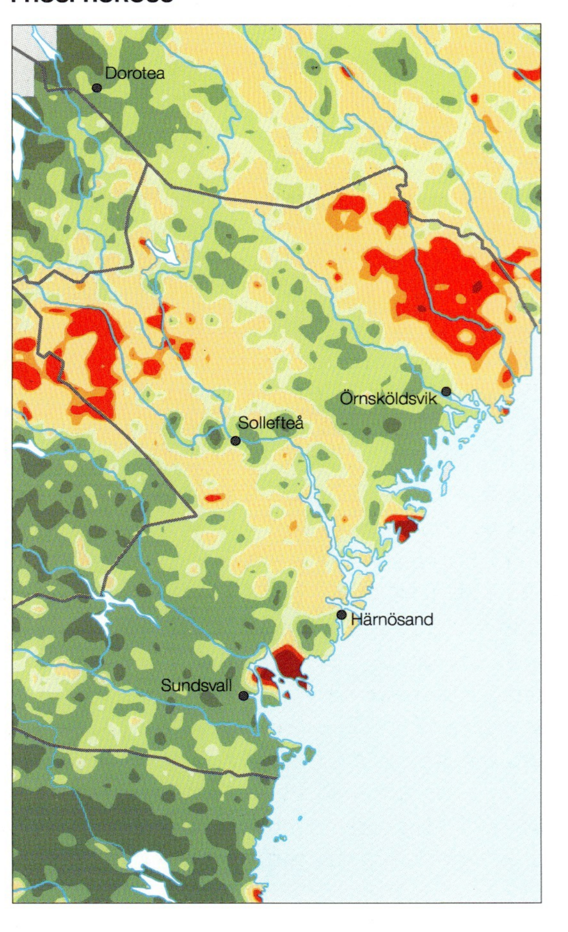

The total content of phosphorus is between 2 and 4 grammes per kilo. The source of phosphorus is the phosphate mineral apatite, which is found in many rocks with varying origins. (M107)

ALUMINIUM, LEACHABILITY

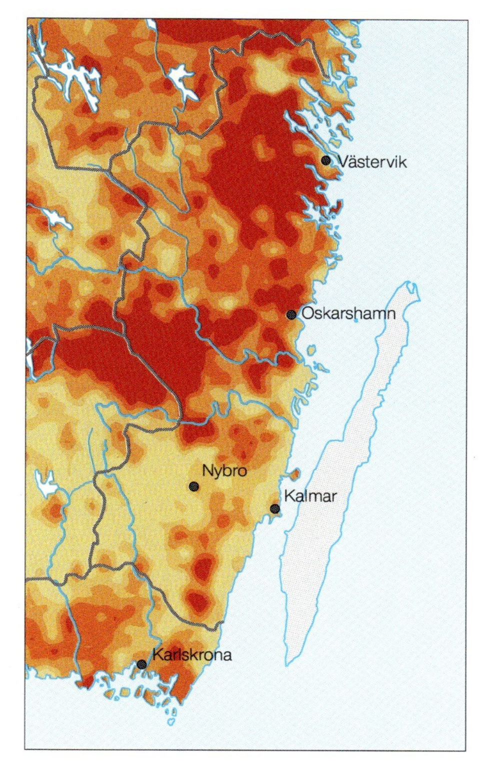

In regions where the leaching ability is high and the pH-value is simultaneously low, there is a large risk that aluminium is released and enters the water system. (M108)

ARSENIC

BERYLLIUM

High content

Low content

No data

Arsenic and beryllium are trace elements that occur naturally in small amounts and are measured in milligrammes (or less) per kilo of till. (M109, M110)

. . . AND HAZARDOUS ELEMENTS

The maps on this page show examples of hazardous elements. *Aluminium* is found abundantly in nature, both in rocks and in till, where it is usually strongly tied in different minerals. However, if aluminium is chemically loosely bound, and if the pH-value is simultaneously low, then acidic percolating water will be neutralised as a result of the aluminium being released and entering the water system. As far as we know, aluminium is not essential for the biological metabolism. Instead, it appears to be a risk to Man, animals and vegetation.

Beryllium occurs naturally in, for example, feldspar, and thus is accidentally spread when feldspar is mined. It is not known whether beryllium has any beneficial effects, but we do know that it influences certain enzymes negatively and causes damage to most organs in the human body. The element is easily absorbed by plants and, thus, is extremely toxic to plants, animals and Man.

Arsenic is another element known only as extremely toxic. It occurs in different rocks as sulphide minerals, frequently together with cobalt, copper and gold. It is easily soluble and can be transported in the surface water.

Cows are allowed to graze in the meadow but can the farmer be sure that the grass is as beneficial as it was earlier, or have the heavy metals, that were earlier bound in the soil, now been absorbed by the vegetation?

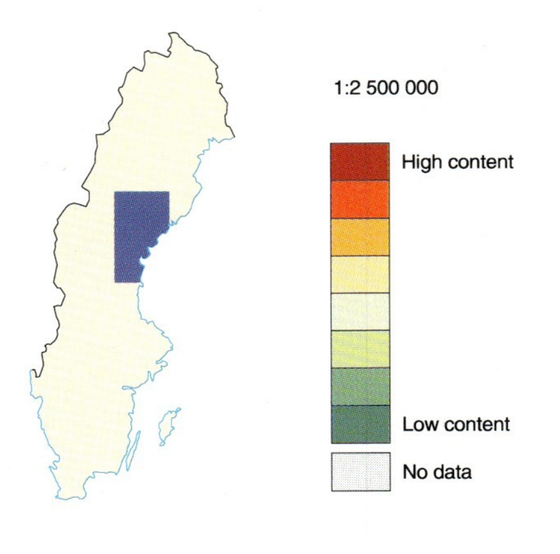

LEAD

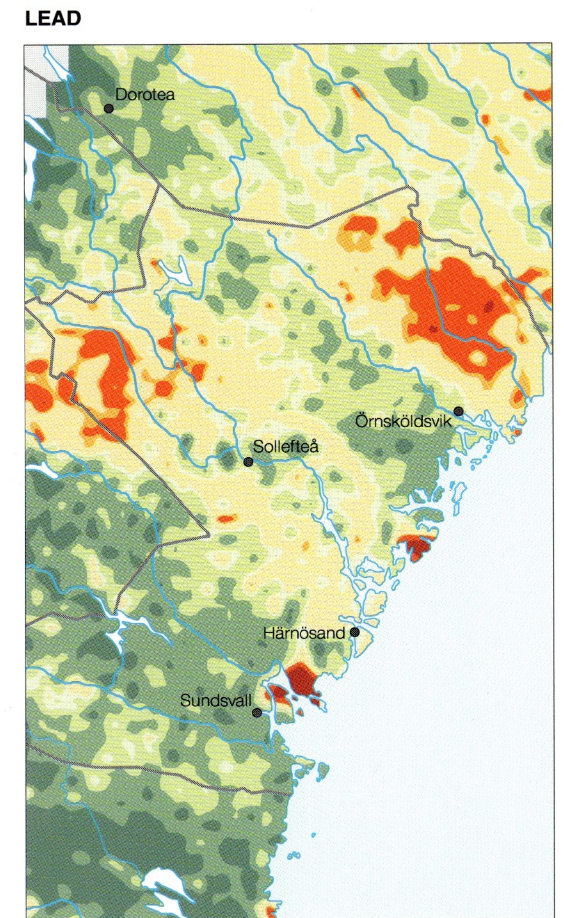

COPPER

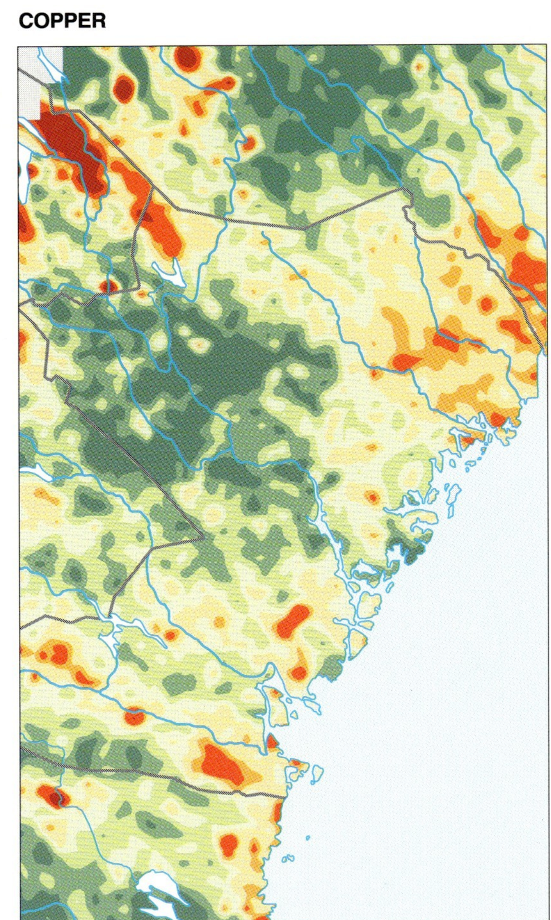

Lead, copper, zinc and *tungsten* have been mined and processed for many years in Sweden. Mineralisations containing lead, silver and zinc are found in, for example, the Cambrian sandstones of the Caledonides. Rock fragments in the corresponding till create the increased metal levels that can be seen in the left-hand part of the map showing lead.

Both the copper map, showing parts of Västernorrland, and the zinc map of southeastern Sweden, show increased levels in areas which have, or have had, mining activities.

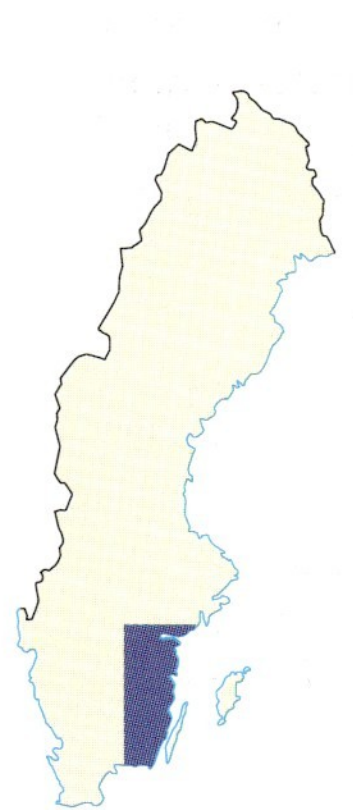

ZINC

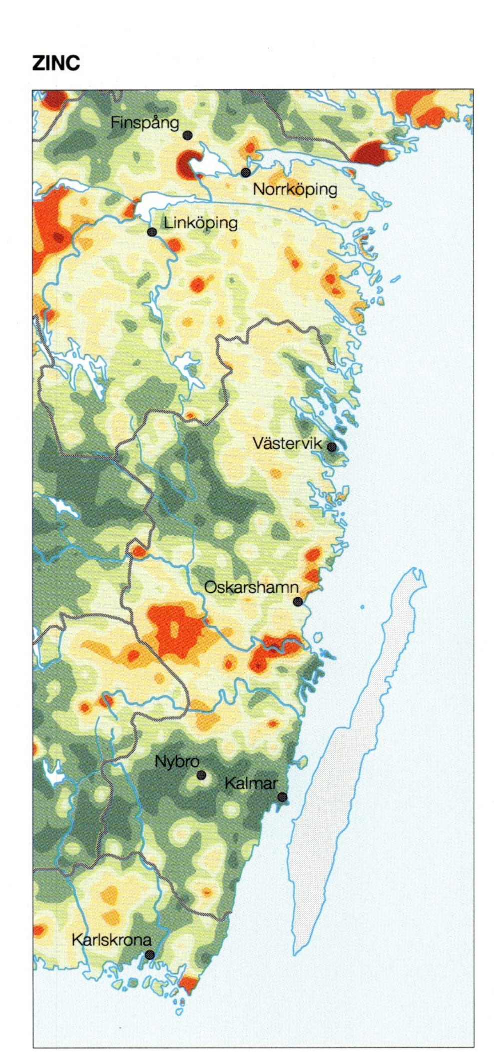

TUNGSTEN

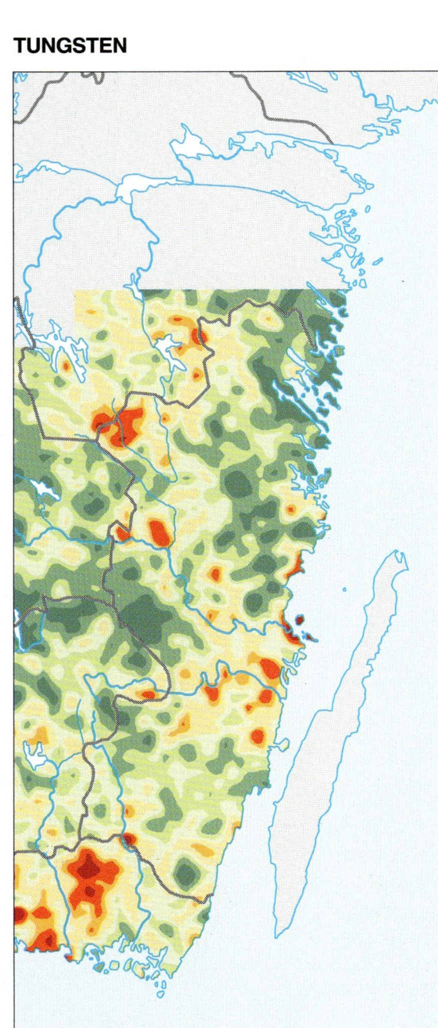

PROSPECTING FOR ORE

In the search for new mineralisations that may result in mining activities, the maps showing the dispersal of metals in till are of particular interest. The continental ice sheet was the massive plane that broke off fragments from a mineralisation, transported them, and dispersed them in a fan-shaped pattern. When the ice margin retreated, it left a layer of till covering the surface of the rock. The till consists not only of ordinary rock fragments collected by the ice, but also of fragments of mineralised bedrock. In this way, the till may have an abnormally high content of metals. When searching for ore, an anomalous region of this kind forms a basis for more detailed investigations.

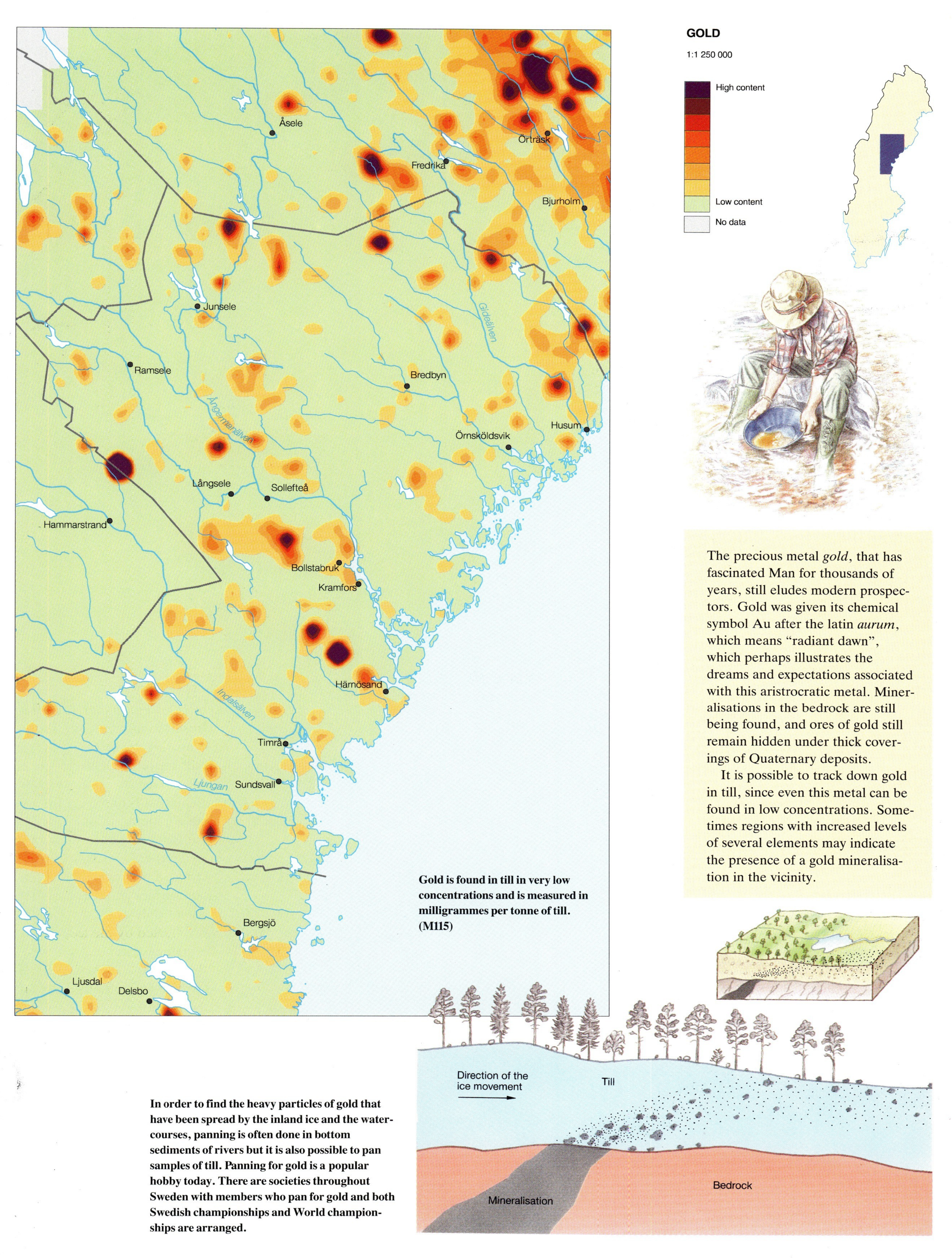

The precious metal *gold*, that has fascinated Man for thousands of years, still eludes modern prospectors. Gold was given its chemical symbol Au after the latin *aurum*, which means "radiant dawn", which perhaps illustrates the dreams and expectations associated with this aristrocratic metal. Mineralisations in the bedrock are still being found, and ores of gold still remain hidden under thick coverings of Quaternary deposits.

It is possible to track down gold in till, since even this metal can be found in low concentrations. Sometimes regions with increased levels of several elements may indicate the presence of a gold mineralisation in the vicinity.

Gold is found in till in very low concentrations and is measured in milligrammes per tonne of till. (M115)

In order to find the heavy particles of gold that have been spread by the inland ice and the watercourses, panning is often done in bottom sediments of rivers but it is also possible to pan samples of till. Panning for gold is a popular hobby today. There are societies throughout Sweden with members who pan for gold and both Swedish championships and World championships are arranged.

Changes to the landscape

Natural processes

Nature is undergoing continuous change. Running water, waves, glaciers and wind reshape the landscape, and tectonic processes such as land uplift, create new areas of land. This transformation has been going on for billions of years and will continue as long as Earth exists. In Sweden the most noticeable events in these natural changes to the landscape are rockfalls, rockslides, and landslides.

When boulders, stones, or particles of gravel and sand move freely, we speak of e.g. *rockfalls*. This occurs in rock-walls, or on slopes made up of gravel or sand. *Landslides* occur in Quaternary deposits made up of fine sand, silt and clay, mainly in quick-clay. Both falls, landslides and debris flows may occur suddenly.

Forces of nature are working to adapt cliffs and slopes to an equilibrium. Land uplift, the climate and anthropological influence alter the stability of the ground.

Falls and landslides are commonly found in connection with snowmelt and the thawing of frozen ground, as well as during periods of intensive

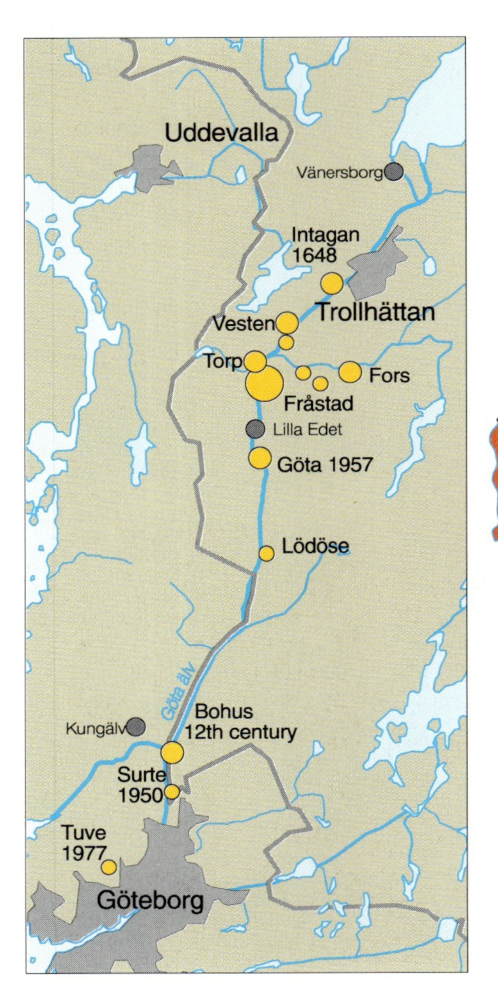

The valley of the River Göta Älv suffers more from landslides than any other area in Sweden. The valley has important communications with extensive shipping, roads and railways. Parts of the valley have a warning system that registers movements in the clay.

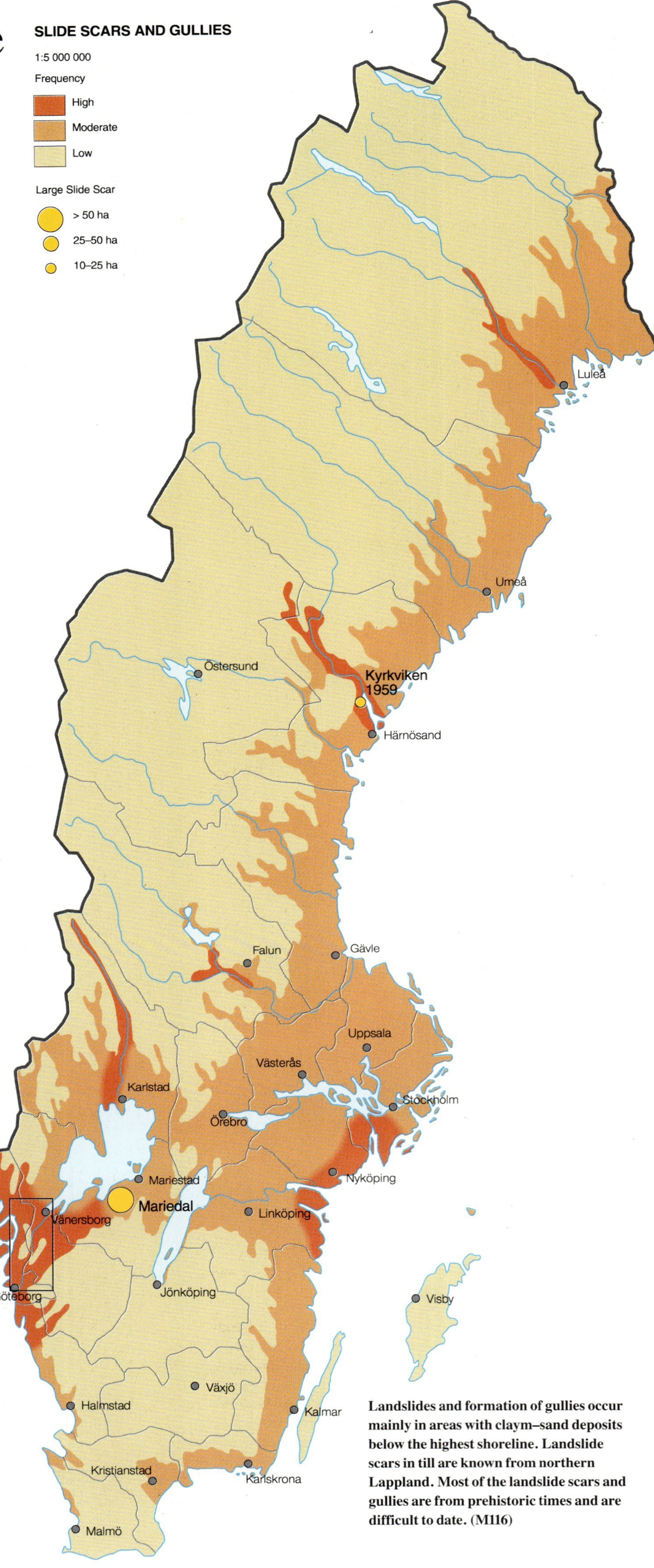

Landslides and formation of gullies occur mainly in areas with claym–sand deposits below the highest shoreline. Landslide scars in till are known from northern Lappland. Most of the landslide scars and gullies are from prehistoric times and are difficult to date. (M116)

Part of the gully system in the River Örekilsälven valley at Gesäter in Dalsland.

In many places in the mountains, the valley sides are subjected to rapid, episodic, sloping processes. The rock slide on the western slope of Viddja, to the south of Lake Torneträsk, has markedly changed the valley's morphology.

rain, i.e., when the water pressure in the ground is high. Large landslides have become increasingly common during the past century, probably as a result of anthropological interference.

In high terrain, there are both rapid and slow erosional processes. Falls of stone and cobbles along rockwalls take place in the autumn and spring as a result of frost weathering. *Avalanches of snow* and *mudflows* on mountain slopes may cause erosion and create new land forms. *Solifluction* is the name of the slow, continuous movements of deposits that can be seen as small, narrow terraces on mountain slopes above the tree limit.

Gullies occur as the result of a slower process than landslides and falls. They are mainly formed in areas where there is fine sand and silt. Gullies generally branch into two or more arms, with abrupt changes of direction.

Landslide is the most feared mass movement. The last major landslide occurred on 30 November 1977 at Tuve, Hisingen. The landslide covered 27 hectares, nine people lost their lives and 65 houses were totally destroyed.

Left: The landslide at Surte on 29 September 1950. The dry crust of the clay broke apart into pillars positioned more or less pell mell.

In the Tuve landslide, the clay masses were pressed out into the Kvillebäcken valley. Today, the area is a park.

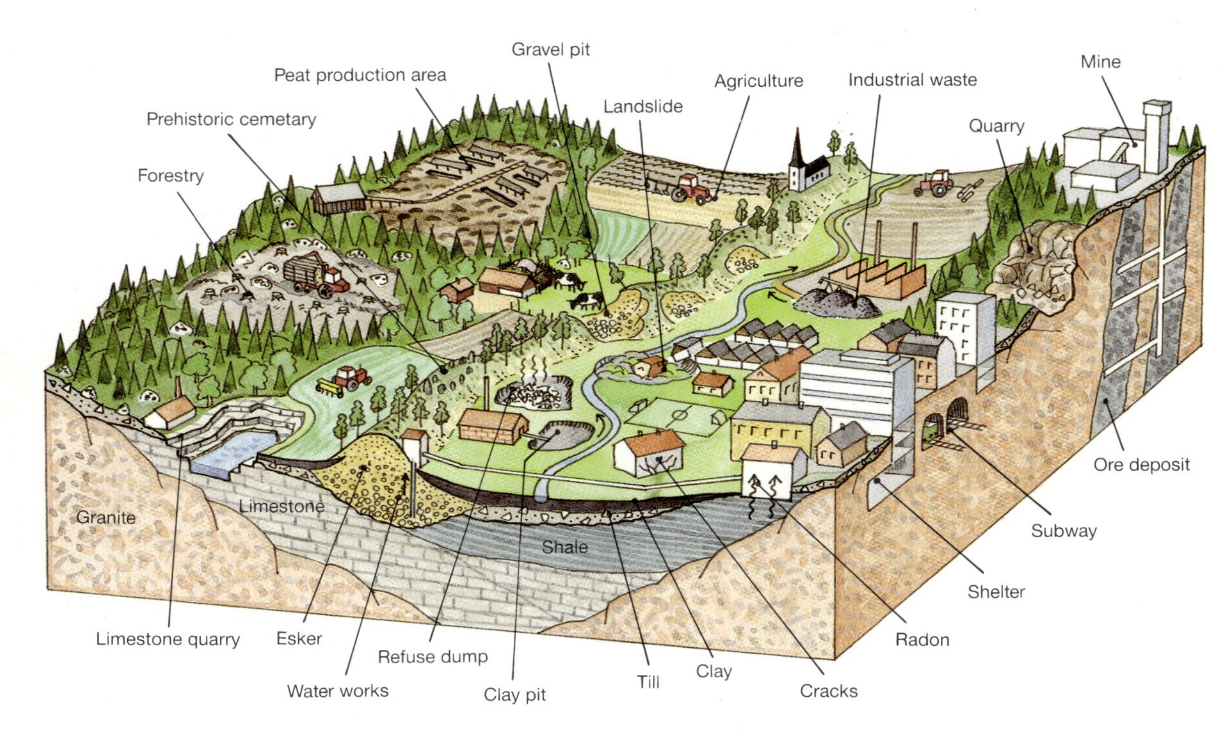

In former times, the landscape was utilized in accordance with the possibilities offered by the soil. Easily tilled ground such as sand and silt were cultivated and dwellings were erected on well-drained land, e.g., on the eskers. However, the natural resources of the landscape gradually became utilized increasingly intensively. All too often, no geological considerations were taken when planning; settlement, cracks, mould and radon gas in the houses, landslides and polluted ground water are expensive examples of this.

An exploited esker can never be recreated, and neither can bedrock. However, an exploited area can be tided up and utilized in other ways. The former limestone quarry at Hällekis, Kinnekulle, has been transformed into a lake for angling.

Natural resources

Geological conditions are of great importance for society. In the small-scale agrarian society, land was used for agriculture and forestry with adaptation to the opportunities offered by the land – mainly depending on the properties, extent and nutrient content of Quaternary deposits – together with the presence and availability of groundwater. Modern possibilities for irrigation, fertilisation and use of efficient machinery for preparing the soil today allow us to cultivate land that was formerly inaccessible.

The *mineral resources* of Sweden have always been of great importance for society. In sectors such as communication, information, energy and building construction, the products are largely based on raw minerals. Energy products such as oil, coal, natural gas, peat, and geothermal energy all have their origin in geological processes.

Water is our most important foodstuff and is a condition of life for humans, animals and plants. Availability and quality of water is determined by prevailing geological conditions. Availability of water in Sweden is good in comparison with most other places in the world. As far as Sweden is concerned, water problems in the future will not concern quantity but quality.

The *eskers* are an important resource for supply of water as well as for recreational activities. However, sand and gravel are also needed in manufacturing concrete, in asphalt and in building roads and other constructions. When an esker or an outcrop of rock is to be exploited, there are almost always conflicts with other interests.

ANTHROPOLOGICAL INFLUENCE ON THE LANDSCAPE

The development of the industrialized society has resulted in stresses on the environment in the form of pollution and waste. Environmental work is today moving from corrective activities to placing greater emphasis on preventative activities and therefore geological knowledge has become increasingly important.

The chemical and physical properties of rocks and Quaternary deposits vary. This is of decisive importance for the natural occurrence of heavy metals, risks for pollution of ground and groundwater, sensitivity to acidification, and the risk of radon gas. Water in the ground is able to transport not only beneficial elements but also pollutions and toxics. Knowledge of different processes in the ground is therefore a condition for solving environmental problems. The threat to the ground from air pollution and acidification places greater demands on agriculture and forestry being conducted with knowledge of the varying properties of deposits.

By using geological information already in the planning work for, e.g., a housing estate, many hazards and unnecessary expenses can be avoided. Radon gas and clays where there is a risk of landslides are examples of threats to human life and health.

Shales for extraction of crude oil have been mined at Kvarntorp in the province of Närke. The waste rock forms an almost 100 m high hill that has become a landmark on the Närke plain.

The winding eskers at Lake Råstojaure in northernmost Lappland are part of a beautiful area, considered to be of national geological interest.

Another area of national geological interest is Valle Parish, to the west of Billingen in Västergötland. This parish is a typical kame area with hills, ridges and depressions.

The glacially polished Bohus granite at Hållö to the south of Smögen contains numerous sculptured forms on the rock surface, for example, smoothly rounded troughs and groove-like channels. The island is an object of national geological interest and protected as a nature reserve.

NATURE CONSERVATION

Before the arrival of Man, the landscape was influenced solely by the forces of nature. The development of humans from the Stone Age to the post-industrial society has been accompanied by increasing anthropological influence on, and a re-shaping of, the landscape. By means of technology, we have become a powerful geological force. Humans are creating new land-forms and changing the natural environment, e.g., with mine shafts, quarries, pits, agriculture and forestry, embankments and dumps.

In nature conservation, attempts are made to restrict the interference with nature and adapt the new forms in a suitable manner. Quarries are being filled up with earth masses, and others are transformed into recreational and angling lakes, sports arenas, parks or forest land for wood production.

The most important instrument that can be used in nature conservation to preserve valuable natural environments and the geological diversity in the landscape is protection. There are different kinds of protection for different areas, all of which are supported by the Law on Nature Conservation:

- National park
- Nature reserve
- Nature conservation area
- Natural monument

In all these protected areas, it is possible to find *geoscientific objects of national and provincial interest*. This refers to phenomena that have few correspondances in the area or in the country as a whole. During the 1980's, the Swedish Environmental Protection Agency reviewed the objects of national interest in Sweden together with the provincial administrations. Geoscientific objects of national interest are referred to the Geological Survey of Sweden for assessment. Sweden also has objects of international interest, mainly reference localities for different fossils.

Radiometric dating methods are used to follow the activities of geological forces at least three billion years backwards in time. A geologist works with long time perspectives. For a bedrock geologist, one million years is an acceptable margin of error, whereas a Quaternary geologist considers 75 years to be a reasonable error.

Research is continuously offering new contributions to geological development, not least with support of atmospheric research, increased knowledge of the circulation of water in nature, physical, chemical and biological processes on Earth and in the inner regions of the Earth, improved geophysical measuring techniques, etc. The movements of the continents have recently been explained and within the youngest part of the history of the Earth, revisions are now being made to the age of the Quaternary deposits since increasing numbers of deposits have been found to have been formed before the last inland ice covered Sweden. Environmental awareness in our dynamic society of today has given geology increasing importance and increased understanding.

"For a couple of hundred generations, we have been living on borrowed time and borrowed land in a green interval between major glaciations"

(*Rolf Edberg*)

Pot-hole

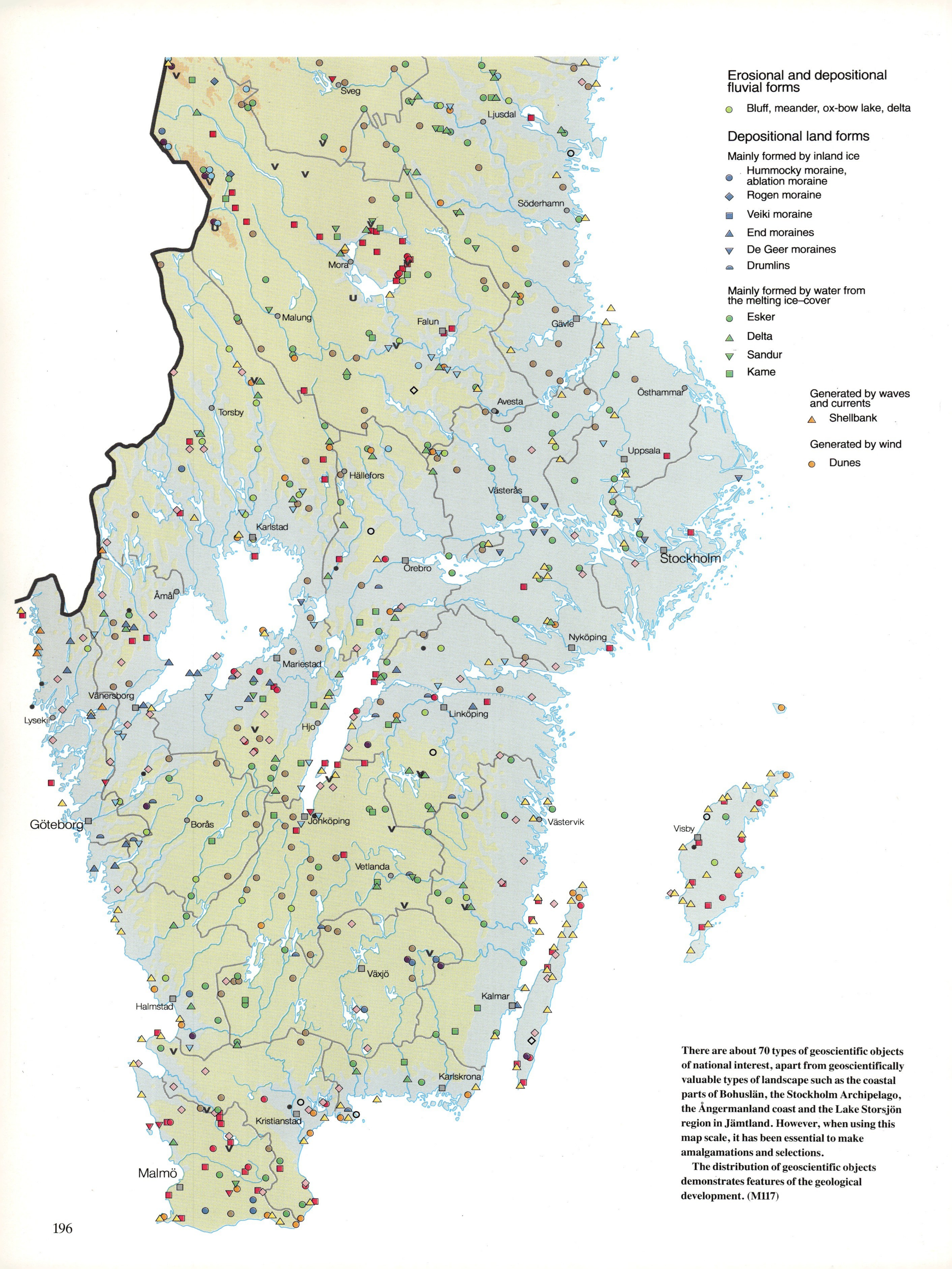

There are about 70 types of geoscientific objects of national interest, apart from geoscientifically valuable types of landscape such as the coastal parts of Bohuslän, the Stockholm Archipelago, the Ångermanland coast and the Lake Storsjön region in Jämtland. However, when using this map scale, it has been essential to make amalgamations and selections.

The distribution of geoscientific objects demonstrates features of the geological development. (M117)

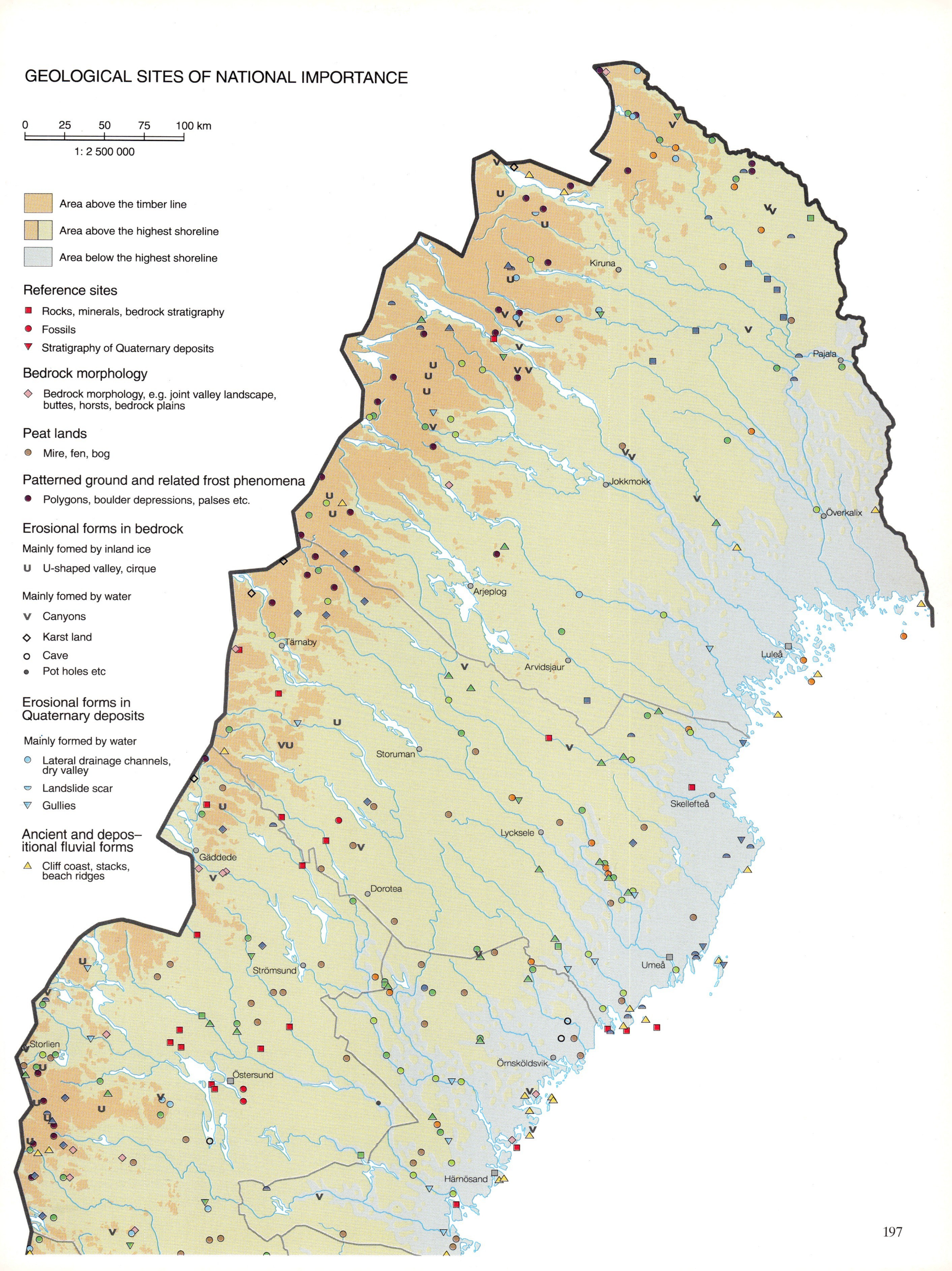
GEOLOGICAL SITES OF NATIONAL IMPORTANCE
0 25 50 75 100 km
1: 2 500 000
Area above the timber line
Area above the highest shoreline
Area below the highest shoreline
Reference sites
Rocks, minerals, bedrock stratigraphy
Fossils
Stratigraphy of Quaternary deposits
Bedrock morphology
Bedrock morphology, e.g. joint valley landscape, buttes, horsts, bedrock plains
Peat lands
Mire, fen, bog
Patterned ground and related frost phenomena
Polygons, boulder depressions, palses etc.
Erosional forms in bedrock
Mainly fomed by inland ice
U-shaped valley, cirque
Mainly fomed by water
Canyons
Karst land
Cave
Pot holes etc
Erosional forms in Quaternary deposits
Mainly formed by water
Lateral drainage channels, dry valley
Landslide scar
Gullies
Ancient and depositional fluvial forms
Cliff coast, stacks, beach ridges
Kiruna
Pajala
Jokkmokk
Överkalix
Arjeplog
Tärnaby
Arvidsjaur
Luleå
Storuman
Skellefteå
Lycksele
Gäddede
Dorotea
Strömsund
Umeå
Storlien
Östersund
Örnsköldsvik
Härnösand

Glossary

The list mainly includes words that may be difficult to find in a conventional dictionary, together with terms that are used repeatedly in this volume. Terms that only occur once or twice can be sought by means of the Register. A more precise illustration of the geological time units is given on the inside of the front cover.

Ablation area. Part of a glacier where melting, ablation, is greater than growth (accumulation).

Accumulation. Deposition of material that has been transported by water, wind or ice; a process that adds snow or ice to a glacier or inland ice sheet.

Acidic rock. Rock with a high content of silica (more than 65%).

Actinolite. Calcium-iron-rich amphibole.

Agate. Compact form of SiO_2 consisting of very small crystals of quartz with pores. The colour alternates in bands or concentric zones.

Albite. See plagioclase.

Alkaline rock. Characterised by high contents of alkali elements (sodium and potassium) in relation to silica and aluminium. Contains minerals such as nepheline, cancrinite, alkali pyroxene and alkali amphibole.

Alkalinity. Ability of water to bind acids.

Allochthonous. Deposits of plant and animal residues that occur outside their original environment; used also to denote bedrock that has been moved from its original position through tectonic movements.

Alum shale. Clay with a high content of organic material (bitumen) and high sulphur content. In Sweden, these shales have earlier been used to produce alum ($KAl(SO_2)_4 \cdot 12\,H_2O$).

Amblygonite. Lithium-aluminium phosphate containing fluorine and hydroxyl.

Ammonites. Cephalopods extinct since the Cretaceous period, usually with spiral tests with folded edges.

Amorphous. Lacks crystalline structure.

Amphibole. A group of hydrous silicates with prismatic crystalline form. The most important are hornblende (calcium-iron-magnesium-aluminium silicate) and actinolite-tremolite (calcium-iron-magnesium silicate).

Andalusite. Aluminium silicate mineral (Al_2SiO_5) occurring in, for example, mica schist and gneiss.

Andesite. Dark, intermediate volcanic rock dominated by plagioclase and iron-magnesium mineral.

Ankerite. Iron-rich dolomite (mineral).

Anorogenic. Rock formed in a relatively quiet period during crustal evolution, following larger orogeneses. The rapakivi granites belong to this group.

Anorthite—see plagioclase.

Anorthosite. A plutonic rock consisting of more than 90% of plagioclase.

Anthraconite. Dark-coloured, bituminous limestone.

Anthropogenic. Originating from human activity.

Anticlinal. A ridge-shaped structure formed by the folding of a stratigraphic sequence.

Apatite. Mineral with the chemical composition $Ca_5\,(OH, F, Cl)\,(PO_4)_3$.

Aquifer. A system in rock or Quaternary deposits capable of yielding water in useful quantities.

Aragonite. Carbonate mineral (a form of $CaCO_3$) included in certain hard constituents of fossils.

Arkose. Sandstone containing small amounts of clay minerals or mica and large amounts of feldspar.

Artesian groundwater. Groundwater that flows up above the ground surface under pressure.

Asthenosphere. Part of the upper mantle of the Earth.

Autochthonous. Deposits of plants and animals that have lived at the place where these remnants are found. Also used to denote rock masses that have not moved from their original site.

Baltic Shield. The flat area in northern Europe where Precambrian bedrock is exposed and is uninfluenced by Phanerozoic orogenesis. Also called the Fennoscandian Shield.

Basalt. Basic volcanic rock consisting of plagioclase, pyroxene and olivine with varying contents of volcanic glass.

Base level. The lowest level to which erosion can proceed.

Basic rock. Magmatic rock with 45–52% by weight of SiO_2.

Batholith. Very large body of plutonic rock without a known lower limit.

Bauxite. Mixture of erosion minerals that largely consist of aluminium hydroxide.

Bentonite. A very fine-grained clay, formed of volcanic ash. Consists largely of montmorillonite.

Beryllium. A green, glassy, mineral. The crystals are hexagonal, usually prismatic and are considered to be gems.

Biogenic. Refers to deposits built up of residues of organisms.

Bitumen. A general term for organic substances, mainly hydrocarbons, in nature.

Black shale. Dark shale containing carbon and sulphides.

Bleached soil. The leached horizon in podzols with lighter colour than the layers above and below.

Bog. Peatland where the Sphagnum vegetation, etc., obtains nutrients exclusively through precipitation.

Boulder. Particles with a size >200 mm.

Boulder depression. Boulders lifted up by frost action and enriched on the surface of the ground in low-lying terrain. Shallow depressions can become covered in this way by boulders.

Brackish. This term is used for water with low salinity (0.05–2%).

Breccia. Rock consisting of angular broken fragments in a fine-grained matrix.

Brucite. Mineral consisting of magnesium hydroxide.

Calcite. Calcium carbonate, $CaCO_3$.

Caledonides. An orogenic belt formed through folding and thrusting that culminated during the Silurian-Devonian with a continent-continent collision. The Caledonides mostly coincide with the present Scandinavian mountain chain.

Cancrinite. A hydrous silicate of sodium, potassium, calcium and aluminium, containing carbonate ions (CO_3^{2-}).

Capillary water. Water present in Quaternary deposits and rocks as a result of surface tension being retained in the smaller cavities. Cannot be removed by pumping.

Carbonate rock. Collective name for limestone and dolomite.

Carbonatite. A magmatic rock dominated by carbonate (calcite and/or dolomite).

Chalcopyrite. Sulphide mineral ($CuFeS_2$), the most important copper mineral in Sweden.

Charnockite. A metamorphic or magmatic rock containing orthopyroxene and sometimes garnet. This rock is formed at high temperature, high pressure and low water content.

Chlorite. Hydrous, mica-like, silicate mineral containing magnesium, iron and aluminium.

C-horizon. Soil horizon below the influenced horizons located above it.

Clastic. Structure formed of fragments of rocks that have become disintegrated through weathering.

Clay. Particles with a size <0.002 mm and a deposit with more than 15% of such particles in the fractions smaller than 20 mm.

Clay gyttja. Quaternary deposit with 6–20% organic material (dry weight).

Clay shale. Non-metamorphic sedimentary shale with a high content of clay minerals.

Cobble. Particles with a size of 20–200 mm.

Coccolith. Microscopic calcareous plate from the outer test of unicellular flagellum-bearing algae.

Colonus shale. Name of a Silurian shale sequence in Skåne. *Colonus* originates from *Monograptus colonus,* a graptolite species.

Complex ore. Sulphide ore containing several extractable metals.

Conglomerate. Sedimentary rock consisting of rounded fragments (pebbles, cobbles, etc.), usually set in a sandy or gravelly matrix.

Constructive plate margin. Margin where the Earth's plates are newly created in spreading zones.

Continental shelf. The part of the continental margin that is located between the shoreline and the continental slope.

Correlation. Determination in different areas of simultaneous occurrences within geological history.

Culmination. A structure where underlying bedrock curves upwards and is exposed beneath overlying rocks or nappes. Also used for a centre in an inland ice from where the ice moves.

Danien. Section of the Danish-Swedish stratigraphic sequence from the oldest Tertiary, formed 63–65 million years ago.

Danish depression. A sedimentation area covering large parts of Denmark, Skåne and the southern Baltic Sea.

Dannemorite. Amphibole containing manganese.

De Geer moraine. A small moraine ridge, usually a couple of hundred metres long and up to 5 m high. Occurs in swarms and is considered to be formed in the marginal zone of the inland ice, in cracks running parallel to it.

Delta. Deposit of gravel, sand and fine sand that has been formed in a lake or sea.

Denudation. Destruction of the land surface by weathering, mass movements and erosion.

Destructive plate margin. Margin where the Earth's plates collide and are broken down.

Diamond. Carbon with a special crystalline structure.

Diatom. A siliceous alga with a test of silica.

Diorite. Plutonic rock consisting of plagioclase, biotite and hornblende, sometimes also pyroxene.

Dolerite. A hypabyssal intrusive rock that often forms more or less steeply-dipping sheets in the bedrock. Mineralogically, diabase contains plagioclase, pyroxene, olivine, titanomagnetite etc. The rock is often characterized by large crystals of pyroxene surrounding tabular grains of plagioclase.

Dolomite. Mineral: Calcium-magnesium carbonate ($CaMg(CO_3)_2$). Rock: Sedimentary rock consisting mainly of the mineral dolomite.

Drumlin. A low, smoothly rounded, elongate oval ridge, partly consisting of glacial till and aligned parallel to the direction of the movement of the inland ice or a glacier.

Dyke. An intrusive rock which is tabular in form and cuts across the planar structures of the surrounding rock. Dykes consist of fillings in fissures and have usually been formed in the upper part of the Earth's crust (hypabyssal).

Earthquake. Sudden motion in the Earth's crust caused by one or more elastic waves occurring as a result of fissure-formation, faulting or other release of strain energy in the Earth's crust or mantle.

Earth's crust. The outermost part of the Earth, extending down to about 5–10 km below the oceans and to about 35–60 km below the continents.

Eclogite. Basic rock metamorphosed under high pressure and temperature. Characterized by pyroxene and garnet.

End moraine. A ridge-like accumulation of till or melt-water sediments deposited at the margin of a glacier or inland ice.

Eolian silt; Eolian sand. Wind-blown sediment consisting mainly of silt or sand.

Eon. The formal geochronological unit of highest rank, next above era.

Epidote. Water-containing silicate with calcium, aluminium and iron.

Era. The formal geochronologic unit next in order of magnitude below an eon.

Esker. Ridge-shaped glaciofluvial deposit extending in approximately the direction the glacier ice has moved.

Eustasy. Fluctuations in sea level.

Fault. Fracture or a zone of fractures along which one block of bedrock has moved in relation to another.

Fayalite—see olivine.

Feldspar. General name for certain rock-forming minerals (aluminium silicates of K, Na and Ca). The most important are potassium feldspar and plagioclase.

Fen. Peatland where the vegetation is highly affected by groundwater and surface water, and which can occasionally be more or less submerged under water.

Fennoscandian Ice-Marginal Zone. Zone of deltas and large terminal moraines in Dalsland, Västergötland, Östergötland and Sörmland. Associated with similar formations in Norway and Finland.

Fennoscandian Marginal Zone. Name of the marginal zone forming the border of the East European Plate form (including the Baltic Shield) towards the southwest. The zone can be followed from the southern Baltic Sea (in the neighbourhood of Bornholm) through Skåne, over the Kattegat and northern Jutland to the North Sea off southwestern Norway.

Ferrimagnetic. The property of a substance to become strongly attracted by a magnet and to possess permanent magnetisation.

Flexure. Open bending of layers without breakage. Also called fold broad dome structure.

Fluorite. Mineral with the chemical composition CaF_2.

Foraminifera. Unicellular animals (protozooans), most of which have a test (0.25–0.55 mm, occasionally larger). The test has frequently been preserved as a fossil and is useful in relative dating of sedimentary rocks and interpretation of their genesis.

Formation. Primary lithostratigraphic unit consisting of a stratigraphic sequence.

Fossil. Fossilized remnants and traces of animals and plants from former times.

Fulvic compound. Water-soluble humus compound.

Gabbro. Plutonic rock consisting of plagioclase, hornblende and pyroxene, sometimes also olivine.

Gal. Gravimetric measure. 1 gal = 1 cm/s^2, named after Galileo Galilei.

Galena. A sulphide mineral of lead (PbS_2), normally occurs in association with sphalerite. Galena is the most important lead mineral and frequently has a considerable silver content.

Gamma radiation. Electromagnetic radiation emitted as the result of a nuclear process.

Garnet. General name for a number of cubic-shaped crystalline minerals of iron, calcium, magnesium, aluminium, silica, etc.

Geomagnetism. The phenomenon connected with the magnetic fields observed at the surface of the Earth and in its atmosphere.

Glacial. Relating to the presence and activities of glacier ice.

Glacial striation. Furrows or lines in solid rock caused by debris transported at the base of a moving glacier or inland ice.

Glaciofluvial. Pertaining to melt-water flowing from wasting glacier ice.

Glauconite. Commonly found sedimentary iron silicate that usually indicates slow sedimentation in a marine environment. The colour is dark blue-green.

Glint. An escarpment or steep cliff produced by erosion, commonly found on the eastern border of the nappes in the Caledonides.

Gneiss. Highly metamorphosed rock with foliated structure, frequently also with band-shaped aggregates of different minerals.

Graben. Elongate depression in which the bedrock has been depressed as a result of faults in relation to the surroundings.

Granite. Plutonic rock characterised by quartz, feldspar (of which at least 35% is microcline or orthoclase and the rest plagioclase), mica and/or hornblende.

Granitoid. General name for quartz-rich plutonic rocks such as granite, granodiorite and tonalite.

Granodiorite. Plutonic rock consisting of quartz, feldspar (of which microcline or orthoclase makes up 10–35%), mica and/or hornblende.

Graphite. Black, shiny, flaky mineral consisting of carbon.

Graptolite. Extinct marine colony-forming group of animals. The preserved parts of the fossil look like straight, curved or rolled-up saw-blades, in some forms branched.

Gravel. Quaternary deposit dominated by a particles with sizes of 2–20 mm.

Gravitation. The mutual attraction between two masses, proportional to the product of the masses and inversely proportional to the square of the distance between them.

Gravity field. Field of force with which, e.g., the planet Earth influences bodies at or near the Earth's surface.

Greenschist. Low-metamorphic rock rich in chlorite, usually a basic lava or tuff.

Greenstone. Basic metamorphic rock of magmatic origin.

Greywacke. A kind of sandstone containing feldspar mixed with clay. The clay content is at least 15%. Greywackes have largely been deposited as a result of turbidity currents on the seabed.

Groundwater. Water that fills cavities in rocks and Quaternary deposits.

Guide fossil. A fossil that is characteristic for strata of a certain age of major importance for the relative dating of sedimentary rocks and interpretation of the environment in which they were formed.

Gypsum. Mineral consisting of hydrous calcium sulphate.

Gyttja. Quaternary deposit containing more than 20% organic material (dry weight).

Gyttja clay. Quaternary deposit with 2–6% (dry weight) organic material.

Hard water. Water with a high content of calcium and magnesium ions.

Heavy metal. Metallic element with a high atomic number, e.g., cadmium, gold, lead.

Hematite. Iron oxide (Fe_2O_3).

Highest shoreline. During the deglaciation large parts of the depressed Earth's crust were covered by the sea. The highest located shoreline markings are called the highest shoreline. These are located at different altitudes in different parts of Sweden depending, for example, on how large the isostatic uplift has been in the area and when it was deglaciated.

Hornblende—see amphibole.

Horst. Elongate ridge bounded by faults and uplifted in relation to the surrounding rock.

Hummocky ablation moraine. Ridges, hummocks and depressions of till in an irregular pattern. Hummocky ablation moraine has formed in a retreating, cracked and stationary part of a glacier or inland ice.

Humus. Layers of partly decomposed raw plant and animal residues that lie below the litter in a podzol profile.

Hypabyssal rock. Igneous intrusion in which the depth of crystallisation is between that of plutonic and volcanic rocks. Often tabular in form, either steeply-dipping (dyke) or flat-lying (sill).

Hyperite. Name of certain dark-coloured doleritic rocks in Värmland and northern Västergötland that consist of two types of pyroxene together with olivine and iron oxide-pigmented plagioclase.

Hyperite dolerite. Black dolerite ("black granite" that usually contains two pyroxenes and iron oxide-pigmented plagioclase. Found in the Protogine zone extending from northern Skåne to Lake Vättern.

Hälleflinta. Name of silica-rich volcanic rocks of the same type as leptites but with a grain size not exceeding 0.05 mm.

Höganäs Formation. Carbon and kaolinite-containing deposits on the Triassic/Jurassic boundary. Frequently called Skåne's coal-containing formation.

Ice-marginal lake. A lake of melt-water dammed-up between an ice barrier and the exposed terrain.

Igneous rock. Rocks formed from magma.

Ignimbrite. Volcanic rock formed by an ash flow. In principle, volcanic ash and pumice deposited at such a high temperature that the particles became welded together.

Ilmenite. Black or black-brown iron-titanium mineral.

Industrial mineral. Mineral of economic value, exclusive of metallic ores and mineral fuels.

Inland ice. The mass of glacier ice covering large parts of a continent.

Interglacial. The period between two successive glacial stages.

Intermediate volcanic rock. Volcanic rock containing 52–65% silica.

Interstadial. The period between two phases within the same glacial stage.

Intrusive. Igneous rock that has penetrated into and solidified in the Earth's crust as dykes or larger bodies.

Inversion. This term has been used in a tectonic context. Tectonic inversion is a result of compressional deformation. The process has the result that sedimentary basins become structurally elevated areas that are exposed to erosion, and that formerly elevated areas can become lowered.

Isostasy. The condition of equilibrium in the Earth's crust.

Jaspilite. A banded siliceous rock consisting of alternate strata of hematite and red jasper.

Jatulian bedrock. General name for certain formations consisting of quartzites, basaltic greenstones, schists and dolomites that were deposited 2,200–2,000 million years ago close to a basement of Archaean bedrock in northern Sweden and in northern and eastern Finland.

Kalevian bedrock. Name of certain formations consisting mainly of phyllites and mica schists in northern-eastern Finland and in northern Sweden. Their age is about 2,000–1,900 million years. Today they are considered to belong to the Svecofennian formation.

Kame. A mound or hummock with marked sides or irregular ridge, mainly consisting of glaciofluvial sediments deposited in contact with a glacier or the inland ice.

Kaolin. The name originates from a Chinese place name. Belongs to a group of weathering minerals of which kaolinite ($Al_4Si_4O_{10}(OH)_8$)is the most important.

Kimmerian. Name of a stage with increased tectonic activity during the Mesozoic.

Kolm. Impure coal, rich in uranium.

Kyanite. A form of Al_2SiO_5 (aluminium silicate) that crystallises with low (triclinic) symmetry, usually at high pressure.

Late orogenic. Refers to granites formed at the end of an orogeny.

Lava. Magma that has extruded onto the surface.

Lazulite. A blue mineral containing water and phosphate of magnesium, iron and aluminium.

Lepidolite. Lithium mica.

Leptite. Silica-rich volcanic rock with a particle size of 0.05–1 mm. Found in the Svecofennian bedrock in Bergslagen and neighbouring areas.

Lias. The oldest of the three series in the Jurassic period.

Limestone. Sedimentary rock consisting mainly of calcite ($CaCO_3$).

Lithology. The description of a rock or Quaternary deposit based on characteristics such as mineral composition and particle size.

Lithosphere. The Earth's crust and the upper part of the mantle.

Lithostratigraphy. The organisation of a sequence of strata on the basis of lithological characteristics.

Litter. General term for loose organic debris composed of freshly fallen or slightly decayed material that has not yet been incorporated into the overburden.

Magma. Molten rock.

Magnesite. Magnesium carbonate ($MgCO_3$).

Magnetite. Magnetic iron oxide mineral (Fe_3O_4).

Mantle. The zone of the Earth below the crust and extending down to a depth of approximately 2,900 km.

Marble. Recrystallised limestone or dolomite as a result of metamorphism.

Marl. Rocks and Quaternary deposits that contain calcium carbonate, used as a soil conditioner.

Marsh ore. Limonite—brown iron hydroxide, $Fe_2O_3 \cdot H_2O$—formed in peatland.

Megacryst. Large crystal, usually of feldspar, occurring in a medium- or coarse-grained matrix, especially in granitic rocks.

Metamorphism. The process whereby new mineral combinations are created when a rock is subjected to changes in pressure and/or temperature.

Mica. Hydrous silicates that crystallize in plates or scales. The most common types are biotite (potassium-iron-magnesium-aluminium silicate) and muscovite (potassium-aluminium silicate).

Mica schist. Mica-rich sedimentary rock formed under medium grade metamorphic conditions.

Microcline—see potassium feldspar.

Microfossil. A fossil that is so small that it cannot be studied without the help of a microscope (e.g., pollen, spores, diatoms, ostracods, foraminifera).

Migmatisation. The process leading to the formation of migmatite, usually a partial melting.

Migmatite. Rock consisting of a mixture of one or more older components (usually gneiss) and one or more younger components (veins, etc.) of granite, granodiorite or pegmatite.

Mineral. Naturally-occurring inorganic material with a composition and crystal structure varying within certain limits.

Mineralisation. The natural enrichment of one or more minerals which are potentially of economic value.

Mire. General term for fen, bog, marshy moorland, swamp.

Moho; Mohorovičić discontinuity. An area of seismic discontinuity between the Earth's crust and the subjacent mantle.

Montmorillonite. Expanding clay mineral belonging to the smectite group.

Monzonite. A plutonic rock consisting of feldspar (of which 35–65% is plagioclase) and dark minerals (mica, hornblende, pyroxene, olivine).

Muscovite—see Mica.

Mylonite. Fine-grained to dense, banded rock formed as a result of very strong ductile deformation in movement zones in the bedrock.

Mylonite Zone. A zone of strong mylonitization in the gneissic bedrock of southwest Sweden.

Nappe. A package of rock that, in the form of a sheet, has been moved over the underlying bedrock along a level surface.

Nepheline. A mineral belonging to the feldspathoid group ($NaAlSiO_4$).

Nunatakk (Eskimo). An isolated hill or ridge emerging above an inland ice or a glacier.

Olivine. Mineral occurring mainly in basic and ultrabasic rocks. Its composition varies from Mg_2SiO_4forsterite) to Fe_2SiO_4 (fayalite).

Orogeny. Folding, faulting, metamorphism and igneous activity. Orogeny is linked with the process mountain-building.

Orthoclase—see potassium feldspar.

Orthogneiss. Gneiss formed from metamorphosed igneous rocks.

Orthopyroxene. A pyroxene with rhombic crystal structure.

Ostracod. A crustacean bivalve. Usually microscopic (0.4–1.5 mm). An important index fossil.

Palsa. A mound, up to 7 m high, in bogs consisting of frozen material, usually peat.

Paragneiss. Gneiss formed from metamorphosed sedimentary rocks.

Paramagnetic. Property of a substance to be weakly attracted by a magnet. The magnetic properties of the substance are completely dependent on an external magnetic field. Paramagnetic minerals are, e.g., biotite, hornblende and olivine.

Parautochthonous. Almost autochthonous.

Peat. Organic material formed in a moist and oxygen-deficient environment by decomposition of dead plant parts.

Pegmatite. A coarsely crystalline, usually granitic rock that forms dykes or small veins. Originates, for example, from residual solutions in granitic melts.

Pellet. Small, less than 1 mm, round aggregates, usually of sediments that are faeces of small marine animals. The term is also used within the ore-processing industry.

Peneplain. An extensive, level and relatively uniform, bedrock surface formed through long erosion in an originally more undulating landscape.

Period. Geological unit of time, e.g., Cambrian, subordinate to an era, in this case the Palaeozoic.

Permafrost. Permanently frozen ground.

Petalite. A mineral similar to feldspar (lithium aluminium silicate).

Phenocryst. Large crystal embedded in a finer-grained matrix, usually a volcanic or hypabyssal rock.

Phlogopite. Magnesium-rich variety of biotite.

Phyllite. Mica-rich sedimentary rock of low metamorphic grade with shiny surfaces.

Pillow lava. Lava that has solidified on the sea-floor usually with basaltic composition. Consists of rounded balls and tubes with a glassy crust.

Pitchblende. A radioactive uranium mineral (oxide).

Plagioclase. Feldspars that consist of a mixed series between albite ($NaAlSi_3O_8$) and anorthite ($CaAl_2Si_2O_8$).

Plateau dolerite. Cover of dolerite that forms the roof of certain buttes or mesas, i.e., mountains with flat top surfaces and steep slopes, e.g., the Västergötland hills.

Plutonic rock. Igneous rock that has crystallized in deep parts of the Earth's crust.

Podzol (Russian). Ash soil, soil type common in coniferous forests.

Porphyrite. Term for phenocryst-bearing basic or intermediate volcanic and dyke rocks.

Porphyry. Name of volcanic and dyke rocks containing phenocrysts and having an acidic composition.

Postorogenic. Name of plutonic rocks that have formed shortly after a period of orogeny.

Potassium feldspar. Feldspar with the formula $KAlSi_3O_8$. The most common forms are microcline (triclinic symmetry) and orthoclase (monoclinic symmetry). Usually found with albite in solid solution or in separated (mixed) form.

Precambrian bedrock. Term covering all Proterozoic and Archaean rocks.

Protogine Zone. A zone running approximately north-south from Skåne to northern Värmland in which the bedrock to the east (the granites etc. of the Trans-Scandinavian Granite-Porphyry Belt) successively change into gneiss towards the west as a result of increasing deformation.

Pyrite. Iron sulphide mineral (FeS_2).

Pyroxene. Silicates with prismatic crystalline form. The most important are augite (calcium-iron-magnesium-aluminium silicate) and hypersthene (iron-magnesium silicate).

Quartz. The most common form of silicon dioxide (SiO_2), crystallized with a trigonal symmetry.

Quartz-banded ore—see jaspilite.

Quartzite. Metamorphosed, quartz-rich sandstone. This term has also been used for certain quartz-rich rocks in alteration zones around, e.g., sulphide ores.

Quartz keratophyre. Sodium-rich, acidic volcanic rock.

Quaternary deposit. Mostly unconsolidated deposits from the Quaternary period.

Quick clay. Clay that loses its shear strength (stability) when water-saturated and stirred.

Radial moraine. A ridge of till that has formed as a result of material being pressed up or falling down into longitudinal cracks in glacier ice.

Radon. A colourless and odourless radioactive gas formed through the radioactive decay of radium.

Raised bog. Bog with a domed surface.

Regression. Here used to describe the withdrawal of the sea with the resulting increase in the land area.

Residual. Remaining, e.g., after weathering.

Residual mountain. Isolated mountain on an erosion surface.

Resistivity. The ratio of field strength and current density.

Rhaetian. The youngest stage in the Triassic.

Rhomb porphyry. Quartz-free porphyry with rhomb-shaped, usually cm long megacrysts of feldspar.

Rhyolite. Volcanic rock which, in composition, corresponds to granite among the plutonic rocks.

Richter Scale. Numerical scale 0–9 indicating the magnitude of the total seismic energy released in an earthquake.

Rogen moraine. Hilly moraine landscape characterized by more or less regular ridges that are mainly oriented at right angles to the direction in which the ice has moved.

Rust soil. An accumulation horizon in a podzol profile.

Rutile. Form of titanium dioxide that crystallizes in prisms with tetragonal symmetry.

Sand. Particles with a size of 0.06–2.0 mm and a deposit dominated by them.

Sandur (Icelandic). Sand and gravel deposits formed by meltwater from glaciers or inland ice. Sandur fields have numerous irregular channels, a braided river system.

Scheelite. Calcium wolframate ($CaWO_4$).

Sedimentary bedrock. Rock formed through sedimentary processes and subsequent lithification.

Sedimentary gneiss. Gneiss of sedimentary origin.

Seismicity. Method implying registration and analysis of controlled seismic waves.

Sericite. Fine-grained, white or light-grey variant of muscovite.

Series. Stratigraphic unit comprising part of a geological system.

Serpentine. Hydrous magnesium-silicate mineral.

Shell deposit. Deposit that largely consists of shells and shell fragments of molluscs.

Shingle. Wave-washed deposit consisting of rounded pebbles, cobbles and sometimes small boulders.

Siderite. Iron-bearing carbonate mineral, relatively common in sedimentary rocks.

Silicate. Refers to minerals containing SiO4 tetrahedra in their crystal structure.

Sillimanite. Silicate mineral (Al_2SiO_5) usually forming needle-like crystals in high-grade gneisses and schists.

Silt. Particles with a size of 0.06–0.002 mm and a deposit dominated by them.

Skarn. Silicates, usually containing calcium, that are found in connection with, for example, the iron ores in Bergslagen. Formed by reactions between silicate rocks and limestone.

Slate. Low-grade, well-cleaved, clay-rich sedimentary rock. See also mica schist.

Smectite. Collective name for swelling clay minerals of mica type.

Soapstone. Rock consisting of chlorite and talc, formed through alteration of olivine-rich rocks.

Soil. Part of the Earth's surface that changes under the influence of climate, vegetation and fauna.

Solifluction. The slow viscous downslope flow of unsorted and saturated surficial material.

Solifluction lobe. Tongue-like mass of solifluction debris.

Soligenous mire. A mire on a sloping surface, usually in mountain terrain or adjacent areas.

Sorted polygon. A ring or polygon of stones and boulders that surround an island of finer material and has been formed through repeated freezing and thawing.

Sorted stripe. A row of stones and boulders down a slope. Formed through freezing processes.

Sphalerite. A yellow, brown or black isometric mineral with the composition ZnS.

Spilite. Basaltic rock consisting of albite, actinolite, epidote, chlorite and calcite, the latter often filling cavities.

Stadial. Colder period during a glaciation with growth of the inland ice.

Stone. Particles with a size of 20–200 mm.

Stone pit. Irregularly-shaped hole in the ground that has been formed as a result of frost action and containing stones and boulders.

Stromatolites. Often domed or pillar-like structures in limestone or dolomite, formed by algae.

Subduction zone. A long, narrow belt in the Earth's crust where the oceanic crust of a lithospheric plate is forced down below another plate.

Sulphide ore. Economically interesting concentration of sulphide minerals of mainly copper, zinc and lead.

Supracrustal rock. Rock formed on the Earth's surface as a result of sedimentary or volcanic processes.

Svecofennian bedrock. Refers to certain Precambrian formations of mainly acidic volcanic and sedimentary rocks in eastern Sweden and southwestern-central Finland.

Svecokarelian or **Svecofennian orogeny**. The orogeny that occurred mainly in what is today eastern Sweden and southwestern and central Finland about 1,800–1,870 million years ago.

Sveconorwegian orogeny. The orogeny that occurred in what is today southwestern Sweden and southern Norway about one billion years ago.

Syenite. A plutonic rock dominated by feldspar (of which max. 35% is plagioclase) and dark minerals (mica, hornblende, pyroxene).

Talc. A hydrous magnesium silicate.

Tectonics. The branch of geology dealing with the large-scale architecture of the outer part of the Earth. Results of tectonic movements are, e.g., cracks, faults, folds and thrusts. Within glacial tectonics, the term also includes very small structures.

Tension movement. Tectonic movement implying pulling apart.

Terminal moraine. Large end moraine that has formed as the result of the ice margin remaining stationary for a long period.

Thrusting. The process whereby sheets of bedrock (nappes) along flat surfaces are displaced over the originally higher layers along gently-dipping surfaces.

Till. Material deposited by glaciers or inland ice. Till has a varying composition of boulders, stones, gravel, sand, fine sand, silt and clay.

Tonalite. Plutonic rock consisting of quartz, feldspar (of which 0–10% is microcline or orthoclase), mica and/or hornblende.

Tornquist Zone. A zone of faulting in a northwest-southeast direction between the Black Sea and the North Sea. The zone cuts across Skåne where it marks the southwestern margin of the Baltic Shield.

Total hardness. The total content of calcium and magnesium in water.

Transgression. When the sea successively spreads or extends over land areas (cf. regression).

Travertine. Limestone formed through chemical or biogenic precipitation of calcium carbonate from solution in surface water and groundwater.

Trilobite. Marine arthropod that lived from the lower Cambrian to the Permian.

Tuffite. Rock of volcanic origin containing ash mixed with normal clastic sediment.

Tundra polygon. Contraction cracking in the ground in areas of permafrost. The cracks form a polygonal pattern with a mesh size sometimes reaching tens of metres.

Ultrabasic rock, ultrabasite. An igneous rock with less than 45% SiO_2.

Unsaturated zone. Area of rock or Quaternary deposits where the cavities not only contain water but also air.

Upper Svecofennian. Supracrustal rocks are included here that were folded during the Svecokarelian orogeny and that are younger than the early orogenic granitoids.

U-valley. Valley with the shape of an U in section, formed through glacial erosion.

Varved clay. Glacial clay deposited in cycles in fresh water. One varve corresponds to sedimentation during one year.

Veiki moraine. Moraine hills similar to plateaus with numerous circular bodies of water. Named after a small farm near Gällivare.

Veined gneiss. A form of migmatite with veined structure.

Visingsö Group. Sedimentary rocks in and around Lake Vättern deposited 700–850 million years ago. The stratigraphy consists of sandstones, arkoses, conglomerates and shales with layers of limestone.

Volcanite. Deposit or rock formed as a result of volcanic processes.

Weathering. Disintegration and conversion of rocks and Quaternary deposits through mechanical and chemical processes. The weathering of the bedrock can be divided into deep weathering and surface weathering.

Wollastonite. Mineral with the chemical composition $CaSiO_3$.

Literature and references

GFF = Geologiska föreningens i Stockholm förhandlingar (A quarterly journal published by the Geological Society of Sweden)
SGU = Sveriges geologiska undersökning (Geological Survey of Sweden)

Agrell, H., 1978, *The Quaternary of Sweden.* SGU C 770.

Ahlberg, P., 1986, *Den svenska kontinentalsockelns berggrund.* SGU Rapporter & meddelanden 47.

Aronsson, M., Hedenäs, L., Lagerbäck, R., Lemdahl, G. and Robertsson, A-M., 1993, *Flora och fauna i Norrbotten för 100 000 år sedan.* Svensk botanisk tidsskrift 87, 241–253.

Axberg, S., 1980, *Seismic stratigraphy and bedrock geology of the Bothnian Sea, northern Baltic.* Stockholm Contrib. Geology. 36:3.

Berglund, B. E., 1979, *The deglaciation of southern Sweden 13,500–10,000 B.P.* Boreas 8, 89–117.

Bergqvist, E., 1981, *Svenska inlandsdyner. Översikt och förslag till dynreservat.* Naturvårdsverket, rapport 1412.

Björck, S., 1981, *A stratigraphic study of Late Weichselian deglaciation, shore displacement and vegetation history in southeastern Sweden.* Fossils and strata 14. Oslo.

Björck, S., 1994, *Review of the history of the Baltic Sea 13.0–8.0 Ka B.P.* Quaternary International 12.

Björck, S. and Digerfeldt, G., 1991, *Alleröd—Younger Dryas sea level changes in southwestern Sweden and their relation to the Baltic Ice Lake development.* Boreas 20, 115–133.

Blundell, D., Freeman, R. and Mueller, S. (eds), 1992, *A continent revealed. The European geotraverse.* Cambridge University Press.

Cato, I., 1987, *On the definitive connection of the Swedish time scale with the present.* SGU Ca 68.

Collet, L., 1927, *The structure of the Alps.* E. Arnold & Co. London.

Denton, G. H. and Hughes, T. J., 1981, *The last great ice sheets.* John Wiley & Sons.

Dyrelius, D., 1980, Aeromagnetic interpretation in a geotraverse area across the central Scandinavian Caledonides. GFF 102, 421–438.

Ekman, M., Eliasson, L., Pettersson, L. and Sjöberg, L. E., 1982, *Bestämning av landhöjningen i Sverige med geodetiska metoder.* Lantmäteriet, tekniska skrifter 1982:13

Elvhage, C. and Lidmar-Bergström, K., 1987, *Some working hypotheses on the geomorphology of Sweden in the light of a new relief map.* Geografiska annaler 69 A, 343–358.

Flodén, T., 1980, *Seismic stratigraphy and bedrock geology of the central Baltic.* Stockholm contrib. geology 35.

Fredén, C., 1988, *Marine life and deglaciation chronology of the Vänern basin southwestern Sweden.* SGU Ca 71.

Fries, M., 1965, *The Late-Quaternary vegetation of Sweden.* Acta Phytogeographica Suecica 50, 269–284.

Fries, M., 1976, *Skogen i förhistorisk tid.* Skogsägaren 2, 23–25.

Frietsch, R., Sundberg, A. and Wik, N.-G., 1991, *Register över svenska fyndigheter av malmmineral och industriella mineral och bergarter.* SGU Rapporter & meddelanden 66.

Frietsch, R., 1980, *The ore deposits of Sweden.* Geological Survey of Finland, Bulletin 306.

Gaál. G. and Gorbatschev, R., 1987, *An outline of the Precambrian evolution of the Baltic Shield.* Precambrian research 35, 15–52.

Gee, D.G., Kumpalainen, R., Roberts, S., Stephens, M.B., Thon, A. and Zachrisson, E., 1985, *Scandinavian Caledonides—Tectonostratigraphic map.* SGU Ba 35.

Gee, D.G. and Sturt, B.A., (eds) 1985, *The Caledonide orogen—Scandinavia and related areas.* John Wiley & Sons.

Geijer, P. and Magnusson, N. H., 1944, *De mellansvenska järnmalmernas geologi.* SGU Ca 35.

Geological surveys of Finland, Norway and Sweden, 1986, *Map of Quaternary geology, sheets 1–3, northern Fennoscandia, 1:1 mill.*

Geological surveys of Finland, Norway and Sweden, 1987, *Geological map. Pre-Quaternary rocks, northern Fennoscandia, 1:1 mill.*

Gorbatschev, R. and Bogdanova, S., 1993, *Frontiers in the Baltic Shield.* Precambrian research 64, 3–21.

Grigelis, A. (ed), 1993, *Pre-Quaternary geology of the Baltic Sea proper.* St. Petersburg Cartography Enterprise.

Grip, E. and Frietsch, R., 1973, *Malm i Sverige 1 och 2.* Almqvist & Wiksell, Stockholm.

Grip, E., 1978, *Sweden. Mineral deposits of Europe, vol.1: Northwest Europe.* The Institute of Mining and Metallurgy, London.

Grip, H. and Rodhe, A., 1988, *Vattnets väg från regn till bäck.* Hallgren & Fallgren, Uppsala.

Gudelis, V. and Königsson, L-K. (eds), 1979, *The Quaternary history of the Baltic.* Acta Univ Ups Symp Univ Ups Annum Quingentesimum Celebrantis 1.

Hallam, A., 1976, *Continental drift and the fossil record. Continents adrift and continents aground.* W. H. Freeman & Co. San Francisco.

Hamilton, W., 1979, *Tectonics of the Indonesian region.* US Geological Survey Prof. paper 1078.

Henkel, H., 1992, *Aeromagnetic and gravity interpretation of three traverses across the Protogine Zone, southern Sweden.* GFF 114, 344–349.

Hoppe, G., 1948, *Isrecessionen från Norrbottens kustland i belysning av de glaciala formelementen.* Geographica 20.

Hoppe, G., 1983, *Fjällens terrängformer.* Naturvårdsverket.

Kabata-Pendias, A. and Pendias, H., 1991, *Trace elements in soils and plants.* CRC Press, Ann Arbor

Kleman, J., 1992, *The palimpsest glacial landscape in northwestern Sweden.* Geografiska annaler 74A, 305–325.

Kleman, J., 1994, *Preservation of landforms under ice sheets and ice caps.* Geomorphology 9, 19–32.

Kleman, J. and Borgström, I., 1990, *The boulder fields of Mt Fulufjället, west-central Sweden.* Geografiska annaler 72A, 63–78.

Knutsson, G. and Morfeldt, C-O., 1993, *Grundvatten, teori & tillämpning.* Svensk byggtjänst. Stockholm.

Kornfält, K. A. and Larsson, K., 1987, *Geological maps and cross-sections of southern Sweden.* SKB Technical report 24.

Lagerbäck, R., 1988, *The Veiki moraines in northern Sweden—widespread evidence of an Early Weichselian deglaciation.* Boreas 17, 469–486.

Lagerbäck, R., 1988, *Periglacial penomena in the wooded areas of Northern Sweden—relicts from the Tärendö Interstadial.* Boreas 17, 487–499.

Lagerbäck, R. and Robertsson, A-M., 1988, *Kettle holes—stratigraphical archives for Weichselian geology and palaeoenvironment in northernmost Sweden*. Boreas 17, 439–468.

Lagerlund, E., 1987, *An alternative Weichselian glaciation model, with special reference to the glacial history of Skåne, south Sweden*. Boreas 16, 433–459.

Levinsson, A. A., 1974, *Introduction to exploration geochemistry*. Applied Publishing Ltd, Wilmette, Ill. USA.

Lidmar-Bergström, K., 1982, *Pre-Quaternary geomorphological evolution in southern Fennoscandia*. SGU C 785.

Lidmar-Bergström, K., 1993, *Denudation surfaces and tectonics in the southernmost part of the Baltic shield*. Precambrian Research 64, 337–345.

Lind, G., 1982, *Gravity interpretation of the crust in southwestern Sweden*. Geologiska institutionen, Göteborgs universitet A 41.

Lindström, M., Lundqvist, J. and Lundqvist, Th., 1991, *Sveriges geologi från urtid till nutid*. Studentlitteratur, Lund.

Loberg, B., 1993, *Geologi. Material, processer och Sveriges berggrund*. 5 ed. Norstedts, Stockholm

Lundegårdh, P.H., 1991, *Stenar i färg*. 9th. ed. Norstedts, Stockholm.

Lundegårdh, P.H. and Laufeld, S., 1984, *Norstedts stora stenbok. Mineral, bergarter, fossil*. Norstedts, Stockholm.

Lundqvist, G., 1940, *Bergslagens minerogena jordarter*. SGU C 433.

Lundqvist, J., 1962, *Patterned ground and related frost phenomena in Sweden*. SGU C 583.

Lundqvist, J., 1972, *Ice-lake types and deglaciation pattern along the Scandinavian mountain range*. Boreas 1, 27–54.

Lundqvist, J., 1988, *Geologi. Processer—landskap—naturresurser*. Studentlitteratur, Lund.

Lundqvist, J., 1992, *Glaciation Stratigraphy in Sweden*. Geological survey of Finland, Spec. paper 15, 43–59.

Mannerfelt, C. M:son, 1945, *Några glacialgeologiska formelement och deras vittnesbörd om inlandsisens avsmältningsmekanik i svensk och norsk fjällterräng*. Geografiska annaler 27.

Möller, P., 1987, *Moraine morphology, till genesis, and deglaciation pattern in the Åsnen area, south-central Småland, Sweden*. Lundqua thesis 20.

Mörner, N-A., 1979, *The Fennoscandian uplift and Late Cenozoic geodynamics*. Geo-Journal 3, 287–318.

Nordberg, L. and Persson, G., 1974, *The national groundwater network of Sweden*. SGU Ca 48.

Norling, E. (ed), 1981, *A Swedish contribution to project Tornquist (IGCP accession no. 86)*. GFF 103, 161–278.

Norling, E. and Bergström, J., 1987, *Mesozoic and Cenozoic tectonic evolution of Scania*. Tectonophysics 137, 7–19. Elsevier.

Palm, H., Gee, D.G., Dyrelius, D. and Björklund, L., 1991, *A reflection seismic image of Caledonian structure in central Sweden*. SGU Ca 75.

Pesonen, L., 1990, *Fennoskandiens kontinentaldrift*. Atlas över Finland 125,36.

Plate tectonics, 1989, *A revolution in the Earth Sciences*. The Open Univ., Science Unit 7–8.

Pozaryski, W., 1959, *Budowa geologiczno Polski*. Warzawa.

Rapp, A., 1960, *Recent development of mountain slopes in Kärkevagge and surroundings, northern Scandinavia*. Geografiska annaler 42, 71–200.

Robertsson, A-M. and García Ambrosiani, K., 1992, *The Pleistocene in Sweden—a review of research, 1960–1990*. SGU Ca 81, 299–306.

Rudberg, S., 1954, *Västerbottens berggrundsmorfologi*. Geografica 25.

Rudberg, S., 1987: *Geology and geomorphology of Norden. I Norden. Man and environment*. Gebrüder Borntraeger. Berlin, Stuttgart.

Statens naturvårdsverk, *Geomorfologiska kartor med beskrivning*.

Stephansson, O., 1983, *Sveriges vandring på jorden*. Liber, Stockholm.

Stephens, M. B. and Gee, D.G., 1985, *A tectonic model for the evolution of the eugeoclinal terraines in the central Scandinavian Caledonides*. I Gee & Sturt (eds) 1985, 953–978.

Stephens, M. B., 1988: *The Scandinavian Caledonides: a complexity of collisions*. Geology Today 4, 20–26.

Stephens, M. B. and Gee, D. G., 1989: *Terraines and polyphase accretionary history in the Scandinavian Caledonides*. Geological Society of America, special paper 230, 17–30.

Strömberg, B., 1989, *Late Weichselian deglaciation and clay varve chronology in east-central Sweden*. SGU Ca 73.

Strömberg, B., 1992, *The final stage of the Baltic Ice Lake*. SGU Ca 81, 347–354.

Svensson, N-O., 1989, *Late Weichselian and Early Holocene shore displacement in the central Baltic, based on stratigraphical and morphological records from eastern Småland and Gotland, Sweden*. Lundqua thesis 25.

Svensson, N-O., 1991, *Late Weichselian and Early Holocene shore displacement in the central Baltic Sea*. Quaternary International 9, 7–26.

Sveriges geologiska undersökning—*Berggrundskartor med beskrivning*. Serie Af, Ba and Ca and *Berggrundskartor* maps serie Ai.—*Bergverksstatistik 1978–1992*. PM-serie.—*Jordartskartor med beskrivning*. Serie Ae, Ak and Ca.—*Grundvattenkartor med beskrivning*. Serie Ag and Ah.—*Maringeologiska kartor med beskrivning*. Serie Am.

Sveriges officiella statistik 1911–1959, *Bergshantering*. Kommerskollegium.

Sveriges officiella statistik 1960–1977, *Bergshantering*. Statistiska Centralbyrån.

Tegengren, F. R. et. al., 1924, *Sveriges ädlare malmer och bergverk*. SGU Ca 17.

Torsvik, T. H., Smethurst, M. A., Van de Voo, R., Trench, A., Abrahamsen, N. and Halvorsen, E., 1992, *Baltica. A synopsis of Vendian—Permian palaeomagnetic data and their palaeotectonic implications*. Earth Science Reviews 33, 133–152.

Varv, 1992, *Geologisk kort over den danske undergrund*. Köbenhavn.

Vasari, Y., 1986, *The Holocene development of the Nordic landscape*. Striae 24, 15–19. Uppsala.

Voipio, A. (ed), 1981, *The Baltic Sea*. Elsevier Oceanographic series.

Wråk, W., 1908, *Bidrag till Skandinaviens reliefkronologi*. Ymer, häfte 2.

Ziegler, P. A., 1982, *Geological Atlas of western and central Europe*. Shell international petroleum maatschappig B.V.

Authors

Aastrup, Mats, 1943, Senior Geologist, SGU

Andersson, Madelen, 1950, Senior Geologist, SGU

Beckholmen, Monica, 1951, Ph.D., Uppsala

Björck, Svante, 1948, Professor, Geologisk institut, Köpenhamn

Bygghammar, Birgitta, 1949, Senior Geologist SGU

Cato, Ingemar, 1946, Reader, Principal Adm Officer, SGU

Ek, John, 1936, F.L., Senior Geologist, SGU

Ekelund, Lena, 1956, Senior Geologist, SGU

Engqvist, Per, 1931, Senior Geologist, SGU

Eriksson, Leif, 1939, Senior Geophysics, SGU

Fredén, Curt, 1937, Reader, Senior Geologist, SGU

Henkel, Herbert, 1942, Senior Lecturer, Royal Institute of Technology

Kjellin, Berndt, 1945, Senior Geologist, SGU

Lidmar-Bergström, Karna, 1940, Reader, University of Stockholm

Lundqvist, Jan, 1926, Professor, University of Stockholm

Lundqvist, Thomas, 1932, Reader, Principal Adm. Officer, SGU

Müllern, Carl-Fredrik, 1939, Senior Geologist, SGU

Norling, Erik, 1931, Reader, SGU

Persson, Christer, 1937, Reader, Senior Geologist, SGU

Robertsson, Ann-Marie, 1941, Reader, Senior Geologist, SGU

Schytt, Anna, 1960, Geologist, Scientific Journalist

Selinus, Olle, 1943, Senior Geologist, SGU

Stephens, Michael, 1948, Ph.D., Senior Geologist, SGU

Svensson, Nils-Olof, 1957 Ph.D, University of Lund

Söderholm, Hans, 1946, Senior Geologist, SGU

Wik, Nils-Gunnar, 1946, Senior Geologist, SGU

Wikström, Anders, 1937, F.L., Senior Geologist, SGU

Zachrisson, Ebbe, 1931, F.L., Principal Adm. Officer, SGU

Åkerman, Christer, 1943, Ph.D., Senior Geologist, SGU

Many persons have been kind enough to place information at our disposal, or in some other way contributed to our work. We would particularly like to mention Sven Aaro, Birgitta Ericsson, Kerstin Finn, Dag Fredriksson, Bengt Holdar, Kajsa Hult, Jan Hultström, Gunilla Johansson, Karl-Axel Kornfält, Anders G Lindén, Jonas Lindgren, Ernest Magnusson, Britt Nordäng, Anders Olsson, Magnus Persson, Lennart Samuelsson, Sten Sandström, Jan Schedin, Lillemor Schultz, Bo Thunholm, Carl-Henric Wahlgren, Rosa Wallgren and Hugo Wikman, all of SGU, and Carl-Erik Johansson, SNV, Magnus Odin, SKB and Tomas Sjöstrand, Helsingborg.

Acknowledgements for illustrations

LMV = Lantmäteriverket (National Survey of Sweden)
N = Naturfotograferna (Agency for Nature Photographers, Sweden)
SGU = Sveriges geologiska undersökning (Geological Survey of Sweden)
SNA = Sveriges Nationalatlas (National Atlas of Sweden)
SKB = Svensk kärnbränslehantering AB (Swedish Nuclear Fuel and Waste Management Company)
Tio = Tiofoto Picture Agency

Page
2 Photo SGU
6 Drawing top Hans Sjögren; Drawing middle Bo Mossberg; Photo Curt Fredén
7 Photo Pål-Nils Nilsson/Tio
8–9 Drawing Hans Sjögren
10 Drawing from Blundell, D., Freeman, R. & Mueller, S., 1992: A continent revealed. The European Geotraverse. Cambridge University Press; Map CGMW/Unesco, 1990: Geological map of the world, 1:25 mill., Paris
11 Drawing Hans Sjögren, top and middle from Hamilton, W., 1979; bottom from Collet, L., 1927.
12 Drawings Hans Sjögren, top left from Plate tectonics Earth Sciences 1989; top middle from Plate tectonics Earth Sciences 1989; bottom from Hallam, A., 1976
13 Drawings Hans Sjögren, top from Plate tectonics 1989; bottom from Torsvik, T.H. et al, 1992 and Pesonen, L., 1990. Photo Tommie Jacobsson/N
14 Drawings Hans Sjögren; Photoes left and right Lars Persson, top Mikael Erlström
15 Drawings Hans Sjögren
16 Map SNA, from Lindström, M., Lundqvist, J. & Lundqvist, T. 1991
17 Drawing Hans Sjögren
18 Photoes from top to bottom: Left Lisbeth Godin, Ingemar Lager, Thomas Lundqvist; Middle Thomas Lundqvist; Right Anders Wikström, Thomas Lundqvist, Thomas Lundqvist
19 Photo top Anders Damberg; Photo middle Thomas Lundqvist; Photo bottom Department of Geology, Uppsala University (Karl Erik Alnavik)
20 Diagram SNA, data SGU; Photoes Lennart Samuelsson
21 Diagram SNA, from Gustaf Lind, 1982; Photo Thomas Lundqvist
22 Map top SNA, from Ziegler, P.A. 1982; Map middle left SNA, from Gee, D.G. et al, 1985; Map middle right SNA, from Stephens, M.B. & Gee, D.G. 1989; Diagram SNA, data SGU
23 Photo LMV; Drawing Hans Sjögren from Gee, D.G. & Sturt, B.A., 1985
24 Photo Claes Grundsten/N; Drawing Hans Sjögren from Stephens, M.B. 1988 and Torsvik, T.H., et al, 1992
25 Map SNA, data SGU; Photoes Anders Damberg; Drawing Hans Sjögren
26 Photo top Erik Norling; Photo middle Anders Damberg; Photo bottom left Christer Bäck; Photo bottom right Anders Damberg
27 Drawing Hans Sjögren; Photo left Anders Damberg; Photo right Göran Kjellström
28 Drawings top and middle Hans Sjögren; Drawing bottom Ingegerd Svensson; Photo Else-Marie Friis
29 Map SNA, data SGU; Photoes Hugo Wikman
30–37 Map SNA, data SGU
38 Map SNA, data SGU
39 Drawings Hans Sjögren; Photo top Staffan Arvegård/N; Photo bottom Lennart Mathiasson/N
40 Maps SNA, data SGU; Photo Bo Sundström
41–43 Map SNA, data SGU from Ahlberg, 1986; Axberg, 1980; Flodén, 1980; Grigelis, 1991; Kornfält & Larsson, 1987; Pozarski, 1959; Varv, 1992; Voipio, 1981
44 Map SNA
45 Maps SNA, data LMV
46 Photo top Lars Jarnemo/N; Photo bottom Tore Hagman/N
47 Photo top Lennart Mathiasson/N; Photo bottom Pål-Nils Nilsson/Tio
48 Photoes Karna Lidmar-Bergström; Drawing Hans Sjögren
49 Drawings Hans Sjögren
50–51 Map SNA
52 Photo Pål-Nils Nilsson/Tio; Drawing Hans Sjögren
53 Drawing Hans Sjögren; Map SNA
54 Map SNA; Photo Claes Grundsten/N
55 Photo top Ulf Risberg/N; Photo middle left Curt Fredén; Photo middle right Anders Damberg; Diagram SNA, data SGU; Photo bottom Axel Ljungquist/N
56–57 Map SNA, data SGU; Photo P. Roland Johansson/N
58 Photo top Boliden Mineral; Photo bottom Göran Hansson/N
59 Photo top Karl-Erik Alnavik; Photo bottom Anders Damberg
60–61 Map SNA, data SGU
62 Photo top Kim Naylor/Tio; Photo middle Boliden Mineral; Photoes bottom Anders Damberg
63 Map SNA, data SGU
64 Photo LMV; Drawing Hans Sjögren
65 Photo top Chad Ehlers/Tio

Photo middle left LKAB
Photo middle right SGU's archive
66 Photo top Tore Hagman/N
Photo middle Lennart Mathiasson/N
67 Drawing Hans Sjögren
68 Photo top Photo Researchers Inc
Photes bottom Anders Damberg
69 Photo top and bottom Anders Damberg
Photo middle Jan Strömberg
70 Photo top Anders Damberg
Photo middle left Hans Wretling/Tio
Photo middle right Anders Damberg
71 Photo top Dalbo Kvartsit AB
Photo middle Anders Lindh/Tio
72 Photo Björn Nordien/Tio
Drawing Hans Sjögren
73 Photo top Per Adlercreutz/Tio
Photo middle left SGU's archive
Photo middle right Anders Damberg
Photo bottom Lennart Nygren, Pica Pressfoto
74 Map SNA, data SGU
75 Maps SNA, data SGU
76 Photo top LMV
Photo bottom National Art Museums
77 Drawing Hans Sjögren
Map SNA, data SGU
78 Maps SNA, data SGU
79 Map SNA, data SGU, LMV, Boliden Mineral AB, Uppsala university
80–87 Map SNA, data SGU, Statens Kärnbränslenämnd, Boliden Mineral AB
88 Photoes Sven Aaro
Drawing Hans Sjögren
89 Map and diagram SNA, data SGU
90–91 Maps SNA, data SGU
Diagram SNA, data Henkel, H., 1992
92 Map left SGU C 812
Map right SGU
93 Photo Claes Grundsten/N
Drawing Hans Sjögren
Map left SGU Af 2
Map right SGU Af 2
94 Map top LMV
Diagram SNA, data Henkel,H.
Map bottom left SGU
Map bottom right SGU
95 Map left SNA, data SGU
Map right SNA, data SGU
96–97 Map SNA, data SGU
98 Map left SGU Ah 12
Map right SGU
99 Map top SNA, data **Seismologic Dept.**, Uppsala University
Map bottom left SNA, data SGU
Map bottom right SNA, data from **Seismologic Dept.**, Uppsala University
100 Diagrammes SNA and Hans Sjögren, top and middle from Palm, H. et.al. 1991; bottom from Dyrelius, D., 1980
101 Photo Curt Fredén
Map left SNA, data SGU
Map right SNA, data Ekman et al 1982
102–103 Drawing Nils Forshed
104 Diagram SNA, data SGU
Drawing Hans Sjögren
Photo Anders Damberg
105 Photo top left Åke W Engman/N
Other photoes Curt Fredén
106 Photo top Claes Grundsten/N
Photo middle Robert Lagerbäck
Photo bottom Esko Daniel
107 Photo top Claes Grundsten/N
Photo middle Kenneth Bengtsson/Naturbild
Drawing Hans Sjögren
Photo bottom left Anders Damberg
Photo bottom middle Bo Strömberg
Photo bottom right Ingemar Cato
108 Photo top Björn Uhr/N
Photo middle Lars Jarnemo/N
Photo bottom left Pål-Nils Nilsson/Tio
Photo bottom right Esko Daniel
109 Photo top Anders Damberg
Photo middle Claes Grundsten/N
Photo bottom left Harald Svensson
Photo bottom right Claes Grundsten/N
110 Maps SNA, data SGU
Photo Göran Hansson/N
111 Diagram SNA, data SGU
Map left SNA, data SGU
Map right SNA, data Peat Producers Ass.
Photo Alf Linderheim/N
112–119 Map SNA, mainly based on maps published by SGU, geomorphological maps of the mountain region published by the Swedish Environmental Protection Agency/University of Stockholm and the Map of Quaternary Geology, sheet 1 Northern Fennoscandia, scale 1:1 mill.
118 Maps SNA, data SGU
120 Photo Olle Melander
Drawing Hans Sjögren
121 Map SNA
Photo left Bo Lind
Photo right Björn E. Berglund
Drawing Hans Sjögren
122 Drawing Hans Sjögren
Photo top left Anders Damberg
Photo top right Jan Lundqvist
Photo bottom Robert Lagerbäck
123 Maps SNA from Jan Lundqvist 1992
124 Map SNA
125 Map SNA from Denton & Hughes 1981
Photo middle Karin Eriksson
Photo bottom Jan Lundqvist
126–127 Map SNA, data SGU
128 Photon top Pål-Nils Nilsson/Tio
Photo middle left Göran Hansson/N
Photo middle right Ingemar Borgström
129 Maps and diagram SNA from Strömberg, B. 1989
Photo top Lars Brunnberg
Photo middle Hans Pettersson
Photo bottom Wibjörn Karlén
130–131 Map SNA
132 Photoes top and bottom Claes Grundsten/N
Photo middle Robert Lagerbäck
133 Drawing Hans Sjögren from Lagerlund, E., 1987
134–135 Map SNA
136 Drawing Hans Sjögren
Maps SNA
Photoes Ann-Marie Robertsson
137 Drawings Hans Sjögren
Photo top left Per-Olov Eriksson/N
Photo top right Ulf Risberg/N
Photo bottom left Claes Grundsten/N
Photo bottom right Tore Hagman/N
138 The large map SNA, data SKB och Nils Olof Svensson
The small map SNA
Photo Nisse Peterson/Tio
139 The large map SNA, data SKB and Nils Olof Svensson
The small map SNA
Photo Pål-Nils Nilsson/Tio
140 The large map SNA, data SKB and Nils Olof Svensson
The small map SNA
Photo Robert Lagerbäck
141 The large map SNA, data SKB and Nils Olof Svensson
The small map SNA
Photo Eva Weiler
142 Map and diagram SNA
Photo Robert Lagerbäck
143 Map SNA
144 Map SGU Ae 100
Photo top Klas Rune/N
Photo bottom Tore Hagman/N
145 Photo top Axel Ljungquist/N
Map SGU Ae 52
Photo bottom Magnus Waller/Tio
146 Map top SGU Ae 17
Map bottom SGU Ae
Photo Bertil K. Johansson/N
147 Map SGU Ca 38
Photo Ulf Simonsson/Tio
148 Alf Linderheim/N
Map SGU Ca 45
149 Photoes Claes Grundsten/N
150 Photo top Mats Carlsson
Photo top middle Boris Winterhalter
Photo bottom middle Mats Carlsson
Photo bottom SGU
151–153 Map SNA, data SGU
154 Drawings Hans Sjögren
155 Photo Claes Grundsten/N
Drawing Hans Sjögren
156 Photo top Jan Töve/N
Drawing Hans Sjögren
Photo bottom left Göran Hansson/N
Photo bottom right SGU
157 Photo Staffan Arvegård/N
Diagram SNA, data SGU
158–165 Map SNA, data SGU
Photo Carl-Fredrik Müllern
166 Diagram SNA, data SGU
Maps SNA, data SGU
167 Photo Bo Brännhage/N
Maps SNA, data SGU
168 Map SNA, data SGU
Drawing Hans Sjögren
169 Maps SNA, data SGU
170 Photo Tore Hagman/N
Maps SNA, data SGU
171 Photo top Axel Ljunquist/N
Photo middle left Curt Fredén
Photo middle right Klas Rune/N
172 Drawing Hans Sjögren
173–175 Maps SNA, data SGU
176 Photo Madelen Andersson
177 Drawing Hans Sjögren
Photo middle SGU
Photo bottom Lennart Norström/N
178–183 Maps SNA, data SGU
184 Photo top Åke W Engman/N
Photo middle Madelen Andersson
Maps SNA, data SGU
185 Maps SNA, data SGU
186 Maps SNA, data SGU
187 Map SNA, data SGU
Drawing Hans Sjögren
188 Maps SNA, data SGU
Drawing Hans Sjögren
189 Maps SNA, data SGU
Photo Jan Töve/N
190 Maps SNA, data SGU
191 Map SNA, data SGU
Drawings Hans Sjögren
192 Maps SNA, data SGU
193 Photo top left Pål-Nils Nilsson/Tio
Photo top right Claes Grundsten/N
Photo middle Curt Fredén
Photo bottom left Gunvor Kristianson's **photo collection**, Surte
Map SGU/LMV
194 Drawing Nils Forshed
Photo middle Jan Johansson/N
Photo bottom Hans Wretling/Tio
195 Photo top Claes Grundsten/N
Photo middle Sture Karlsson/Tio
Photo bottom left Tore Hagman/N
Photo bottom right Alf Linderheim/N
196–197 Map SNA, data SGU and The National Environmental Protection Agency

Thematic maps

Index

National Atlas of Sweden

A geographical description of the landscape, society and culture of Sweden in 17 volumes

MAPS AND MAPPING

From historic maps of great cultural significance to modern mapping methods using the latest advanced technology. What you didn't already know about maps you can learn here. A unique place-name map (1:700,000) gives a bird's-eye view of Sweden. Editors: Professor Ulf Sporrong, geographer, Stockholm University, and Hans-Fredrik Wennström, economist, National Land Survey, Gävle.

THE FORESTS

Sweden has more forestland than almost any other country in Europe. This volume describes how the forests have developed and how forestry works: ecological cycles, climatic influences, its importance for the economy etc. One of many maps shows, on the scale of 1:1.25 million, the distribution of the forests today. Editor: Professor Nils-Erik Nilsson, forester, National Board of Forestry, Jönköping.

THE POPULATION

Will migration to the towns continue, or shall we see a new "green wave"? This volume highlights most sides of Swedish life: how Swedes live, education, health, family life, private economy etc. Political life, the population pyramid and immigration are given special attention. Editor: Professor Sture Öberg, geographer, Uppsala University, and Senior Administrative Officer Peter Springfeldt, geographer, Statistics Sweden, Stockholm.

THE ENVIRONMENT

More and more people are concerning themselves with environmental issues and nature conservancy. This book shows how Sweden is being affected by pollution, and what remedies are being applied. Maps of protected areas, future perspectives and international comparisons. Editors: Dr Claes Bernes and Claes Grundsten, geographer, National Environment Protection Agency, Stockholm.

AGRICULTURE

From horse-drawn plough to the highly-mechanized production of foodstuffs. A volume devoted to the development of Swedish agriculture and its position today. Facts about the parameters of farming, what is cultivated where, the workforce, financial aspects etc. Editor: Birger Granström, state agronomist, and Åke Clason, managing director of Research Information Centre, Swedish University of Agricultural Sciences, Uppsala.

The work of producing the National Atlas of Sweden is spread throughout the country.

THE INFRASTRUCTURE

Sweden's welfare is dependent on an efficient infrastructure, everything from roads and railways to energy production and public administration. If you are professionally involved, this book will provide you with a coherent survey of Sweden's infrastructure. Other readers will find a broad explanation of how Swedish society is built up and how it functions. Editor: Dr Reinhold Castensson, geographer, Linköping University.

SEA AND COAST

The Swedes have a deep-rooted love for the sea and the coast. This volume describes the waters that surround Sweden and how they have changed with the evolution of the Baltic. Facts about types of coastline, oceanography, marine geology and ecology, including comparisons with the oceans of the world. Editor: Björn Sjöberg, oceanographer, Swedish Meteorological and Hydrological Institute, Göteborg.

CULTURAL LIFE, RECREATION AND TOURISM

An amateur drama production in Hässleholm or a new play at the Royal Dramatic Theatre in Stockholm? Both fill an important function. This volume describes the wide variety of culture activities available in Sweden (museums, cinemas, libraries etc), sports and the various tourist areas in Sweden. Editor: Dr Hans Aldskogius, geographer, Uppsala University.

SWEDEN IN THE WORLD

Sweden is the home of many successful export companies. But Sweden has many other relations with the rest of the world. Cultural and scientific interchange, foreign investment, aid to the Third World, tourism etc. are described in a historical perspective. Editor: Professor Gunnar Törnqvist, geographer, Lund University.

WORK AND LEISURE

Describes how Swedes divide their time between work and play, with regional, social and age-group variations. The authors show who does what, the role of income, etc, and make some predictions about the future. Editor: Dr Kurt V Abrahamsson, geographer, Umeå University.

CULTURAL HERITAGE AND PRESERVATION

Sweden is rich in prehistoric monuments and historical buildings, which are presented here on maps. What is being done to preserve our cultural heritage? This volume reviews modern cultural heritage policies. Editor: Reader Klas-Göran Selinge, archeologist, Central Board of National Antiquities, Stockholm. Ass. Editor:
1st Antiquarian Marit Åhlén, runologist, Central Board of National Antiquities, Stockholm.

GEOLOGY

Maps are used to present Sweden's geology—the bedrock, soils, land forms, ground water. How and where are Sweden's natural geological resources utilised? Editor: Curt Fredén, state geologist, Geological Survey of Sweden, Uppsala.

LANDSCAPE AND SETTLEMENTS

How has the Swedish landscape evolved over the centuries? What traces of old landscapes can still be seen? What regional differences are there? This volume also treats the present landscape, settlements, towns and cities, as well as urban and regional planning. Editor: Professor Staffan Helmfrid, geographer, Stockholm University.

CLIMATE, LAKES AND RIVERS

What causes the climate to change? Why does Sweden have fewer natural disasters than other countries? This volume deals with the natural cycle of water and with Sweden's many lakes and rivers. Climatic variations are also presented in map form. Editors: Birgitta Raab, state hydrologist, and Haldo Vedin, state meteorologist, Swedish Meteorological and Hydrological Institute, Norrköping.

MANUFACTURING, SERVICES AND TRADE

Heavy industry is traditionally located in certain parts of Sweden, while other types of industry are spread all over the country. This volume contains a geographical description of Swedish manufacturing and service industries and foreign trade. Editor: Dr Claes Göran Alvstam, geographer, Göteborg University.

GEOGRAPHY OF PLANTS AND ANIMALS

Climatic and geographical variations in Sweden create great geographical differences in plant and animal life. This volume presents the geographical distribution of Sweden's fauna and flora and explains how and why they have changed over the years. There is a special section on game hunting. Editors: Professor Ingemar Ahlén and Dr Lena Gustafsson, Swedish University of Agricultural Sciences, Uppsala.

THE GEOGRAPHY OF SWEDEN

A comprehensive picture of the geography of Sweden, containing excerpts from other volumes but also completely new, summarizing articles. The most important maps in the whole series are included. Indispensable for educational purposes. Editors: The editorial board of the National Atlas of Sweden, Stockholm.